Lecture Notes in Computer S

Commenced Publication in 1973
Founding and Former Series Editors:
Gerhard Goos, Juris Hartmanis, and Jan van Lee

Editorial Board

Costas S. Iliopoulos William F. Smyth (Eds.)

Combinatorial Algorithms

21st International Workshop, IWOCA 2010
London, UK, July 26-28, 2010
Revised Selected Papers

Volume Editors

Costas S. Iliopoulos
University of London, King's College
Department of Computer Science
The Strand, London, WC2R 2LS, UK
E-mail: csi@dcs.kcl.ac.uk

William F. Smyth
McMaster University, Department of Computing and Software
1280 Main Street West, Hamilton, ON L8S 4K1, Canada
E-mail: bill@arg.cas.mcmaster.ca

ISSN 0302-9743 e-ISSN 1611-3349
ISBN 978-3-642-19221-0 e-ISBN 978-3-642-19222-7
DOI 10.1007/978-3-642-19222-7
Springer Heidelberg Dordrecht London New York

Library of Congress Control Number: 2011920712

CR Subject Classification (1998): G.2.1, G.2.2, I.1, F.2, E.1, E.4, H.1

LNCS Sublibrary: SL 1 – Theoretical Computer Science and General Issues

Typesetting: Camera-ready by author, data conversion by Scientific Publishing Services, Chennai, India

Printed on acid-free paper

Springer is part of Springer Science+Business Media (www.springer.com)

Preface

This volume contains the papers presented at IWOCA 2010: The 21st International Workshop on Combinatorial Algorithms. The 21st IWOCA was held in the Great Hall of the Strand Campus, King's College London (KCL), July 26–28, 2010. The meeting was sponsored and supported financially by KCL and the London Mathematical Society; it was hosted by the Algorithm Design Group in the KCL Department of Informatics. The local Organizing Committee, co-chaired by Tomasz Radzik and German Tischler, did an outstanding job; the Program Committee was co-chaired by Costas Iliopoulos and Bill Smyth.

The EasyChair system was used to facilitate management of submissions and refereeing, with three referees selected from the Program Committee assigned to each paper. Out of 85 contributed papers, a total of 31 were accepted, subject to revision, for presentation at the workshop and publication in the LNCS proceedings. An additional 13 papers were accepted for poster presentation, of which eight are included as four-page papers in the proceedings. Authors and titles of all 44 accepted papers are available at the IWOCA 2010 website. The workshop also featured a problem session, organized by Joe Ryan, and four invited talks by Alan Frieze, Gregory Kucherov, Mirka Miller, and Dorothea Wagner.

The number of participants in IWOCA 2010 more than doubled the number of speakers: altogether 75 researchers attended, affiliated to institutions in 20 different countries on five continents.

Here are brief summaries of the invited talks, in order of their presentation:

(1) Mirka Miller, University of Newcastle, Australia: "Constructions of Large Graphs and Digraphs of Given Diameter and Maximum Degree" — a survey of recent progress in a research area with numerous practical applications, together with a knowledgeable outline and analysis of research directions likely to be significant in the future.
(2) Dorothea Wagner, University of Karlsruhe, Germany: "Clustering of Static and Temporal Graph" — a survey of a research area with growing applications to networks of all kinds (for example, in the social sciences and biology), with particular focus on algorithmic aspects of quality measures for graph clustering.
(3) Alan Frieze, Carnegie Mellon University, USA: "The Karp-Sipser Matching Algorithm and Refinements" — an analysis of a well-known greedy graph matching algorithm in the context of sparse random graphs, presenting new performance results and describing its application to finding a maximum matching.
(4) Gregory Kucherov, J.-V. Poncelet Laboratory, Russia and Laboratoire d'Informatique Fondamentale de Lille, France: "Seeding Methods for Biosequence Search: Algorithmic Ideas and Applications" — a discussion of the use of spaced seeds as a technique for searching DNA sequences, describing a new

method for designing seeds adapted to different search situations, together with applications to protein search and mapping reads from high-throughput sequencing technology.

It was a warm summer in London, and the weather continued to cooperate throughout a delightful Thames cruise on the riverboat *Viscountess*, where the conference banquet was held.

Finally we would like to thank the members of the Program Committee and the subreferees for their fine work, their thorough reviews, their constructive and helpful comments.

December 2010

Costas S. Iliopoulos
William F. Smyth

Organization

Program Chairs

Costas S. Iliopoulos
William F. Smyth

Program Committee

Subramanian Arumugam
Hideo Bannai
Ljiljana Brankovic
Gerth Stølting Brodal
Charles Colbourn
Maxime Crochemore
Diane Donovan
Michael Fellows
Jiri Fiala
Aviezri Fraenkel
Alan Frieze
Dalibor Froncek
Roberto Grossi
Jan Holub
Seok-Hee Hong
Ralf Klasing
Rao Kosaraju
Jan Kratochvil
Marcin Kubica
Laurent Mouchard
Mirka Miller
Ian Munro
Kunsoo Park
Simon Puglisi
Sohel Rahman
Rajeev Raman
Michiel Smid
Iain Stewart
Eli Upfal
Koichi Wada
Dorothea Wagner
Sue Whitesides
Christos Zaroliagis
Michal Ziv-Ukelson
Lynette van Zijl

Organizing Committee

Golnaz Badkobeh
Kaichun Chang
Colin Cooper
Maxime Crochemore
Costas Iliopoulos
Leonidas Kapsokalivas
Laurent Mouchard
Solon Pissis
Tomasz Radzik (Co-chair)
Kathleen Steinhöfel
German Tischler (Co-chair)

External Reviewers

Mohamed Abouelhoda
Jawaherul Alam
Sivan Albagli
Mitchell Archibald
Md. Faizul Bari
Nicolas Bedon
Cedric Benz
Philip Bille
Guillaume Blin
Hans-Joachim Böckenhauer
Hajo Broersma
Manolis Christodoulakis
Paul Christophe
Shane Culpepper
Andrzej Dudek
Robert Elsasser
Vladimir Estivill-Castro
Frantisek Franek
Alan Frieze
Stanley Fung
Travis Gagie
Robert Görke
Serge Gaspers
Sylvain Guillemot
Tanja Hartmann
Danny Hermelin
Pim van't Hof
Michael Hoffmann
Peter Horak
Louis Ibarra
Shunsuke Inenaga
Taisuke Izumi
Tomoko Izumi
Matthew Johnson
Tomas Kaiser
Yoshiaki Katayama
Bastian Katz
Josef Kolar
Adrian Kosowski
Dieter Kratsch
Marcus Krug
Alessio Langiu
Thierry Lecroq
Christos Makris
Bodo Manthey
Luke Mathieson
Sascha Meinert
Daniel Meister
Debajyoti Mondal
Tanaeem M Moosa
Rolf Niedermeier
Sotiris Nikoletseas
Hirotaka Ono
Sang-il Oum
Daniel Paulusma
Hebert Perez-Roses
Oudone Phanalasy
Guillermo Pineda-Villavicencio
Solon Pissis
Tomasz Radzik
Dror Rawitz
Alex Rosa
Ignaz Rutter
Thomas Sauerwald
Ignasi Sau
Andrea Schumm
Francesc Sebe
Steven Skiena
Bettina Speckmann
Daniel Spielman
Kathleen Steinhöfel
Jan Stoklasa
Jan Arne Telle
Christopher Thraves
Dikel Tsur
Walter Unger
Yushi Uno
Pavel Valtr
Kristina Vuskovic
MD. Wasiur Rahman
Oren Weimann
David R. Wood
Shay Zakov
Shira Zucker

Table of Contents

Parameterized Algorithms for the Independent Set Problem in Some Hereditary Graph Classes

Konrad Dabrowski[1], Vadim Lozin[1], Haiko Müller[2], and Dieter Rautenbach[3]

[1] DIMAP, University of Warwick, Coventry CV4 7AL, UK
[2] School of Computing, University of Leeds, Leeds, LS2 9JT, UK
[3] Institut für Optimierung und Operations Research, Universität Ulm, D-89069 Ulm, Germany

Abstract. The maximum independent set problem is NP-complete for graphs in general, but becomes solvable in polynomial time when restricted to graphs in many special classes. The problem is also intractable from a parameterized point of view. However, very little is known about parameterized complexity of the problem in restricted graph classes. In the present paper, we analyse two techniques that have previously been used to solve the problem in polynomial time for graphs in particular classes and apply these techniques to develop fpt-algorithms for graphs in some classes where the problem remains NP-complete.

Keywords: Independent set; Fixed-parameter tractability; Augmenting graph; Modular decomposition; Hereditary class of graphs.

AMS subject classification: 05C69.

1 Introduction

We study simple undirected graphs without loops or multiple edges. In a graph, an independent set is a subset of vertices, no two of which are adjacent and a clique is a subset of pairwise adjacent vertices. The size of a maximum independent set in a graph G is called the *independence number* of G and is denoted $\alpha(G)$, while the size of a maximum clique is called the *clique number* of G and is denoted $\omega(G)$.

The MAXIMUM INDEPENDENT SET problem is that of finding an independent set of maximum size in a graph. From a computational point of view, this is a difficult problem, i.e. it is NP-hard. Moreover, it remains NP-hard under substantial restrictions, for instance, for triangle-free graphs [27] and for graphs of vertex degree bounded by d, where $d \geq 3$. On the other hand, in many special graph classes the problem admits polynomial-time algorithms, which is the case for perfect graphs [16], claw-free graphs [24], and graphs of bounded clique-width [7].

In this paper, we study the following parameterization of the MAXIMUM INDEPENDENT SET problem:

C.S. Iliopoulos and W.F. Smyth (Eds.): IWOCA 2010, LNCS 6460, pp. 1–9, 2011.

k-INDEPENDENT SET
Instance: A graph G and a positive integer k.
Parameter: k.
Problem: Decide whether G has an independent set of size k and find such a set if it exists.

An approach to deal with NP-complete problems in practice is to split a problem that contains a parameter as part of the input into sub-problems for each value of this parameter. A parameterized problem is said to be *fixed-parameter tractable (fpt)* if it can be solved in time $f(k)p(n)$ on instances of input size n, where $f(k)$ is an efficiently computable function, depending only on the value of the parameter k and $p(n)$ is a polynomial independent of k.

Unfortunately, the MAXIMUM INDEPENDENT SET problem remains difficult even under this relaxation. More formally, it is W[1]-hard [10]. However, for graphs in some restricted classes the problem becomes fixed-parameter tractable. In particular, this is the case for graphs without large cliques, which follows from a Ramsey argument (see e.g. [29]). This argument alone implies fixed-parameter tractability of the problem for graphs of bounded degree, of bounded degeneracy, of bounded chromatic number, in all proper minor-closed graph classes (which includes, in particular, classes of graphs excluding single-crossing graphs as minors [8]), all proper classes closed under taking subgraphs (not necessarily induced). Beyond this argument, very little is known on the parameterized complexity of the problem in restricted graph families. Other classes where the problem is known to be fixed-parameter tractable are the complements of t-multiple-interval graphs [13], segment intersection graphs with a bounded number of directions [19] and graphs whose vertices can be partitioned into two subsets of which one induces a graph of bounded clique number and the other induces a graph of bounded independence number [20].

In search of new fpt results, we analyse algorithmic techniques which are traditionally used for obtaining polynomial-time solutions for the MAXIMUM INDEPENDENT SET problem on graphs in special classes. In particular, we study the augmenting graph approach and the modular decomposition technique and apply them to develop fpt-algorithms that solve the problem in several new classes of graphs, generalising some of the previously known results.

All classes considered in this paper are hereditary, in the sense that for any graph G in such a class, all induced subgraphs of G are also in the class. It is known that a class of graphs is hereditary if and only if it can be characterised by a set of forbidden induced subgraphs. We denote the set of graphs containing no induced subgraphs from a set M by $Free(M)$ and call graphs in this class M-free.

For a graph G we denote the vertex set and the edge set of G by $V(G)$ and $E(G)$ respectively. If v is a vertex of G, then $N(v)$ is the neighbourhood of v (i.e. the set of vertices adjacent to v) and $N[v] = N(v) \cup \{v\}$ is the closed neighbourhood of v. For a subset $U \subseteq V(G)$ we let $G[U]$ be the subgraph of G induced by U, and $N(U)$ be the neighbourhood of U, i.e. the set of vertices outside U that have at least one neighbour in U. By $R(r, s)$ we denote the Ramsey number, i.e. the minimum number n such that every graph with at least n vertices has either an independent set of size r or a clique of size s. As usual, K_n, C_n and P_n denote the complete graph, the chordless cycle and the chordless path on n vertices, respectively. We denote the graph obtained from K_n by deleting an edge by $K_n - e$.

2 The Augmenting Graph Technique

The idea of augmenting graphs was proposed by Berge [3] and then implemented by Edmonds [11] to solve the maximum matching problem, which is equivalent to the MAXIMUM INDEPENDENT SET problem restricted to the class of line graphs. With this restriction, the idea reduces to finding augmenting chains. However, the notion of an augmenting graph lies in the basis of a general approach to solve the problem, which can be described as follows:

Let G be a graph and S an independent set in G. We shall call the vertices of S *white* and the remaining vertices of G *black.*

Definition 1. *An* augmenting graph for S in G *is an induced bipartite subgraph $H = (W, B, E)$ of G, where $W \cup B$ is the bipartition of its vertex set and E its edge set, such that:*

- $|B| > |W|$,
- $W \subseteq S$,
- $B \subseteq V(G) \backslash S$, *and*
- $N(B) \cap S \subseteq W$.

Clearly if $H = (W, B, E)$ is an augmenting graph for S, then S is not a maximum independent set in G, since the set $S' = (S \backslash W) \cup B$ is independent and $|S'| > |S|$. Conversely, if S is not a maximum independent set, and S' is a larger independent set, then the subgraph of G induced by the set $(S \setminus S') \cup (S' \setminus S)$ is augmenting for S. Thus we have the following theorem:

Theorem of Augmenting Graphs. *An independent set S in a graph G is maximum if and only if there are no augmenting graphs for S in G.*

This theorem suggests the following general approach to find a maximum independent set in a graph G: Begin with any independent set S in G and, as long as S admits an augmenting graph H, augment S as above. This approach has proven to be a useful tool to develop approximate solutions to the problem [18], to compute bounds on the independence number [9], and to solve the problem in polynomial time for graphs in special classes such as claw-free graphs [24], fork-free graphs [1] and some others [2, 4, 22, 26]. In the present paper, we use the idea of augmenting graphs to derive the following fpt result:

Theorem 1. *The k-*INDEPENDENT SET *problem can be solved for $(K_r - e)$-free n-vertex graphs in time $f(k, r)p(n)$, where $f(k, r)$ is a function of k and r only and $p(n)$ is a polynomial independent of k and r.*

Proof. Let G be a $(K_r - e)$-free graph with n vertices and let S be an independent set in G. We assume that S is maximal with respect to set-inclusion and admits no augmenting P_3. Obviously such a set can be found in polynomial time. If $|S| \geq k$, we are done. Therefore, we suppose that $|S| < k$ and explain how to determine whether G admits an augmenting graph in time $g(k,r)p(n)$, which clearly implies the desired result for $f(k,r) = kg(k,r)$. We split the process of finding an augmenting graph into the following general steps.

Step 1: Partition the set of black vertices of G into subsets called *node classes* putting two vertices in the same node class if and only if they have the same neighbourhood in S. Note that there are at most $2^{|S|} - 1 < 2^k$ node classes. We call a node class *light* if its neighbourhood in S contains exactly 1 vertex and *heavy* otherwise. Since S admits no augmenting P_3, every light node class is a clique. Clearly, this step can be implemented in polynomial time.

Step 2: Consider a heavy node class C. Since G is $(K_r - e)$-free, the subgraph of G induced by C must be K_{r-2}-free. Therefore, if $|C| \geq R(k, r-2)$, then C necessarily has an independent set of size k. Thus, in this case we can arbitrarily choose $R(k, r-2)$ vertices in the class and then find an independent set of size k among them in time bounded by a function of k and r. Otherwise, the size of every heavy node class is less than $R(k, r-2)$, in which case the total number of vertices in heavy node classes is bounded by a function of k and r.

Step 3: Generate all independent sets contained in the union of the heavy node classes. From the previous step it follows that the number of such sets and the time needed to generate all of them is bounded by a function of k and r. For each independent set T found in this step, execute Step 3.1.

Step 3.1: If the size of T is strictly larger than the number of its white neighbours, then $G[T \cup (N(T) \cap S)]$ is an augmenting graph. Otherwise, extend T by adding to it some vertices from the light node classes. To this end, delete from the light node classes those vertices that have neighbours in T and then split the thus-modified light classes into two groups: those containing at most kr vertices, we call them *small*, and those containing more than kr vertices, called *large*. Let s be the number of small classes and l the number of large classes. Obviously, $s + l < k$.

Extend T to a larger independent set by adding to it some vertices from small node classes. Since the number of small node classes is at most k and each of them contains at most kr vertices, the number of such extensions and the time needed to find all of them is bounded by a function of k and r. For each such an extension T', execute Step 3.1.1.

Step 3.1.1: If the size of T' is strictly larger than the number of its white neighbours, then $G[T' \cup (N(T') \cap S)]$ is an augmenting graph. Otherwise, extend T' by adding to it vertices from the large node classes. To this end, delete from the large node classes those vertices that have neighbours in T'. Since every light class (small or large) is a clique,

- T' contains at most one vertex in each light class,
- no vertex u from a light node class has more than $r-3$ neighbours in another light node class, since otherwise a $K_r - e$ arises using u, $r-2$ neighbours of u in another light node class and their only neighbour in S.

Therefore, deleting from the large node classes those vertices that have neighbours in T' leaves at least lr vertices in each large node class. Consequently, the set of vertices left in the large node classes necessarily contains an independent set L of size l. This set can be constructed iteratively by picking an arbitrary vertex, deleting its neighbours, and so on. Now we add L to T' and check if $G[(T' \cup L) \cup (N(T' \cup L) \cap S)]$ is an augmenting graph. Observe that it does not matter how we choose L, since for any choice of this set, its neighbourhood in S coincides with the neighbourhood of the large node classes in S.

Summarising, we conclude that determining whether S admits an augmenting graph can be done in time $g(k,r)p(n)$. The number of augmentations to solve the problem is at most k. Therefore, the result follows. □

3 Modular Decomposition

The idea of modular decomposition was first introduced in the 1960s by Gallai [15], and also appeared in the literature under various other names, such as *prime tree decomposition* [12], *X-join decomposition* [17], and *substitution decomposition* [25]. To describe this idea, let us fix some terminology.

Given a graph $G = (V, E)$, a subset of vertices $U \subseteq V$ and a vertex $x \in V$ outside U, we say that x *distinguishes* U if x has both a neighbour and a non-neighbour in U. A subset $U \subseteq V$ is called a *module* of G if no vertex in $V \setminus U$ distinguishes U. A module U is *nontrivial* if $1 < |U| < |V|$, otherwise it is *trivial.* A graph is called *prime* if it has only trivial modules.

An important property of maximal modules is that if G and the complement of G are both connected, then the maximal modules of G are pairwise disjoint. Moreover, from the above definition it follows that if U and W are maximal modules, then either all possible edges between them are present, or none of them are. This property is useful when we deal with the weighted version of the maximum independent set problem.

We say that G is a weighted graph if each vertex v of G is assigned a positive integer $w(v)$, the weight of the vertex. The MAXIMUM WEIGHT INDEPENDENT SET problem is that of finding an independent set of maximum total weight in a weighted graph. This maximum total weight is denoted $\alpha_w(G)$. By using the properties of maximal modules we can find a maximum weight independent set in G by:

(1) recursively solving the problem in the subgraphs of G induced by maximal modules,
(2) contracting each maximal module M to a single vertex and assigning to it the weight $\alpha_w(G[M])$, obtaining in this way a new graph G^0,
(3) solving the problem for the graph G^0.

The graph G^0 constructed in step 2 of the outlined procedure is prime. So the procedure reduces the MAXIMUM WEIGHT INDEPENDENT SET problem from any hereditary class X to prime graphs in X. This reduction can be implemented in polynomial time (see e.g. [23]). In this section we show that this is also an fpt-reduction, i.e. it preserves fixed-parameter tractability. In case of weighted graphs we parameterize the problem by the weight W of a solution. Without loss of generality we will assume that if the input graph has no independent set of weight at least W, the problem asks for an independent set of maximum weight. This generalisation increases the complexity of any algorithm solving the problem at most W times and therefore preserves fixed-parameter tractability. More formally, we consider the following parameterization of the MAXIMUM WEIGHT INDEPENDENT SET problem:

w-INDEPENDENT SET
Instance: A weighted graph G with weight function $w : V(G) \to \{1, 2, 3, \ldots\}$ and a positive integer W.
Parameter: W.
Problem: Decide whether G has an independent set of weight at least W and find such a set if it exists. If no such set exists, find an independent set of weight $\alpha_w(G)$ instead.

Theorem 2. *Let $\mathcal{X}$ be a hereditary class of graphs and let $\mathcal{X}_0$ denote the class of prime graphs in $\mathcal{X}$. If the w-INDEPENDENT SET problem is fixed-parameter tractable in $\mathcal{X}_0$, then it is fixed-parameter tractable in $\mathcal{X}$.*

Proof. Let (G, W) be an instance of the w-INDEPENDENT SET problem with $G \in \mathcal{X}$. Let n denote the number of vertices of G. Recall that the modular decomposition tree T of G can be determined in linear time [23, 6] and that the set of leaves of T equals the vertex set V of G. To each node v of T we associate the subgraph G_v of G induced by the leaves of the subtree of T rooted at v. Processing the vertices of T in an order of non-increasing height, for each node v of T we will find an independent set I_v of G_v such that $w(I_v) \geq \min\{W, \alpha_w(G_v)\}$. If the weight of I_v is at least W, we stop the procedure and output I_v. Otherwise, to each node v we assign an independent set I_v of weight $\alpha_w(G_v)$. The procedure starts by assigning to each leaf v of T the independent set $I_v = \{v\}$. Now let v be an inner node of T.

If G_v is disconnected, then the children $v_1, v_2, \ldots, v_l$ of v correspond to the connected components of G_v. In this case, we let $I_v = I_{v_1} \cup I_{v_2} \cup \ldots I_{v_l}$.

If the complement of G_v is disconnected, then the children $v_1, v_2, \ldots, v_l$ of v correspond to the connected components of the complement of G_v. In this case, we let $I_v = I_{v_i}$ where $w(I_{v_i}) = \max\{w(I_{v_1}), w(I_{v_2}), \ldots, w(I_{v_l})\}$.

Finally, if both G_v and its complement are connected, then the children $v_1, \ldots, v_l$ of v correspond to the subgraphs of G_v induced by the maximal modules $U_1, U_2, \ldots, U_l$ of G_v, which partition the vertex set of G_v. Let the graph G_v^0 arise from G_v by contracting each maximal module U_i of G_v into a single vertex, denoted i, to which we assign the weight $w(i) = w(I_{v_i})$. Since G_v^0 belongs to $\mathcal{X}_0$, there is an algorithm $\mathcal{A}$ that solves w-INDEPENDENT SET on the instance (G_v^0, W) in time $f(W)l^c \leq f(W)n^c$, where c is a constant. If I is the output of $\mathcal{A}$, then let $I_v = \bigcup_{i \in I} I_{v_i}$. It is not difficult to see that the set assigned to the root of T correctly solves w-INDEPENDENT SET on the instance (G, W). Since T has $O(n)$ vertices, the overall time complexity is at most $f(W)n^{c+1}$. $\square$

Theorem 2 reduces the w-INDEPENDENT SET problem from general graphs to prime graphs. The corresponding result for the non-parameterized problem is well-known. Our next result shows that the problem can be further reduced to prime graphs containing a clique K_r for a certain value of r.

Theorem 3. *For any $r \in \mathbb{N}$, the w-INDEPENDENT SET problem is fixed-parameter tractable in the class of K_r-free graphs.*

Proof. Let (G, W) be an instance of the w-INDEPENDENT SET problem with $G = (V, E)$ being a K_r-free graph on n vertices. Since the weight of each vertex is a positive integer, the weight of every independent set is at least its size. Therefore, if G has at least $R(r, W)$ vertices it necessarily has an independent set of size (and therefore of weight) at least W. If the number of vertices of G is strictly more than $R(r, W)$, we can delete, without loss of generality, any $n - R(r, W)$ vertices of G, since the remaining vertices of the graph still necessarily have an independent set of size (of weight at least) W.

When the number of vertices of G is bounded by $R(r, W)$, the problem can be solved in time independent of n. This completes the proof. □

To illustrate Theorems 2 and 3, we apply them to solve the w-INDEPENDENT SET problem in the class of $(house, bull)$-free graphs. The *house* and *bull* graphs are shown in Fig. 1. Observe that this class contains (C_3, C_4)-free graphs, where the MAXIMUM INDEPENDENT SET problem is NP-hard [27].

Fig. 1. The house and the bull graphs

Theorem 4. *The w-INDEPENDENT SET problem is fixed-parameter tractable in the class of (house,bull)-free graphs.*

Proof: Our proof is based on a characterisation of $(house, bull)$-free graphs from [28]: every prime $(house, bull)$-free graph is either triangle-free or the complement of a bipartite chain graph. (A bipartite graph is a bipartite chain graph if the vertices in both parts of the bipartition are linearly ordered by inclusion of neighbourhoods.) Obviously, for the complements of bipartite graphs, the MAXIMUM WEIGHT INDEPENDENT SET problem can be solved in polynomial time, since the size of any independent set in such a graph is at most 2. Also, by Theorem 3, the w-INDEPENDENT SET problem is fixed-parameter tractable in the class of triangle-free graphs. Therefore, by Theorem 2, it is fixed-parameter tractable in the class of $(house, bull)$-free graphs. □

4 Concluding Remarks and Open Problems

In this paper, we used the augmenting graph technique and modular decomposition to obtain new results on the parameterized complexity of the MAXIMUM INDEPENDENT SET problem in hereditary classes of graphs. The new results together with some previously known results allow us to conclude, in particular, that the problem is fixed-parameter tractable in all hereditary classes defined by a single forbidden induced subgraph G with at most 4 vertices, except for $G = C_4$. Finding the parameterized complexity of the problem in the class of C_4-free graphs is a challenging open problem. In addition to the two techniques studied in this paper, some other approaches can be useful for finding an answer to the above question, such as graph transformations [21], separating cliques [5] and split decomposition [30].

References

1. Alekseev, V.E.: Polynomial algorithm for finding the largest independent sets in graphs without forks. Discrete Appl. Math. 135(1-3), 3–16 (2004)
2. Alekseev, V.E., Lozin, V.V.: Augmenting graphs for independent sets. Discrete Appl. Math. 145(1), 3–10 (2004)
3. Berge, C.: Two theorems in graph theory. Proc. Nat. Acad. Sci. USA 43(9), 842–844 (1957)
4. Boliac, R., Lozin, V.V.: An augmenting graph approach to the stable set problem in P_5-free graphs. Discrete Appl. Math. 131(3), 567–575 (2003)
5. Brandstädt, A., Hoàng, C.T.: On clique separators, nearly chordal graphs, and the maximum weight stable set problem. Theoret. Comput. Sci. 389(1-2), 295–306 (2007)
6. Corneil, D., Habib, M., Paul, C., Tedder, M.: Simpler Linear-Time Modular Decomposition Via Recursive Factorizing Permutations. In: Aceto, L., Damgård, I., Goldberg, L.A., Halldórsson, M.M., Ingólfsdóttir, A., Walukiewicz, I. (eds.) ICALP 2008, Part I. LNCS, vol. 5125, pp. 634–645. Springer, Heidelberg (2008)
7. Courcelle, B., Makowsky, J.A., Rotics, U.: Linear time solvable optimization problems on graphs of bounded clique-width. Theory of Computing Systems 33(2), 125–150 (2000)
8. Demaine, E.D., Hajiaghayi, M., Thilikos, D.M.: Exponential speedup of fixed-parameter algorithms for classes of graphs excluding single-crossing graphs as minors. Algorithmica 41(4), 245–267 (2005)
9. Denley, T.: The independence number of graphs with large odd girth. Electron. J. Comb. 1, R9 (1994)
10. Downey, R.G., Fellows, M.R.: Parameterized complexity. Monographs in Computer Science. Springer, New York (1999)
11. Edmonds, J.: Paths, trees, and flowers. Canadian J. Math. 17, 449–467 (1965)
12. Ehrenfeucht, A., Rozenberg, G.: Primitivity is hereditary for 2-structures. Theoret. Comput. Sci. 70(3), 343–358 (1990)
13. Fellows, M.R., Hermelin, D., Rosamond, F., Vialette, S.: On the parameterized complexity of multiple-interval graph problems. Theoret. Comput. Sci. 410(1), 53–61 (2009)

14. Flum, J., Grohe, M.: Parameterized complexity theory. Texts in Theoretical Computer Science. Springer, Berlin (2006)
15. Gallai, T.: Transitiv orientierbare graphen. Acta Math. Acad. Sci. Hungar. 18, 25–66 (1967)
16. Grötschel, M., Lovász, L., Schrijver, A.: Geometric Algorithms and Combinatorial Optimization. Springer, Berlin (1988)
17. Habib, M., Maurer, M.C.: On the X-join decomposition for undirected graphs. Discrete Appl. Math. 1(3), 201–207 (1979)
18. Halldórsson, M.M.: Approximating discrete collections via local improvements. In: Proceedings of the Sixth SAIM-ACM Symposium on Discrete Algorithms, San Francisco, CA, pp. 160–169 (1995)
19. Kára, J., Kratochvíl, J.: Fixed parameter tractability of Independent Set in segment intersection graphs. In: Bodlaender, H.L., Langston, M.A. (eds.) IWPEC 2006. LNCS, vol. 4169, pp. 166–174. Springer, Heidelberg (2006)
20. Lozin, V.V.: Parameterized complexity of the maximum independent set problem and the speed of hereditary properties. Electronic Notes in Discrete Mathematics 34, 127–131 (2009)
21. Lozin, V.V.: Stability preserving transformations of graphs. Annals of Operations Research (to appear) doi: 10.1007/s10479-008-0395-1
22. Lozin, V.V., Milanič, M.: On finding augmenting graphs. Discrete Appl. Math. 156(13), 2517–2529 (2008)
23. McConnell, R.M., Spinrad, J.P.: Modular decomposition and transitive orientation. Discrete Math. 201(1-3), 189–241 (1999)
24. Minty, G.J.: On maximal independent sets of vertices in claw-free graphs. J. Combin. Theory Ser. B 28(3), 284–304 (1980)
25. Möhring, R.H.: Algorithmic aspects of comparability graphs and interval graphs. In: Rival, I. (ed.) Graphs and Orders, pp. 41–101. D. Reidel, Boston (1985)
26. Mosca, R.: Stable sets in certain P_6-free graphs. Discrete Appl. Math. 92(2-3), 177–191 (1999)
27. Murphy, O.J.: Computing independent sets in graphs with large girth. Discrete Appl. Math. 35(2), 167–170 (1992)
28. Olariu, S.: On the homogeneous representations of interval graphs. J. Graph Theory 15(1), 65–80 (1991)
29. Raman, V., Saurabh, S.: Triangles, 4-Cycles and Parameterized (In-)Tractability. In: Arge, L., Freivalds, R. (eds.) SWAT 2006. LNCS, vol. 4059, pp. 304–315. Springer, Heidelberg (2006)
30. Rao, M.: Solving some NP-complete problems using split decomposition. Discrete Appl. Math. 156(14), 2768–2780 (2008)

On the Maximal Sum of Exponents of Runs in a String

Maxime Crochemore[1,3], Marcin Kubica[2], Jakub Radoszewski[2,*], Wojciech Rytter[2,4], and Tomasz Waleń[2]

[1] King's College London, London WC2R 2LS, UK
maxime.crochemore@kcl.ac.uk
[2] Dept. of Mathematics, Computer Science and Mechanics, University of Warsaw, Warsaw, Poland
[3] Université Paris-Est, France
[4] Dept. of Math. and Informatics, Copernicus University, Toruń, Poland
{kubica,jrad,rytter,walen}@mimuw.edu.pl

Abstract. A run is an inclusion maximal occurrence in a string (as a subinterval) of a repetition v with a period p such that $2p \leq |v|$. The exponent of a run is defined as $|v|/p$ and is ≥ 2. We show new bounds on the maximal sum of exponents of runs in a string of length n. Our upper bound of $4.1\,n$ is better than the best previously known proven bound of $5.6\,n$ by Crochemore & Ilie (2008). The lower bound of $2.035\,n$, obtained using a family of binary words, contradicts the conjecture of Kolpakov & Kucherov (1999) that the maximal sum of exponents of runs in a string of length n is smaller than $2n$.

1 Introduction

Repetitions and periodicities in strings are one of the fundamental topics in combinatorics on words [1,14]. They are also important in other areas: lossless compression, word representation, computational biology, etc. In this paper we consider bounds on the sum of exponents of repetitions that a string of a given length may contain. In general, repetitions are studied also from other points of view, like: the classification of words (both finite and infinite) not containing repetitions of a given exponent, efficient identification of factors being repetitions of different types and computing the bounds on the number of various types of repetitions occurring in a string. More results and motivation can be found in a survey by Crochemore et al. [4].

The concept of runs (also called maximal repetitions) has been introduced to represent all repetitions in a string in a succinct manner. The crucial property of runs is that their maximal number in a string of length n (denoted as $\rho(n)$) is $O(n)$, see Kolpakov & Kucherov [10]. This fact is the cornerstone of

* Corresponding author. Some parts of this paper were written during the corresponding author's Erasmus exchange at King's College London.

C.S. Iliopoulos and W.F. Smyth (Eds.): IWOCA 2010, LNCS 6460, pp. 10–19, 2011.

any algorithm computing all repetitions in strings of length n in $O(n)$ time. Due to the work of many people, much better bounds on $\rho(n)$ have been obtained. The lower bound $0.927\,n$ was first proved by Franek & Yang [7]. Afterwards, it was improved by Kusano et al. [13] to $0.944565\,n$ employing computer experiments, and very recently by Simpson [18] to $0.944575712\,n$. On the other hand, the first explicit upper bound $5\,n$ was settled by Rytter [16], afterwards it was systematically improved to $3.48\,n$ by Puglisi et al. [15], $3.44\,n$ by Rytter [17], $1.6\,n$ by Crochemore & Ilie [2,3] and $1.52\,n$ by Giraud [8]. The best known result $\rho(n) \le 1.029\,n$ is due to Crochemore et al. [5], but it is conjectured [10] that $\rho(n) < n$. Some results are known also for repetitions of exponent higher than 2. For instance, the maximal number of cubic runs (maximal repetitions with exponent at least 3) in a string of length n (denoted $\rho_{cubic}(n)$) is known to be between $0.406\,n$ and $0.5\,n$, see Crochemore et al. [6].

A stronger property of runs is that the maximal sum of their exponents in a string of length n (notation: $\sigma(n)$) is linear in terms of n, see final remarks in Kolpakov & Kucherov [12]. This fact has applications to the analysis of various algorithms, such as computing branching tandem repeats: the linearity of the sum of exponents solves a conjecture of [9] concerning the linearity of the number of maximal tandem repeats and implies that all can be found in linear time. For other applications, we refer to [12]. The proof that $\sigma(n) < cn$ in Kolpakov and Kucherov's paper [12] is very complex and does not provide any particular value for the constant c. A bound can be derived from the proof of Rytter [16] but the paper mentions only that the obtained bound is "unsatisfactory" (it seems to be $25\,n$). The first explicit bound $5.6\,n$ for $\sigma(n)$ was provided by Crochemore and Ilie [3], who claim that it could be improved to $2.9\,n$ employing computer experiments. As for the lower bound on $\sigma(n)$, no exact values were previously known and it was conjectured [11,12] that $\sigma(n) < 2n$.

In this paper we provide an upper bound of $4.1\,n$ on the maximal sum of exponents of runs in a string of length n and also a stronger upper bound of $2.5\,n$ for the maximal sum of exponents of cubic runs in a string of length n. As for the lower bound, we bring down the conjecture $\sigma(n) < 2n$ by providing an infinite family of binary strings for which the sum of exponents of runs is greater than $2.035\,n$.

2 Preliminaries

We consider *words* (*strings*) u over a finite alphabet Σ, $u \in \Sigma^*$; the empty word is denoted by ε; the positions in u are numbered from 1 to $|u|$. For $u = u_1 u_2 \ldots u_m$, let us denote by $u[i\,..\,j]$ a *factor* of u equal to $u_i \ldots u_j$ (in particular $u[i] = u[i\,..\,i]$). Words $u[1\,..\,i]$ are called prefixes of u, and words $u[i\,..\,|u|]$ suffixes of u.

We say that an integer p is the (shortest) *period* of a word $u = u_1 \ldots u_m$ (notation: $p = \mathsf{per}(u)$) if p is the smallest positive integer such that $u_i = u_{i+p}$ holds for all $1 \le i \le m - p$. We say that words u and v are cyclically equivalent (or that one of them is a cyclic rotation of the other) if $u = xy$ and $v = yx$ for some $x, y \in \Sigma^*$.

A *run* (also called a maximal repetition) in a string u is an interval $[i\,..\,j]$ such that:

- the period p of the associated factor $u[i\,..\,j]$ satisfies $2p \le j - i + 1$,
- the interval cannot be extended to the right nor to the left, without violating the above property, that is, $u[i-1] \neq u[i+p-1]$ and $u[j-p+1] \neq u[j+1]$, provided that the respective letters exist.

A *cubic run* is a run $[i\,..\,j]$ for which the shortest period p satisfies $3p \le j-i+1$. For simplicity, in the rest of the text we sometimes refer to runs and cubic runs as to occurrences of the corresponding factors of u. The (fractional) *exponent* of a run v, denoted $\mathsf{exp}(v)$, is defined as $(j-i+1)/p$.

For a given word $u \in \Sigma^*$, we introduce the following notation:

- $\rho(u)$ and $\rho_{cubic}(u)$ are the numbers of runs and cubic runs in u resp.
- $\sigma(u)$ and $\sigma_{cubic}(u)$ are the sums of exponents of runs and cubic runs in u resp.

For a non-negative integer n, we use the same notations $\rho(n)$, $\rho_{cubic}(n)$, $\sigma(n)$ and $\sigma_{cubic}(n)$ to denote the maximal value of the respective function for a word of length n.

3 Upper Bounds for $\sigma(n)$ and $\sigma_{cubic}(n)$

In this section we utilize the concept of *handles* of runs as defined in [6]. The original definition refers only to cubic runs, but here we extend it also to ordinary runs.

Let $u \in \Sigma^*$ be a word of length n. Let us denote by $P = \{p_1, p_2, \ldots, p_{n-1}\}$ the set of inter-positions in u that are located *between* pairs of consecutive letters of u. We define a function H assigning to each run v in u a set of some inter-positions within v (called later on *handles*) — H is a mapping from the set of runs occurring in u to the set 2^P of subsets of P. Let v be a run with period p and let w be the prefix of v of length p. Let $w_{\min}$ and $w_{\max}$ be the minimal and maximal words (in lexicographical order) cyclically equivalent to w. $H(v)$ is defined as follows:

a) if $w_{\min} = w_{\max}$ then $H(v)$ contains all inter-positions within v,
b) if $w_{\min} \neq w_{\max}$ then $H(v)$ contains inter-positions from the middle of any occurrence of $w_{\min}^2$ or $w_{\max}^2$ in v.

Note that $H(v)$ can be empty for a non-cubic-run v.

Proofs of the following properties of handles of runs can be found in [6]:

1. Case (a) in the definition of $H(v)$ implies that $|w_{\min}| = 1$.
2. $H(v_1) \cap H(v_2) = \emptyset$ for any two distinct runs v_1 and v_2 in u.

To prove the upper bound for $\sigma(n)$, we need to state an additional property of handles of runs. Let $\mathcal{R}(u)$ be the set of all runs in a word u, and let $\mathcal{R}_1(u)$ and $\mathcal{R}_{\ge 2}(u)$ be the sets of runs with period 1 and at least 2 respectively.

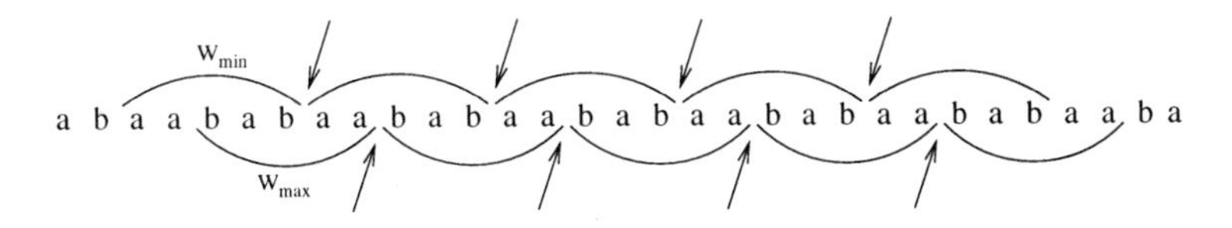

Fig. 1. An example of a run $v = (abaab)^6 a$ with exponent $\mathsf{exp}(v) = 6.2$. It contains $\lfloor 6.2 \rfloor - 1 = 5$ occurrences of each of the words $w_{\min} = aabab$ and $w_{\max} = babaa$. The set $H(v)$ contains $2 \cdot (\lfloor 6.2 \rfloor - 2) = 8$ inter-positions, pointed by arrows in the figure

Lemma 1
If $v \in \mathcal{R}_1(u)$ *then* $\mathsf{exp}(v) = |H(v)| + 1$.
If $v \in \mathcal{R}_{\geq 2}(u)$ *then* $\lceil \mathsf{exp}(v) \rceil \leq \frac{|H(v)|}{2} + 3$.

Proof. For the case of $v \in \mathcal{R}_1(u)$, the proof is straightforward from the definition of handles. Assume now that $v \in \mathcal{R}_{\geq 2}(u)$ and let w be a prefix of v of length $\mathsf{per}(v)$. Then the word w^k for $k = \lfloor \mathsf{exp}(v) \rfloor$ is a prefix of v, and therefore both words $w_{\min}^{k-1}$ and $w_{\max}^{k-1}$ are factors of v. Each of the words provides $k - 2$ distinct handles for v. Hence,

$$|H(v)| \geq 2 \cdot (\lfloor \mathsf{exp}(v) \rfloor - 2) .$$

□

Now we are ready to prove the upper bound for $\sigma(n)$. In the proof we use the bound $\rho(n) \leq 1.029\, n$ on the number of runs from [5].

Theorem 1. *The sum of the exponents of runs in a string of length* n *is less than* $4.1\, n$.

Proof. Let u be a word of length n. Using Lemma 1, we obtain:

$$\begin{aligned}
\sigma(u) &= \sum_{v \in \mathcal{R}_1(u)} \mathsf{exp}(v) + \sum_{v \in \mathcal{R}_{\geq 2}(u)} \mathsf{exp}(v) \\
&\leq \sum_{v \in \mathcal{R}_1(u)} (|H(v)| + 1) + \sum_{v \in \mathcal{R}_{\geq 2}(u)} \left(\frac{|H(v)|}{2} + 3 \right) \\
&= \sum_{v \in \mathcal{R}_1(u)} |H(v)| + |\mathcal{R}_1(u)| + \sum_{v \in \mathcal{R}_{\geq 2}(u)} \frac{|H(v)|}{2} + 3 \cdot |\mathcal{R}_{\geq 2}(u)| \\
&\leq 3 \cdot |\mathcal{R}(u)| + A + B/2, \qquad (1)
\end{aligned}$$

where $A = \sum_{v \in \mathcal{R}_1(u)} |H(v)|$ and $B = \sum_{v \in \mathcal{R}_{\geq 2}(u)} |H(v)|$. Due to the disjointness of handles of runs (the second property of handles), $A + B < n$, and thus, $A + B/2 < n$. Combining this with (1), we obtain:

$$\sigma(u) < 3 \cdot |\mathcal{R}(u)| + n \leq 3 \cdot \rho(n) + n \leq 3 \cdot 1.029\, n + n < 4.1\, n .$$

□

A similar approach for cubic runs, this time using the bound of $0.5\, n$ for $\rho_{cubic}(n)$ from [6], enables us to immediately provide a stronger upper bound for the function $\sigma_{cubic}(n)$.

Theorem 2. *The sum of the exponents of cubic runs in a string of length n is less than $2.5\,n$.*

Proof. Let u be a word of length n. Using same inequalities as in the proof of Theorem 1, we obtain:

$$\sigma_{cubic}(u) \;<\; 3 \cdot |\mathcal{R}_{cubic}(u)| + n \;\leq\; 3 \cdot \rho_{cubic}(n) + n \;\leq\; 3 \cdot 0.5\,n + n \;=\; 2.5\,n\;,$$

where $\mathcal{R}_{cubic}(u)$ denotes the set of all cubic runs of u. □

4 Lower Bound for $\sigma(n)$

Let us start by investigating the sums of exponents of runs for words of two known families that contain a large number of runs. We consider first the words defined by Franek & Yang [7], then the Padovan words defined by Simpson [18]. They give large sums of exponents, however below $2n$. Then we construct a new family of words which breaks the barrier of $2n$.

Table 1. Number of runs and sum of exponents of runs in Franek & Yang's [7] words x_i

i	$\lvert x_i\rvert$	$\rho(x_i)/\lvert x_i\rvert$	$\sigma(x_i)$	$\sigma(x_i)/\lvert x_i\rvert$
1	6	0.3333	4.00	0.6667
2	27	0.7037	39.18	1.4510
3	116	0.8534	209.70	1.8078
4	493	0.9047	954.27	1.9356
5	2090	0.9206	4130.66	1.9764
6	8855	0.9252	17608.48	1.9885
7	37512	0.9266	74723.85	1.9920
8	158905	0.9269	316690.85	1.9930
9	673134	0.9270	1341701.95	1.9932

Let $\circ$ be a special concatenation operator defined as:

$$x[1\mathinner{..}n] \circ y[1\mathinner{..}m] = \begin{cases} x[1\mathinner{..}n]y[2\mathinner{..}m] = x[1\mathinner{..}n-1]y[1\mathinner{..}m] & \text{if } x[n] = y[1], \\ x[1\mathinner{..}n-1]y[2\mathinner{..}m] & \text{if } x[n] \neq y[1]. \end{cases}$$

Also let g be a morphism defined as:

$$g(x) = \begin{cases} 010010 & \text{if } x = 0, \\ 101101 & \text{if } x = 1, \\ g(x[1\mathinner{..}n]) = g(x[1]) \circ g(x[2]) \circ \ldots \circ g(x[n]) & \text{if } |x| > 1. \end{cases}$$

Then $x_i = g^i(0)$ is the family of words described by Franek and Yang [7], which gives the lower bound $\rho(n) \geq 0.927\,n$, conjectured for some time to be optimal. The sums of exponents of runs of several first terms of the sequence x_i are listed in Table 1.

Define a mapping $\delta(x) = R(f(x))$, where $R(x)$ is the reverse of x and f is the morphism

$$f(a) = aacab,\ f(b) = acab,\ f(c) = ac\ .$$

Let y'_i be a sequence of words defined for $i > 5$ recursively using $y'_{i+5} = \delta(y'_i)$. The first 5 elements of the sequence y'_i are:

$$b,\ a,\ ac,\ ba,\ aca\ .$$

The strings y'_i are called modified Padovan words. If we apply the following morphism h:

$$h(a) = 1010010110010100101101000,$$

$$h(b) = 1010010110100,\quad h(c) = 10100101$$

to y'_i, we obtain a sequence of run-rich strings y_i defined by Simpson [18], which gives the best known lower bound $\rho(n) \geq 0.944575712\,n$. Table 2 lists the sums of exponents of runs of selected words from the sequence y_i.

Table 2. Number of runs and sum of exponents of runs in Simpson's [18] modified Padovan words y_i

i	$\lvert y_i \rvert$	$\rho(y_i)/\lvert y_i \rvert$	$\sigma(y_i)$	$\sigma(y_i)/\lvert y_i \rvert$
1	13	0.6154	16.00	1.2308
6	69	0.7971	114.49	1.6593
11	287	0.8990	542.72	1.8910
16	1172	0.9309	2303.21	1.9652
21	4781	0.9406	9504.38	1.9879
26	19504	0.9434	38903.64	1.9946
31	79568	0.9443	158862.94	1.9966
36	324605	0.9445	648270.74	1.9971
41	1324257	0.9446	2644879.01	1.9973

The values in Tables 1 and 2 have been computed experimentally. They suggest that for the families of words x_i and y_i the maximal sum of exponents could be less than $2n$. We show, however, a lower bound for $\sigma(n)$ that is greater than $2n$.

Theorem 3. *There are infinitely many binary strings w such that*

$$\frac{\sigma(w)}{|w|} > 2.035\ .$$

Proof. Let us define two morphisms $\phi : \{a,b,c\} \mapsto \{a,b,c\}$ and $\psi : \{a,b,c\} \mapsto \{0,1\}$ as follows:

$$\phi(a) = baaba,\quad \phi(b) = ca,\quad \phi(c) = bca$$

$$\psi(a) = 01011,\quad \psi(b) = \psi(c) = 01001011\ .$$

Table 3. Sums of exponents of runs in words w_i

i	$\|w_i\|$	$\sigma(w_i)$	$\sigma(w_i)/\|w_i\|$
1	31	47.10	1.5194
2	119	222.26	1.8677
3	461	911.68	1.9776
4	1751	3533.34	2.0179
5	6647	13498.20	2.0307
6	25205	51264.37	2.0339
7	95567	194470.30	2.0349
8	362327	737393.11	2.0352
9	1373693	2795792.39	2.0352
10	5208071	10599765.15	2.0353

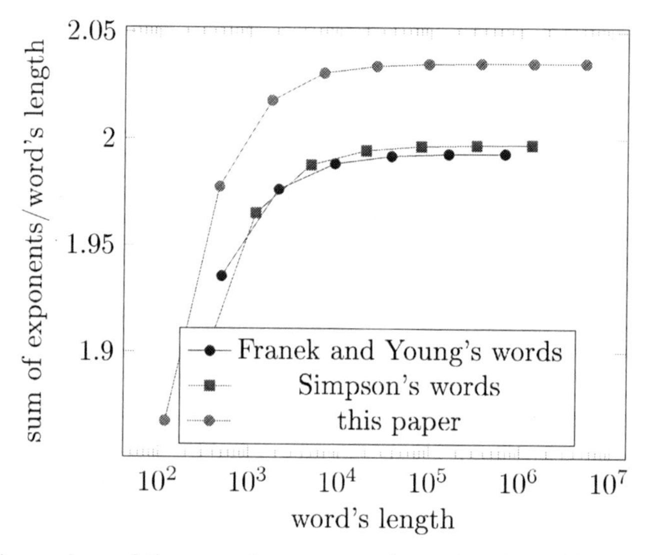

Fig. 2. Comparison of the sum of exponents of runs in selected families of words

We define $w_i = \psi(\phi^i(a))$. Table 3 and Figure 2 show the sums of exponents of runs in words $w_1, \dots, w_{10}$, computed experimentally.

Clearly, for any word $w = (w_8)^k$, $k \geq 1$, we have

$$\frac{\sigma(w)}{|w|} > 2.035 .$$

□

5 Relating the Upper Bound for $\sigma(n)$ to Semicubic Runs

Recall that $1.029\,n$ is the best known upper bound for $\rho(n)$ [5]. On the other hand, the best known corresponding upper bound for cubic runs, for which the exponent is at least 3, is much smaller: $0.5\,n$ [6].

This suggests that the upper bound for the maximal number of runs with an intermediate exponent, e.g. at least 2.5, in a string of length n could be smaller than the general bound for exponent at least 2. Let us call such runs with exponent at least 2.5 *semicubic* runs.

Observation
The number of semicubic runs in Fibonacci strings is relatively small, it can be proved that in case of these strings every semicubic run is also cubic (if exponent is at least 2.5 then it is at least 3).

Let $\rho_{semic}(u)$ be the number of semicubic runs in the string u and let $\rho_{semic}(n)$ denote the maximum of $\rho_{semic}(u)$ over all strings u of length n.

Using extensive computer experiments, we have found the following family of binary words $z_i = \nu(\mu^i(a))$, where:

$$\mu(a) = ba \qquad \nu(a) = 0010100010$$
$$\mu(b) = aba \qquad \nu(b) = 001010$$

that contain, for sufficiently large i, at least $0.52\,n$ semi-cubic runs, see Table 4.

Table 4. The number of semicubic runs in the words $z_i = \nu(\mu^i(a))$

i	$\lvert z_i \rvert$	$\rho_{semic}(z_i)$	$\rho_{semic}(z_i)/\lvert z_i \rvert$
1	16	5	0.3125
2	42	17	0.4048
3	100	46	0.4600
4	242	118	0.4876
5	584	296	0.5069
6	1410	724	0.5135
7	3404	1762	0.5176
8	8218	4266	0.5191
9	19840	10316	0.5200
10	47898	24920	0.5203
11	115636	60182	0.5204
12	279170	145310	0.5205
13	673976	350832	0.5205
14	1627122	847004	0.5206

Using the same program we managed to construct strings with $\rho(n) \geq 0.944575\,n$, that is, very close to the best known lower bound for this function and also close to the best known upper bound. This suggests that the data from the program are good approximation for semicubic runs as well. Therefore we conjecture the following.

Conjecture 1. **[Semicubic-Runs Conjecture]**
$\rho_{semic}(n) \leq 0.6\,n.$

If the above conjecture holds, it lets us instantly improve the upper bound for $\sigma(n)$.

Theorem 4. *If Conjecture 1 is true then* $\sigma(n) \leq 3.9\,n$.

Proof. Let $\mathcal{R}_{semic}(u)$ denote the set of all semicubic runs in a string u. Conjecture 1 lets us improve the part of the proof of Theorem 1 related to the term $\sum_{v\in\mathcal{R}_{\geq 2}(u)} \mathsf{exp}(v)$:

$$\begin{aligned}
\sigma(u) &= \sum_{v\in\mathcal{R}_1(u)} \mathsf{exp}(v) + \sum_{v\in\mathcal{R}_{\geq 2}(u)} \mathsf{exp}(v) \\
&= \sum_{v\in\mathcal{R}_1(u)} \mathsf{exp}(v) + \sum_{v\in\mathcal{R}_{\geq 2}(u)\setminus\mathcal{R}_{semic}(u)} \mathsf{exp}(v) + \sum_{v\in\mathcal{R}_{semic}(u)} \mathsf{exp}(v) \\
&\leq \sum_{v\in\mathcal{R}_1(u)} (|H(v)|+1) + \sum_{v\in\mathcal{R}_{\geq 2}(u)\setminus\mathcal{R}_{semic}(u)} 2.5 + \sum_{v\in\mathcal{R}_{semic}(u)} \left(\frac{|H(v)|}{2}+3\right) \\
&\leq \sum_{v\in\mathcal{R}(u)} |H(v)| + |\mathcal{R}_1(u)| + 2.5\cdot|\mathcal{R}_{\geq 2}(u)\setminus\mathcal{R}_{semic}(u)| + 3\cdot|\mathcal{R}_{semic}(u)| \\
&\leq n + 2.5\cdot(1.029\,n - 0.6\,n) + 3\cdot 0.6\,n \;<\; n + 1.1\,n + 1.8\,n \;=\; 3.9\,n\ .
\end{aligned}$$

In the above formula, in order to obtain an upper bound we used a greedy approach to distribute $1.029\,n$ among the sizes of the sets $\mathcal{R}_1(u)$, $\mathcal{R}_{\geq 2}(u)\setminus\mathcal{R}_{semic}(u)$ and $\mathcal{R}_{semic}(u)$. □

6 Conclusions

In this paper we have provided an upper bound of $4.1\,n$ on the maximal sum of exponents of runs in a string of length n and also a stronger upper bound of $2.5\,n$ for the maximal sum of exponents of cubic runs in a string of length n. As for the lower bound, we bring down the conjecture $\sigma(n) < 2n$ by providing an infinite family of binary strings for which the sum of exponents of runs is greater than $2.035\,n$.

A natural open problem is to tighten these bounds. One of the possible directions for this improvement, presented in this paper, consists in finding bounds for the maximal number of runs with exponent at least f, where $f \in (2,3)$, in a string of length n.

References

1. Berstel, J., Karhumaki, J.: Combinatorics on words: a tutorial. Bulletin of the EATCS 79, 178–228 (2003)
2. Crochemore, M., Ilie, L.: Analysis of maximal repetitions in strings. In: Kucera, L., Kucera, A. (eds.) MFCS 2007. LNCS, vol. 4708, pp. 465–476. Springer, Heidelberg (2007)
3. Crochemore, M., Ilie, L.: Maximal repetitions in strings. J. Comput. Syst. Sci. 74(5), 796–807 (2008)
4. Crochemore, M., Ilie, L., Rytter, W.: Repetitions in strings: Algorithms and combinatorics. Theor. Comput. Sci. 410(50), 5227–5235 (2009)

5. Crochemore, M., Ilie, L., Tinta, L.: Towards a solution to the "runs" conjecture. In: Ferragina, P., Landau, G.M. (eds.) CPM 2008. LNCS, vol. 5029, pp. 290–302. Springer, Heidelberg (2008)
6. Crochemore, M., Iliopoulos, C.S., Kubica, M., Radoszewski, J., Rytter, W., Walen, T.: On the maximal number of cubic runs in a string. In: Dediu, A.H., Fernau, H., Martín-Vide, C. (eds.) LATA 2010. LNCS, vol. 6031, pp. 227–238. Springer, Heidelberg (2010)
7. Franek, F., Yang, Q.: An asymptotic lower bound for the maximal number of runs in a string. Int. J. Found. Comput. Sci. 19(1), 195–203 (2008)
8. Giraud, M.: Not so many runs in strings. In: Martín-Vide, C., Otto, F., Fernau, H. (eds.) LATA 2008. LNCS, vol. 5196, pp. 232–239. Springer, Heidelberg (2008)
9. Gusfield, D., Stoye, J.: Simple and flexible detection of contiguous repeats using a suffix tree (preliminary version). In: Farach-Colton, M. (ed.) CPM 1998. LNCS, vol. 1448, pp. 140–152. Springer, Heidelberg (1998)
10. Kolpakov, R.M., Kucherov, G.: Finding maximal repetitions in a word in linear time. In: Proceedings of the 40th Symposium on Foundations of Computer Science, pp. 596–604 (1999)
11. Kolpakov, R.M., Kucherov, G.: On maximal repetitions in words. J. of Discr. Alg. 1, 159–186 (1999)
12. Kolpakov, R.M., Kucherov, G.: On the sum of exponents of maximal repetitions in a word. Tech. Report 99-R-034, LORIA (1999)
13. Kusano, K., Matsubara, W., Ishino, A., Bannai, H., Shinohara, A.: New lower bounds for the maximum number of runs in a string. CoRR, abs/0804.1214 (2008)
14. Lothaire, M.: Combinatorics on Words. Addison-Wesley, Reading (1983)
15. Puglisi, S.J., Simpson, J., Smyth, W.F.: How many runs can a string contain? Theor. Comput. Sci. 401(1-3), 165–171 (2008)
16. Rytter, W.: The number of runs in a string: Improved analysis of the linear upper bound. In: Durand, B., Thomas, W. (eds.) STACS 2006. LNCS, vol. 3884, pp. 184–195. Springer, Heidelberg (2006)
17. Rytter, W.: The number of runs in a string. Inf. Comput. 205(9), 1459–1469 (2007)
18. Simpson, J.: Modified Padovan words and the maximum number of runs in a word. Australasian J. of Comb. 46, 129–145 (2010)

Path-Based Supports for Hypergraphs

Ulrik Brandes[1], Sabine Cornelsen[1], Barbara Pampel[1], and Arnaud Sallaberry[2]

[1] Fachbereich Informatik & Informationswissenschaft, Universität Konstanz
{Ulrik.Brandes,Sabine.Cornelsen,Barbara.Pampel}@uni-konstanz.de
[2] CNRS UMR 5800 LaBRI, INRIA Bordeaux - Sud Ouest, Pikko
arnaud.sallaberry@labri.fr

Abstract. A path-based support of a hypergraph H is a graph with the same vertex set as H in which each hyperedge induces a Hamiltonian subgraph. While it is $\mathcal{NP}$-complete to compute a path-based support with the minimum number of edges or to decide whether there is a planar path-based support, we show that a path-based tree support can be computed in polynomial time if it exists.

1 Introduction

A *hypergraph* is a pair $H = (V, A)$ where V is a finite set and A is a (multi-)set of non-empty subsets of V. The elements of V are called *vertices* and the elements of A are called *hyperedges*. A *support* (or *host graph*) of a hypergraph $H = (V, A)$ is a graph $G = (V, E)$ such that each hyperedge of H induces a connected subgraph of G, i.e., such that the graph $G[h] := (h, \{e \in E, e \subseteq h\})$ is connected for every $h \in A$. See Fig. 1(b) for an example.

Applications for supports of hypergraphs are, e.g., in hypergraph coloring [10, 4], databases [1], or hypergraph drawing [7, 8, 3, 12]. E.g., see Fig. 1 for an application of a support for designing Euler diagrams. An *Euler diagram* of a hypergraph $H = (V, A)$ is a drawing of H in the plane in which the vertices are drawn as points and each hyperedge $h \in A$ is drawn as a simple closed region containing the points representing the vertices in h and not the points representing the vertices in $V \setminus h$. There are various well-formedness conditions for Euler diagrams, see e.g. [5, 12].

Recently the problem of deciding which classes of hypergraphs admit what kind of supports became of interest again. It can be tested in linear time whether a hypergraph has a support that is a tree [13], a path or a cycle [3]. It can be decided in polynomial time whether a hypergraph has a tree support with bounded degrees [3] or a cactus support [2]. A minimum weighted tree support can be computed in polynomial time [9]. It is $\mathcal{NP}$-complete to decide whether a hypergraph has a planar support [7], a compact support [7, 8] or a 2-outerplanar support [3]. A support with the minimum number of edges can be computed in polynomial time if the hypergraph is closed under intersections [3]. If the set of hyperedges is closed under intersections and differences, it can be decided in polynomial time whether the hypergraph has an outerplanar support [2].

C.S. Iliopoulos and W.F. Smyth (Eds.): IWOCA 2010, LNCS 6460, pp. 20–33, 2011.

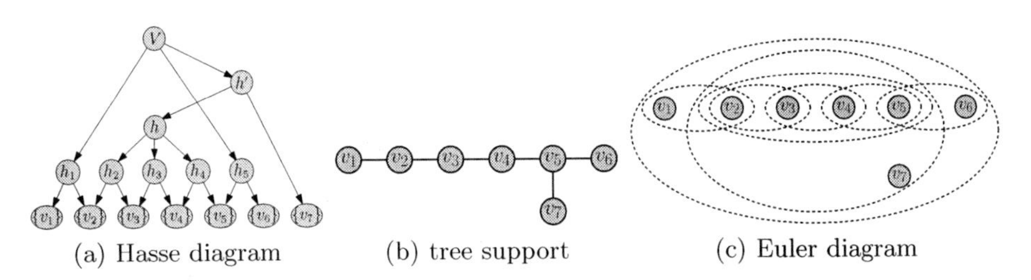

(a) Hasse diagram (b) tree support (c) Euler diagram

Fig. 1. Three representations of the hypergraph $H = (V, A)$ with hyperedges $h_1 = \{v_1, v_2\}$, $h_2 = \{v_2, v_3\}$, $h_3 = \{v_3, v_4\}$, $h_4 = \{v_4, v_5\}$, $h_5 = \{v_5, v_6\}$, $h = \{v_2, v_3, v_4, v_5\}$, $h' = \{v_2, v_3, v_4, v_5, v_7\}$, and $V = \{v_1, \ldots, v_7\}$

In this paper, we consider a restriction on the subgraphs of a support that are induced by the hyperedges. A support G of a hypergraph $H = (V, A)$ is called *path-based* if the subgraph $G[h]$ contains a *Hamiltonian path* for each hyperedge $h \in A$, i.e., $G[h]$ contains a path that contains each vertex of h. This approach was on one hand motivated by hypergraph drawing and on the other hand by the aesthetics of metro map layouts. I.e., the hyperedges could be visualized as lines along the Hamiltonian path in the induced subgraph of the support like the metro lines in a metro map. See Fig. 2 for examples of metro maps and Fig. 3(c) for a representation of some hyperedges in such a metro map like drawing. For metro map layout algorithms see, e.g., [11, 14].

We briefly consider planar path-based supports and minimum path-based supports. Our main result is a characterization of those hypergraphs that have a path-based tree support and a polynomial time algorithm for constructing path-based tree supports if they exist. E.g., Fig. 1 shows an example of a hypergraph $H = (V, A)$ that has a tree support but no path-based tree support. However, the tree support in Fig. 1(b) is a path-based tree support for $(V, A \setminus \{V\})$.

The contribution of this paper is as follows. In Section 2, we give the necessary definitions. We then briefly mention in Section 3 that finding a minimum path-based support or deciding whether there is a planar path-based support, respectively, is $\mathcal{NP}$-complete. We consider path-based tree supports in Sect. 4. In Section 4.1, we review a method for computing tree supports using the Hasse diagram. In Section 4.2, we show how to apply this method to test whether a hypergraph has a path-based tree support and if so how to compute one in polynomial time. Finally, in Section 4.3 we discuss the run time of our method.

2 Preliminaries

In this section, we give the necessary definitions that were not already given in the introduction. Throughout this paper let $H = (V, A)$ be a hypergraph. We denote by $n = |V|$ the number of vertices, $m = |A|$ the number of hyperedges, and $N = \sum_{h \in A} |h|$ the sum of the sizes of all hyperedges of a hypergraph H. The *size of the hypergraph* H is then $N + n + m$. A hypergraph is a *graph* if

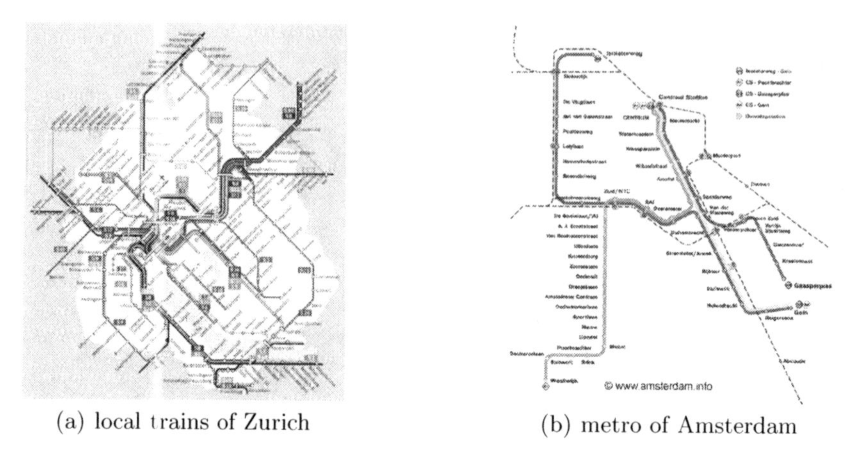

(a) local trains of Zurich (b) metro of Amsterdam

Fig. 2. Local train map of Zurich (`www.zvv.ch`) and the metro map of Amsterdam (`www.amsterdam.info`). In (b) the union of all lines forms a tree.

all hyperedges contain exactly two vertices. A hypergraph $H = (V, A)$ is *closed under intersections* if $h_1 \cap h_2 \in A \cup \{\emptyset\}$ for $h_1, h_2 \in A$.

The *Hasse diagram* of a hypergraph $H = (V, A)$ is the directed acyclic graph with vertex set $A \cup \{\{v\}; v \in V\}$ and there is an edge (h_1, h_2) if and only if $h_2 \subsetneq h_1$ and there is no set $h \in A$ with $h_2 \subsetneq h \subsetneq h_1$. Fig. 1(a) shows an example of a Hasse diagram. Let (v, w) be an edge of a directed acyclic graph. Then we say that w is a *child* of v and v a *parent* of w. For a *descendant* d of v there is a directed path from v to d while for an *ancestor* a of v there is a directed path from a to v. A *source* does not have any parents, a *sink* no children and an *inner vertex* has at least one parent and one child.

3 Minimum and Planar Path-Based Supports

Assuming that each hyperedge contains at least one vertex, each hypergraph $H = (V, A)$ has a path-based support $G = (V, E)$ with at most $N - m$ edges: Order the vertices arbitrarily. For each hyperedge $\{v_1, \dots, v_k\} \in A$ with $v_1 < \dots < v_k$ with respect to that ordering the edge set E contains $\{v_{i-1}, v_i\}, i = 1, \dots, k$. It is, however, $\mathcal{NP}$-complete to find an ordering of the vertices that minimizes the number of edges of the thus constructed path-based support of H [6]. Moreover, even if we had an ordering of the vertices that had minimized the number of the thus constructed path-based support, this support still does not have to yield the minimum number of edges in any path-based support of H. E.g., consider the hypergraph with hyperedges $\{1, 2, 4\}$, $\{1, 3, 4\}$, and $\{2, 3, 4\}$. Nevertheless, we have the following theorem.

Theorem 1. *It is $\mathcal{NP}$-complete to minimize the number of edges in a path-based support of a hypergraph – even if it is closed under intersections.*

Proof. Reduction from Hamiltonian path. Let $G = (V, E)$ be a graph. Let $H = (V, E \cup \{V\} \cup \{\{v\};\ v \in V\})$ and $K = |E|$. Then G contains a Hamiltonian path if and only if H has a path-based support with at most K edges. □

For the application of Euler diagram like drawings, planar supports are of special interests. However, like for general planar supports, the problem of testing whether there is a path-based planar support is hard.

Theorem 2. *It is $\mathcal{NP}$-complete to decide whether a hypergraph – even if it is closed under intersections – has a path-based planar support.*

Proof. The support that Johnson and Pollak [7] constructed to prove that it is $\mathcal{NP}$-complete to decide whether there is a planar support was already path-based. □

4 Path-Based Tree Supports

In this section we show how to decide in polynomial time whether a given hypergraph has a path-based tree support. If such a support exists, it is at the same time a path-based support of minimum size and a planar path-based support. So far it is known how to decide in linear time whether there is a path-based tree support if $V \in A$ [3].

4.1 Constructing a Tree Support from the Hasse Diagram

A support with the minimum number of edges and, hence, a tree support if one exists can easily be constructed from the Hasse diagram if the hypergraph is closed under intersections [3].

To construct a tree support of an arbitrary hypergraph, it suffices to consider the *augmented Hasse diagram* – a representation of "necessary" intersections of hyperedges. The definition is as follows. First consider the smallest set $\overline{A}$ of subsets of V that contains A and that is closed under intersections. Consider the Hasse diagram $\overline{D}$ of $\overline{H} = (V, \overline{A})$. Note that any tree support of H is also a tree support of $\overline{H}$. Let $h_1, \ldots, h_k$ be the children of a hyperedge h in $\overline{D}$. The hyperedge $h \in \overline{A}$ is *implied* if the hypergraph $(h_1 \cup \cdots \cup h_k, \{h_1, \ldots, h_k\})$ is connected and *non-implied* otherwise. Let $\{h_1, \ldots, h_k\}$ be a maximal subset of the children of a non-implied hyperedge in $\overline{A}$ such that $(h_1 \cup \cdots \cup h_k, \{h_1, \ldots, h_k\})$ is connected. Then $h_1 \cup \cdots \cup h_k$ is a *summary hyperedge*. Note that a summary hyperedge does not have to be in $\overline{A}$. Let A' be the set of subsets of V containing the summary hyperedges, the hyperedges in $\overline{A}$ that are not implied, and the sources of $\overline{D}$. E.g., for the hypergraph in Fig. 1 it holds that $A' = A$. In this example, the hyperedge h is a summary hyperedge, h' is not implied, and V is a source.

The augmented Hasse diagram of H is the Hasse diagram D' of $H' = (V, A')$. If H has a tree support then the augmented Hasse diagram has $\mathcal{O}(n+m)$ vertices and can be constructed in $\mathcal{O}(n^3 m)$ time [3]. Further note that if H has a tree support and $h \in A'$ is non-implied then all children of h in D' are disjoint.

If a tree support $G = (V, E)$ of H exists it can be constructed as follows [3]. Starting with an empty graph G, we proceed from the sinks to the sources of D'. If $h \in A'$ is not implied, choose an arbitrary ordering $h_1, \ldots, h_k$ of the children of h in D'. We assume that at this stage, $G[h_i], i = 1, \ldots, k$ are already connected subgraphs of G. For $j = 2, \ldots, k$, choose vertices $v_j \in \bigcup_{i=1}^{j-1} h_i$, $w_j \in h_j$ and add edges $\{v_j, w_j\}$ to E.

If we want to construct a path-based tree support, then $G[h_j], j = 1, \ldots, k$ are paths and as vertices v_{j+1} and w_j for the edges connecting $G[h_j]$ to the other paths, we choose the end vertices of $G[h_j]$. The only choices that remain is the ordering of the children of h and the choice of which end vertex of $G[h_j]$ is w_j and which one is v_{j+1}. The implied hyperedges give restrictions on how these choices might be done.

4.2 Choosing the Connections: A Characterization

When we want to apply the general method introduced in Sect. 4.1 to construct a path-based tree support G, we have to make sure that we do not create vertices of degree greater than 2 in $G[h]$ when processing non-implied hyperedges contained in an implied hyperedge h.

Let $h', h'' \in A'$. We say that h', h'' overlap if $h' \cap h'' \neq \emptyset$, $h' \not\subseteq h''$, and $h'' \not\subseteq h'$. Two overlapping hyperedges $h', h'' \in A'$ have a *conflict* if there is some hyperedge in A' that contains h' and h''. Two overlapping hyperedges $h', h'' \in A'$ have a *conflict with respect to* $h \in A'$ if h' has a conflict with h'', $h' \cap h'' \subseteq h$ and h is a child of h' or h''. In that case we say that h' and h'' are *conflicting* hyperedges of h. Let A'_h be the set of conflicting hyperedges of h. Let A^c_h be the set of children h_i of h such that $h \in A'_{h_i}$.

Assume now that H has a path-based tree support G and let $h, h', h'' \in A'$ be such that h' and h'' have a conflict with respect to h. We have three types of restrictions on the connections of the paths.

1. $G[h' \setminus h]$ and $G[h'' \setminus h]$ are paths that are attached to different end vertices of $G[h]$. Otherwise $G[h_a]$ contains a vertex of degree higher than 2 for any hyperedge $h_a \supseteq h' \cup h''$.
2. Assume further that $h_1 \in A^c_h$. For all hyperedges $h^1 \in A'_h$ that have a conflict with h with respect to h_1 it holds that $G[h^1 \setminus h]$ has to be appended to the end vertex of $G[h]$ that is also an end vertex of $G[h_1]$. Hence, all these paths $G[h^1 \setminus h]$ have to be appended to the same end vertex of $G[h]$.
3. Assume further that $h_2 \in A^c_h, h_2 \neq h_1$. Let $h^i \in A'_h$ have a conflict with h with respect to $h_i, i = 1, 2$, respectively. Then $G[h^i \setminus h]$ has to be appended to the end vertex of $G[h]$ that is also an end vertex of $G[h_i]$. Hence, $G[h^1 \setminus h]$ and $G[h^2 \setminus h]$ have to be appended to different end vertices of $G[h]$.

E.g., consider the hypergraph $H = (V, A)$ in Fig. 1. Then on one hand, h' has a conflict with h_1 and h_5 with respect to h. Hence, by the first type of restrictions $G[h_1 \setminus h]$ and $G[h_5 \setminus h]$ have to be appended to the same end vertex of $G[h]$, i.e. the end vertex of $G[h]$ to which $G[h' \setminus h]$ is not appended. On the other hand, h_1 and h have a conflict with respect to h_2 while h_5 and h have a conflict with respect to h_4. Hence, by the third type of restrictions it follows that $G[h_1 \setminus h]$ and $G[h_5 \setminus h]$ have to be appended to different end vertices of $G[h]$. Hence, there is no path-based tree support for H.

This motivates the following definition of conflict graphs. The *conflict graph* $C_h, h \in A'$ is a graph on the vertex set $A'_h \cup A^c_h$. The conflict graph C_h contains the following three types of edges.

1. $\{h', h''\}, h', h'' \in A'_h$ if h' and h'' have a conflict with respect to h.
2. $\{h', h_1\}, h' \in A'_h, h_1 \in A^c_h$ if $h' \in A'_{h_1}$ and h' and h have a conflict with respect to h_1.
3. $\{h_1, h_2\}, h_1, h_2 \in A^c_h, h_1 \neq h_2$.

E.g., consider the hypergraph $H = (V, A)$ in Fig. 1. Then the conflict graph C_h contains the edges $\{h', h_5\}$ and $\{h', h_1\}$ of type one, the edges $\{h_2, h_1\}$ and $\{h_4, h_5\}$ of type 2 and the edge $\{h_2, h_4\}$ of type 3. Hence, C_h contains a cycle of odd length, reflecting that there is no suitable assignment of the end vertices of $G[h]$ to h_1, h_5 and h'.

Theorem 3. *A hypergraph $H = (V, A)$ has a path-based tree support if and only if*

1. *H has a tree support,*
2. *no hyperedge contains three pairwise overlapping hyperedges $h_1, h_2, h_3 \in A'$ with $h_1 \cap h_2 = h_2 \cap h_3 = h_1 \cap h_3$, and*
3. *all conflict graphs $C_h, h \in A', |h| > 1$ are bipartite.*

From the observations before the definition of the conflict graph it is clear that the conditions of Theorem 3 are necessary for a path-based tree support. In the remainder of this section, we prove that the conditions are also sufficient.

In the following assume that the conditions of Theorem 3 are fulfilled. We show in Algorithm 1 how to construct a path-based tree support G of H. We consider the vertices of the augmented Hasse diagram D' from the sinks to the sources in a *reversed topological order*, i.e., we consider a hyperedge only if all its children in D' have already been considered. During the algorithm, a conflicting hyperedge h' of a hyperedge h is labeled with the end vertex v of $G[h]$ if the path $G[h' \setminus h]$ will be appended to v. We will call this label $\text{side}_h(h')$. Concerning Step 2a, the sets $A^c_h, h \in A'$ contain at most two hyperedges – otherwise the subgraph of C_h induced by A^c_h contains a triangle and, hence, is not bipartite.

Algorithm 1 constructs a tree support G of H [3]. Before we show that G is a path-based tree support, we illustrate the algorithm with an example. Consider the hypergraph H in Fig. 3. We show how the algorithm proceeds h^5_1 and all its descendants in D'. For the hyperedges h^1_3, h^1_4, h^1_6, and h^1_8 the conflict graphs are empty while for the other leaves we have $\text{side}_{h^1_5}(h^2_2) = \text{side}_{h^1_5}(h^2_3) = \text{side}_{h^1_5}(h^3_1) = \text{side}_{h^1_5}(h^4_2) = v_5$, $\text{side}_{h^1_7}(h^2_4) = \text{side}_{h^1_7}(h^3_1) = v_7$, and $\text{side}_{h^1_9}(h^2_4) = \text{side}_{h^1_9}(h^4_1) = \text{side}_{h^1_9}(h^2_5) = \text{side}_{h^1_9}(h^2_6) = \text{side}_{h^1_9}(h^2_7) = v_9$. When operating h^2_2 and h^2_3, respectively, we add edges $\{v_4, v_5\}$ and $\{v_5, v_6\}$, respectively, to G. While the conflict graph of h_2^2 does only contain h_5^1 with $\text{side}_{h_2^2}(h_5^1) = v_4$, in $C_{h_3^2}$ we set $\text{side}_{h_3^2}(h_5^1) = \text{side}_{h_3^2}(h_1^3) = v_6$, and $\text{side}_{h_3^2}(h_2^2) = v_5$. h_4^2 has a conflict with respect to h_7^1 and h_9^1. Hence, we add edges $\{v_7, v_8\}$ and $\{v_8, v_9\}$ to G. Further, $\text{side}_{h_4^2}(h_7^1) = \text{side}_{h_4^2}(h_5^2) = v_9$ and $\text{side}_{h_4^2}(h_9^1) = \text{side}_{h_4^2}(h_1^4) = v_7$. When operating h_1^3 we can choose $h_1 = h_3^2$ and $h_2 = h_7^1$, since $\text{side}_{h_3^2}(h_3^1) = v_6$ and $\text{side}_{h_7^1}(h_3^1) = v_7$. We add the edge $\{v_6, v_7\}$ to G. The conflict graph $C_{h_1^3}$ is shown in Fig. 3(b). The hyperedge h_1^4 is implied and we set $\text{side}_{h_1^4}(h_4^2) = v_4$. We can finally connect v_3 to v_4 or v_9 when operating h_1^5.

To prove the correctness of Algorithm 1, it remains to show that all hyperedges of H induce a path in G. Since we included all inclusion maximal hyperedges of H in A', it suffices to show this property for all hyperedges in A'. We start with a technical lemma.

Lemma 1. *Let h' and h'' be two overlapping hyperedges and let h' be not implied. Then there is a hyperedge $h \in A'$ with $h' \cap h'' \subseteq h \subsetneq h'$.*

Proof. Let $h_c \in \overline{A}$ be maximal with $h' \cap h'' \subseteq h_c \subsetneq h'$. The hyperedge h_c is a child of the non-implied hyperedge h' in $\overline{D}$. Consider the summary hyperedge h with $h_c \subseteq h \subsetneq h'$. By definition of A' it follows that $h \in A'$. □

For an edge $\{v, w\}$ of G let h_{vw} be the intersection of all hyperedges of A' that contain v and w. Note that then h_{vw} is not implied since v and w cannot both be contained in a subset of h_{vw}. Hence, $h_{vw} \in A'$.

Algorithm 1. Path-based tree support

Let $E = \emptyset$.
For $h \in A'$ in a reversed topological order of D'.
1. If $h = \{v\}$ for some $v \in V$
   (a) set $\text{side}_h(h') = v$ for all vertices h' of C_h.
2. Else
   (a) let $h_1, \ldots, h_k$ be the children of h such that $h_2, \ldots, h_{k-1} \notin A_h^c$.
   (b) If h is non-implied
      i. let $w_i, v_{i+1}, i = 1, \ldots, k$ be the end vertices of $G[h_i]$ such that
         A. $\text{side}_{h_1}(h) = v_2$ if $h \in A'_{h_1}$ and
         B. $\text{side}_{h_k}(h) = w_k$ if $h \in A'_{h_k}$.
      ii. Add the edges $\{v_i, w_i\}, i = 2, \ldots, k$ to E.
   (c) Else let $w_1 \neq v_{k+1}$ be the end vertices of $G[h]$ such that
      i. $v_{k+1} \notin h_1$ and
      ii. $w_1 \notin h_k$.
   (d) If $h_1 \in A_h^c$ set $\text{side}_h(h_1) = v_{k+1}$.
   (e) If $h_k \in A_h^c$ set $\text{side}_h(h_k) = w_1$.
   (f) Label the remaining vertices of C_h with v_{k+1} or w_1 such that no two adjacent vertices have the same label.

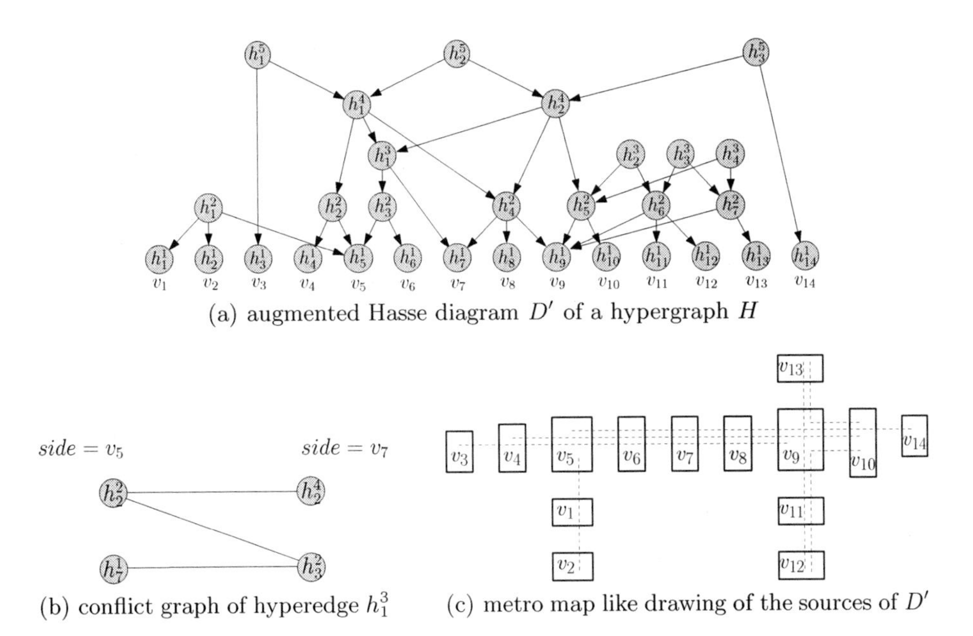

(a) augmented Hasse diagram D' of a hypergraph H

(b) conflict graph of hyperedge h_1^3

(c) metro map like drawing of the sources of D'

Fig. 3. Illustration of Algorithm 1

Lemma 2. *Let Conditions 1-3 of Theorem 3 be fulfilled and let* $G = (V, E)$ *be the graph computed in Algorithm 1. Let* $h', h'' \in A'$ *have a conflict with respect to a child* h *of* h' *and let* $G[h']$ *and* $G[h'']$ *be paths. Then*

1. $\text{side}_g(h'') = \text{side}_h(h'')$ *for all* $g \in A'$ *with* $h' \cap h'' \subseteq g \subseteq h$,
2. $\text{side}_h(h'') \in h''$,
3. $\text{side}_h(h'')$ *is an end vertex of* $G[h']$,
4. $G[h' \setminus h'']$ *is a path, and*
5. $\text{side}_h(h'')$ *is adjacent in* G *to a vertex of* $h'' \setminus h'$.

Proof. We prove the lemma by induction on the sum of the steps in which h' and h'' were considered in Algorithm 1. If h' and h'' had been considered in the first two steps, then at least one of them is a leaf of D' and, hence, h' and h'' have no conflict. So there is nothing to show. Let now h' and h'' be considered in later steps. Let $h'' \in A'$ have a conflict with h' with respect to a child h of h' and let $G[h']$ and $G[h'']$ be paths.

1. + 2. if $h' \cap h'' \in A'$**:** There is nothing to show if $h = h' \cap h''$. So let h_1 be the child of h with $h_1 \supseteq h' \cap h''$. Then h, h'' have a conflict with respect to h_1. Hence, C_h contains the path h', h'', h_1. By the inductive hypothesis on Property 3, it follows that $\text{side}_{h_1}(h'')$ is an end vertex of $G[h]$, and, especially that h_1 and h share an end vertex. By construction, it follows that $\text{side}_h(h_1)$ is the end vertex of h that is not in h_1. Hence, $\text{side}_h(h'') \in h_1$ and $\text{side}_{h_1}(h'') = \text{side}_h(h'')$. By the inductive hypothesis it follows that $\text{side}_g(h'') = \text{side}_h(h'')$ for $h \cap h'' \subseteq g \subseteq h_1$. Since the labels in $\text{side}_{h' \cap h''}(.)$ are the end vertices of $G[h' \cap h'']$, it follows that $\text{side}_h(h'') \in h' \cap h'' \subset h''$.

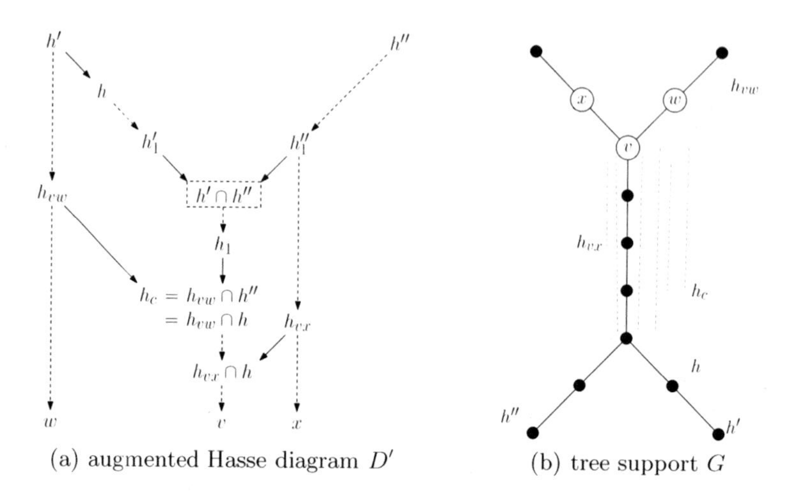

Fig. 4. Illustration of the proof of Lemma 2.3

1. + 2. + 5. if $h' \cap h'' \notin A'$: Let $h''_1 \subseteq h''$ be minimal with $h' \cap h'' \subset h''_1$. Since h' and h''_1 overlap there is an edge $\{v, w\} \in E$ such that $v \in h' \cap h''$ and $w \in h''_1 \setminus h'$. We show that $\text{side}_h(h'') = v$.

By Lemma 1 there is a child h_c of h_{vw} that contains $h \cap h_{vw}$. Since $v \in h \cap h_{vw}$ it follows that $w \notin h_c$ and, hence, v is an end vertex of h_c.

Note that by the minimality of h''_1 it follows that $h' \cap h'' \not\subseteq h_{vw}$. Since $G[h''], G[h']$ are paths, it follows that $h_c \subsetneq h$ and, hence, $h_c = h \cap h_{vw}$. Let h_p be minimal with $h_c \subsetneq h_p \subseteq h$. Then h_p, h_{vw} have a conflict with respect to h_c and it follows from the inductive hypothesis on Property 5 that $\text{side}_{h_c}(h_{vw}) = v$. Let h'_c be maximal with $h_c \subseteq h'_c \subsetneq h$. By the inductive hypothesis on Property 1 it follows that $\text{side}_{h'_c}(h_{vw}) = v$. Since h, h_{vw} have a conflict with respect to h'_c it follows by the inductive hypothesis on Property 3 that v is an end vertex of h. In C_h there is the path h'_c, h_{vw}, h', h''. By construction, $\text{side}_h(h'_c)$ is the end vertex of h that is not in h'_c. Hence, $\text{side}_h(h_{vw}) = \text{side}_h(h'') = v$.

3.: Let $v = \text{side}_h(h'')$. By the construction in Algorithm 1, v is an end vertex of $G[h']$ if h' is non-implied. So assume that h' is implied and that v is not an end vertex of $G[h']$. Let $w \in h' \setminus h$ be a neighbor of v in G. By Property 2, it follows that $v \in h''$. Let h_c be the child of h_{vw} that contains $h_{vw} \cap h''$. By the inductive hypothesis on Property 4, it follows that $G[h_{vw} \setminus h'']$ is a path that contains w but not v. Hence, $h_c = h_{vw} \cap h'' = h_{vw} \cap h$.

Let $h'_1, h''_1 \in A'$, respectively, be minimal with $h' \supseteq h'_1 \supsetneq h' \cap h''$ and $h'' \supseteq h''_1 \supsetneq h' \cap h''$. Assume first that $h' \cap h'' \in A'$. Then $C_{h' \cap h''}$ contains the triangle $h_{vw}, h'_1, h''_1, h_{vw}$ and, hence, is not bipartite.

Assume now that $h' \cap h'' \notin A'$. By the already proven part of Property 5 it follows that there is an edge $\{v, x\}$ of G with $x \in h''_1 \setminus h$. We have $h_c = h_{vw} \cap h'' \supseteq h_{vw} \cap h_{vx}$. Further, the child of h_{vx} that contains $h_{vx} \cap h$ equals $h_{vx} \cap h$. Since h''_1 is implied and h_{vx} not, it follows that $h''_1 \neq h_{vx}$ and, hence,

$h_{vx} \not\supseteq h' \cap h''$. Hence, either $h_{vx} \cap h \subseteq h_{vw} \cap h$ or $h_{vw} \cap h \subsetneq h_{vx} \cap h \subsetneq h' \cap h''$. In the first case let $h_1 \in A'$ be minimal with $h_{vw} \cap h \subsetneq h_1 \subseteq h$. Then there is the triangle $h_{vw}, h_{vx}, h_1, h_{vw}$ in $C_{h \cap h_{vw}}$. In the latter case let $h_1 \in A'$ be minimal with $h_{vx} \cap h \subsetneq h_1 \subseteq h$. Then there is the triangle $h_{vw}, h_{vx}, h_1, h_{vw}$ in $C_{h \cap h_{vx}}$.

4.: By the inductive hypothesis $G[h \setminus h'']$ is a path. Further, h and h' share $\text{side}_h(h'') \in h''$ as a common end vertex. By the precondition of the lemma, $G[h']$ is a path. Hence, $G[h' \setminus h'']$ is a path.

5. if $h' \cap h'' \in A'$: If $h \neq h' \cap h''$ let h_1 be the child of h with $h' \cap h'' \subseteq h_1$. By the inductive hypothesis $\text{side}_{h_1}(h'')$ is adjacent in G to a vertex of $h'' \setminus h = h'' \setminus h'$ and by Property 1 $\text{side}_{h_1}(h'') = \text{side}_h(h'')$.

If $h = h' \cap h''$, let $h_1'' \in A'$ be minimal with $h \subsetneq h_1'' \subseteq h''$. Applying Property 3 with h_1'' as "h'" and h' as "h''" reveals that $\text{side}_h(h')$ is an end vertex of $G[h_1'']$. Since $G[h_1'']$ is a path it follows that some vertex of $h_1'' \setminus h$ is adjacent to $\text{side}_h(h'')$. □

Lemma 3. *If Conditions 1-3 of Theorem 3 are fulfilled then all hyperedges in A' induce a path in the graph G constructed in Algorithm 1.*

Proof. Again, we prove the lemma by induction on the step in which h was considered in Algorithm 1. There is nothing to show if h had been considered in the first step. So assume that $h \in A'$ and that $G[h]$ contains a vertex v of degree greater than two.

Let u_1, u_2, u_3 be the first three vertices connected to v in G. Let $h_i = h_{vu_i}, i = 1, 2, 3$. Then h_1, h_2, h_3 are all three contained in h and its intersection contains v. Hence, any two of them have a conflict if and only if one of them is not contained in the other. A case distinction reveals that we wouldn't have appended all three, u_1, u_2 and u_3, to v.

$h_2 = h_3$: Since h_3 contains no vertex of degree higher than two, it follows that $u_1 \notin h_3$, $h_3 \cap h_1 = \{v\}$. Hence, h_1 and h_3 have a conflict with respect to the common child $\{v\}$, contradicting that v is added in the middle of h_3.

$h_1 = h_2$ or $h_1 = h_3$: These cases are analogous to the first case.

$h_1 \subsetneq h_3$: Like in the first case it follows that $u_2 \notin h_3$. Let $h_i', i = 2, 3$ be the child of h_i that contains v. Then h_2 and h_3 have a conflict with respect to $h_i', i = 2, 3$. Since we add the edge $\{v, u_i\}$ to G when we process h_i it follows on one hand that $\text{side}_{h_i'}(h_i) = v$. On the other hand, since h_1 is contained in h_3 and $v \in h_1$ it follows that $h_1 \subseteq h_3'$. Hence, h_3' has more than one vertex. If $h_3' \neq h_3 \cap h_2$ then v is the only end vertex of $G[h_3']$ that is contained in h_2. By Lemma 2 Property 2 it follows that $\text{side}_{h_3'}(h_2) = v$ and hence, $\text{side}_{h_3'}(h_3) \neq v$. If $h_3' = h_3 \cap h_2$ let $v' \neq v$ be the other end vertex of h_2'. Since we know that $\text{side}_{h_2'}(h_2) = v$ it follows that $\text{side}_{h_2'}(h_3) = v'$. Hence, by Lemma 2 Property 1, we can conclude that $\text{side}_{h_3'}(h_3) = v'$. In both cases, we have a contradiction.

$h_1 \subsetneq h_2$ or $h_2 \subsetneq h_3$: These cases are analogous to the third case.

h_1, h_2, h_3 pairwise overlapping: Then $h_1 \cap h_2 = h_2 \cap h_3 = h_1 \cap h_3 = \{v\}$. Hence, Condition 2 of Theorem 3 is not fulfilled. □

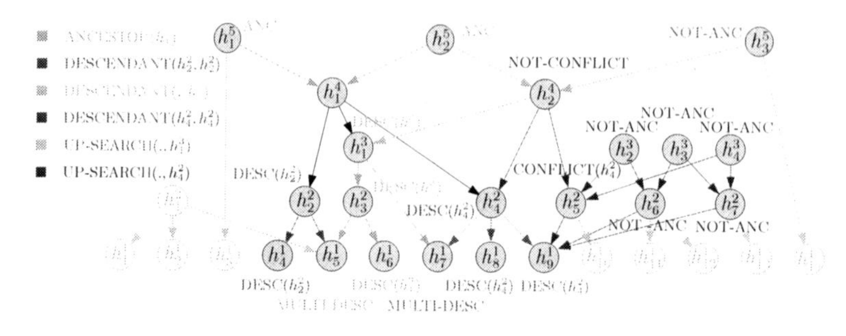

Fig. 5. Computation of the potential conflicts for h_1^4

This completes the proof of Theorem 3. We conclude this section with the following corollary.

Corollary 1. *Algorithm 1 computes a path-based tree support of a hypergraph H if H has a path-based tree support, i.e., if and only if the conditions of Theorem 3 are fulfilled.*

4.3 Conflict Computation and Run Time

In this section we show how to efficiently compute the conflicts and give an upper bound for the run time of testing whether a hypergraph has a path-based tree support and of constructing one, if it exists.

Representing the hyperedges as sorted lists of their elements, all conflicts can be determined straight-forwardly in $\mathcal{O}(n^3(n+m))$ time. In the following, we show how this time bound can be improved.

We first compute candidates for conflicting pairs of hyperedges, which in the case of hypergraphs having a path-based tree support turn out to be a superset of the set of all conflicts. The idea is, that all potential conflicts lie on a path from an ancestor of h to one of h's descendants. The method can be found as pseudocode in Algorithm 2.

We illustrate Algorithm 2 with an example. Figure 5 shows the computation of potential conflicts for the hyperedge h_1^4 of the hypergraph H from Figure 3(a). The different methods are colored. h_5^2 is the only hyperedge that can be in conflict with h_1^4 with respect to a child of h_1^4 and if so, with respect to h_4^2.

Lemma 4. *Let D' be the augmented Hasse diagram of a hypergraph that has a path-based tree support and let h' and h have a conflict with respect to a child h_c of h. Then Algorithm 2 applied to D' and h labels h' with* CONFLICT(h_c).

Proof. Let G be a path-based tree support of a hypergraph and let h' and h have a conflict with respect to a child h_c of h.

1. Let v be the end vertex of $G[h]$ that is contained in h'. Then v and all its ancestors on the path from v to h_c are labeled DESC(h_c) (and not MULTI-DESC).

Algorithm 2. Conflict Computation.

Input : augmented Hasse diagram D' of a hypergraph, vertex h
Output : vertices h' with LABEL(h') = CONFLICT(h_c) for all children h_c of h
Data : there are the following vertex labels

LABEL(h') = ANC iff $h \subsetneq h'$
LABEL(h') = NOT-ANC only if $h \cup h'$ not contained in any source of D'
LABEL(h') = DESC(h_c) iff $h' \subseteq h_c$ for exactly one child h_c of h
LABEL(h') = MULTI-DESC iff h' is contained in more than one child of h
LABEL(h') = NOT-CONFLICT only if $h \cap h'$ not contained in any child of h and $h \cup h'$ contained in some source of D'
LABEL(h') = CONFLICT(h_c) only if $h_c \cap h' \neq \emptyset$ for a child h_c of h and $h \cup h'$ contained in some source of D'

```
ANCESTOR(vertex h') begin
    foreach parent h'' of h' do
        LABEL(h'') ← ANC;
        ANCESTOR(h'');
end

DESCENDANT(vertex h', vertex h_c) begin
    if LABEL(h') = DESC(h'_c), h_c ≠ h'_c then
        LABEL(h') ← MULTI-DESC;
    else
        LABEL(h') ← DESC(h_c);
    foreach child h'' of h' do
        if LABEL(h'') ≠ MULTI-DESC then
            DESCENDANT(h'', h_c);
end

UP-SEARCH(vertex h', vertex h_c) begin
    foreach parent h'' of h' do
        if LABEL(h'') ∈ {∅, CONFLICT(h'_c), h'_c ≠ h_c} then
            UP-SEARCH(h'', h_c);
        if LABEL(h') = CONFLICT(h'_c), h_c ≠ h'_c then
            LABEL(h') ← NOT-CONFLICT;
        else if LABEL(h') ≠ DESC(h_c) then
            if LABEL(h'') ∈ {CONFLICT(h_c), ANC, NOT-CONFLICT} then
                LABEL(h') ← CONFLICT(h_c);
            if LABEL(h') ≠ CONFLICT(h_c) then
                LABEL(h') ← NOT-ANC;
end

begin
    Clear all labels;
    LABEL(h) ← NOT-CONFLICT;
    ANCESTOR(h);
    foreach child h_c of h do
        DESCENDANT(h_c, h_c);
    foreach vertex h' of D' with LABEL(h') ∈ {DESC(h_c); h_c child of h} do
        UP-SEARCH(h', h_c);
end
```

2. If there was a descendant of h' labeled $\text{DESC}(h'_c)$ for a child $h'_c \neq h_c$ of h, then h_c does not contain $h \cap h'$ contradicting that h and h' have a conflict with respect to h_c.

Hence, Algorithm 2 labels h' with $\text{CONFLICT}(h_c)$. □

Theorem 4. *It can be tested in $\mathcal{O}(n^3 m)$ time whether a hypergraph has a path-based tree support and if so such a support can be constructed within the same time bounds.*

Proof. Let H be a hypergraph. First test in linear time whether there is a tree support for H [13]. Let D' be the augmented Hasse diagram of H. The method works in four steps.

1. Start with an empty array conflict indexed with pairs of inner vertices of D'. Set $\text{conflict}_{h,h'} \leftarrow h_c$ if and only if h' is labeled $\text{CONFLICT}(h_c)$ in Algorithm 2 applied to D' and h.
2. For each pair h, h' of inner vertices of D' test whether $\text{conflict}_{h,h'}$ contains $h \cap h'$. Otherwise set $\text{conflict}_{h,h'} \leftarrow \emptyset$. Now, if H has a path-based tree support then h, h' has a conflict with respect to the child h_c of h if and only if $h_c = \text{conflict}_{h,h'}$.
3. Apply Algorithm 1 to compute a support G. If the algorithm stops without computing a support then H does not have a path-based tree support.
4. Test whether every hyperedge induces a path in G. If not, H does not have a path-based tree support.

D' has $\mathcal{O}(n+m)$ vertices, $\mathcal{O}(n^2+nm)$ edges and can be computed in $\mathcal{O}(n^3 m)$ time if H has a tree support [3]. Algorithm 2 visits every edge of D' at most twice and, hence, runs in $\mathcal{O}(n^2+nm)$ time for each of the $\mathcal{O}(n)$ inner vertices of D'.

We may assume that the hyperedges are given as sorted lists of their elements. If not given in advance, these lists could straight forwardly be computed from D' in $\mathcal{O}(n^3+mn^2)$ time by doing a graph search from each leaf. Now, for each of the $\mathcal{O}(n^2)$ pairs h, h' of inner vertices it can be tested in $\mathcal{O}(n)$ time whether $\text{conflict}_{h,h'}$ contains $h \cap h'$.

The sum of the sizes of all conflict graphs is in $\mathcal{O}(n^2)$. Hence, Algorithm 1 runs in $\mathcal{O}(n^2+mn)$ time. For each of the $\mathcal{O}(m)$ hyperedges h it can be tested in $\mathcal{O}(n)$ time, whether $G[h]$ is a path. Hence, the overall run time is dominated by the computation of the augmented Hasse diagram and is in $\mathcal{O}(n^3 m)$. □

5 Conclusion

We have introduced path-based supports for hypergraphs. Hence, as a new model, we considered a restriction on the appearance of those subgraphs of a support that are induced by the hyperedges. We have shown that it is $\mathcal{NP}$-complete to find the minimum number of edges of a path-based support or to decide whether there is a planar path-based support. Further, we characterized those hypergraphs that have a path-based tree support and we gave an algorithm that computes a path-based tree support in $\mathcal{O}(n^3 m)$ run time if it exists. Our algorithm completed the paths for the hyperedges in the order in which they appeared in a reversed topological ordering of the augmented Hasse diagram. To connect the subpaths in the right order, we introduced a conflict graph for each hyperedge h and colored the vertices of this conflict graph with the end vertices of the path induced by h.

References

1. Beeri, C., Fagin, R., Maier, D., Yannakakis, M.: On the desirability of acyclic database schemes. Journal of the Association for Computing Mashinery 30(4), 479–513 (1983)
2. Brandes, U., Cornelsen, S., Pampel, B., Sallaberry, A.: Hypergraphs and outerplanarity. In: Iliopoulos, C.S., Smyth, W.F. (eds.) IWOCA 2010. LNCS, vol. 6460, pp. 201–211. Springer, Heidelberg (2010)
3. Buchin, K., van Kreveld, M., Meijer, H., Speckmann, B., Verbeek, K.: On planar supports for hypergraphs. In: Eppstein, D., Gansner, E.R. (eds.) GD 2009. LNCS, vol. 5849, pp. 345–356. Springer, Heidelberg (2010)
4. Bujtás, C., Tuza, Z.: Color-bounded hypergraphs, II: Interval hypergraphs and hypertrees. Discrete Mathematics 309, 6391–6401 (2009)
5. Flower, J., Fish, A., Howse, J.: Euler diagram generation. Journal on Visual Languages and Computing 19(6), 675–694 (2008)
6. Johnson, D.S., Krishnan, S., Chhugani, J., Kumar, S., Venkatasubramanian, S.: Compressing large boolean matrices using reordering techniques. In: Nascimento, M.A., Özsu, M.T., Kossmann, D., Miller, R.J., Blakeley, J.A., Schiefer, K.B. (eds.) Proceedings of the 13th International Conference on Very Large Data Bases (VLDB 2004), pp. 13–23. Morgan Kaufmann, San Francisco (2004)
7. Johnson, D.S., Pollak, H.O.: Hypergraph planarity and the complexity of drawing Venn diagrams. Journal of Graph Theory 11(3), 309–325 (1987)
8. Kaufmann, M., van Kreveld, M., Speckmann, B.: Subdivision drawings of hypergraphs. In: Tollis, I.G., Patrignani, M. (eds.) GD 2008. LNCS, vol. 5417, pp. 396–407. Springer, Heidelberg (2009)
9. Korach, E., Stern, M.: The clustering matroid and the optimal clustering tree. Mathematical Programming, Series B 98, 385–414 (2003)
10. Král', D., Kratochvíl, J., Voss, H.-J.: Mixed hypercacti. Discrete Mathematics 286, 99–113 (2004)
11. Nöllenburg, M.: An improved algorithm for the metro-line crossing minimization problem. In: Eppstein, D., Gansner, E.R. (eds.) GD 2009. LNCS, vol. 5849, pp. 381–392. Springer, Heidelberg (2010)
12. Simonetto, P., Auber, D., Archambault, D.: Fully automatic visualisation of overlapping sets. Computer Graphics Forum 28(3), 967–974 (2009)
13. Tarjan, R.E., Yannakakis, M.: Simple linear-time algorithms to test chordality of graphs, test acyclicity of hypergraphs, and selectively reduce acyclic hypergraphs. SIAM Journal on Computing 13(3), 566–579 (1984)
14. Wolff, A.: Drawing subway maps: A survey. Informatik-Forschung und Entwicklung 22, 23–44 (1970)

On Improved Exact Algorithms for $L(2,1)$-Labeling of Graphs

Konstanty Junosza-Szaniawski and Paweł Rzążewski

Warsaw University of Technology
Faculty of Mathematics and Information Science
Pl. Politechniki 1 / 207, 00-661 Warsaw, Poland
{k.szaniawski,p.rzazewski}@mini.pw.edu.pl

Abstract. $L(2,1)$-labeling is graph labeling model where adjacent vertices get labels that differ by at least 2 and vertices in distance 2 get different labels. In this paper we present an algorithm for finding an optimal $L(2,1)$-labeling (i.e. an $L(2,1)$-labeling in which largest label is the least possible) of a graph with time complexity $O^*(3.5616^n)$, which improves a previous best result: $O^*(3.8739^n)$.

1 Introduction

$L(2,1)$-labeling is inspired by a channel assignment problem in telecommunication. It asks for such a labeling with non-negative integers, that no vertices in distance 2 get the same labels and labels of adjacent vertices differ by at least 2.

The k-$L(2,1)$-labeling problem is to determine if there exists an $L(2,1)$-labeling of a graph with no label greater than k. $L(2,1)(G)$ denotes $L(2,1)$-span, which is the smallest value of k for which there exists a k-$L(2,1)$-labeling of G.

For any fixed $k \geq 4$, the k-$L(2,1)$-labeling problem is NP-complete [1].

Havet *et al.* [2] presented an algorithm for finding $L(2,1)(G)$ in time $O^*(3.8739^n)$.

In this paper we present and analyze an improved version of this algorithm. The time complexity bound $O^*(3.5616^n)$ of our algorithm is substantially better than the time complexity of the original one. The difference lies in a better bound on the number of 2-packings, a smaller number of triples considered in the main loop of the algorithm and more carefully formulated conditions for this loop.

2 Preliminaries

The number of vertices of a graph is called the *order* of the graph, denoted in this paper by n.

Let $dist_G(x,y)$ be the *distance* between vertices x and y in a graph G.

Formally a $L(2,1)$-*labeling* is defined as a function $f\colon V(G) \to \mathbb{N}$, such that
$\forall x,y \in V(G)\ dist_G(x,y) = 1 \Rightarrow |f(x) - f(y)| \geq 2$ and
$\forall x,y \in V(G)\ dist_G(x,y) = 2 \Rightarrow |f(x) - f(y)| \geq 1$.

A set $X \subseteq V(G)$ is a *2-packing* in $G \Leftrightarrow \forall x,y \in X\ dist_G(x,y) > 2$.

C.S. Iliopoulos and W.F. Smyth (Eds.): IWOCA 2010, LNCS 6460, pp. 34–37, 2011.

Let $N(v) = \{u : uv \in E(G)\}$ denote the *neighborhood* of a vertex v. The set $N[v] = N(v) \cup \{v\}$ denotes the *closed neighborhood* of v. Let $N^2[v]$ denote the set of all vertices in distance at most 2 from v.

By the *neighborhood* of a set X of vertices in G we mean the set $N(X) = \bigcup_{v \in X} N(v)$ and by the *closed neighborhood* of X – the set $N[X] = N(X) \cup X$.

We write $f(n) = O^*(g(n))$ if there exists a polynomial $p(n)$ such that $f(n) \leq p(n) \cdot g(n)$ for all n greater than some n_0.

3 Improved Exact Algorithm for $L(2,1)$ - Labeling

3.1 Generating 2-Packings

In this section we present an algorithm for generating all 2-packings of a specified size in a connected graph G.

In the beginning, the algorithm finds T – a spanning tree of G and then P, which is the longest path in T. Let v denote the endvertex of P, u denote the neighbor of v on P and w denote second neighbor of u on P (other than v).

The algorithm constructs 2-packings by branching on a vertex v. The vertex v is either included or not in a constructed 2-packing and we call the algorithm recursively for a graph of smaller order. Since P is the longest path in T, we can delete vertices from its end without a risk of losing connectivity.

We also keep the set of active vertices $\hat{V}$, which contains vertices that can belong to a 2-packing constructed in current recursive call. Since by deleting some vertex we could lose information about vertices in distance 2, we have to keep the original graph G and use it to find $N^2_G[v]$ in line 14. A set S denotes currently constructed 2-packing and is initialized by the emptyset $\emptyset$.

Algorithm 1. G2P

Arguments: Graph G, Graph $\hat{G}$, Int k, Set $\hat{V}$, Set S, Family of sets X
1 **if** $k = 0$ **then** $X \leftarrow X \cup \{S\}$ and **return**
2 **if** $|\hat{V}| < k$ **then return**
3 **if** $\Delta(\hat{G}) = |V(\hat{G})| - 1$ **then**
4   **if** $k = 1$ **then** $X \leftarrow X \cup \{S \cup \{v\} : v \in \hat{V}\}$
5   **return**
6 $P \leftarrow$ longest path in spanning tree of $\hat{G}$
7 $(v, u, w) \leftarrow$ three consecutive vertices from the end of P
8 **G2P**$(G, \hat{G} - v, k, \hat{V}, S, X)$
9 **if** $v \in \hat{V}$ **then G2P**$(G, \hat{G} - (N_T[u] - \{w\}), k - 1, \hat{V} \setminus N^2_G[v], S \cup \{v\}, X)$

To generate all k-element 2-packings in a graph G, we call the algorithm: **G2P**$(G, G, k, V(G), \emptyset, \emptyset)$.

Lemma 1. *The algorithm **G2P** generates each 2-packing of size k exactly once.*

Proof. (Sketch) All the 2-packings not containing v are generated in the call in line 8, while all the 2-packing containing v are generated in the call in line 9. Since $v \in S$, the algorithm is to include $k-1$ vertices. Notice that none of vertices in $N_G^2[v]$ can belong to any 2-packing containing v, so they have to be removed from $\hat{V}$. □

Theorem 1. *The algorithm **G2P** works in time $O^*(\binom{n-k+1}{k})$.*

Proof. Let $T(n,k)$ be the number of recursive calls of the algorithm. We observe that $T(n,k) \leq f(n,k)$, where f is given by the following recursion:

$$\begin{cases} f(n,0)=1,\ f(1,1)=1,\ f(n,k)=0 \text{ for } k>n \\ f(n,k) = \underbrace{f(n-1,k)}_{\text{vertex } v \text{ does not belong to } S} + \underbrace{f(n-2,k-1)}_{\text{vertex } v \text{ belongs to } S} \quad \text{for } n\geq 2, k\geq 1 \end{cases}$$

Solving this recursive equation we obtain $T(n,k) \leq f(n,k) = \binom{n-k+1}{k}$. □

Corollary 1. *The number of 2-packings with exactly k vertices in a connected graph is at most $f(n,k) = \binom{n-k+1}{k}$.*

Corollary 2. *The number of all 2-packings in a connected graph can be bounded by $\sum_{k=0}^{n} f(n,k) = O^*((\frac{1+\sqrt{5}}{2})^n) = O^*(1.6181^n)$.*

3.2 Improved Algorithm

This subsection describes the improved algorithm for finding $L(2,1)$-span.

The algorithm iteratively marks all subsets of $V(G)$ that can be labeled with no label greater than i for $i=0,\ldots,2n$. Note here that $L(2,1)(G) \leq 2n$, since labeling the vertices by distinct even integers is always a valid $L(2,1)$-labeling.

We introduce Boolean variables $Lab[X,Y,i]$ for all pairs X,Y, where Y is a 2-packing in G and $X \subseteq V(G) \setminus Y$ and $i=0,\ldots,2n$. The value of $Lab[X,Y,i]$ is set *true* if and only if there exists a partial $L(2,1)$-labeling with no label exceeding i for X, such that no vertex in $N(Y) \cap X$ has label i. The values of $Lab[X,Y,i]$ are computed by dynamic programming using the implication:

If $Lab[A,U,i-1]$ is *true* for a 2-packing U and $A \subseteq V(G) \setminus U$, then $Lab[U \cup A, Y, i]$ is *true* for any $Y \subseteq V(G) \setminus (A \cup U)$, disjoint with $N(U)$.

Algorithm 2. IFL(2,1)-Span

```
1 foreach Y - 2-packing in G, X ⊆ V(G) \ Y, i = 0,...,2n do
2 |  Lab[X,Y.i] ← false
3 foreach X - 2-packing in G, Y - 2-packing in G − N[X] do
4 |  Lab[X,Y.0] ← true
5 for i ← 1 to 2n do
6 |  foreach U - 2-packing in G, Y - 2-packing in G − N[U] do
7 |  |  foreach A ⊆ V(G) \ (U ∪ Y) do
8 |  |  |  if Lab[A,U,i − 1] then Lab[U ∪ A,Y,i] ← true
9 |  |  |  if U ∪ A = V(G) then return "L(2,1)(G) = i"
```

Lemma 2. *The algorithm **IFL(2,1)-Span** finds $L(2,1)$-span of any graph.*

Proof. To prove the correctness of the algorithm, let us prove the statement:

After i iterations of the main loop, all values of $Lab[X, Y, i]$ are set correctly.

Proof by induction on i. For $i = 0$ values of $Lab[X, Y, i]$ are set in the loop in lines 3-5. It is easy to prove, that they are set correctly.

Assume that statement is correct for $i-1$. Let Y be a 2-packing in G and $X \subseteq V(G) \setminus Y$. Suppose there exists i-$L(2,1)$-labeling f of X, such that $f(u) \leq i-1$ for every $u \in N(Y) \cap X$. Let $U = \{v \in X \colon f(v) = i\}$ and $A = X \setminus U$. Then U must be a 2-packing in G and $N(Y) \cap U = \emptyset$. Since $X \cap Y = \emptyset$, U forms a 2-packing in graph induced by set of vertices $V(G) - N[Y]$. Moreover, every labeled vertex from $N(U)$ must have label at most $i-2$. Thus labeling f restricted to A satisfies the requirements for setting $Lab[A, U, i-1] \leftarrow true$. By the inductive assumption, this value is set correctly, so $Lab[X, Y, i]$ is also set $true$ in line 9.

Extending f - a partial $(i-1)$-$L(2,1)$-labeling of A by setting $f(u) = i$ for all $u \in U$ gives a labeling of X satisfying our requirements, if $N[Y] \cap U = \emptyset$.

This justifies the computation of $Lab[X, Y, i]$ by dynamic programming. □

Theorem 2. *The algorithm **IFL(2,1)-Span** finds $L(2,1)$-span of any connected graph in time $O^*(3.5616^n)$ and space $O^*(2.7321^n)$.*

Proof. Notice that $|N(U)| \geq |U|$ since every vertex in U has at least one neighbor and there is no common neighbor for any two vertices in U.

Let us bound the number of desired triples U, A, Y. By Corollary 1 there are at most $\binom{n-k+1}{k}$ 2-packings of size k. For every such 2-packing U we can choose $Y \subseteq V(G) \setminus N[U]$ (notice that we cannot use the bound from Corrolary 1, because the graph $G - N[U]$ may be disconected). Let Y have j elements. Each of the remaining $n-k-j$ vertices is either in A or not. Hence the total number of desired triples is at most: $A(n) = \sum_{k=0}^{n} \binom{n-k+1}{k} \sum_{j=0}^{n-2k} \binom{n-2k}{j} 2^{n-k-j}$. From this formula we obtain our result: $A(n) = O^*\left(\left(\frac{3+\sqrt{17}}{2}\right)^n\right) = O^*(3.5616^n)$.

Space complexity is determined by the number of pairs X, Y for which we have to store values of $Lab[X, Y, i]$. For every fixed k-element 2-packing Y ($k \in \{0, \ldots n\}$), X can be any subset of $V(G) \setminus Y$. Hence, by Corollary 1, the total number of pairs X, Y is bounded by $O^*(\sum_{k=0}^{n} \binom{n-k+1}{k} 2^{n-k})) = O^*((1+\sqrt{3})^n) = O^*(2.7321^n)$. □

References

1. Fiala, J., Kratochvíl, J., Kloks, T.: Fixed-parameter complexity of λ-labelings. Discrete Applied Mathematics 113(1), 59–72 (2001)
2. Havet, F., Klazar, M., Kratochvíl, J., Kratsch, D., Liedloff, M.: Exact Algorithms for $L(2,1)$-Labeling of Graphs. Algorithmica, doi: 10.1007/s00453-009-9302-7

Thread Graphs, Linear Rank-Width and Their Algorithmic Applications

Robert Ganian*

Faculty of Informatics, Masaryk University
Botanická 68a, Brno, Czech Republic
xganian1@fi.muni.cz

1 Introduction

The introduction of tree-width by Robertson and Seymour [7] was a breakthrough in the design of graph algorithms. A lot of research since then has focused on obtaining a width measure which would be more general and still allowed efficient algorithms for a wide range of NP-hard problems on graphs of bounded width. To this end, Oum and Seymour have proposed rank-width, which allows the solution of many such hard problems on a less restricted graph classes (see e.g. [3,4]). But what about problems which are NP-hard even on graphs of bounded tree-width or even on trees? The parameter used most often for these exceptionally hard problems is path-width, however it is extremely restrictive – for example the graphs of path-width 1 are exactly paths.

In the article we study a new width measure called linear rank-width, defined by an additional requirement on the rank-decomposition of graphs analogous to the requirement path-width imposes on tree-decompositions. The goal is to obtain a width measure which on one hand is less restrictive than path-width and yet on the other hand allows efficient algorithms for problems which are hard on graphs of bounded rank-width or even tree-width. We first provide a constructive characterization of graphs having linear rank-width 1 (further referred to as thread graphs), and then continue by providing positive algorithmic results on this class of graphs.

The algorithmic section contains three new polynomial algorithms on thread graphs. Due to space restrictions, we assume that the reader is familiar with rank-width and also refer to the full version of this paper for all proofs [2].

2 Linear Rank-Width

The most popular width parameters are subject to a certain hierarchy. Rank-width is the most general of the three parameters, though if some problem is hard on rank-width one can try solving it on graphs of bounded tree-width, and if that again fails then there is path-width. This relationship is illustrated in the following table:

* This research has been supported by the Czech research grants 201/09/J021 and MUNI/E/0059/2009.

C.S. Iliopoulos and W.F. Smyth (Eds.): IWOCA 2010, LNCS 6460, pp. 38–42, 2011.

	paths	trees	cliques
path-width	bounded	unbounded	unbounded
tree-width	bounded	bounded	unbounded
rank-width	bounded	bounded	bounded

The catch here is that many of the problems which are hard on tree-width and trees tend to be solvable on cliques as well, not just paths. It is a natural question to ask whether there exists a width parameter which remains capable of solving problems hard on tree-width, but at the same time relaxes the restrictions of path-width and achieves low values also on cliques. This is strong motivation for linear rank-width.

Definition 2.1. *A rank-decomposition (T, μ) is* linear *if T is a caterpillar (i.e. a path with pendant vertices). The linear rank-width of a graph G is the minimum of the width of all linear rank-decompositions of G.*

Theorem 2.2. *The linear rank-width of paths and cliques is* 1, *and the linear rank-width of trees is not bounded by any constant.*

3 Thread Graphs

The classes of graphs of rank-width 1, tree-width 1 or path-width 1 each possess interesting structural properties. For rank-width these are called distance hereditary graphs, while for tree-width and path-width we speak of forests and disjoint unions of paths respectively. In this section we introduce a new graph class called *thread graphs* and prove that this is exactly the class of graphs which have linear rank-width 1, answering a question asked by Oum at GROW 2009.

Definition 3.1. *A thread graph is a graph which can be constructed by sequentially creating vertices. Every new vertex is created with 3 attributes, as follows:*

1. *Passive ($\mathcal{P}$) or Active ($\mathcal{A}$);*
2. *Disconnect ($\mathcal{D}$) or Join ($\mathcal{J}$);*
3. *Normal or Reset ($\mathcal{R}$);*

Each new vertex is either $\mathcal{P}$ or $\mathcal{A}$, either $\mathcal{D}$ or $\mathcal{J}$ and may or may not be $\mathcal{R}$.

A $\mathcal{D}$ vertex is created without any incident edges. A $\mathcal{J}$ vertex on the other hand is created with incident edges to all vertices which are currently $\mathcal{A}$.

Finally, an $\mathcal{R}$ vertex changes all previous $\mathcal{A}$ vertices to $\mathcal{P}$. Every vertex is normal (not $\mathcal{R}$) unless explicitly said otherwise.

Notice that for connected thread graphs it is enough to consider $\mathcal{AJ}$, $\mathcal{AD}$, $\mathcal{PJ}$, $\mathcal{AJR}$ vertices (any other type of vertices disconnects the graph) and that any thread graph can be created by disjoint union of connected thread graphs. Also, given a thread graph it is possible to reconstruct its creation sequence [2].

Theorem 3.2. *A graph G has linear rank-width 1 if and only if G is a thread graph.*

4 Algorithms on Thread Graphs

4.1 Computing Bandwidth

Definition 4.1. *Given a graph G and a one-to-one mapping $f : V \rightarrow \{1, \dots |V|\}$, the bandwidth of f is defined as the maximum difference between the labels of vertices sharing an edge. The bandwidth of G, denoted by $bwd(G)$ is then the minimum bandwidth over all such f.*

Bandwidth has many applications in theory as well as practice, ranging from networking to biology (see e.g. [8] or the dedicated survey [1]). Unfortunately, it turns out that computing the bandwidth of graphs is extremely hard. Even on trees, approximating bandwidth within some constant factor is NP-hard and the best known polynomial-time approximation bound is $O(log^{2.5}n)$ [5].

Theorem 4.2. *There exists a polynomial time algorithm for 2-approximation of bandwidth on thread graphs.*

$\mathcal{PJ}^{i}$ $\mathcal{R}_i$ $\mathcal{PJ}^{i+1}$ $\mathcal{R}_{i+1}$

$\mathcal{AD}^{i}$ & $\mathcal{AJ}^{i}$

Fig. 1. Mapping order of vertices for bandwidth 2-approximation

4.2 Dominating Bandwidth

While bandwidth is a well-known problem, in this subsection we introduce a related problem called dominating banwidth. This may have practical applications in communication (i.e. constructing an array of communicating relays with bandwidth restrictions, each relay covering the surrounding areas), but our main goal here is to show that there exist interesting problems which are NP-hard on trees and at the same time polynomially solvable on thread graphs.

Definition 4.3. *The dominating bandwidth problem for a given graph G and a minimum dominating set $X \subseteq V(G)$ of G is the problem of computing a mapping $f : V \rightarrow \{1, \dots |X|\}$ such that:*

1. *each $v \in X$ receives a unique label.*
2. *each $u \in V(G) - X$ receives the same label as some u-neighbour $v \in X$.*
3. *the bandwidth of f (defined as the maximum difference between the labels of vertices sharing an edge) is minimized.*

Theorem 4.4. *The dominating bandwidth problem is NP-hard on trees.*

Theorem 4.5. *The dominating bandwidth of thread graphs is 1.*

4.3 The Path-Width Problem

The final algorithm in this section is a polynomial time algorithm for computing the path-width of thread graphs.

Definition 4.6. *A path-decomposition of a graph* $G = (V, E)$ *is a path* $P = (T, A)$ *where the nodes* T *are subsets of* V *(also called bags) such that the following holds:*

1. *Each vertex* $v \in V$ *appears in some bag.*
2. *For every edge* $\{v, w\}$ *there exists a bag containing both* v *and* w.
3. *For every vertex* $v \in V$, *the bags containing* v *induce a subpath in* P *(the interpolation property).*
   The width of the path-decomposition $P = (T, A)$ *equals the cardinality of the largest bag in* T *minus one. The path-width of* G, *denoted by* $pwd(G)$, *is the minimum width over all path decompositions of* G.

Path-width itself is a powerful (albeit extremely restrictive) width parameter. However, computing path-width is a hard problem – it remains NP-hard even when restricted to weighted trees and distance hereditary graphs (graphs of rank-width 1) [6].

Theorem 4.7. *There exists a polynomial time algorithm for computing the path-width of thread graphs.*

5 Conclusion

The main contribution of the article may be summarized in two main points. First, it gives a constructive characterization of graphs of linear rank-width 1 and provides insight into the structure of such graphs, which we call thread graphs. This new graph class contains paths and cliques but also many other graphs. Second, the article uses the obtained results in the design of new polynomial algorithms for bandwidth, dominating bandwidth and path-width on thread graphs. Each of these problems remains hard on other well-known classes of graphs, such as distance hereditary graphs and trees. Further research in this area should focus on possible parameterized algorithms on linear rank-width – it is not clear whether or how our polynomial algorithms might be extended to graphs of bounded linear rank-width.

Acknowledgment

The author wishes to thank Jan Obdržálek for his contribution and help in characterizing graphs of linear rank-width 1.

References

1. Chinn, P., Chvátalová, J., Dewdney, A., Gibbs, N.: The bandwidth problem for graphs and matrices—a survey. J. Graph Theory 6, 223–254 (1982)
2. Ganian, R.: Thread graphs, linear rank-width and their algorithmic applications (manuscript), `http://is.muni.cz/www/99352/threadgraphs.pdf`

3. Ganian, R., Hliněný, P.: Better polynomial algorithms on graphs of bounded rank-width. In: Fiala, J., Kratochvíl, J., Miller, M. (eds.) IWOCA 2009. LNCS, vol. 5874, pp. 266–277. Springer, Heidelberg (2009)
4. Ganian, R., Hliněný, P.: On parse trees and Myhill–Nerode–type tools for handling graphs of bounded rank-width. Discrete Appl. Math. (2009) (to appear)
5. Gupta, A.: Improved bandwidth approximation for trees and chordal graphs. J. Algorithms 40(1), 24–36 (2001)
6. Mihai, R., Todinca, I.: Pathwidth is np-hard for weighted trees. In: Deng, X., Hopcroft, J.E., Xue, J. (eds.) FAW 2009. LNCS, vol. 5598, pp. 181–195. Springer, Heidelberg (2009)
7. Robertson, N., Seymour, P.D.: Graph minors. II. Algorithmic aspects of tree-width. Journal of Algorithms 7(3), 309–322 (1986)
8. Zhu, Q., Adam, Z., Choi, V., Sankoff, D.: Generalized gene adjacencies, graph bandwidth, and clusters in yeast evolution. IEEE/ACM Trans. Comput. Biol. Bioinformatics 6(2), 213–220 (2009)

Minimum Number of Holes in Unavoidable Sets of Partial Words of Size Three*

Francine Blanchet-Sadri[1], Bob Chen[2], and Aleksandar Chakarov[3]

[1] Department of Computer Science, University of North Carolina, P.O. Box 26170, Greensboro, NC 27402–6170, USA
blanchet@uncg.edu

[2] Department of Mathematics, University of California, San Diego, 9500 Gilman Drive, Dept 0112, LaJolla, CA 92093–0112, USA

[3] Department of Computer Science, University of Colorado at Boulder, 430 UCB, Boulder, CO 80309–0430, USA

Abstract. Partial words are sequences over a finite alphabet that may contain some undefined positions called holes. In this paper, we consider unavoidable sets of partial words of equal length. We compute the minimum number of holes in sets of size three over a binary alphabet (summed over all partial words in the sets). We also construct all sets that achieve this minimum. This is a step towards the difficult problem of fully characterizing all unavoidable sets of partial words of size three.

1 Introduction

An *unavoidable* set of (full) words X over a finite alphabet A is one for which any two-sided infinite word over A has a factor in X. For example, the set $X = \{aa, ba, bb\}$ is unavoidable over the alphabet $\{a, b\}$, since avoiding aa and bb forces a word to be an alternating sequence of a's and b's. This fundamental concept was explicitly introduced in 1983 in connection with an attempt to characterize the rational languages among the context-free ones [1]. Since then it has been consistently studied by researchers in both mathematics and theoretical computer science (see for example [2–9]).

Partial words are sequences that may contain some undefined positions called holes, denoted by $\diamond$'s, that match every letter of the alphabet (we also say that $\diamond$ is *compatible* with each letter of the alphabet). For instance, $a\diamond bca\diamond b$ is a partial word with two holes over $\{a, b, c\}$. A set of partial words X over A is unavoidable if any two-sided infinite full word over A has a factor compatible with an element in X. The problem of deciding the avoidability of finite sets of

* This material is based upon work supported by the National Science Foundation under Grant No. DMS–0754154. The Department of Defense is gratefully acknowledged. The authors would also like to acknowledge Sean Simmons from the Department of Mathematics of the University of Texas at Austin for pointing out an approach to proving our characterization of the $D_m(i, j)$ unavoidable sets. We thank him for his contributions and insightful suggestions.

C.S. Iliopoulos and W.F. Smyth (Eds.): IWOCA 2010, LNCS 6460, pp. 43–55, 2011.

partial words turns out to be NP-hard [10, 11], which is in constrast with the well known feasibility results for finite sets of full words [12, 13].

Unavoidable sets of partial words were introduced in [14], where the problem of characterizing such sets of cardinality n over a k-letter alphabet was initiated. Note that if X is unavoidable, then every infinite unary word has a factor compatible with a member of X; thus X cannot have fewer elements than the alphabet, and so $k \leq n$ (note that the cases $n = 1$ and $k = 1$ are trivial). The characterization of *all* unavoidable sets of cardinality $n = 2$ was settled recently in [15] using deep arguments related to Cayley graphs. So our next long-term goal is to characterize unavoidable sets of cardinality $n = 3$. Since in [14], all such sets over a three-letter alphabet were completely characterized (in fact, there are no nontrivial such sets), we need to focus on sets over a two-letter alphabet.

In [14], a complete characterization of all three-word unavoidable sets over a binary alphabet where each partial word has at most two defined positions was given, and some special cases where one partial word has more than two defined positions were discussed, but general criteria for these sets had not been found. In this paper, among other things, we answer affirmatively a conjecture that was left open there. Our main goal however is to make another step towards the full $n = 3$ characterization by computing the minimum number of holes in any unavoidable set of partial words of equal length and of cardinality three over a binary alphabet. We also construct all sets that achieve this minimum.

2 Preliminaries

Let A be a fixed non-empty finite set called an *alphabet* whose elements we refer to as *letters*. In this paper we restrict our attention to the binary alphabet $\{a, b\}$. Hence, we may refer to a and b as complements of each other, so that $\overline{a} = b$ and $\overline{b} = a$. A *full word* w over A is a finite sequence of elements of A. We write $|w|$ to denote the length of w, and $w(i)$ to denote the ith symbol. By convention, we begin indexing with 0, so a word w of length m can be represented as $w = w(0)w(1) \cdots w(m-1)$. Formally, a finite word of length m is a function $w : \{0, \ldots, m-1\} \to A$. The number of occurrences of the letter a (resp. b) in w is denoted by $|w|_a$ (resp. $|w|_b$).

A *two-sided infinite full word* (hereafter *infinite word*) w is a function $w : \mathbb{Z} \to A$. We call w p-periodic, or of period p, if for some positive integer p $w(i) = w(i+p)$ for all $i \in \mathbb{Z}$. We say w is periodic if it has a period. On the other hand, w is p-alternating if $w(i) = \overline{w(i+p)}$ for all $i \in \mathbb{Z}$. Note that if w is p-alternating, it is also $2p$-periodic. If v is a non-empty finite word, then $v^{\mathbb{Z}}$ denotes $w = \cdots vvvv \cdots$. A finite word u is a *factor* of w if some $i \in \mathbb{Z}$ satisfies $u = w(i) \cdots w(i + |u| - 1)$. An *m-factor* is a factor of length m.

A *partial word* u of length m over A is a function $u : \{0, \ldots, m-1\} \to A_\diamond$, where $A_\diamond = A \cup \{\diamond\}$ with $\diamond \notin A$ called a *hole*. For $0 \leq i < |u|$, if $u(i) \in A$, then i belongs to the *domain* of u, denoted $D(u)$. Note that full words are simply partial words whose domain is the entire set $\{0, \ldots, |u| - 1\}$. Two partial words u and v of equal length are *compatible*, denoted $u \uparrow v$, if $u(i) = v(i)$ whenever $i \in D(u) \cap D(v)$. We denote by $h(u)$ the number of holes in u, thus, $h(u) = |u| - |D(u)|$.

Let w be an infinite word and let u be a partial word. We say w *meets* u if w has a factor compatible with u, and w *avoids* u otherwise. Now, w meets a set of partial words X if it meets some $u \in X$, and w avoids X otherwise. If X is avoided by some infinite word, then X is *avoidable*; otherwise, X is *unavoidable*. We say X is *m-uniform* if every partial word in X has length m.

The partial word u is *contained* in the partial word v, denoted $u \subset v$, if $|u| = |v|$ and $u(i) = v(i)$, for all $i \in D(u)$. We say that v is a *strengthening* of u if v has a factor containing u, and write $v \succ u$. We also say that u is a *weakening* of v. We use strengthening and weakening as operations performed on partial words: to strengthen a partial word u is to replace an instance of $\diamond$ with any letter in A, while to weaken a partial word u is to set $u(i) = \diamond$ for some $i \in D(u)$. Note that if an infinite word w meets the partial word u, it also meets every weakening of u, and if w avoids u then w avoids all strengthenings of u.

Let X, Y be sets of partial words. We extend the notions of strengthening and weakening as follows. We say that X is a strengthening of Y (written as $X \succ Y$) if, for each $x \in X$, there exists $y \in Y$ such that $x \succ y$. We also say that Y is a weakening of X. It is not hard to see that if the infinite word w meets X, then it also meets every weakening of X, and if w avoids X then it avoids any strengthening of X. Hence if X is unavoidable, so are all weakenings of X, and if X is avoidable all strengthenings of X are avoidable.

Two partial words u and v are *conjugate*, denoted $u \sim v$, if there exist partial words x, y such that $u \subset xy$ and $v \subset yx$. It is well-known that conjugacy on full words is an equivalence relation, and we use $c(m, k)$ to denote the number of conjugacy classes of words of length m over a k-letter alphabet. However, in the case of partial words, conjugacy is no longer an equivalence relation [16]. Therefore, two partial words u, v are *hole-conjugate* if there exist partial words x, y such that $u = xy$ and $v = yx$; in this case we write $u \sim_\diamond v$.

We conclude with some number theoretic notation used in this paper. We write $a \mid b$ if a divides b. Next, let p be a prime and let $e, m \in \mathbb{N}$. We write $p^e \parallel m$ if p^e *maximally divides* m, that is, if $p^e \mid m$ but $p^{e+1} \nmid m$. Finally, we write $i \equiv_m j$ if i is congruent to j modulo m.

3 Minimum Number of Holes in Unavoidable Sets

We denote by $H_{m,n}$ the minimum number of holes in any unavoidable m-uniform set (summed over all partial words in the set) of cardinality n over a binary alphabet. To have words of "real length" m, we require that $D(u) \ni 0, m-1$ for each u in any such set.

An unavoidable set of full words of equal length m over a k-letter alphabet A has to contain at least one word of each conjugacy class of words of length m over A. Thus the minimum number $\alpha(m, k)$ of elements in an unavoidable set of full words of length m over A is greater than or equal to the number

$c(m,k)$ of conjugacy classes of words of length m over A. It has been shown by Schützenberger that it is asymptotically true that $\alpha(m,k) \sim c(m,k)$ (both numbers are asymptotically equivalent to $\frac{k^m}{m}$) [17]. Later on, answering a conjecture of Golomb, Mykkeltveit proved that actually, $\alpha(m,k) = c(m,k)$ [18]. A new proof of this equality, namely that for all integers $m \geq 1$ and $k \geq 1$, there exists an unavoidable set of full words of length m over an alphabet of size k with $c(m,k)$ elements was given in [2]. Thus, $H_{m,c(m,2)} = 0$ for $m \geq 1$.

We recall that strengthenings of avoidable sets are always avoidable, though the same is not true in general for unavoidable sets. As we are interested in unavoidable sets with the minimum number of holes, and strengthenings do not contain more holes than the original set, it is reasonable to investigate "maximal strength" unavoidable sets. So let X be an unavoidable set. If, for all $Y \succ X$, Y is avoidable, then we say X is *maximal*.

Proposition 1. *If every m-uniform unavoidable set of cardinality n having a total of h holes is maximal, then $H_{m,n} \geq h$.*

Proof. If $h = 0$ then the claim is clear, so assume $h \geq 1$. Suppose that $H_{m,n} < h$, and let Y be an m-uniform unavoidable set of cardinality n with $h' < h$ holes for some $h' \in \mathbb{N}$. Now add holes to words in Y arbitrarily until the new set, Y', has h holes. Since $Y' \prec Y$, Y' is also unavoidable. Hence Y' is an m-uniform unavoidable set that is not maximal. □

We now state the main result and focus of this paper.

Theorem 1. *For $m \geq 4$, $H_{m,3} = 2m - 5$ if m is even, and $H_{m,3} = 2m - 6$ if m is odd.*

Remark 1. As long as we are discussing an m-uniform unavoidable set of size three, say $X = \{x_1, x_2, x_3\}$, we may always assume, without loss of generality, $x_1(0) = x_1(m-1) = a, x_2(0) = x_2(m-1) = b, x_3(0) = b$ and $x_3(m-1) = a$, and require that $h(x_1) \leq h(x_2)$. Moreover, only a's and $\diamond$'s appear in x_1, and only b's and $\diamond$'s appear in x_2. We call this the *standard form* of an m-uniform three-element unavoidable set of partial words. The presence of x_1, x_2 is justified since any unavoidable set over $\{a,b\}$ must contain words compatible with $a^{\mathbb{Z}}$ and $b^{\mathbb{Z}}$, respectively. Now, x_3 must have complementary ends, since otherwise $X \succ \{a\diamond^{m-2}a, b\diamond^{m-2}b\}$ and as the latter set is avoidable so is X. Next, if $h(x_1) > h(x_2)$, we may consider instead the set $\{\overline{x_1}, \overline{x_2}, \overline{x_3}\}$. This "switches" the identity of x_1 and x_2 so that $h(x_1) \leq h(x_2)$. Finally, we may fix the orientation of x_3 by taking the reverse of each word, if necessary.

In the next two sections, we give constructions of sets that achieve the proposed minimum of Theorem 1.

4 The C-Sets

In this section, we define and completely characterize the unavoidable *C-sets*.

Definition 1. *Let $\Lambda \subset \{1, \ldots, m-2\}$. We denote by $C_m(\Lambda)$ the m-uniform set $\{x_1, x_2, x_3\}$ where $x_1 = a^m$, $x_2 = b\diamond^{m-2}b$, and x_3 is defined as follows: $x_3(i) = b$ if $i = 0$, $x_3(i) = a$ if $i \in \Lambda \cup \{m-1\}$, and $x_3(i) = \diamond$ otherwise.*

Remark 2. If $\Lambda = \{i_1, i_2, \ldots, i_s\}$, we often write $C_m(i_1, i_2, \ldots, i_s)$ instead of $C_m(\{i_1, i_2, \ldots, i_s\})$. By convention, we order the arguments of $C_m(i_1, i_2, \ldots, i_s)$ in increasing order, so that $i_1 < i_2 < \cdots < i_s$.

Remark 3. We have $C_m(\Lambda) \prec C_m(\Gamma)$ precisely when $\Lambda \subset \Gamma$.

Proposition 2. *The set $C_m(i)$ is unavoidable if and only if $i \mid m-1$.*

Proof. Suppose $i \mid m-1$ with $li = m-1$ for some $l \in \mathbb{N}$, and suppose to the contrary that w is an infinite word that avoids $X = C_m(i)$. The word w must contain a b in order to avoid x_1; say, without loss of generality, that $w(0) = b$. To avoid x_2, it must be that $w(m-1) = a$. This, however, forces $w(i) = b$, or else w meets x_3. We may repeat the argument to conclude that $w(l'i) = b$ for all $l' \in \mathbb{N}$. This yields a contradiction, as we claimed that $w(li) = w(m-1) = a$. Conversely, if $i \nmid m-1$, then let $w = (ba^{i-1})^{\mathbb{Z}}$. Now, w clearly avoids x_1 and x_3 as it is i-periodic. Finally, all indices containing b are congruent to each other modulo i. Thus, w does not meet x_2, since any two positions $m-1$ apart are not congruent modulo i, and so cannot both be b. Hence, X is avoidable. □

Proposition 3. *The set $C_m(i, j)$ is unavoidable iff $i, j \mid m-1$ and $2i = j$.*

Proof. Suppose $i, j \mid m-1$ with $li = m-1$ for some $l \in \mathbb{N}$ and $2i = j$, and suppose to the contrary that w is an infinite word that avoids $X = C_m(i, j)$. Note that every b in w must be followed by an a after $m-1$ positions (to avoid x_2), and be followed by a b after either i or j positions (to avoid x_3). It is impossible that every consecutive pair of b's be separated by j positions, for if so w meets x_2 (as $j \mid m-1$). Hence, some pair of b's are separated by i positions; say $w(0) = w(i) = b$. This implies that $w(m-1) = w(m-1+i) = a$. Now, if $w(m-1-i) = b$, then w meets x_3 (since that b has a's both i and $2i = j$ positions later). This argument cascades backwards since we once again have a's separated by i positions. Thus $w(m-1-l'i) = a$ for all $l' \in \mathbb{N}$, but this is a contradiction since $w(m-1-li) = w(0) = b$. Hence no word w avoids X.

On the other hand, if $i \nmid m-1$ then $C_m(i, j) \succ C_m(i)$, where the latter set is avoidable by Proposition 2, and so $C_m(i, j)$ is also avoidable (similarly, for the case when $j \nmid m-1$). Finally, if $2i \neq j$ and $i, j \mid m-1$, put $lj = m-1$ for some $l \in \mathbb{N}$. Let $u = ba^{i-1}(ba^{j-1})^{l-1}$. Then we claim $w = u^{\mathbb{Z}}$ is an infinite word avoiding X. Clearly w avoids x_1 and x_3 (for every b is followed by another one after either i or j positions). Now let v be any m-factor of w with $v(0) = b$. We claim that $v(m-1) = a$ and so w avoids x_2. Note that b's appear in positions congruent to 0 modulo j until the first factor of ba^{i-1} appears, after which they appear in positions congruent to i modulo j. The next time a factor of ba^{i-1} appears, b's start appearing in indices congruent to $2i$ modulo j, and so on.

Now, recall that $i < j$, and so $m = lj + 1 > lj + i - j = (l-1)j + i = |u|$. Furthermore, since $j < m-1$, we know that $l \geq 2$. It follows that

$$m < m - 1 + 2i \leq m - 1 + 2i + (l-2)j = lj + 2i + lj - 2j = 2((l-1)j + i) = 2|u|$$

Therefore, any m-factor v of w contains more than one but less than two full copies of u. Hence there are either one or two occurrences of ba^{i-1} (which appear once per u). So b's appear at the end of v in positions congruent to i or $2i$ modulo j. Now, the only way for $v(m-1) = b$ is if $m-1 \equiv_j i$ or $m-1 \equiv_j 2i$. But $j \mid m-1$, so $m-1 \equiv_j 0$. It is easy to see that $i \equiv_j 0$ is impossible since $i < j$, and $2i \equiv_j 0$ implies $2i = lj$ for some l. As $i < j$, this forces $l = 1$ and so $2i = j$, contrary to hypothesis. Hence if v is an m-factor of w with $v(0) = b$, then $v(m-1) = a$. So, w avoids x_2 and hence the set X. □

Corollary 1. *Let $\Lambda \subset \{1, \ldots, m-2\}$ with $|\Lambda| \geq 3$. Then $C_m(\Lambda)$ is avoidable.*

Proof. Put $\Lambda = \{i_1, \ldots, i_s\}$ with $s \geq 3$. Now, $C_m(\Lambda) \succ C_m(i_1, i_2)$ and $C_m(\Lambda) \succ C_m(i_1, i_3)$, and since $i_2 \neq i_3$ at least one of $C_m(i_1, i_2)$ and $C_m(i_1, i_3)$ is avoidable by Proposition 3. Hence, so is the set $C_m(\Lambda)$. □

5 The D-Sets

In this section, we define and completely characterize the unavoidable D-*sets.*

Definition 2. *Let $\Lambda \subset \{1, \ldots, m-2\}$. We denote by $D_m(\Lambda)$ the m-uniform set $\{x_1, x_2, x_3\}$ where $x_1 = a\diamond^{m-2}a$, $x_2 = b\diamond^{m-2}b$, and x_3 is defined as follows: $x_3(i) = b$ if $i = 0$, $x_3(i) = a$ if $i \in \Lambda \cup \{m-1\}$, and $x_3(i) = \diamond$ otherwise.*

As before, if $\Lambda = \{i_1, i_2, \ldots, i_s\}$, we often write $D_m(i_1, i_2, \ldots, i_s)$ instead of $D_m(\{i_1, i_2, \ldots, i_s\})$, and we order the arguments of $D_m(i_1, i_2, \ldots, i_s)$ in increasing order, so that $i_1 < i_2 < \cdots < i_s$.

We now characterize the unavoidable D-sets with one position filled in. However, this process is much more difficult than the corresponding task for C-sets, owing to the stricter requirements imposed by x_1.

Lemma 1 **([14]).** *Let $X = \{a\diamond^m a, b\diamond^n b\}$. Put $2^s \parallel m+1$ and $2^t \parallel n+1$. Then X is unavoidable if and only if $s \neq t$.*

Lemma 2. *The sets $X = \{a\diamond^{m-2}a, b\diamond^{n-2}b\}$, $Y = \{a\diamond^{m-2}a, b\diamond^{n-2}b, a\diamond^{n-2}a\}$ have the same avoidability.*

Proof. Suppose X is avoidable, say by the infinite word w. Suppose that w meets $a\diamond^{n-2}a$, so that $w(i) = w(i+n-1) = a$ for some $i \in \mathbb{Z}$. Then $w(i+m-1) = w(i+n-1+m-1) = b$, since w avoids $a\diamond^{m-2}a$, but this contradicts the fact that w avoids $b\diamond^{n-2}b$. Hence w avoids $a\diamond^{n-2}a$ and so avoids Y. But clearly $X \succ Y$, and so if X is unavoidable so is Y. □

Proposition 4. *Let $2^s \parallel m-1$ and $2^t \parallel i$. Then the set $D_m(i)$ is unavoidable if and only if $t \leq s$.*

Proof. Let $X = \{b\diamond^{m-2}b, a\diamond^{m-2-i}a\}$. We first show that X has the same avoidability as $D_m(i)$. For suppose X is avoidable. Then so is $Y = X \cup \{a\diamond^{m-2}a\}$, by Lemma 2. As Y is an avoidable weakening of $D_m(i)$, we conclude that $D_m(i)$

is avoidable. On the other hand, suppose X is unavoidable. Let w be any infinite word. If w meets $b\diamond^{m-2}b$, then it also meets $D_m(i)$. If it does not, then $w(j) = w(j-m+1+i) = a$ for some $j \in \mathbb{Z}$. Now, if $w(j-m+1) = a$, then w meets x_1, and if $w(j-m+1) = b$, it meets x_3. In either case, w meets $D_m(i)$, and so $D_m(i)$ is unavoidable. Hence X has the same avoidability as $D_m(i)$.

Next, let $2^s \parallel m-1, 2^t \parallel i, 2^r \parallel m-1-i$. We show that $r \neq s$ if and only if $t \leq s$. Set $2^s p = m-1, 2^t q = i$ for odd p, q. Now, if $t < s$, then $2^{s-t}p - q$ is odd, and so $2^t \parallel 2^t(2^{s-t}p - q) = 2^s p - 2^t q = m-1-i$ and $r = t \neq s$. If $t = s$, then, since $p - q$ is even, we have $2^{s+1} \mid 2^s(p-q) = 2^s p - 2^t q = m-1-i$. Thus $r \geq s+1$ and so r cannot be equal to s. Finally, if $t > s$, then $p - 2^{t-s}q$ is odd. It follows that $2^s \parallel 2^s(p - 2^{t-s}q) = 2^s p - 2^t q = m-1-i$ and so $r = s$. Hence $r \neq s$ if and only if $t \leq s$. Recall that by Lemma 1, X is unavoidable if and only if $r \neq s$. Therefore, $D_m(i)$ is unavoidable if and only if $t \leq s$. □

We now turn our attention to D-sets with two positions filled in. A previous result gives necessary conditions for the unavoidability of $D_m(i,j)$, provided that $i, j, m-1$ are relatively prime.

Theorem 2 ([15]). *Let l, n_1, n_2 be nonnegative integers such that $n_1 \leq n_2$ and $\gcd(l+1, n_1+1, n_2+1) = 1$. If the set $\{a\diamond^l a, b\diamond^l b, a\diamond^{n_1}a\diamond^{n_2}a, b\diamond^{n_1}b\diamond^{n_2}b\}$ is unavoidable, then at least one of the following conditions hold:*

(i) $l = 6$ *and* $(n_1, n_2) \in \{(1,3), (3,7), (1,7)\}$;
(ii) $n_1 + 1 \equiv_{2l+2} 0$;
(iii) $n_2 + 1 \equiv_{2l+2} 0$;
(iv) $n_1 + n_2 + 2 \equiv_{2l+2} 0$;
(v) $2n_1 + n_2 + 3 \equiv_{2l+2} l + 1$;
(vi) $2n_2 + n_1 + 3 \equiv_{2l+2} l + 1$;
(vii) $n_2 - n_1 \equiv_{2l+2} l + 1$.

Corollary 2. *If $D_m(i,j)$ is unavoidable and* $\gcd(m-1, i, j) = 1$, *then $j = 2i$, or $i + j = m - 1$, or the three conditions $m = 8, i = 1$, and $j \in \{3, 5\}$ hold.*

Proof. Suppose $D_m(i,j)$ is unavoidable. Put $l = m-2, n_1 = j-i-1, n_2 = m-j-2$ and let $Y = \{a\diamond^l a, b\diamond^l b, a\diamond^{n_1}a\diamond^{n_2}a, b\diamond^{n_1}b\diamond^{n_2}b\}$. Note that Y is also unavoidable since $Y \prec D_m(i,j) = \{a\diamond^l a, b\diamond^l b, b\diamond^{i-1}a\diamond^{n_1}a\diamond^{n_2}a\}$; moreover, $\gcd(l+1, n_1+1, n_2+1) = 1$. Hence, l, n_1, n_2 must satisfy one of the conditions given in Theorem 2. However, as $i > 0$ we have that $n_1 + n_2 + 1 < l$; this forces one of (i), (v), or (vi) to hold. It is easy to verify that these conditions are equivalent to the ones stated about m, i, j. □

The following proposition shows that we do not gain any new unavoidable sets by considering cases where $m-1, i, j$ are not relatively prime. Thus we may extend the above result to all i, j, m.

Proposition 5. *For any $\Lambda = \{i_1, \ldots, i_s\}$, let $d\Lambda = \{di \mid i \in \Lambda\}$. Then $D_m(\Lambda)$ is avoidable if and only if $D_{d(m-1)+1}(d\Lambda)$ is.*

Proof. Let $\Lambda = \{i_1, \ldots, i_s\} \subset \{1, \ldots, m-2\}$. Let $Y = D_m(\Lambda) = \{y_1, y_2, y_3\}$ and $Z = D_{d(m-1)+1}(d\Lambda) = \{z_1, z_2, z_3\}$, where $y_1 = a\diamond^{m-2}a, y_2 = b\diamond^{m-2}b, z_1 = a\diamond^{d(m-1)-1}a, z_2 = b\diamond^{d(m-1)-1}b$. If w is a word avoiding Y, then we claim the word $w' = \cdots w(-1)^d w(0)^d w(1)^d \cdots$ avoids Z. To see this, note that as w is $(m-1)$-alternating, w' is $d(m-1)$-alternating and so avoids z_1, z_2. Now, if w' meets z_3, then there exists l such that $w'(l) = b, w'(l+di_1) = \cdots = w(l+di_s) = w(l+d(m-1)) = a$. But if we put $h = \lfloor \frac{l}{d} \rfloor$, then $w(h) = b, w(h+i_1) = \cdots = w(h+i_s) = w(h+m-1) = a$ so w meets y_3. This is a contradiction, so w' in fact avoids z_3 and hence Z. The reverse direction is analogous, except that if w is a word avoiding Z, then the word $w' = \cdots w(-d)w(0)w(d) \cdots$ avoids Y. □

Corollary 3. *If $D_m(i,j)$ is unavoidable, then $j = 2i$, or $i + j = m - 1$, or both $m = 7i + 1$ and $j \in \{3i, 5i\}$.*

Proof. This is an immediate consequence of Corollary 2 and Proposition 5. □

We now show that the above conditions are sufficient.

Lemma 3. *Let $m, n \in \mathbb{N}$, $2^s \parallel m$ and $2^t \parallel n$. If $s \geq t$, $\gcd(m,n) = \gcd(2m,n)$.*

Proof. Since $s \geq t$, we know that the power of 2 maximally dividing $\gcd(m,n)$ is just $\min(s,t) = t$. But the power of 2 maximally dividing $\gcd(2m,n)$ is $\min(s+1,t) = t$. It is clear that the other prime factors of $\gcd(m,n)$ are unaffected by doubling m, and the result follows. □

Proposition 6. *Let $2^s \parallel m-1, 2^t \parallel i$, and $2^r \parallel j$. Then the set $D_m(i,j)$ is unavoidable if and only if (iv) holds in addition to one of (i), (ii), or (iii):*

- *(i)* $j = 2i$;
- *(ii)* $i + j = m - 1$;
- *(iii)* $m = 7i + 1$ *and* $j \in \{3i, 5i\}$;
- *(iv)* $s \geq t, r$.

Proof. If $t > s$, then $D_m(i)$ is avoidable by Proposition 4. Hence $D_m(i,j)$ is avoidable, as $D_m(i,j) \succ D_m(i)$. A similar argument applies if $r > s$. Together with Corollary 3, we have one direction of the proof.

It remains to show that the above conditions are sufficient. We assume for the remainder of the proof that (iv) holds.

Suppose (i) holds, and that w is a word avoiding $D_m(i,j)$. We show that this leads to a contradiction. Since w avoids x_1, we have $|w|_b \geq 1$ and we may take without loss of generality $w(0) = b$. To avoid x_2, $w(m-1) = a$, and to avoid x_3, $w(i) = b$ or $w(j) = b$. Similarly, for every b, there must be a b that occurs i or $j = 2i$ positions later. Suppose that $w(i) = b$. Then $w(m-1+i) = a$. Now, note that $w(m-1-i) = a$, for there are a's that occur i positions and $j = 2i$ positions after $m-1-i$. Thus $w(-i) = b$. Since we have another two a's separated by i positions (at $m-1$ and $m-1-i$), we may apply the same argument to conclude that $w(-2i) = b$. We may repeat this to get $w(li) = b$ for all $l \leq 0$. Now, w is $(m-1)$-alternating since it avoids $\{x_1, x_2\}$, and so it is $(2m-2)$-periodic. Hence $w(x) = b$ whenever $x \equiv_{2m-2} li$ for some $l \leq 0$.

Let $d = \gcd(m-1, i)$. Then $d \mid m-1$, say with $dq = m-1$, and furthermore $d = \gcd(2m-2, i)$ by Lemma 3. By Bezout's theorem, we may write $d = xi + y(2m-2)$ for some $x, y \in \mathbb{Z}$ (x negative). Hence $xi \equiv_{2m-2} d$. It follows that $w(m-1) = w(dq) = b$, as $dq \equiv_{2m-2} xqi$. This contradicts our previous assertion that $w(m-1) = a$.

It remains to consider the case where b appears in every position congruent to lj modulo $2m-2$ for some $l \in \mathbb{Z}$ (that is, when no two b's are separated by i positions), but this leads to a contradiction in the same way, since $r \leq s$. Hence we may represent $m-1$ as a multiple of j modulo $2m-2$ and so reach a contradiction. We conclude that $D_m(i,j)$ is unavoidable when (i) holds.

Now suppose (ii) holds. Again, let w be a word that avoids $D_m(i,j)$, and take without loss of generality $w(0) = b$. Suppose that $w(i) = b$. Then $w(m-1) = w(m-1+i) = a$. Now, the b in position i already has an a $m-1-i = j$ positions later, so it must have a b i positions later. Hence $w(2i) = b$, and now $w(m-1+2i) = a$. Repeating this argument gives us that $w(li) = b$ for all $l \geq 0$. Since w is $(2m-2)$-periodic, we have $w(x) = b$ whenever $x \equiv_{2m-2} li$ for some l. A contradiction is obtained in a manner identical to the previous case, since (iv) holds. Hence $D_m(i,j)$ is unavoidable when (ii) holds. Finally, note that there are only a finite number of words that are $(m-1)$-alternating, for any fixed m. Thus we may show the unavoidability of $D_8(1,3)$ and $D_8(1,5)$ (and hence the unavoidability of $D_{7i+1}(i,3i)$ and $D_{7i+1}(i,5i)$, by Proposition 5) via an exhaustive search. It follows that $D_m(i,j)$ is unavoidable if (iii) holds. □

Finally, we show that, like the C-sets, the D-sets are always avoidable when x_3 has at least three positions filled in.

Proposition 7. *Let $\Lambda \subset \{1, \ldots, m-2\}$ with $|\Lambda| \geq 3$. Then $D_m(\Lambda)$ is avoidable.*

Proof. It suffices to show that $D_m(i,j,l)$ is avoidable, as if $|\Lambda| > 3$ we can choose a weakening with exactly three positions filled in x_3. Moreover, by Proposition 5, we only need to consider the cases when $\gcd(m-1, i, j, l) = 1$.

If $D_m(i,j,l)$ is unavoidable, then it is necessary that each of the sets $D_m(i,j)$, $D_m(j,l)$, and $D_m(i,l)$ be unavoidable. Hence each weakening must satisfy Proposition 6. Suppose some of these three weakenings satisfies (iii). If $m = 8$ it is easy to see that one of the above weakenings of $D_m(i,j,l)$ is avoidable, as $D_8(1,3)$ and $D_8(1,5)$ are the only unavoidable D-sets. On the other hand, suppose $m = 7d+1$ with $d > 1$. If $D_m(i,j)$ satisfies (iii), then l is also a multiple of d regardless of which condition $D_m(i,l)$ satisfies. This contradicts our claim of relative primeness. An analogous argument shows that $D_m(i,l)$ cannot satisfy (iii).

Now suppose $D_m(j,l)$ satisfies (iii). Then $j = d$ and $l = pd$ for $p \in \{3,5\}$. If $D_m(i,j)$ satisfies (ii) then again i is a multiple of d and we have a contradiction. Hence $D_m(i,j)$ satisfies (i) and $j = 2i$. If $i > 1$ we again contradict relative primeness (since $\gcd(m-1,i,j,l) = i$), and if $i = 1$, we have $d = 2$. But both $D_{15}(1,6), D_{15}(1,10)$ are avoidable, so $D_m(i,j,l)$ has the avoidable weakening $D_m(i,l)$. Hence if any of the three weakenings satisfy (iii), $D_m(i,j,l)$ is avoidable.

Next suppose none of the three weakenings satisfies (iii). Set $2^s \parallel m-1, 2^t \parallel i, 2^r \parallel l$. It is impossible that all three weakenings satisfy (i), just as it is impossible for more than one weakening to satisfy (ii). Hence it must be that two weakenings satisfy (i) and one weakening satisfies (ii). It is easy to see that we must have $j = 2i, l = 2j$, and $i + l = m - 1$. But this implies $l = 4i$, and so $5i = m - 1$. It follows that $s = t$. Hence we have $r > s$, which is a contradiction as we assumed (iv) holds. Therefore, $D_m(i, j, l)$ is avoidable. □

6 Proof of Our Main Result

With our characterization of unavoidable C-sets and D-sets, we may begin to prove Theorem 1. We first prove Conjecture 2 from [14] (see Corollary 4).

Lemma 4. *Let $i_1 < \cdots < i_s < j_1 < \cdots < j_r$ be elements of $\{1, \ldots, m-2\}$. Let x be defined as follows: $x(i) = b$ if $i \in \{0, i_1, \ldots, i_s\}$, $x(i) = a$ if $i \in \{j_1, \ldots, j_r, m-1\}$, and $x(i) = \diamond$ otherwise. Then the set $X = \{a\diamond^{m-2}a, b\diamond^{m-2}b, x\}$ has the same avoidability as some D-set $D_m(\Lambda)$ with $|\Lambda| = s + r$.*

Proof. We proceed by induction on s. The base case of $s = 0$ is trivial as then X is itself a D-set. Now let $s \geq 1$. Note that a word w meets x if and only if it meets x' defined as

$$b\diamond^{i_2-i_1-1}b\cdots b\diamond^{i_s-i_{s-1}-1}b\diamond^{j_1-i_s-1}a\diamond^{j_2-j_1-1}a\cdots a\diamond^{j_r-j_{r-1}-1}a\diamond^{m-1-j_r-1}a\diamond^{i_1-1}a$$

since w must be $(m-1)$-alternating. Hence X has the same avoidability as $X' = \{a\diamond^{m-2}a, b\diamond^{m-2}b, x'\}$ which has one fewer b. Applying the induction hypothesis to X' yields the claim. □

Corollary 4. *If the set $X = \{a\diamond^{m-2}a, b\diamond^{m-2}b, x\}$ is unavoidable, where $x \uparrow b\diamond^{m-2}a$, then x has at most two interior defined positions.*

Proof. If x has any a appearing before a b, then the set X is avoided by $(b^{m-1}a^{m-1})^{\mathbb{Z}}$. Otherwise, if x has at least three interior defined positions, then by Lemma 4 it has the same avoidability as some set $D_m(\Lambda)$ with $|\Lambda| \geq 3$. But all such D-sets are avoidable, by Proposition 7, and so X is avoidable. □

Next, we show that the C-sets are the only unavoidable sets with the minimum number of holes. We divide the sets into multiple cases, conditioning on the quantity $h(x_1) + h(x_2)$.

Corollary 5. *Let m be odd (resp. even). Let X be an m-uniform set of size three of the form described in Remark 1. Suppose $h(x_1) + h(x_2) > m - 2$ (resp. $m - 1$). Then if X has $2m - 6$ (resp. $2m - 5$) holes in total, X is avoidable.*

Proof. There are at most $m - 5$ holes in x_3, and so x_3 has at least three positions other than 0 and $m - 1$ defined. Then we may weaken x_1, x_2 to $a\diamond^{m-2}a, b\diamond^{m-2}b$. The resulting set is avoidable by Corollary 4, and therefore so is X. □

Note that we did not treat the case where $h(x_1) + h(x_2) = m - 1$ for even m. This case is covered by the following proposition.

Proposition 8. *Let $m \geq 4$ be even, and let X be an m-uniform set of size three of the form described in Remark 1 with $h(x_1) + h(x_2) = m - 1$. Then if X has $2m - 5$ holes in total, X is avoidable.*

Proof. First, suppose that $h(x_1) > 1$. Assume that $m \geq 8$. We find a two sided-infinite word w with period $m-1$ that avoids X. Since w is $(m-1)$-periodic, any m-factor of w begins and ends with the same letter, and so w immediately avoids x_3. Moreover, we only have to consider whether w meets $x'_1 = x_1(0) \cdots x_1(m-2)$ (and $x'_2 = x_2(0) \cdots x_2(m-2)$), as any m-factor v with $v(0) = a$ necessarily has $v(m-1) = a$ (analogously, every m-factor that begins with b has to end with b).

Now consider the set B, which contains all conjugacy classes of length $m-1$ over $\{a, b\}$, with exactly $h(x_1)$ b's and $h(x_2)$ a's. Since $m \geq 8$, it follows that $|B| > 2$. Choose a representative u of a conjugacy class not covered by x'_1 and x'_2. By considering the number of a's and b's in u, we see that if $w = u^{\mathbb{Z}}$ were to meet x'_1 via the $(m-1)$-factor v, the $\diamond$'s in x'_1 need to align with the b's in v. However, for any factor v of w this is impossible, since $u \not\sim x'_1$ and $v \sim u$. Thus, it follows that v cannot be compatible with x'_1. A similar argument shows that w avoids x'_2. Hence w avoids x_1 and x_2, and therefore avoids X. We may check the cases for $m \leq 6$ easily via a computer program.

Now, suppose that $h(x_1) = 1$. In this case we know that $x_1 \sim_\diamond a^{m-1}\diamond$ and $x_2 = b\diamond^{m-2}b$. Moreover, x_3 has precisely two interior positions defined. First, if both the interior positions have letter b, then the word $w_1 = (baba^{m-3})^{\mathbb{Z}}$ avoids X since each m-factor of w_1 contains exactly two occurrences of the letter b, and so cannot be compatible with either x_1 or x_3. The word w_1 avoids x_2 as well since both m-factors that begin with b end with a. Second, if the interior positions have letters, from left to right, a, b, then the word $(b^{m-1}a^{m-1})^{\mathbb{Z}}$ avoids X. Third, if the interior positions have letters, from left to right, b, a, and the b occurs in position 1, then $(baba^{m-3})^{\mathbb{Z}}$ avoids X. Otherwise, the word $(bba^{m-1})^{\mathbb{Z}}$ avoids X, since in any m-factor which contains two instances of b, these letters appear in consecutive positions, and so cannot be compatible with x_2 or x_3.

Finally, if both the interior positions i, j, $i < j$, have letter a, then we proceed as follows. If $i, j \mid m - 1$, then, since $m - 1$ is odd it cannot be that $j = 2i$. Therefore the word $w_2 = (ba^{i-1}(ba^{j-1})^{l-1})^{\mathbb{Z}}$ (where $jl = m - 1$) avoids the set $C_m(i, j)$ by Proposition 3, and so avoids x_2 and x_3. Since w_2 has at least two occurrences of b in each m-factor, w_2 avoids x_1 as well. Hence w_2 avoids X.

If i and j do not simultaneously divide $m - 1$, let $l \in \{i, j\}$ be an index that does not divide $m-1$. Now, $(ba^{l-1})^{\mathbb{Z}}$ avoids x_2 and x_3, but it might meet x_1 if the number of a's on either side of the $\diamond$ in x_1 are both less than l. This can happen only if $l > \frac{m}{2}$, which in turn implies that $j > \frac{m}{2}$ (either $l = j$ or $l = i < j$). Hence $j \nmid m - 1$ as well. Then the j-periodic word $w_3 = (bba^{j-2})^{\mathbb{Z}}$ avoids x_1 and x_3 (consider the number of instances of b in w_3 and its period, respectively). Unless either $j + 1 = m - 1$ or $2j - 1 = m - 1$, the word w_3 avoids x_2 as well. However, in both of these last cases the word $(baba^{j-3})^{\mathbb{Z}}$ avoids X. □

Proposition 9. *Let X be an m-uniform set of three partial words of the form described in Remark 1. If $h(x_1) + h(x_2) = m - 2$, then either X is a C-set or X is avoidable.*

Proof. Suppose $h(x_1) = 0$. Then if $|x_3|_b \geq 2$, the infinite word $w = (ba^{m-1})^{\mathbb{Z}}$ avoids X; otherwise, X is a C-set. Therefore, for the remainder of this proof we may assume that $h(x_1) \geq 1$. For brevity, let $h(x_1) = i - 2$. Then $h(x_2) = m - i$.

First, suppose that $x_2 \not\sim_\diamond b^i \diamond^{m-i}$. The word $w = (b^{i-1}a^{m-i})^{\mathbb{Z}}$ avoids X. Note that w is $(m-1)$-periodic, so w does not meet x_3 (any m-factor of w has the same symbol in its first and last position). Since any m-factor of w has at least $i-1$ b's, while x_1 contains only $i-2$ $\diamond$'s, we can conclude that w avoids x_1. Finally, let v be any m-factor of w with $v(0) = b$. Then $v(m-1) = b$ as w is $(m-1)$-periodic, and $v(0) \cdots v(m-2) \sim b^{i-1}a^{m-i}$. This implies that there exists a contiguous block of $m-i$ a's within v. It is now clear that $v \not\uparrow x_2$, as x_2 has precisely $m-i$ $\diamond$'s to match the a's, but they do not form a contiguous block. By assumption v is any m-factor of w that begins with a b, we can therefore conclude that w avoids x_2 and hence the set X.

Now, suppose that $x_2 \sim_\diamond b^i \diamond^{m-i}$. The word $w_1 = (b^{i-2}aba^{m-i-1})^{\mathbb{Z}}$ avoids X. It avoids x_1 and x_3 for the same reasons w does. Now, if v is any m-factor of w_1 beginning (and ending) with b, then $v(0) \cdots v(m-2) \sim b^{i-2}aba^{m-i-1}$. This implies that there are $m-i$ occurrences of a in v, not situated in a contiguous block. It is now clear that $v \not\uparrow x_2$, as x_2 has only $m-i$ $\diamond$'s to align with the a's, however, all appearing in a single contiguous block. Thus w_1 avoids x_2. □

Corollary 6. *Let X be an m-uniform set of three partial words of the form described in Remark 1. If $h(x_1) + h(x_2) < m - 2$, then X is avoidable.*

Proof. Insert holes into x_1, x_2 so that $1 \leq h(x_1) \leq h(x_2), h(x_1) + h(x_2) = m - 2$. The new set, X', is still in standard form, and is not a C-set since $h(x_1) \geq 1$. Hence it is avoidable by Proposition 9, and thus so is $X \succ X'$. □

Before we apply Proposition 1 to prove Theorem 1, it remains to show that the unavoidable C-sets are maximal.

Proposition 10. *If m is even (resp. odd), then the unavoidable C-sets described in Proposition 2 (resp. Proposition 3) are maximal.*

Proof. Let m be even, and let $X = C_m(i)$ be an unavoidable C-set. We cannot strengthen x_2, for the resulting set would be avoidable by Corollary 6. If we strengthen x_3 with a b, then the resulting set is avoidable by Proposition 9 (as it is no longer a C-set). Finally, suppose we strengthen x_3 with an a in position j. Let $i' = \min(i, j)$ and $j' = \max(i, j)$. Then $C_m(i', j')$ is avoidable by Proposition 3, since either $j' \neq 2i'$, or $j' = 2i' \nmid m - 1$ (since $m - 1$ is odd). Hence X is maximal. Now let m be odd, and let $Y = C_m(\Lambda)$ an unavoidable C-set where $|\Lambda| = 2$. Again, we cannot strengthen x_2 at all, nor can we strengthen x_3 with a b. Now suppose we strenghten x_3 with an a. Then the resulting set is of the form $C_m(i, j, l)$, which is avoidable by Corollary 1. Hence Y is maximal. □

We now complete the proof of Theorem 1.

Proof (of Theorem 1). Let m be odd (resp. even), and let X be an m-uniform unavoidable set of three partial words, with $2m - 6$ (resp. $2m - 5$) total holes. Now, Corollaries 5 and 6 (resp. along with Proposition 8) together tell us that $h(x_1) + h(x_2) = m - 2$, and moreover Proposition 9 gives that X is necessarily a C-set. But we know that unavoidable C-sets with $2m-6$ (resp. $2m-5$) holes are maximal, by Proposition 10, and hence X is. Therefore, $H_{m,n} \geq 2m-6$ (resp. $2m-5$) by application of Proposition 1. On the other hand, $C_m(1,2)$ (resp. $C_m(1)$) is always unavoidable, and so we can in fact achieve $2m-6$ (resp. $2m-5$) holes in an unavoidable set. This yields the reverse inequality. □

References

1. Ehrenfeucht, A., Haussler, D., Rozenberg, G.: On regularity of context-free languages. Theoretical Computer Science 27, 311–332 (1983)
2. Champarnaud, J.M., Hansel, G., Perrin, D.: Unavoidable sets of constant length. International Journal of Algebra and Computation 14, 241–251 (2004)
3. Crochemore, M., Le Rest, M., Wender, P.: An optimal test on finite unavoidable sets of words. Information Processing Letters 16, 179–180 (1983)
4. Choffrut, C., Culik II, K.: On extendibility of unavoidable sets. Discrete Applied Mathematics 9, 125–137 (1984)
5. Evdokimov, A., Kitaev, S.: Crucial words and the complexity of some extremal problems for sets of prohibited words. Journal of Combinatorial Theory, Series A 105, 273–289 (2004)
6. Higgins, P.M., Saker, C.J.: Unavoidable sets. Theoretical Computer Science 359, 231–238 (2006)
7. Rosaz, L.: Unavoidable languages, cuts and innocent sets of words. RAIRO-Theoretical Informatics and Applications 29, 339–382 (1995)
8. Rosaz, L.: Inventories of unavoidable languages and the word-extension conjecture. Theoretical Computer Science 201, 151–170 (1998)
9. Saker, C.J., Higgins, P.M.: Unavoidable sets of words of uniform length. Information and Computation 173, 222–226 (2002)
10. Blanchet-Sadri, F., Jungers, R.M., Palumbo, J.: Testing avoidability on sets of partial words is hard. Theoretical Computer Science 410, 968–972 (2009)
11. Blakeley, B., Blanchet-Sadri, F., Gunter, J., Rampersad, N.: On the complexity of deciding avoidability of sets of partial words. In: Diekert, V., Nowotka, D. (eds.) DLT 2009. LNCS, vol. 5583, pp. 113–124. Springer, Heidelberg (2009)
12. Choffrut, C., Karhumäki, J.: Combinatorics of Words. In: Rozenberg, G., Salomaa, A. (eds.) Handbook of Formal Languages, vol. 1, pp. 329–438. Springer, Berlin (1997)
13. Lothaire, M.: Algebraic Combinatorics on Words. Cambridge University Press, Cambridge (2002)
14. Blanchet-Sadri, F., Brownstein, N.C., Kalcic, A., Palumbo, J., Weyand, T.: Unavoidable sets of partial words. Theory of Computing Systems 45, 381–406 (2009)
15. Blanchet-Sadri, F., Blakeley, B., Gunter, J., Simmons, S., Weissenstein, E.: Classifying All Avoidable Sets of Partial Words of Size Two. In: Martín-Vide, C. (ed.) Scientific Applications of Language Methods. Mathematics, Computing, Language, and Life: Frontiers in Mathematical Linguistics and Language Theory, pp. 59–101. Imperial College Press, London (2010)
16. Blanchet-Sadri, F.: Algorithmic Combinatorics on Partial Words. Chapman & Hall/CRC Press, Boca Raton (2008)
17. Schützenberger, M.P.: On the synchronizing properties of certain prefix codes. Information and Control 7, 23–36 (1964)
18. Mykkeltveit, J.: A proof of Golomb's conjecture for the de Bruijn graph. Journal of Combinatorial Theory, Series B 13, 40–45 (1972)

Shortest Paths between Shortest Paths and Independent Sets

Marcin Kamiński[1,⋆], Paul Medvedev[2], and Martin Milanič[3,⋆⋆]

[1] Département d'Informatique, Université Libre de Bruxelles, Brussels, Belgium
Marcin.Kaminski@ulb.ac.be
[2] Department of Computer Science, University of Toronto, Toronto, Canada
pashadag@cs.toronto.edu
[3] FAMNIT and PINT, University of Primorska, Koper, Slovenia
martin.milanic@upr.si

Abstract. We study problems of reconfiguration of shortest paths in graphs. We prove that the shortest reconfiguration sequence can be exponential in the size of the graph and that it is NP-hard to compute the shortest reconfiguration sequence even when we know that the sequence has polynomial length. Moreover, we also study reconfiguration of independent sets in three different models and analyze relationships between these models, observing that shortest path reconfiguration is a special case of independent set reconfiguration in perfect graphs, under any of the three models. Finally, we give polynomial results for restricted classes of graphs (even-hole-free and P_4-free graphs).

1 Introduction

One of the biggest impacts of algorithmic graph theory has been its usefulness in modeling real-world problems, where the domain of the problem is modeled as a graph and the constraints on the solution define feasible solutions. For example, consider the problem of routing a certain commodity between two nodes in a transportation network, using as few hops as possible. The transportation network can be modeled as a graph, each route can be modeled as a path, and the feasible solutions are all the shortest paths between the two nodes. Traditionally, the real-world user first defines a problem instance and then uses an algorithm to find a feasible solution which she then "implements" in the real world. However, some real-world situations do not follow this simple paradigm and are more dynamic because they allow the solution to "evolve" over time. For example, consider the situation where the commodity is already being transferred along a shortest route, but the operator has been instructed to use a different route, which is also a shortest path. She can physically switch the route only one node

⋆ Chargé de Recherches du FRS-FNRS.
⋆⋆ Supported in part by "Agencija za raziskovalno dejavnost Republike Slovenije", research program P1-0285.

C.S. Iliopoulos and W.F. Smyth (Eds.): IWOCA 2010, LNCS 6460, pp. 56–67, 2011.

at a time, but does not wish to interrupt the transfer. Thus, she would like to switch between the two routes in as few steps as possible, while maintaining a shortest path route at every intermediate step.

In general, this type of situation gives rise to a *reconfiguration* framework, where we consider an algorithmic problem $\mathcal{P}$ and a way of transforming one feasible solution of an instance I of $\mathcal{P}$ to another (*reconfiguration rule*). Given two feasible solutions s_1, s_k of I, we want to find a *reconfiguration sequence* $s_1, \ldots, s_k$ such that each s_i $(1 \leq i \leq k)$ is a feasible solution of I, and the transition between s_i and s_{i+1} is allowed by the reconfiguration rule. An alternate definition is via the *reconfiguration graph*, where the vertices are the feasible solutions of I, and two solutions are adjacent if and only if one can be obtained from the other by the reconfiguration rule. The reconfiguration sequence is then a path between s_1 and s_k in the reconfiguration graph. We can then ask for the shortest reconfiguration sequence, or, in the *reconfigurability problem*, to simply check if the two solutions are *reconfigurable* (i.e., if such a sequence exists).

The reconfiguration framework has recently been applied in a number of settings, including vertex coloring [3,4,2,1], list-edge coloring [12], clique, set cover, integer programming, matching, spanning tree, matroid bases [11], block puzzles [10], independent set [10,11], and satisfiability [9]. In the well-studied vertex coloring problem, for example, we are given two k-colorings of a graph, and the reconfiguration rule allows to change the color of a single vertex. In a different example, we are given two independent sets, which we imagine to be two sets of tokens placed on the vertices, and the reconfiguration rule is to slide a single token along an edge (*token sliding*).

Though the complexities of each of the many reconfiguration problems may each be studied independently, a fundamental question is whether there exists any systematic relationship between the complexity of the original problem and that of its reconfigurability problem. To this end, current studies have revealed a pattern where most "natural" problems in P have their reconfigurability problems in P as well, while problems whose reconfigurability versions are at least NP-hard are NP-complete. For example, spanning tree, matching, and matroid problems in general (all in P) lead to polynomially solvable reconfigurability problems, while the reconfigurability of independent set, set cover, and integer programming (all NP-complete) are PSPACE-complete [11]. Another example is satisfiability, where Gopalan et al. [9] showed that reconfigurability instances arising from tight relations—a class for which it is easy to determine if the formula is satisfiable—can be solved in linear time; on the other hand, reconfigurability is PSPACE-complete for the class of formulas arising from non-tight relations.

Ito et al. [11] have conjectured that this relationship is not true in general, and that there exist problems in P which give rise, in a natural way, to NP-hard reconfigurability problems. Indeed, the problem of deciding whether two k-colorings are reconfigurable is PSPACE-complete for (i) bipartite graphs and $k \geq 4$, and (ii) planar graphs, for $4 \leq k \leq 6$ [1]. Clearly, 4-coloring of bipartite or planar graphs is in P. However, these are not "natural" problems in the sense that the colorings are not optimal. It is interesting to ask if there exists a "natural" problem in P whose reconfiguration version is NP-hard.

Another systematic relationship that has been pursued is between the complexity of a reconfigurability problem and the diameter of the reconfiguration graph. When the diameter is polynomial, a reconfiguration sequence is a trivial certificate for the reconfigurability of two instances, guaranteeing that the problem is in NP. However, current evidence further suggests that for reconfigurability problems that are solvable in polynomial time, the diameter is also polynomial. In the study of k-coloring, it was found that for $k \leq 3$, the reconfigurability problem is solvable in polynomial time and the diameter of the reconfiguration graph is at most quadratic in the number of vertices of the colored graph. For satisfiability, the formulas built from tight relations (whose reconfigurability is polynomial) lead to reconfiguration graphs with linear diameter [9]. We are not aware of any natural problems with the property that the diameter can be exponential while reconfigurability can be decided in polynomial time[1]; however, such an example, if found, would indicate that the diameter cannot serve as a reliable indicator of the reconfigurability complexity.

In this paper, we introduce the reconfiguration version of the shortest path problem (Section 2), which arises naturally, such as in the routing example above. We show that the reconfiguration graph can have exponential diameter, implying that the shortest path reconfiguration problem probably breaks one of the two established patterns described above. On the one hand, if reconfigurability of shortest paths can be decided in polynomial time, then it is the first example of a reconfigurability problem in P with exponential diameter. On the other hand, if it is NP-hard, it is the first example of a "natural" problem in P whose reconfigurability version is NP-hard. For these reasons, we believe that shortest path reconfiguration is an important problem to study, not only for its practical application but also for our understanding of the systematic relationship between the hardness of a problem, the diameter of its reconfiguration graph, and the hardness of its reconfigurability problem. Towards this end, we give a non-trivial reduction from SAT to show that it is NP-hard to find the shortest reconfiguration sequence between two shortest paths (however, the complexity status of the reconfigurability problem remains open).

We also study reconfiguration of independent sets, where, unlike many other problems, there is more than one natural reconfiguration rule. In particular applications, for example, a threshold is specified that bounds the cardinality of the intermediate feasible solutions. Based on this idea, Ito et al. [11] considered an alternative to token sliding called *token addition and removal*, where one is allowed to either add or remove a token as long as there are at least $k-1$ tokens at any given time, for some k. In this paper, we introduce *token jumping*, where one is allowed to move a single token to any other vertex. The token

---

[1] For a very artificial one, consider the problem in which instances are n-bit words and two instances are adjacent if they differ by 1 modulo 2^n. The diameter of the reconfiguration graph is 2^{n-1} but all pairs of instances are reconfigurable.

jumping reconfiguration graph is often easier to analyze than the token addition and removal one, since the cardinalities of two adjacent token sets are always the same. However, we show that the two models are polynomially equivalent, allowing for an easier way to analyze token addition and removal reconfiguration graphs (Section 3).

Finally, we show that reconfiguration of independent sets is a generalization of the reconfiguration of shortest paths; our hardness result for shortest paths then implies that it is NP-hard to find the shortest reconfiguration sequence between two independent sets, even in perfect graphs (Section 4). We also identify two restricted graph classes where reconfigurability is easy – even-hole-free graphs under token jumping, for which the reconfiguration graph is always connected, and P_4-free graphs under token sliding (Section 5). Due to the space constraint, some proofs are omitted but can be found in the full version [13].

2 Shortest Path Reconfiguration

We define the reconfiguration rule for shortest paths in the natural way: two shortest (s,t)-paths are adjacent in the reconfiguration graph of shortest (s,t)-paths if and only if they differ, as sequences, in exactly one vertex.

2.1 Instances with Exponential Diameter

We now present a family of graphs G^k whose size is linear in k but the diameter of the reconfiguration graph is $\Omega(2^k)$. The graph G^1 contains vertices $\{x_i^1 \mid 1 \leq i \leq 7\} \cup \{y_i^1 \mid 1 \leq i \leq 6\} \cup \{s,t\}$ and edges $\{(x_i^1, y_i^1), (x_{i+1}^1, y_i^1), (y_i^1, t) \mid i \leq 6\} \cup \{(s, x_i^1) \mid 1 \leq i \leq 7\}$. The graph G^k is defined recursively with vertices $\{x_i^k \mid 1 \leq i \leq 7\} \cup \{y_i^k \mid 1 \leq i \leq 6\} \cup V(G^{k-1})$ and the edges $\{(x_i^k, y_i^k), (x_{i+1}^k, y_i^k) \mid i \leq 6\} \cup \{(y_i^k, x_j^{k-1}) \mid i \in \{1,3,5\}, j \leq 7\} \cup \{(y_2^k, x_1^{k-1}), (y_4^k, x_7^{k-1}), (y_6^k, x_1^{k-1})\} \cup E(G^{k-1} \setminus \{s\}) \cup \{(s, x_i^k) \mid 1 \leq i \leq 7\}$ (see Figure 1). Let $p_b^k = s, x_1^k, y_1^k, \ldots, x_1^1, y_1^1, t$, and let $p_e^k = s, x_7^k, y_6^k, x_1^{k-1}, x_1^{k-1}, \ldots, x_1^1, y_1^1, t$. We will consider the problem of reconfiguring p_b^k to p_e^k in G^k.

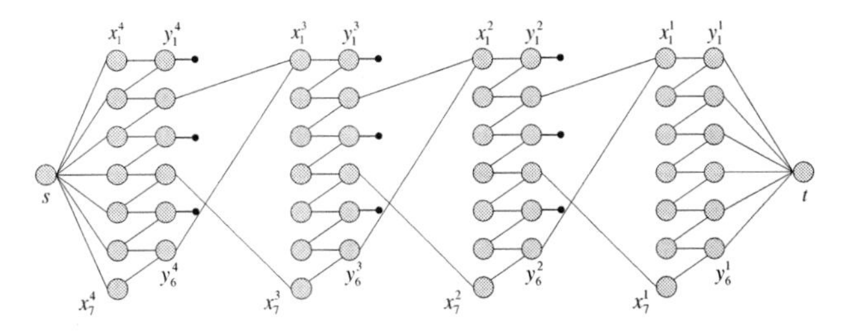

Fig. 1. The graph G^k for $k = 4$, where the reconfiguration distance between $p_b^k = s, x_1^k, y_1^k, \ldots, x_1^1, y_1^1, t$ and $p_e^k = s, x_7^k, y_6^k, x_1^{k-1}, x_1^{k-1}, \ldots, x_1^1, y_1^1, t$ is $\Theta(2^k)$. An edge with a circle end means that the vertex is connected to all the vertices in the next layer.

Lemma 1. *Let p be a shortest path in G^k that goes through y_1^k, and let q be a path that goes through y_6^k. Then the reconfiguration distance between p and q is at least $9(2^k - 1)$.*

Proof. We prove by induction on k, where the base case is clear. Let $\rho = p_1, \ldots, p_n$ be the shortest reconfiguration sequence between p and q. First, let i' be the smallest integer such that $p_{i'+1}$ contains y_4^k, and let $i \leq i'$ be the smallest integer such that every path $p_i, \ldots, p_{i'}$ contains y_3^k. By construction, we know that p_{i-1}, and hence p_i, contains y_1^{k-1} and $p_{i'+1}$, and hence $p_{i'}$, contains y_6^{k-1}. Hence, by the induction hypothesis, the length of this first phase, $i' - i + 1$, is at least $9(2^{k-1} - 1)$.

Next, let j' be the smallest integer such that $p_{j'+1}$ contains y_6^k, and let $j \leq j'$ be the smallest integer such that every path $p_j, \ldots, p_{j'}$ contains y_5^k. By construction, we know that p_{j-1}, and hence p_j, contains y_6^{k-1} and $p_{j'+1}$, and hence $p_{j'}$, contains y_1^{k-1}. Hence, by the induction hypothesis, the length of this second phase, $j' - j + 1$, is at least $9(2^{k-1} - 1)$.

Observe from the graph construction that ρ must always visit y_{x-1}^k before visiting y_x^k, hence $i' < j$, and so the length of ρ is at least the sum of the two phases plus the moves of the first and second vertex necessary to percolate y_1^k down to y_6^k, proving the lemma. □

On the other hand, there exists a reconfiguration sequence of length $11(2^k - 1)$ (see the full version [13] for a proof), giving the following theorem:

Theorem 1. *The reconfiguration distance in G^k between p_b^k and p_e^k is $\Theta(2^k)$.*

2.2 NP-Hardness of Min-SPR

Given (G, s, t, p_b, p_e, k), where p_b and p_e are shortest (s,t)-paths and k is an integer, the Min-SPR problem is to determine whether there is a reconfiguration sequence between p_b and p_e of length at most k. Let ϕ be a formula with variables $x_1, \ldots, x_n$ and clauses $C_1, \ldots, C_m$. We will create an instance $(G_\phi, s, t, p_b, p_e, 2m(n+2))$ and show that ϕ is satisfiable if and only this instance is in Min-SPR. For ease of presentation, the graph G_ϕ will be directed. However, our result holds for undirected graphs because the directed shortest (s,t)-paths in G_ϕ are exactly the shortest paths in the undirected version of G_ϕ.

For every variable x_i and its possible value $vs \in \{0, 1\}$, we build a gadget $G(i, vs)$. The vertex set is $\{v(i, vs, cs, j) \mid cs \in \{0, 1\}, 1 \leq j \leq 2m\}$. The values i, vs, cs, and j for a vertex are referred to as its *level*, *v-state*, *c-state*, and *depth*, and denoted by $l(v)$, $vs(v)$, $cs(v)$, and $d(v)$, respectively. For every $1 \leq j \leq 2m - 1$, and every cs, there is an edge from $v(i, vs, cs, j)$ to $v(i, vs, cs, j + 1)$. For all $1 \leq j \leq m - 1$, there is an edge from $v(i, vs, 0, 2j)$ to $v(i, vs, 1, 2j + 1)$, and from $v(i, vs, 1, 2j)$ to $v(i, vs, 0, 2j + 1)$. We also add edges, called formula edges, that are formula dependent. For all j, if $x_i = vs$ satisfies C_j, we add an edge from $v(i, vs, 1, 2j - 1)$ to $v(i, vs, 0, 2j)$. This gadget is shown in Figure 2A.

We now connect some of these gadgets together. The gadgets we connect are $G(i, vs)$ to $G(i + 1, 0)$ and to $G(i + 1, 1)$, for all $i \leq n - 1$ and all vs. Given two

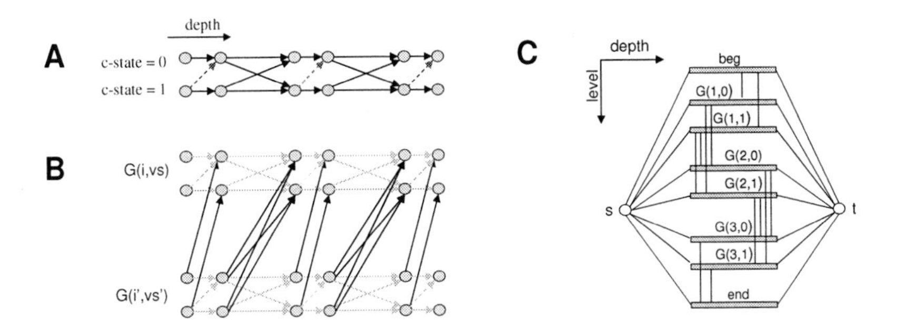

Fig. 2. The reduction from a formula ϕ to a graph G_ϕ for the case of three clauses and three variables. Panel A shows the internal connections of a gadget, with the potential formula edges that depend on ϕ given in red (dashed). Panel B shows the way we connect two given gadgets, while C shows the structure of the whole graph. Each of the rectangles represents a gadget, with the lines showing which parts are connected together.

gadgets, $G(i, vs)$ and $G(i', vs')$, the meaning of connecting $G(i, vs)$ to $G(i', vs')$ is given as follows (shown in Figure 2BC). For all $j \leq 2m - 1$ and cs, there is an edge from $v(i', vs', cs, j - 1)$ to $v(i, vs, cs, j)$. Also, for all $j \leq m - 1$ and cs, there is an edge from $v(i', vs', cs, 2j)$ to $v(i, vs, 1 - cs, 2j + 1)$.

We next add a begin and end gadget to the graph, consisting of vertices beg_j and end_j, respectively, for $1 \leq j \leq 2m$. These are connected in a path, with edges (beg_j, beg_{j+1}) and (end_j, end_{j+1}) for $j \leq 2m - 1$. The level of the vertices in the begin (end) gadget is 0 $(i + 1)$, the c-state is 0 (1), and the depth of beg_j or end_j is j. For all vs, $j \leq 2m - 1$, there is an edge from $v(1, vs, 0, j)$ to beg_{j+1}, and from end_j to $v(n, vs, 1, j + 1)$.

Finally, we add a s and t vertex to the graph, and make an edge from s to every depth 1 vertex, and from every depth $2m$ vertex to t. The depth of $s(t)$ is defined to be 0 $(2m + 1)$. We call the resulting directed graph G_ϕ. Let $p_b = s, beg_1, \ldots, beg_{2m}, t$ and $p_e = s, end_1, \ldots, end_{2m}, t$ be two paths in this graph. Then, $(G_\phi, s, t, p_b, p_e, 2m(n+2))$ is the instance of the MIN-SPR problem that we will consider here.

The intuition behind the reduction is that in order for the path to percolate down from p_b to p_e in a minimal number of steps, it must pass consecutively through exactly one of $G(i, 0)$ or $G(i, 1)$ for every variable x_i. The choice of which one corresponds to assigning x_i the corresponding value. Furthermore, each shortest path that goes through a gadget can visit the vertex at depth $2j$ with a c-state of 0 or 1. This corresponds to having the jth clause satisfied or not. Initially, the path goes only through vertices with c-state 0, and the only way to switch the c-state at a given depth is via a formula edge. By going through a gadget $G(i, vs)$, there is an opportunity to use the formula edges to switch the c-state of all clauses that $x_i = vs$ would satisfy. In order to reach the final path p_e, the c-state of all the vertices must be 1, hence all the clauses must be satisfied.

First, we will show that the reduction is sound. Each edge (a, b) is considered to be either *odd* or *even*, depending on the parity of $d(a)$. We call edges that connect vertices on the same level (exactly those that belong to the same gadget) as *intra-level*, while the edges that connect vertices on different levels are called *inter-level*. We say that a reconfiguration sequence *visits* a vertex if there exists $p \in \rho$ that contains that vertex.

Fact 1. *Let $e = (a, b)$ be an edge in G_ϕ. The following follows directly from construction:*

1. $l(b) \leq l(a) \leq l(b) + 1$.
2. *If e is an intra-level odd edge, $cs(a) = 0$ implies that $cs(b) = 0$.*
3. *If e is an inter-level odd edge, then $cs(a) = cs(b)$.*
4. *If e is a non-formula odd edge, then $cs(a) = cs(b)$.*
5. *If e is intra-level, then $vs(a) = vs(b)$.*

These facts about G_ϕ capture most of the properties of the reduction that are needed to prove correctness of the following Lemmas (2, 3, 4, 5, and 6), whose proofs can be found in the full version [13].

Lemma 2. *The length of a reconfiguration sequence is at least $2m(n+2)$. Moreover, each move in an sequence that has this length must either increase the c-state or the level of the switched vertex, but not both.*

Lemma 3. *No path can contain two vertices with the same level but different v-state.*

Lemma 4. *Suppose there exists a reconfiguration sequence ρ of length $2m(n+2)$. Then ρ visits at least one vertex at every level, and all the vertices that it visits at a given level have the same v-state.*

Suppose there exists a reconfiguration sequence ρ of length $2m(n+2)$. Lemma 4 allows us to build an assignment θ by assigning θ_i the v-state of the vertices of level i in ρ.

Lemma 5. *The assignment θ is satisfying for ϕ.*

We also show that the reduction is complete.

Lemma 6. *If ϕ is satisfiable, then there exists a reconfiguration sequence of length at most $2m(n + 2)$.*

Combining Lemma 5 and Lemma 6 together with the fact that the reduction can be clearly done in polynomial time, we have the following theorem.

Theorem 2. *The* Min-SPR *problem is NP-hard, even if k is polynomial in $|V(G)|$.*

3 Independent Set Reconfiguration: The Models

We now turn our attention to reconfiguring independent sets. Consider the reconfiguration rule in which a move from one valid configuration to another is made by a *token jump*: moving a token from one vertex to an unoccupied vertex (not necessarily a neighbor of it), such that the resulting set is independent. The token sliding, token jumping, and token addition and removal reconfiguration rules give rise to the following three reconfigurability problems.

Token sliding (TS) / token jumping (TJ): Given a graph G and two independent sets A, B in G, determine if A can be reconfigured into B via a sequence of independent sets, each of which results from the previous one by a single token slide (for TS) or jump (for TJ).

Token addition and removal (TAR): Given a graph G, an integer k and two independent sets A, B in G, both of size $\geq k$, is there a way to transform A into B via independent sets, each of which results from the previous one by adding or removing one vertex of G, without ever going through an independent set of size less than $k-1$?

We say that A and B are TS- (TAR-, TJ-) reconfigurable if they belong to the same connected component of the TS- (TAR-, TJ-) reconfiguration graph. We now establish the equivalence between the TJ and TAR problems (the proof can be found in the full version [13]).

Theorem 3. *Two independent sets A and B of size s in a graph G are TJ-reconfigurable if and only if they are TAR-reconfigurable with parameter $k = s$. Moreover, $dist_{\mathrm{TAR}}(A, B) = 2dist_{\mathrm{TJ}}(A, B)$, and there exists an algorithm that, given a reconfiguration sequence between two independent sets in one of these two models outputs a reconfiguration sequence connecting the two sets in the other model in time polynomial in the length of the sequence. The algorithm maps every shortest TAR-sequence to a shortest TJ-sequence, and vice versa.*

Theorem 3 immediately implies that results holding for the TAR model can be transferred to the TJ model. New results can also be derived via this relationship:

Corollary 1. *There exists a polynomial-time algorithm for the TJ problem in line graphs.*

Proof. By Theorem 3, the TJ problem in line graphs is polynomially reducible to the TAR problem in line graphs. Due to the correspondence between matchings in a graph and independent sets in its line graph, the problem is polynomially equivalent to the MATCHING RECONFIGURATION problem. For a polynomial-time algorithm for this problem, see Ito et al. [11]. □

4 Hardness of Independent Set Reconfiguration

TS, TAR, and TJ reconfiguration problems are all PSPACE-complete in general graphs. For the TS problem, this was announced in [10] (see also [1]) without an explicit proof. For the TAR problem, this was shown by Ito et al. [11]. In fact, their proof uses only token slides (which are done by token additions and removals), also implying that TS is PSPACE-complete. Theorem 3 immediately implies PSPACE-completeness for the TJ model. The hardness of these reconfigurability problems of course implies the hardness of the more difficult problems of finding the length of the shortest reconfiguration sequence. However, in this section we show that these related variants remain NP-hard even when the graph is restricted to be perfect.

We use a reduction from the reconfiguration of shortest paths. Given a graph G and two vertices s and t of G at a distance k apart, let G_1 denote the graph obtained from G by deleting from it all the vertices and edges not appearing on any shortest (s,t)-path. For $i \in \{0, 1, \ldots, k\}$, let D_i be the set of vertices in G_1 at distance i from s and $k - i$ from t. The graph G' is the graph obtained from G_1 by turning every set D_i into a clique, and complementing the edges of G_1 between every pair of consecutive layers D_i and D_{i+1}. Formally, $V(G') = V(G_1)$ and $E(G') = \{uv : u \neq v, \exists i \text{ such that } u, v \in D_i\} \cup \bigcup_{i=0}^{k-1} \{uv : u \in D_i, v \in D_{i+1}, uv \notin E(G_1)\}$. The idea of the construction is that there is a bijective correspondence between shortest (s,t)-paths in G and independent sets of size $k + 1$ in G'. This gives the following theorem (the proof is straightforward):

Theorem 4. *For every graph G, there is a polynomially computable length-preserving bijection (length-doubling for TAR) between shortest reconfiguration sequences in the shortest path reconfiguration graph of G and those in the TS- (TJ-, TAR-) reconfiguration graph for G'.*

The following corollary is a direct consequence of Theorem 2 and the fact that graph G' contains no odd holes or their complements and hence is perfect [5]. Recall that a *hole* in a graph is a chordless cycle with at least four vertices, and a hole is even (odd) if it has an even (odd) number of vertices.

Corollary 2. *Let G be a perfect graph, A, B two independent sets in G, and k an integer. It is NP-hard to determine if there exists a reconfiguration sequence of length at most k between A and B in the TS, TJ, and TAR models,even if k is polynomial in $|V(G)|$.*

5 Positive Results for Independent Set Reconfiguration

In this section we identify two restrictions on the input graphs which make the reconfigurability of independents sets easy to solve.

5.1 Even-Hole-Free Graphs in the TJ Model

We will show that two token sets of the same size in any even-hole-free graph are TJ-reconfigurable. Given a graph G and two independent sets A and B in G of the same size, the *Piran graph* $\Pi(A, B)$ of A and B is the subgraph of G induced by the vertex set $(A \backslash B) \cup (B \backslash A)$. The following simple lemma gives a sufficient condition under which it is always possible to jump a token from A to B.

Lemma 7. *Let A and B be two independent sets of the same size in a graph G. If the Piran graph $\Pi(A, B)$ is even-hole-free then there exists a token in $B \setminus A$ with at most one neighbor in $A \setminus B$.*

Proof. The Piran graph is bipartite, and as such, it does not contain odd cycles. If in addition, $\Pi(A, B)$ is even-hole-free, then it must be a forest. Since $|A \setminus B| = |B \setminus A|$, the number of edges in $\Pi(A, B)$ is in fact at most $|A \setminus B| + |B \setminus A| - 1 = 2|B \setminus A| - 1$. Therefore there exists a vertex in $B \setminus A$ with at most one neighbor in $A \setminus B$. □

A consequence of Lemma 7 is the following result (a proof of which can be found in the full version [13]).

Theorem 5. *Let A and B be two independent sets of the same size in a graph G. If the Piran graph $\Pi(A, B)$ is even-hole-free, then A and B are TJ-reconfigurable. Moreover, there exists an algorithm running in time $O(|A|)$ that (if the Piran graph is even-hole-free) finds a shortest TJ-path between the two sets.*

The class of even-hole-free graphs includes the well known class of chordal graphs (hence also trees and interval graphs). The structure of even-hole-free graphs is understood and membership in this class can be decided in polynomial time [6]. Notice that if the input graph is even-hole-free, so is the Piran graph. Due to Theorem 5 we can easily solve the TJ reconfiguration problem for the class of even-hole-free graphs. Interestingly, determining the complexity of computing the maximum size of an independent set in an even-hole-free graph is, to the best of our knowledge, an open problem.

The example of the claw $G = K_{1,3}$ with leaves $\{v_1, v_2, v_3\}$, and the independent sets $A = \{v_1, v_2\}$, $B = \{v_1, v_3\}$, shows that the analogue of Theorem 5 does not hold for the TS model for the whole class of even-hole-free graphs. We leave it as an open question to determine whether the analogue holds for the class of (claw, even-hole)-free graphs.

5.2 P_4-Free Graphs in the TS Model

In this subsection we give a polynomial time algorithm to solve the TS problem in P_4-free graphs. P_4-free graphs (also known as cographs) are graphs without an induced subgraph isomorphic to a 4-vertex path. A polynomial-time algorithm for token sliding in P_4-free graphs can be developed based on the following well-known characterization of P_4-free graphs [7]: a graph G is P_4-free if and only if for every induced subgraph F of G with at least two vertices, either F or the complement to F is disconnected. A *co-component* of a graph $G = (V, E)$ is the subgraph of G induced by the vertex set of a connected component of the complementary graph $\overline{G} = (V, \{uv \mid u, v \in V,\ u \neq v,\ uv \notin E\})$.

Theorem 6. *The TS problem is solvable in time $O(|V| + |E|)$ if the input graph $G = (V, E)$ is P_4-free. Moreover, a shortest reconfiguration sequence, if it exists, can be found in time $O(|V| + |E|)$.*

Algorithm 1. TS-reconfiguration of independent sets in P_4-free graphs

Input: A P_4-free graph $G = (V, E)$ and two independent sets A, B.
Output: A shortest (A, B)-path in the TS-graph, if one exists, NO otherwise.

1: **if** $|V(G)| = 1$ **then** return the trivial TS-path if $|A| = |B|$, and NO otherwise.
2: **if** G is disconnected, with connected components $C_1, \ldots, C_m$ **then**
3:     **if** there is an $i \in \{1, \ldots, m\}$ such that $|A \cap C_i| \neq |B \cap C_i|$ **then** return NO.
4:     **else** solve the problem recursively for the connected components $C_1, \ldots, C_m$ with respective token sets $(A \cap C_1, B \cap C_1), \ldots (A \cap C_m, B \cap C_m)$.
5:         **if** one of the outputs is NO **then** return NO.
6:         **else** merge the corresponding $(A \cap C_i, B \cap C_i)$-paths into an (A, B) TS-path P, return P.
7: **else**
8:     **if** $|A| = |B| = 1$ **then** return an (A, B) TS-path corresponding to a shortest (A, B)-path in G.
9:     **else**
10:         **if** A and B are in the same co-component of G **then** solve the problem for A and B recursively on that co-component and return the output.
11:         **else** return NO.

Proof. We claim that Algorithm 1 below solves the TS problem on P_4-free graphs.

The correctness of the algorithm is straightforward, using the above-mentioned characterization of P_4-free graphs [7]. Using the result of Corneil et al. [8] showing that the decomposition of a P_4-free graph $G = (V, E)$ into one-vertex graphs by means of taking components or co-components can be computed in time $O(|V| + |E|)$ [8], it is also easy to see that the algorithm can be implemented so that it runs in linear time. □

Theorem 6 can be used to prove that the TS problem is solvable in polynomial time if the input graph is (claw, paw)-free (recall that the claw is $K_{1,3}$ and the paw is the graph obtained from the claw by adding one edge). This is due to the observation that the only connected (claw, paw)-free graph containing an induced P_4 are (long enough) paths and cycles.

6 Concluding Remarks

In this paper, we studied the reconfiguration variants of the shortest path and independent set problems. We believe that the major open problem is to determine the complexity of deciding whether two shortest paths are reconfigurable. If the problem is NP-hard, then it will be the first example of an efficiently solvable reconfigurability problem with reconfiguration graphs of large diameter. If the problem is polynomially solvable, then it will be the first example of a "natural" problem in P whose reconfigurability version is NP-hard.

Acknowledgements. We are grateful to Paul Bonsma, Takehiro Ito and Daniel Pellicer for interesting and fruitful discussions. We also thank Paul Bonsma for pointing us to problems in P whose reconfiguration versions are PSPACE-complete.

References

1. Bonsma, P.S., Cereceda, L.: Finding paths between graph colourings: PSPACE-completeness and superpolynomial distances. Theor. Comput. Sci. 410(50), 5215–5226 (2009)
2. Bonsma, P.S., Cereceda, L., van den Heuvel, J., Johnson, M.: Finding paths between graph colourings: Computational complexity and possible distances. Electronic Notes in Discrete Mathematics 29, 463–469 (2007)
3. Cereceda, L., van den Heuvel, J., Johnson, M.: Connectedness of the graph of vertex-colourings. Discrete Mathematics 308(5-6), 913–919 (2008)
4. Cereceda, L., van den Heuvel, J., Johnson, M.: Mixing 3-colourings in bipartite graphs. European Journal of Combinatorics 30, 1593–1606 (2009)
5. Chudnovsky, M., Robertson, N., Seymour, P., Thomas, R.: The strong perfect graph theorem. Ann. of Math. 164, 51–229 (2006)
6. Conforti, M., Cornuéjols, G., Kapoor, A., Vušković, K.: Even-hole-free graphs part II: Recognition algorithm. J. Graph Theory 40, 238–266 (2002)
7. Corneil, D.G., Lerchs, H., Stewart Burlingham, L.: Complement reducible graphs. Discrete Applied Mathematics 3(3), 163–174 (1981)
8. Corneil, D.G., Perl, Y., Stewart, L.K.: A linear recognition algorithm for cographs. SIAM J. Comput. 14(4), 926–934 (1985)
9. Gopalan, P., Kolaitis, P.G., Maneva, E.N., Papadimitriou, C.H.: The connectivity of Boolean satisfiability: Computational and structural dichotomies. SIAM J. Comput. 38(6), 2330–2355 (2009)
10. Hearn, R.A., Demaine, E.D.: PSPACE-completeness of sliding-block puzzles and other problems through the nondeterministic constraint logic model of computation. Theor. Comput. Sci. 343(1-2), 72–96 (2005)
11. Ito, T., Demaine, E.D., Harvey, N.J.A., Papadimitriou, C.H., Sideri, M., Uehara, R., Uno, Y.: On the complexity of reconfiguration problems. In: Hong, S.-H., Nagamochi, H., Fukunaga, T. (eds.) ISAAC 2008. LNCS, vol. 5369, pp. 28–39. Springer, Heidelberg (2008)
12. Ito, T., Kamiński, M., Demaine, E.D.: Reconfiguration of list edge-colorings in a graph. In: Dehne, F., Gavrilova, M., Sack, J.-R., Tóth, C.D. (eds.) WADS 2009. LNCS, vol. 5664, pp. 375–386. Springer, Heidelberg (2009)
13. Kaminski, M., Medvedev, P., Milanic, M.: Shortest paths between shortest paths and independent sets. CoRR, abs/1008.4563 (2010)

Faster Bit-Parallel Algorithms for Unordered Pseudo-tree Matching and Tree Homeomorphism

Yusaku Kaneta and Hiroki Arimura

Hokkaido University, N14, W9, Sapporo 060-0814, Japan
{y-kaneta,arim}@ist.hokudai.ac.jp

Abstract. In this paper, we consider the unordered pseudo-tree matching problem, which is a problem of, given two unordered labeled trees P and T, finding all occurrences of P in T via such many-one embeddings that preserve node labels and parent-child relationship. This problem is closely related to tree pattern matching problem for XPath queries with child axis only. If $m > w$, we present an efficient algorithm that solves the problem in $O(nm\log(w)/w)$ time using $O(hm/w + m\log(w)/w)$ space and $O(m\log(w))$ preprocessing on a unit-cost arithmetic RAM model with addition, where m is the number of nodes in P, n is the number of nodes in T, h is the height of T, and w is the word length. We also discuss a modification of our algorithm for the unordered tree homeomorphism problem, which corresponds to a tree pattern matching problem for XPath queries with descendant axis only.

1 Introduction

Tree matching is a fundamental problem in computer science, and it has a wide range of applications in XML/Web database, schema validation, information extraction, document analysis, image processing, and semi-structured data processing. In particular, tree matching and tree inclusion problems have attracted much attention and have been extensively studied [2,8,9,14]. In this paper, we study a non-standard version of the unordered tree matching and inclusion problems, called the unordered pseudo-tree matching problem (UPTM) [16] and the unordered tree homeomorphism problem (UTH) [4], respectively, where embedding mappings can be many-one (See Fig. 1).

As main results, we present an efficient algorithm that solves UPTM problem with the following complexities (Theorem 1):

- $O(n\lceil m/w\rceil\log w)$ time using $O(h\lceil m/w\rceil + \lceil m/w\rceil\log w)$ space and $O(m\log w)$ preprocessing if $m > w$. (the large pattern case)
- $O(n\log m)$ time using $O(h + \log m)$ space and $O(m\log m)$ preprocessing if $m \leq w$. (the small pattern case)

where m and n are the sizes of pattern tree P and text tree T, h is the height of T, and w is the word length of RAM. We also show that UTH problem is solvable in the same time and space complexities as above (Theorem 2).

C.S. Iliopoulos and W.F. Smyth (Eds.): IWOCA 2010, LNCS 6460, pp. 68–81, 2011.

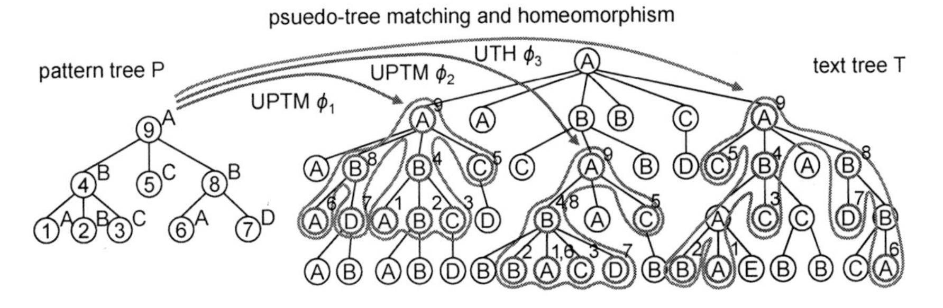

Fig. 1. The unordered pseudo-tree matching problem (UPTM) and the unordered tree homeomorphism problem (UTH)

A key of our algorithm is a data structure for the small pattern case, where $m \leq w$, based on bit-parallel computation of set operations, including *tree aggregation* that checks the branching of internal nodes. Developing bit-assignment technique based on separator trees, we improve the complexity of the tree aggregation from $O(m)$ time and space to $O(\log m)$ time and space. Combining this result to dynamic programming tree matching algorithms and a module decomposition technique of [10], we have claimed results for both UPTM and UTH.

For the UPTM, our $O(n\lceil m/w\rceil \log w)$ time and $O(h\lceil m/w\rceil \log w + \lceil m/w\rceil \log w)$ space algorithm improves the complexity of the previous $O(n{\cdot}rep(P)l(P)d(P)/w) = O(nm^3/w)$ time and $O(n{\cdot}l(P)d(P)/w) = O(nm^2/w)$ space algorithm[1] by Yamamoto and Takenouchi [16] in the worst case. For the UTH, our algorithm is one of the first bit-parallel algorithm for the problem and slightly faster than the previous $O(nm \cdot d(P)) = O(nm^2)$ time and $O(d(T)b(T)) = O(n^2)$ space algorithm[1] by Götz, Koch, and Martens [4]. These results for UPTM and UTH correspond to evaluation of fragments of Core XPath queries consisting with child axis only and with descendant axis only [4], respectively.

Tree matching problems with many-one embeddings have been studied in the area of FO and MSO logics over combinatorial structures such as strings, trees, and graphs as well as in database and Web systems [4]. These problems have less constraints than the other tree matching problems, but this does not necessarily mean that many-one matching problems are easiest among them. Hence, we hope that these results become steps towards development of efficient query mechanism for such data intensive applications.

Organization of this paper is as follows. Section 2 prepares definitions and notations. Section 3 shows a fast bit-parallel algorithm for UPTM. Section 4 gives an extension to UTH. In Section 5, we conclude.

2 Preliminaries

In this section, we give basic definitions and notations on our unordered tree matching problems according to [4,8,16]. For a set S, we denote by $|S|$ the

[1] In the results, $l(P)$, $d(P)$, and $b(P) = O(m)$ are the number of leaves, the maximum depth, and the maximum branching in a tree P, respectively. The parameter $rep(P) = O(m)$ is the maximum number of the same label on paths in P.

cardinality of S. Let $\mathbf{N}_+ = \{1, 2, \ldots\}$. We define an *interval from i to j* by $[i..j] = \{i, i+1, \ldots, j\} \subseteq \mathbf{N}_+$, where $i \leq j$. We define the *smallest interval* including set $S \subseteq \mathbf{N}_+$ by $Int(S) = [\min S, \max S] \subseteq \mathbf{N}_+$. For an array $A = A[1] \cdots A[n]$ and $i \leq j$, we define $A[i..j] = A[i] \cdots A[j]$. For a binary relation $R \subseteq A^2$ on a set A, we denote by $R^+ \subseteq A^2$ the *transitive closure* of R.

2.1 Unordered Trees

Let $\Sigma = \{a, b, a_1, a_2, \ldots\}$ be a finite alphabet of *labels*. In this paper, we will mainly consider *unordered trees*, which are the labeled, rooted trees, where the ordering among their siblings is irrelevant.

Let P be an unordered tree of m nodes whose labels are drawn from Σ. We denote by $V(P)$ the *node set*, by $E(P)$ the *edge set*, and by $root(P)$ the *root* of P. For each node x, $label_P(x) \in \Sigma$ denotes the *label* of x in P, and $P(x)$ denotes the *subtree* of P rooted at x.

If $(x, y) \in E(P)$ then we say that x is the *parent* of y and y is a *child* of x. If there exists some downward path from x to y, i.e., $(x, y) \in E(P)^*$, then we say that x is an *ancestor* of y and y is a *descendant* of x and write $x \preceq y$. If $x \preceq y$ and $x \neq y$ then we say that x is a *proper ancestor* of y and y is a *proper descendant* of x and write $x \prec y$. If both of $x \not\preceq y$ and $y \not\preceq x$ hold then x and y are *incomparable* each other and write $x \sharp y$. For nodes x and y in P, if $x \sharp y$ and x precedes y in the preorder traversal of P, then we say that x *precedes* y in P (or, *x is to the left of y*) and write $x \lhd y$. If $x \sharp y$ then either $x \lhd y$ or $y \lhd x$ holds.

For unordered tree P, we denote by $|P|$ and by $height(P)$ the number of nodes in P and the height of P. We denote the sets of all leaves and all internal nodes in P, respectively, by $leaves(P)$ and $internal(P)$. Clearly, $V(P) = internal(P) \cup leaves(P)$. Let x be any node in P. The *arity* of x, denote by $\alpha(x) \geq 0$, is the number of children of x. For every $1 \leq i \leq \alpha(x)$, we denote the i-th child of node x by $x[i]$, and the list of the children of x by $children(x) = x[1] \cdots x[\alpha(x)]$.

2.2 Unordered Tree Matching Problem

Let $P = P[1..m]$ be an unordered tree of size m, called a *pattern tree*, and $T = T[1..n]$ be an unordered tree of size n, called a *text tree*. In this subsection, we introduce the unordered pseudo-tree matching and unordered tree homeomorphism problems. For other variations of tree matching problems as in [2,4,8,9,14,16], please consult the full paper [6].

Definition 1 (conditions for tree matching and inclusion). For any (possibly many-one) mapping $\phi : V(P) \to V(T)$, we define the following conditions:

(E0) ϕ preserves node labels. That is, for every node $x \in V(P)$, $label_P(x) = label_T(\phi(x))$ holds.

(E1) ϕ preserves the parent-child relationship. That is, for every node $x, y \in V(P)$, $(x, y) \in E(P) \Rightarrow (\phi(x), \phi(y)) \in E(T)$ holds.

(E1') ϕ preserves the ancestor-descendant relationship. That is, for every node $x, y \in V(P)$, $(x, y) \in E(P) \Rightarrow \phi(x) \prec \phi(y)$ holds.

Let $\mathcal{F}$ be a class of mappings. Then, a pattern tree P *maps to* a node $v \in V(T)$ in T w.r.t. class $\mathcal{F}$ if $\phi(root(P)) = v$ for some $\phi \in \mathcal{F}$. Then, the node v is called an *occurrence* of P in T w.r.t. class $\mathcal{F}$. Then, the tree pattern matching problem w.r.t. $\mathcal{F}$ ($\mathcal{F}$-matching problem) is the problem of, given a pattern tree P and a text tree T, finding all occurrences of P in T w.r.t. class $\mathcal{F}$.

An embedding from P to T is a possibly many-one mapping $\phi : V(P) \to V(T)$ with (E0). An *unordered pseudo-tree matching* (UPTM) [16] is a many-one version of unordered tree matching, i.e., an embedding ϕ with (E0) and (E1). A *unordered tree homeomorphism* (UTH) [4] is a many-one version of unordered tree inclusion, i.e., an embedding ϕ with (E0) and (E1'). We denote by $UPTM(P,T)$ and $UTH(P,T)$ the sets of all pseudo-tree matching and all tree homeomorphism from P to T. The *unordered pseudo-tree matching problem* (UPTM) and the *unordered tree homeomorphism problem* (UTH) are tree matching problem related to the above classes of mappings.

3 Faster Bit-Parallel Algorithm for Unordered Pseudo tree Matching

In this section, we present efficient algorithm BP-MatchUPTM based on bit-parallel pattern matching method for the pseudo-tree matching problem. Let $P = P[1..m]$ be a pattern tree of size m and $T = T[1..n]$ be a text tree of size n.

3.1 Decomposition Formula and a Bottom-Up Algorithm

In Fig. 2, we show an algorithm MatchUPTM for Unordered Pseudo Tree Matching. Our matching algorithm computes, for every text node v in T, the set $Emb^{P,T}(v)$ of integers in $V(P) = [1..m]$, called the *embedding set*, defined by:

$$Emb^{P,T}(v) = \{\, x \in [1..m] \mid (\exists \phi)\ \phi \in UPTM(P(x), T) \wedge \phi(x) = v \,\}, \quad (1)$$

where $P(x)$ is the subtree of P rooted at the pattern node $x \in V(P)$. Clearly, for every pattern node $x \in [1..m]$, $x \in Emb^{P,T}(v)$ if and only if the corresponding subtree $P(x)$ has an occurrence at the current text node v by some UPTM ϕ. By definition, we see that P matches T at node v iff $root(P) \in Emb^{P,T}(v)$. Now, we have the next lemma, which is crucial to the correctness of our algorithm.

Lemma 1 (decomposition formula for UPTM). *For every $x \in V(P)$ and $v \in V(T)$, $x \in Emb^{P,T}(v)$ if and only if*

(i) $label_P(x) = label_T(v)$, and
(ii) $children(x) \subseteq \bigcup_{1 \le j \le \alpha(v)} Emb^{P,T}(v[j])$.

From Lemma 1 above, we show in Fig. 2 a bottom-up procedure VisitUPTM to compute $Emb^{P,T}(v)$ by using the post-order traversal of T. To describe the procedure VisitUPTM, we need the following operators.

---

algorithm MatchUPTM($P[1..m]$: a pattern tree, $T[1..n]$: a text tree):
Global Variables: P and T;
Output: all occurrences of P in T w.r.t. unordered pseudo-tree matching (UPTM);
1: VisitUPTM($root(T)$);

procedure VisitUPTM(v: a text node):
Return Value: $R = Emb^{P,T}(v)$;
2: $S \leftarrow$ Constant($\emptyset$); {See Definition 2}
3: **for** $i = 1, \dots \alpha(v)$ **do**
4: $\quad S \leftarrow$ Union(S, VisitUPTM($v[i]$));
5: $R \leftarrow$ Constant($[1..m]$);
6: $R \leftarrow$ LabelMatch$_P$($R, label_T(v)$); {See Definition 2}
7: $R \leftarrow$ TreeAggr$_P$(R, S); {See Definition 2}
8: **if** Member($R.root(P)$) **then** {See Definition 2}
9: $\quad$ **output** "*A match is found at a node v.*";
10: **return** R;

Fig. 2. An algorithm for the unordered pseudo-tree matching problem

Definition 2 (set manipulation operators). We define operators Constant, Union, Member, LabelMatch (label matching), and TreeAggr (tree aggregation) on subsets of $[1..m]$ as follows, where $R, S \subseteq [1..m]$, $x \in [1..m]$, and $\alpha \in \Sigma$:

- Constant(S) $\equiv$ S. This operation returns the set S itself.
- Union(R, S) $\equiv$ $R \cup S$. This returns the set-union of R and $S \subseteq [1..m]$.
- Member(R, x) $\equiv$ $x \in R$. Given a set R and an element x, this operation returns "*yes*" if $x \in R$ and "*no*" otherwise.
- LabelMatch$_P$(R, α) $\equiv$ $\{ k \in R \mid label_P(k) = \alpha \}$. Given any set R and label α, this operation returns the set of elements in R satisfying (i) of Lemma 1.
- TreeAggr$_P$(R, S) $\equiv$ $\{ k \in R \mid children_P(k) \subseteq S \}$. Given any sets R, S, this operation returns the set of elements in R satisfying (ii) of Lemma 1.

In the procedure VisitUPTM, we use the last two operators LabelMatch and TreeAggr to check (i) and (ii) of Lemma 1. Later, the above set operations will be implemented in bit-parallel manner in Sec. 3.2.

By representing sets R and $S \subseteq [1..m]$ in lists of integers, it is easy to see that these operators can be implemented to run in $O(m)$ time and space. Then, we have the following lemma.

Lemma 2. *For the unordered pseudo-tree matching problem, and for every pattern tree P and a text tree T, the algorithm* MatchUPTM *in Fig. 2 correctly finds all occurrences of P in T. Moreover, the algorithm can be implemented to run in $O(nm)$ time and $O(hm)$ additional space, where m is the size of P, n and h are the size and the height of T, respectively.*

The algorithm MatchUPTM can run in streaming setting using a stack of length $O(hm)$, where T is given as an input stream consisting of a sequence of balanced open and close parentheses on alphabet $\Sigma \cup \{ \bar{a} \mid a \in \Sigma \}$ as in XML databases [4,12,13].

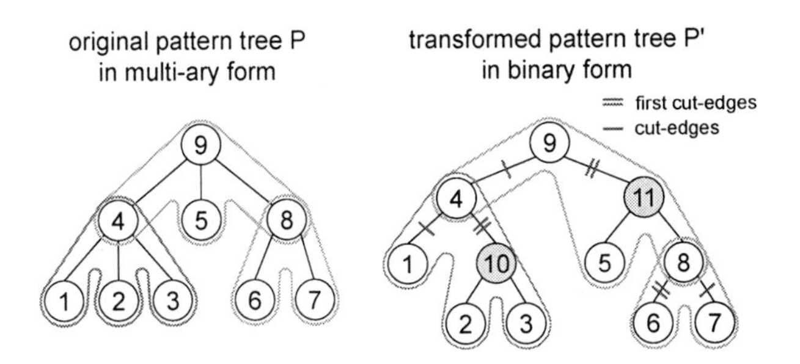

Fig. 3. An original pattern tree P and its binarization P', where white and shadowed circles indicate original (real) nodes and dummy (virtual) nodes, respectively. The number in each circle indicates the node id.

3.2 Bit-Parallel Implementation: Overview

In the following subsections, we give the bit-parallel version of the algorithm MatchUPTM, called BP-MatchUPTM, that runs in $O(n\lceil m/w\rceil \log w)$ time and $O(h\lceil m/w\rceil + \lceil m/w\rceil \log w)$ space, where m is the size of pattern tree P, n and h are the size and the height of text tree T, and w is the word length of underlying computer. Let us fix a pattern tree $P = P[1..m]$ of size m on a finite alphabet Σ. In what follows, we assume that $|\Sigma| = O(1)$.

In the bit-parallel implementation of MatchUPTM, we introduce a data structure $\mathcal{A}$ for representing a subset S of the universe $[1..m]$ that efficiently supports the collection of set manipulation operators in Definition 2 in Sec. 3.1. In $\mathcal{A}$, we represent any subsets of $V(P)$ by bitmasks $X \in \{0,1\}^m$ with length m as m-bit integers from 0 to $2^m - 1$. To do this, we need an assignment $Bit : V(P) \to [1..m]$ of the unique bit-position $Bit(x)$ in the interval $[1..m]$ to each node x in P. Since Bit is one-one, we define the inverse mapping $Node : [1..m] \to V(P)$ as $Node(i) = x$ if and only if $Bit(x) = i$ for any bit-position i and pattern node $x \in V(P)$. At this moment, we leave Bit undefined and the appropriate definition for Bit will be given later in the next subsection.

Basic set operations. Once the assignment Bit is given, for any node set $S \subseteq V(P)$, we extend this Bit by $BIT(S) = \{\, Bit(x) \,|\, x \in S \,\} \subseteq [1..m]$. For any subset $X \subseteq [1..m]$, we define $NUM(X) \in \{0,1\}^m$ to be the bitmask for X. Among the set operators in Definition 2, the following operators are easy to implement.

Lemma 3. *Let $S \subseteq V(P)$ be any node sets, and $X, Y \in \{0.1\}^m$ be the corresponding bitmasks, respectively. Then, the following codes correctly implement the operators. Moreover, all operations are executed in $O(1)$ time if $m \leq w$.*

- *Preprocess:* Constant$(S) \equiv NUM(BIT(S))$;
- *Runtime:* Union$_P(X, Y) \equiv (X \mid Y)$;
- *Runtime:* Member$_P(X, x) \equiv$ **if** $(X \,\&\, NUM(BIT(\{x\}))) > 0$ **then** 1 **else** 0;

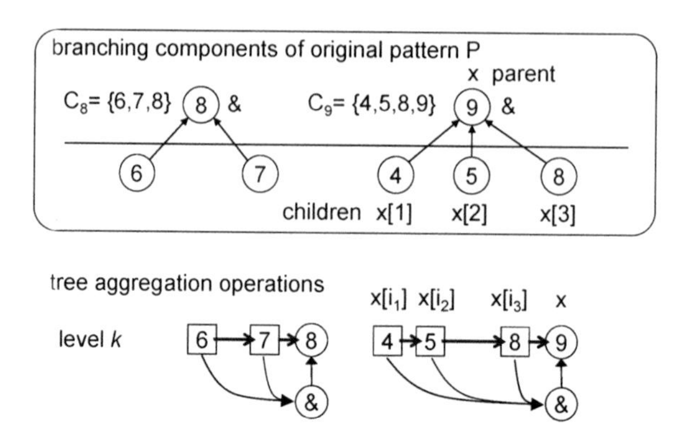

Fig. 4. Branching components of a pattern tree P and the corresponding tree aggregation operation in bit-parallel computation

Label matching operation. The label matching operation can be implemented using a set of character masks as in SHIFT-AND method for exact match [1,15,11] or in Move operation for regular expression match [3]

Lemma 4. *The operator* LabelMatch *can be correctly implemented by the following codes, where* $\{ LAB[\alpha] \in \{0,1\}^m \,|\, \alpha \in \Sigma \}$ *is a set of bitmasks for* P*. Moreover the operation can be executed in* $O(1)$ *time if* $m \leq w$*.*

- *Preprocess: For each* $\alpha \in \Sigma$*,* $LAB[\alpha] = |_{x \in V(P), label_P(x)=\alpha} NUM(Bit(x))$*;*
- *Runtime:* $\mathsf{LabelMatch}_P(X, \alpha) \equiv (X \,\&\, LAB[\alpha])$*;*

Tree aggregation operation. Remaining task is to show how to efficiently implement TreeAggr operation in bit-parallel computation. For each node x in P, we define the *branching component* for x in P by the connected component $C_x = \{x\} \cup children(x)$ of P consisting of parent x and its children. If no confusion arises, we identify C_x and the induced depth-one subtree $P(C_x)$ rooted at x, called a *branching tree*. We denote by $\mathcal{C}_P = \{ C_x \,|\, x \in internal(P) \}$ the set of all branching components of P. For example, pattern tree P of Fig. 3 has three branching components $C_4 = \{1, 2, 3, 4\}$, $C_8 = \{6, 7, 8\}$, and $C_9 = \{4, 5, 8, 9\}$. The upper half of Fig. 4 shows C_8 and C_9 with their branching trees.

Then, the tree aggregation operation means gathering the values of children of x and then copying the value of their conjunction to parent x (See Fig. 4). We want to compute tree aggregations simultaneously for all internal node x in P.

First, to implement the tree aggregation operation in correct and efficient way, we require the assignment Bit to have the following properties:

Definition 3 (monotone bit-assignment). A bit-assignment mapping $Bit : V(P) \to [1..m]$ is said to be *monotone* w.r.t. the ancestor relation $\preceq$ for P if for any $x, y \in V(P)$, $(x \succ y) \Rightarrow (Bit(x) < Bit(y))$ holds.

Next, we introduce an overlap-free decomposition $\mathcal{C}_P = \mathcal{C}[1] + \cdots + \mathcal{C}[K]$ of $\mathcal{C}_P$ as follows, where $K \geq 0$. For any component $C_x \in \mathcal{C}_P$, we assign to C_x the interval $I_x = Int(BIT(C_x))$ in $[1..m]$, which is the smallest interval containing all bit-positions of C_x. Then, two components C_x and C_y $(x \neq y)$ *overlap* if $I_x \cap I_y \neq \emptyset$. A subset $\mathcal{D} \subseteq \mathcal{C}_P$ is said to be *overlap-free* if there are no pairs of components in $\mathcal{D}$ that overlap each other.

Definition 4 (overlap-free decomposition). A partition $\mathcal{C}_P = \mathcal{C}[1] + \cdots + \mathcal{C}[K]$ for some $K \geq 1$ is said to be an *overlap-free decomposition* of $\mathcal{C}_P$ w.r.t. Bit if for every $k = 1, \ldots, K$, the k-th subset $\mathcal{C}[k]$ is overlap-free w.r.t. Bit. Then, K is called the *height* and $\mathcal{C}[k]$ is called the *k-th level* of the partition.

Suppose that there exists some monotone bit-assignment Bit and some overlap-free decomposition $\mathcal{C}_P = \mathcal{C}[1] + \cdots + \mathcal{C}[K]$ of $\mathcal{C}_P$ for some Bit. Then, tree aggregation is implemented in bit-parallel way as follows.

Definition 5 (Preprocess). We first precompute the following bitmasks:

- $LEAF = |_{x \in leaves(P)} NUM(BIT(\{x\}))$.
- For every level $k = 1, \ldots, K$, and for each C_x in $\mathcal{C}[k]$, we define
  - $DST[k][x] = NUM(BIT(\{x\})))$. The position of parent x.
  - $SRC[k][x] = NUM(BIT(children(x)))$. The positions of $children(x)$.
  - $INT[k][x] = NUM(Int(BIT(C_x)))$. The interval for C_x.
  - $SEED[k][x] = NUM(\min BIT(C_x))$. The "seed" position.
- For every level $k = 1, \ldots, K$, and for each $Mask \in \{DST, SRC, INT, SEED\}$,
  - $Mask[k] = |_{C_x \in \mathcal{C}[k]} Mask[k][x]$.

Lemma 5 (Runtime). *Suppose that $\mathcal{C}_P = \mathcal{C}[1] + \cdots + \mathcal{C}[K]$ of $\mathcal{C}_P$ is an overlap-free decomposition with height $K \geq 1$ for a monotone bit-assignment Bit w.r.t. $\preceq$. Then, the following code correctly implements the tree aggregation operator for the component C_x. Moreover, this procedure runs in $O(K)$ time if $m \leq w$.*

- **procedure** $\mathsf{TreeAggr}_P(X, Y) \equiv$
  - $Z \leftarrow \mathsf{Constant}(\emptyset)$;
  - *For every level* $k = 1, \ldots, K$ *do:*
    $BLK \leftarrow (Y \,\&\, SRC[k]) \mid (INT[k] \,\&\, (\sim (SRC[k] \mid DST[k])))$;
    $Z \leftarrow Z \mid ((BLK + SEED[k]) \,\&\, DST[k])$;
  - $Z \leftarrow X \,\&\, (Z \mid LEAF)$;
  - **return** Z;

Therefore, the remaining thing is how to find a good overlap-free decomposition $\mathcal{C}_P$ with small height as well as a monotone bit-assignment Bit. We discuss this issue in the next subsection.

3.3 Construction of a Monotone Bit-Assignment and an Overlap-Free Decomposition Based on Separator Trees

In this subsection, we show how to find both a monotone bit-assignment Bit and an overlap-free decomposition $\mathcal{C}_P$ with height $O(\log m)$. For this purpose, we use a data structure called a *separator tree*.

Binarization of P. Let P be a pattern tree of m nodes. We note that P is a multi-ary tree whose internal node x may have arity $\alpha(x) > 2$. First, before constructing separator tree composition, we apply a standard transformation, called *binarization* to P for obtaining a binary version P' of P. The binarization transforms each branching component $C_x = \{x\} \cup \{x[1], \cdots, x[\alpha]\}$ with the root x and $\alpha(x)$ children into a new component C'_x a binary subtree with the same root and the same number of children by inserting $\alpha(x) - 2$ dummy internal nodes of out-degree two. In general, the resulting binary tree P' has size at most $2m$. In what follows, let $m' = O(m)$ be the size of P'.

Fig. 3 shows an example of the binarization P' of the original pattern P, where the component $C_4 = \{1, 2, 3, 4\}$ with root 4 is transformed into $C'_4 = \{1, 2, 3, 4, 10\}$ with the same root 4.

Construction of a separator tree for P. Secondly, we build a separator tree $\mathcal{M}$ from a binarization P' of pattern tree. A separator tree is a binary tree obtained from P' by iteratively removing edges in $E(P')$. The following well-known lemma is sufficient for our purpose:

Lemma 6 (Jordan [5]). *Let S be a binary tree. Then, there exists a node in S such that $|S(v)| \leq (2/3)|S|$ and $|S(\bar{v})| \leq (2/3)|S|$, where $S(v)$ is the subtree of S rooted at v and $S(\bar{v})$ is the tree obtained by pruning $S(v)$ from S.*

Suppose that each node w of $\mathcal{M}$ has the fields U_w for a subset of $V(P')$, and e_w for a cut-edge. Applying the above theorem recursively, we construct a separator tree $\mathcal{M}$ from P' as follows.

- We start with a new node as the root of $\mathcal{M}$. We associate $V(P')$ to the root by setting $U_{root(\mathcal{M})} = V(P')$. We visit $w = root(\mathcal{M})$, and repeat the following process at each node w.
- Suppose that $|U_w| = 1$. Then, the associated node set U_w is a singleton $\{x\}$, and the current node w is a leaf. We stop the process.
- Otherwise, $|U_w| > 1$. Then, we find an edge $e = (x, y) \in E(P')$ according to Lemma 6 so that removal of e splits the associated component U_w into two subcomponents U_w^1 and U_w^0 of almost equal sizes no more than $(2/3)|U_w|$, where U_w^1 is the subcomponent containing $root(P')$ and U_w^0 is the other subcomponent. Record the *cut-edge* e for w as e_w. Create the left and the right children, w_L and w_R of w, and associate the component U_w^1 to w_R and U_w^0 to w_L. Then, recursively visit both of w_L and w_R as w.

Lemma 6 ensures that the height of $\mathcal{M}$ is $O(\log m)$ and the construction requires $O(m \log m)$ time. Furthermore, we can observe that (1) there exists a one-one correspondence between $internal(\mathcal{M})$ and $E(P')$, and (2) there exists a one-one correspondence between $leaves(\mathcal{M})$ and $V(P')$.

Now, we compute a bit-assignment $Bit : V(P') \rightarrow [1..m]$ as follows. We order $leaves(\mathcal{M})$ from left to right. Then, we number all leaves in $leaves(\mathcal{M})$, which are original (real) nodes in P, from left to right consecutively from 1 to m (not m'). We just skip and unnumber dummy (virtual) nodes included in binarization.

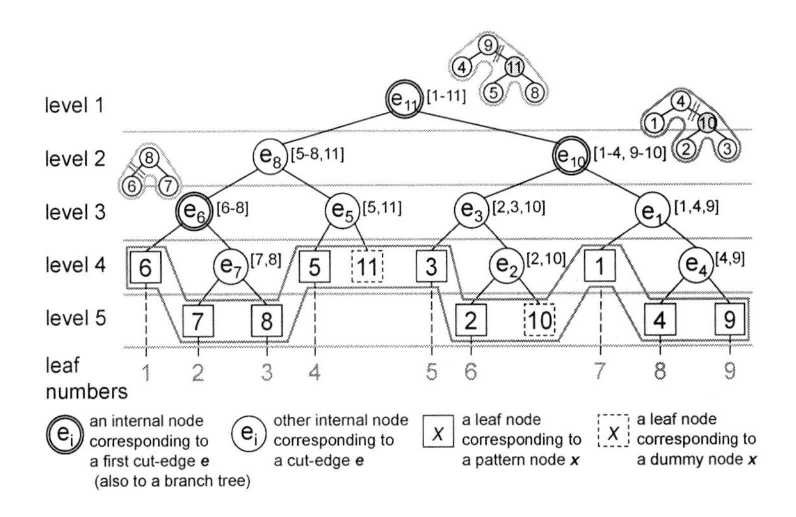

Fig. 5. A separator tree $\mathcal{M}$ for the binarization P' of a pattern tree P. Circles and squares indicate internal nodes and leaf nodes in $\mathcal{M}$, respectively. At each node w, the associated set of numbers in a pair of brackets indicates the connected component associated to w. Each edge (x, y) is denoted as e_y indexed by its lower end point y.

Then, we define bit-assignment $Bit : V(P) \to [1..m]$ as follows. For each node x in P, if x is the i-th leaf in this listing over $leaves(\mathcal{M})$, then we define $Bit(x) = i$ and $Node(i) = x$. For the proof of the next lemma, see the full paper [6].

Lemma 7. *The bit-assignment Bit constructed above from $\mathcal{M}$ is monotone w.r.t. the ancestor relation $\preceq$ for P.*

Proof. Suppose that $x \succeq y$ and both of them are included in some component U in $\mathcal{M}$. Then, there exists an upward path π from x to y, and eventually, some edge in π becomes a cut-edge at some node w in $\mathcal{M}$. This split U_w into U_w^0 and U_w^1 such that $x \in U_w^0$ and $y \in U_w^1$ since the latter locates the upper part. We see that U_w^0 precedes U_w^1 in interval $[1..m]$ by Bit, and that $Bit(x) < Bit(y)$. Hence, we see that $x \succeq y$ implies $Bit(x) < Bit(y)$. □

By using Bit based on leaf numbering of $\mathcal{M}$, we associate an interval I_w to each node w in $\mathcal{M}$ by $I_w = Int(BIT(U_w))$. Then, we give a technical lemma.

Lemma 8. *For any nodes u, w in $\mathcal{M}$, the following properties hold:*

(1) $BIT(U_w) \subseteq I_w$.
(2) If $u \succeq w$ *then* $I_u \subseteq I_w$.
(3) If $u \sharp w$ *then* $I_u \cap I_w = \emptyset$.

Proof. (1) The result is obvious since Int is expanding, i.e., $S \subseteq Int(S)$ for any subset $S \subseteq [1..m]$. (2) By construction, if $u \succeq w$ then $U_u \subseteq U_w$ holds, and thus $I_u \subseteq I_w$ holds from the definition of I_u. (3) Suppose on the contrary that $I_u \cap I_w \neq \emptyset$, that is, I_u and I_w overlap each other. From the tree structure of $\mathcal{M}$ that this implies that either $u \succeq v$ or $v \succeq u$. This shows the claim. □

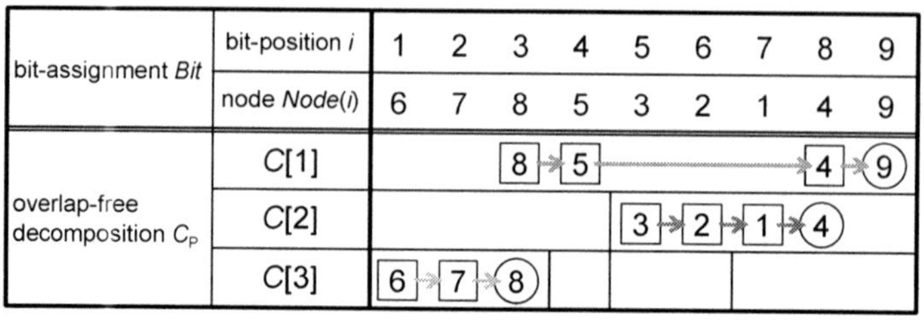

bit-assignment *Bit*	bit-position *i*	1	2	3	4	5	6	7	8	9
	node *Node*(*i*)	6	7	8	5	3	2	1	4	9
overlap-free decomposition C_P	C[1]			8	5				4	9
	C[2]					3	2	1	4	
	C[3]	6	7	8						

Fig. 6. The monotone bit-assignment *Bit* and an overlap-free decomposition w.r.t. *Bit*, for the pattern tree P in Fig. 3

Finally, we construct an overlap-free decomposition for $\mathcal{C}_P$ as follows. We traverse the separator tree $\mathcal{M}$ from the root to leaves top-down. Initially, we visit *root* and U_{root} contains the whole $\mathcal{C}_P$. Assume that we are going down and visiting a node w in $\mathcal{M}$. Let $C_x \in \mathcal{C}_P$ be the unique component in $\mathcal{C}_P$ that contains e_w in the induced tree $P(C_x)$. Then, there are two cases. If this happens at the first time with C_x, that is, e_w is the *first cut-edge* for C_x on the path from the root to w, then we mark the current node w and associate to w the C_x by $Comp(w) = C_x$. Otherwise, this is at least second time cut or so. Then, we skip w and continue the traversal for descendants. After the traversal, we perform breadth-first search for level $k = 1$ to $K = depth(\mathcal{M}) + 1$. Then, we construct a decomposition $\mathcal{C}_P = \mathcal{C}[1] + \cdots + \mathcal{C}[K]$ such that $\mathcal{C}[k] = \{ C_x \mid C_x = Comp(w), w \text{ is a marked internal node in } \mathcal{M} \text{ of depth } k \}$ for each $k = 1, \ldots, K$. See Fig. 5 for an example of a separator tree $\mathcal{M}$ and a bit-assignment *Bit* based on leaf numbering of $\mathcal{M}$.

In Fig. 6, we show the monotone bit-assignment *Bit* and an overlap-free decomposition w.r.t. *Bit*, for the pattern tree P in Fig. 3. In the figure, the node $Node(i)$ of pattern tree P is assigned the bit-position i for $i = 1, \ldots, 9$. The decomposition $\mathcal{C}_P$ contains three branching components $\mathcal{C}_9 = \{4, 5, 8, 9\}$ at level 1, $\mathcal{C}_4 = \{1, 2, 3, 4\}$ at level 2, and $\mathcal{C}_8 = \{6, 7, 8\}$ at level 3.

For the proof of the next lemma, see the full paper [6].

Lemma 9. *An overlap-free decomposition* $\mathcal{C}_P = \mathcal{C}[1] + \cdots + \mathcal{C}[K]$ *of* $\mathcal{C}_P$ *w.r.t. Bit can be computed by the above procedure in* $O(m \log m)$ *time and* $O(\log m)$ *space.*

3.4 Complexity Analysis

Combining the algorithm MatchUPTM of Fig. 2 in Sec. 3.1 and the bit-parallel implementation of the set manipulation operations in Sec. 3.2, and results in Sec. 3.3, we now have a modified version of the algorithm, called BP-MatchUPTM for the UPTM problem.

Lemma 10. *The operation* $\mathsf{TreeAggr}_P$ *can be implemented to run* $O(\log m)$ *time using* $O(m \log m)$ *preprocessing and* $O(\log m)$ *space if* $m \leq w$.

Proof. The claim follows from Lemma 5, Lemma 7, and Lemma 9. □

By applying the module decomposition technique of Myers [10] and Bille [3], we have the main theorem of this paper below:

Theorem 1 (complexity of UPTM). *The algorithm* BP-MatchUPTM *solves the unordered pseudo-tree matching problem with the following complexities:*

- *In the large pattern case* ($m > w$)*:* $O(n\lceil m/w\rceil \log w)$ *time using* $O(h\lceil m/w\rceil + \lceil m/w\rceil \log w)$ *space and* $O(m \log w)$ *preprocessing.*
- *In the small pattern case* ($m \leq w$)*:* $O(n \log m)$ *time using* $O(h + \log m)$ *space and* $O(m \log m)$ *preprocessing.*

where m *and* n *are the sizes of pattern tree* P *and text tree* T*,* h *is the height of* T*, and* w *is the length of computer word.*

4 Extension for Unordered Tree Homeomorphism

In this section, we give a modified algorithm for the unordered tree homeomorphism problem (UTH). Let v be any node in T. Then, the set $\mathit{Desc\text{-}Emb}^{P,T}(v)$, called the *descendant embedding set* and the auxiliary set $\mathit{Sub\text{-}Emb}^{P,T}(v)$ are defined by:

$$\mathit{Desc\text{-}Emb}^{P,T}(v) = \{\, x \in [1..m] \mid (\exists \phi \in \mathit{UTH}(P(x), T)\ \phi(x) = v \,\}. \quad (2)$$
$$\mathit{Sub\text{-}Emb}^{P,T}(v) = \{\, x \in [1..m] \mid (\exists \phi \in \mathit{UTH}(P(x), T)(\exists w \succeq v)\ \phi(x) = w \,\}.$$

Lemma 11. *For any* P *and* T*, we have the following properties:*

(1) For every $x \in V(P)$ *and* $v \in V(T)$*,* $x \in \mathit{Desc\text{-}Emb}^{P,T}(v)$ *if and only if (i)* $\mathit{label}_P(x) = \mathit{label}_T(v)$*, and (ii)* $\mathit{children}(x) \subseteq \bigcup_{1 \leq j \leq \alpha(v)} \mathit{Sub\text{-}Emb}^{P,T}(v[j])$*.*
(2) For any v *in* T*,* $\mathit{Sub\text{-}Emb}^{P,T}(v) = \mathit{Desc\text{-}Emb}^{P,T}(v) \cup \bigcup_{1 \leq j \leq \alpha(v)} \mathit{Sub\text{-}Emb}^{P,T}(v)$*.*

From the above decomposition lemma, we can develop a bit-parallel algorithm, called BP-MatchUTH, for the unordered tree homeomorphism problem based on a dynamic programming algorithm similar to MatchUPTM in Sec. 3 and the bit-parallel implementation of operators including LabelMatch and TreeAggr. We can obtain an algorithm MatchUTH for the unordered tree homeomorphism problem (UTH) from the algorithm MatchUPTM by replacing Line 10 of the recursive subprocedure VisitUPTM with the following line:

10: **return** Union(R, S); $\{R \cup S = \mathit{Sub\text{-}Emb}^{P,T}(v)\}$

Then, we have the following theorem. For details, please consult the full paper [6].

Theorem 2 (complexity of the unordered tree homeomorphism problem). *A modified version of the algorithm,* BP-MatchUTH*, solves the unordered tree homeomorphism problem (UTH) with the following complexities:*

- *In the large pattern case* ($m > w$)*:* $O(n\lceil m/w\rceil \log w)$ *time using* $O(h\lceil m/w\rceil + \lceil m/w\rceil \log w)$ *additional space and* $O(m \log w)$ *preprocessing.*
- *In the small pattern case* ($m \leq w$)*:* $O(n \log m)$ *time using* $O(h + \log m)$ *additional space and* $O(m \log m)$ *preprocessing.*

where m *and* n *are the sizes of pattern tree and text tree, and* w *is the length of computer word.*

5 Conclusion

In this paper, we consider the unordered pseudo-tree matching problem and the unordered tree homeomorphism problem. As results, we present efficient algorithms for both problems that run in $O(n\lceil m/w\rceil \log w)$ time using $O(h\lceil m/w\rceil + \lceil m/w\rceil \log w)$ space and $O(m \log w)$ preprocessing with $m > w$ on a unit-cost arithmetic RAM model with integer addition. As future work, applications to tree pattern matching for practical subclasses of XPath and XQuery queries are interesting problems. Kaneta *et al.* [7] presented a bit-parallel pattern matching algorithm on RAM model with integer addtion for the class of network and regular expressions. Combination of the techiniques in this paper and one in [7] will be another future problem.

Acknowledgments. The authors would like to thank Tatsuya Akutsu, Kouichi Hirata, Takuya Kida, Shin-ichi Minato, Hiroshi Sakamoto, Shinichi Shimozono, Kilho Shin, Akihiro Yamamoto, and Thomas Zeugmann for their fruitful discussions and valuable comments on the earlier version of this paper. This research was partly supported by MEXT Grant-in-Aid for Scientific Research (A), 20240014, FY2008–2011, and MEXT/JSPS Global COE Program, FY2007–2011.

References

1. Baeza-Yates, R.A., Gonnet, G.H.: A new approach to text searching. Communications of the ACM 35(10), 74–82 (1992)
2. Bille, P., Gørtz, I.L.: The tree inclusion problem: In optimal space and faster. In: Caires, L., Italiano, G.F., Monteiro, L., Palamidessi, C., Yung, M. (eds.) ICALP 2005. LNCS, vol. 3580, pp. 66–77. Springer, Heidelberg (2005)
3. Bille, P.: New algorithms for regular expression matching. In: Bugliesi, M., Preneel, B., Sassone, V., Wegener, I. (eds.) ICALP 2006. LNCS, vol. 4051, pp. 643–654. Springer, Heidelberg (2006)
4. Götz, M., Koch, C., Martens, W.: Efficient algorithms for descendant-only tree pattern queries. Inf. Syst. 34(7), 602–623 (2009)
5. Jordan, C.: Sur les assemblages de lignes. Journal für die Reine und Angewandte Mathematik 70, 185–190 (1869)
6. Kaneta, Y., Arimura, H.: Fast bit-parallel algorithm for unordered pseudo-tree matching and tree homeomorphism. TCS-TR-A-10-43, Hokkaido University (2010)
7. Kaneta, Y., Minato, S., Arimura, H.: Fast bit-parallel matching for network and regular expressions. In: Chavez, E., Lonardi, S. (eds.) SPIRE 2010. LNCS, vol. 6393, pp. 372–384. Springer, Heidelberg (2010)
8. Kilpeläinen, P.: Tree matching problems with applications to structured text databases. Ph.D Thesis, Report A-1992-6, DCS, University of Helsinki (1992)
9. Kilpeläinen, P., Mannila, H.: Ordered and unordered tree inclusion. SIAM Journal on Computing 24(2), 340–356 (1995)
10. Myers, E.W.: A four Russian algorithm for regular expression pattern matching. Journal of the ACM 39(2), 430–448 (1992)
11. Navarro, G., Raffinot, M.: Flexible Pattern Matching in Strings: Practical On-Line Search Algorithms for Texts and Biological Sequences, Cambridge (2002)

12. Tsuji, H., Ishino, A., Takeda, M.: A bit-parallel tree matching algorithm for patterns with horizontal VLDC's. In: Consens, M., Navarro, G. (eds.) SPIRE 2005. LNCS, vol. 3772, pp. 388–398. Springer, Heidelberg (2005)
13. W3C, Extensive Markup Language (XML) 1.0 (Second Edition), W3C Recommendation (October 2000), `http://www.w3.org/TR/REC-xml`
14. Valiente, G.: Constrained tree inclusion. Journal of Discrete Algorithms 3(2-4), 431–447 (2005)
15. Wu, S., Manber, U.: Fast text searching: allowing errors. Communications of the ACM 35(10), 83–91 (1992)
16. Yamamoto, H., Takenouchi, D.: Bit-parallel tree pattern matching algorithms for unordered labeled trees. In: Dehne, F., Gavrilova, M., Sack, J.-R., Tóth, C.D. (eds.) WADS 2009. LNCS, vol. 5664, pp. 554–565. Springer, Heidelberg (2009)

Dichotomy for Coloring of Dart Graphs*

Martin Kochol[1] and Riste Škrekovski[2]

[1] MÚ SAV, Štefánikova 49, 814 73 Bratislava 1, Slovakia
martin.kochol@mat.savba.sk
[2] Department of Mathematics, University of Ljubljana,
Jadranska 19, 1111 Ljubljana, Slovenia
skrekovski@gmail.com

Abstract. We study a $(k+1)$-coloring problem in a class of (k,s)-dart graphs, $k, s \geq 2$, where each vertex of degree at least $k+2$ belongs to a (k,i)-diamond, $i \leq s$. We prove that dichotomy holds, that means the problem is either NP-complete (if $k < s$), or can be solved in linear time (if $k \geq s$). In particular, in the latter case we generalize the classical Brooks Theorem, that means we prove that a (k,s)-dart graph, $k \geq \max\{2, s\}$, is $(k+1)$-colorable unless it contains a component isomorphic to K_{k+2}.

1 Introduction

An *r-coloring* of a graph is a mapping from the set of vertices to $\{1, \dots, r\}$ such that any two adjacent vertices have different colors. The decision problem whether a given graph G has an r-coloring is a classical NP-complete problem for every fixed $r \geq 3$ (see [3,4]).

The aim of this paper is to study $(k+1)$-coloring problem in a class of (k,s)-dart graphs, $k, s \geq 2$, where each vertex of degree at least $k+2$ belongs to a (k,i)-diamond, $i \leq s$ (formal definition we introduce in the following section). The main result of the paper is that dichotomy holds: the $(k+1)$-coloring problem for a (k,s)-dart graph is either

(a) NP-complete for $k < s$, or
(b) can be solved in linear time for $k \geq s$.

Subject to the assumption P $\neq$ NP both cases exclude each other. For the linear-time cases we present an algorithm, which not only decides existence, but also finds a $(k+1)$-coloring, if there is one. Moreover we generalize the classical Brooks theorem [1] (every graph with the maximum vertex degree at most $r \geq 3$ and without a component isomorphic to K_{r+1} has an r-coloring) and show that a (k,s)-dart graph, $k \geq \max\{2, s\}$, is $(k+1)$-colorable unless it contains a component isomorphic to K_{k+2}. Notice that for the case $k = 2$ this statement follows directly from Kochol, Lozin, and Randerath [6, Theorem 4.3].

* Supported by grants VEGA 2/0118/10 and ARRS Research Program P1-0297.

C.S. Iliopoulos and W.F. Smyth (Eds.): IWOCA 2010, LNCS 6460, pp. 82–89, 2011.

2 Definitions

In this paper we consider simple graphs, i.e., without multiple edges and loops. If G is a graph, then $V(G)$ and $E(G)$ denote the vertex and the edge sets of G, respectively.

Let G be a graph and x, y two vertices of G. Then $G + xy$ denotes the graph constructed from G by adding an edge xy. Since we consider simple graphs, $G + xy = G$ if x, y are adjacent in G. For a vertex v of G, let $d_G(v)$ denote the degree of v in G. Let H, G be two graphs such that H is not a subgraph of G. Then we use to say that G is a H-*free* graph.

A (k, s)-*diamond* is a join of a clique of size $k \geq 1$ and an independent set of size $s \geq 1$. These graphs are also known as split graphs. In a (k, s)-diamond D, vertices that belong to the independent set are called *pick* vertices, and the remaining (i.e. those in the k-clique) are called *central* vertices. Denote by $C(D)$ and $P(D)$ the sets of central vertices and pick vertices of D, respectively. A $(4, 2)$-diamond D with $C(D) = \{c_1, \ldots, c_4\}$ and $P(D) = \{p_1, p_2\}$ is in Fig. 1.

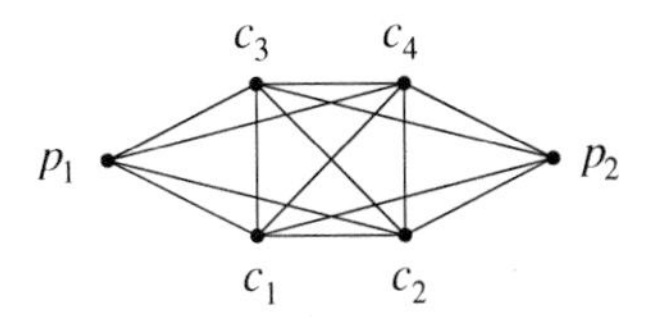

Fig. 1. A $(4, 2)$-diamond

Note that a $(k, 1)$-diamond is isomorphic to K_{k+1}; in this case the unique pick vertex does not distinguish from the central vertices but in such a situation this is irrelevant for us.

Definition 1. *A graph G is a* (k, s)-dart *if each vertex of G of degree $\geq k + 2$ is a central vertex of some (k, i)-diamond D as an induced subgraph of G with $i \leq s$, for which*

(a) *$d_D(x) \geq d_G(x) - 1$ for each $x \in V(D)$;*
(b) *no two vertices of $C(D)$ have a common neighbor in $G - D$.*

Every graph of maximum degree $\leq k+1$ is a $(k, 1)$-dart graph since in the above definition, we only prescribe the structure on the neighborhood of vertices of higher degree. Also, every (k, s_1)-dart is a (k, s_2)-dart if $s_1 \leq s_2$.

Note that the assumption that x is of degree $\geq k + 2$ implies that $i \geq 2$. In a (k, s)-dart graph G, every vertex of degree at least $k + 2$ belongs to an induced (k, i)-diamond with $2 \leq i \leq s$. Denote by $\mathcal{D}(G)$ the set of all induced maximal (k, i)-diamonds of G with $i \geq 2$. Observe that we do not require that a diamond of $\mathcal{D}(G)$ must contain a vertex of degree $k + 2$ or more, just to satisfy conditions (a) and (b) of Definition 1.

We say that a vertex of a dart G is *central* if it is a central vertex of a diamond of $\mathcal{D}(G)$. Similarly define a *pick* vertex of G. Denote the sets of central vertices and pick vertices by $C(G)$ and $P(G)$, respectively.

Let G be a (k,s)-dart and $D \in \mathcal{D}(G)$. Then, each central vertex $x \in C(D)$ is adjacent to at most one vertex v' from $G - D$. In this case, v' is called *isolated* neighbor of v. The set of all isolated neighbors of the central vertices of D is denoted by $I(D)$. Possibility $I(D) = \emptyset$ is not excluded.

We remark that the following observations for a (k,s)-dart G hold:

(1) A central vertex v of a (k,s)-dart G is not necessarily of degree at least $k+2$. This happens only if v is a central vertex of a $(k,2)$-diamond $D \in \mathcal{D}(G)$ and it has no neighbor in $G - D$. Then, v is of degree $k+1$.
(2) If K_{k+2} is a subgraph of a (k,s)-dart G, then it must be a component of G. Thus a copy of K_{k+2} in G is disjoint from diamonds of $\mathcal{D}(G)$.
(3) No two pick vertices of the same diamond from $\mathcal{D}(G)$ are adjacent.

3 Properties of Dart Graphs

The following lemma is an easy observation.

Lemma 1. *Let G be a (k,s)-dart graph and $D \in \mathcal{D}(G)$. Let λ be a proper $(k+1)$-coloring of $G - C(D)$ such that all pick vertices $P(D)$ are assigned the same color a. Then, λ can be extended to G unless every central vertex of D has an isolated neighbor and λ assigns the same color $c \neq a$ to all vertices of $I(D)$.*

Proof. Let $L(v) \subset \{1,\ldots,k+1\}$ be the set of available colors for a central vertex $v \in V(G)$ regarding λ. Notice that $k \geq |L(v)| \geq k-1$. And, $|L(v)| = k-1$ if and only if v has a pendant neighbor v' and $\lambda(v') \neq a$. Thus, each central vertex of D has an isolated neighbor and all vertices of $I(D)$ are assigned a same color $c \neq a$, if and only if the unions of all $L(v)$'s is of size $k-1$. Now the proof follows by Hall's theorem.

Next lemma assures that diamonds in a dart graph are vertex disjoint:

Lemma 2. *Let G be a (k,s)-dart graph with $k \geq 3$ and with no K_{k+2} as a subgraph. Then*

(a) $V(D_1) \cap V(D_2) = \emptyset$, for every two distinct diamonds $D_1, D_2 \in \mathcal{D}(G)$.
(b) $C(G) \cap P(G) = \emptyset$; in particular each pick vertex is of degree k or $k+1$.

Proof. We prove (a). Suppose that v is a vertex of two distinct diamonds $D_1, D_2 \in \mathcal{D}(G)$.

Assume that $v \in C(D_1) \cap C(D_2)$. If $C(D_1) = C(D_2)$, then by Definition 1(*b*) we obtain that $P(D_1) = P(D_2)$, whence $D_1 = D_2$. Thus $C(D_1) \neq C(D_2)$.

Suppose first $|C(D_1) \cap C(D_2)| = 1$, i.e., $C(D_1) \cap C(D_2) = \{v\}$. Then by Definition 1, either $k-1$ or $k-2$ vertices of $C(D_2)$ (resp. $C(D_1)$) are pick vertices of D_1 (resp. D_2). But then for $k \geq 4$, we obtain also two adjacent pick vertices i.e. we obtain K_{k+2}. So we may assume that $k = 3$, $C(D_1) = \{u_1, w_1, v\}$, $C(D_2) = \{u_2, w_2, v\}$, and u_1 (resp. u_2) are pick vertices of D_2 (resp. D_1). As the graph is K_5-free, w_1 (resp. w_2) is not a pick vertex of D_2 (resp. D_1). Then $w_1 \in I(D_2)$ (resp. $w_2 \in I(D_1)$) is a common neighbor of $v, u_2 \in C(D_2)$ (resp. $v, u_1 \in C(D_1)$), a contradiction with Definition 1(b).

Suppose now $|C(D_1) \cap C(D_2)| \geq 2$. Then each vertex $u \in C(D_1) \setminus C(D_2)$ is a neighbor of at least two vertices from $C(D_2)$, whence by item (b) of Definition 1, $u \in P(D_2)$ and thus $C(D_1) \setminus C(D_2) \subseteq P(D_2)$. Similarly $C(D_2) \setminus C(D_1) \subseteq P(D_1)$. Thus the subgraph of G induced by $C(D_1) \cup C(D_2)$ is a clique, whence $|C(D_1) \cup C(D_2)| = k + 1$, and so $|C(D_1) \cap C(D_2)| = k - 1$. By assumptions, D_1 is a (k, s_1)-dart, $s \geq s_1 \geq 2$. Thus there exists $x_1 \in P(D_1) \setminus C(D_2)$. As G is K_{k+2}-free, we insert that $x_1 \in I(D_2)$ but then it is a common neighbor of at least two vertices from $C(D_2)$, a contradiction with Definition 1(b).

By the above two paragraphs, we can assume that $C(D_1) \cap C(D_2) = \emptyset$. If $v \in V(D_1) \cap P(D_2)$, then $d_{D_2}(v) \leq d_G(v) + 2$, a contradiction with Definition 1(a). Similarly if $v \in V(D_2) \cap P(D_1)$. This proves claim (a). Claim (b) is an easy consequence of (a).

Next lemmas assures that removing small vertices or diamnods in dart graphs we preserve the class of dart graphs.

Lemma 3. *Let G be a (k, s)-dart graph with $k \geq 3$. Then*

(a) *if v is a vertex of degree $\leq k$, then $G' = G - v$ is a (k, s)-dart graph,*
(b) *if $D \in \mathcal{D}(G)$, then $G' = G - D$ is a (k, s)-dart graph.*

Moreover, in both cases, $\mathcal{D}(G')$ can be determined from $\mathcal{D}(G)$ in a constant time.

Proof. We first show that in both cases G' is a (k, s)-dart graph. Suppose that u' is an arbitrary vertex of degree $\geq k+2$ in G'. Then, it is also of degree $\geq k+2$ in G, and hence it belongs to a (k, i)-diamond $D' \in \mathcal{D}(G)$ with $2 \leq i \leq s$. In case (b), by Lemma 2, diamonds D and D' are disjoint, and hence D' is an induced (k, s)-diamond in G'. Consider now case (a). If D' is an induced subgraph of G', then we are done. Otherwise, $v \in V(D')$ is a pick vertex of D'. Since u' is of degree $\geq k+2$ in G', it follows that $i \geq 3$, and hence $D' - v$ is a $(k, i-1)$-diamond in G'.

Regarding $\mathcal{D}(G')$ and $\mathcal{D}(G)$, in case (b), Lemma 2 assure that $\mathcal{D}(G)$ consists of D and $\mathcal{D}(G')$. In case (a), $\mathcal{D}(G)$ may change only if v is a pick vertex of some $(k, 2)$-diamond D' of G. But then we have that $\mathcal{D}(G)$ is either $\mathcal{D}(G')$ or $\mathcal{D}(G') \cup \{D\}$.

In the next few lemmas, we study properties of a graph G' obtained from G be applying some local changes.

Lemma 4. *Let G be a (k, s)-dart graph with $k \geq 3$ and let a_1, a_2 be two central vertices of a diamond $D \in \mathcal{D}(G)$. Suppose that x_1 and x_2 are the isolated neighbors of a_1 and a_2, respectively. Then, each $(k+1)$-coloring λ^* of $G^* := G - x_1a_1 - x_2a_2 + x_1x_2$ can be modified into a $(k+1)$-coloring of G in a constant time.*

Proof. Clearly $\lambda^*(a_1) \neq \lambda^*(a_2)$ and $\lambda^*(x_1) \neq \lambda^*(x_2)$. By Definition 1, a_1 and x_2 are non-adjacent, and similarly a_2 and x_1 are non-adjacent. Notice that λ^* is not a coloring of G if and only if $\lambda^*(a_1) = \lambda^*(x_1)$ or $\lambda^*(a_2) = \lambda^*(x_2)$. But in that case, we can simply interchange the colors between a_1 and a_2, and obtain a proper $(k+1)$-coloring of G.

Lemma 5. *Let G be a K_{k+2}-free (k,s)-dart graph and $D \in \mathcal{D}(G)$. Let a_1, a_2 be two central vertices of D and let x_1, x_2 be their isolated neighbors, respectively. Then the graph $G' = G - x_1a_1 - x_2a_2 + x_1x_2$ is a K_{k+2}-free graph unless x_1, x_2 are pick vertices of a diamond of $\mathcal{D}(G)$.*

Proof. Suppose that G' contains a copy H of K_{k+2}. Then, x_1, x_2 are vertices of H, thus cannot be adjacent in G and there is a set S of k common neighbors of x_1 and x_2 in G, which induce a clique. Notice that $|S| = k$ and $d_G(x_1), d_G(x_2) \geq k+1$.

Suppose that $d_G(x_1) \geq k+2$. Then, x_1 is a central vertex of some diamond $D' \in \mathcal{D}(G)$, whence by item (b) of definition 1, $S \subseteq V(D')$ and clearly, $|S \cap C(D')| \geq k-1 \geq 2$. Then X_2 has at least 2 neighbors in $C(D')$, whence x_2 belongs to D' and is adjacent with x_1 in G, a contradiction.

Thus $d(x_1) = k+1$ and analogously $d(x_2) = k+1$. Then x_1, x_2 and S belong to a diamond of $D' \in \mathcal{D}(G)$ in which $x_1, x_2 \in P(D')$ and $S = C(D')$.

Lemma 6. *Let G be a (k,s)-dart graph and $D \in \mathcal{D}(G)$. Let a_1, a_2 be two central vertices of D and let x_1, x_2 be their isolated neigbors, respectively. Then graph $G' = G - x_1a_1 - a_2x_2 + x_1x_2$ is a (k,s)-dart graph unless one of the following conditions occurs:*

(a) x_1, x_2 are pick vertices of a diamond of $\mathcal{D}(G)$;
(b) there exists a diamond $D' \in \mathcal{D}(G)$ and $i \in \{1,2\}$ such that $x_i \in C(D')$ and x_{3-i} is an isolated neighbor of a central vertex from D', which is distinct from x_i.

Proof. Suppose that G' is not a (k,s)-dart graph. First notice that each vertex preserve its degree from G except a_1, a_2, which belong to D and it is a diamond in G' as well. If there is some $D' \in \mathcal{D}(G)$ that is not induced diamond of G', then x_1 and x_2 must be pick vertices of D', which is the excluded case (a). Next observe that each diamond of $\mathcal{D}(G)$ satisfies condition (a) of Definition 1 in G'. Finally, if condition (b) Definition 1 is not satisfied for some $D' \in \mathcal{D}(G)$ in G', then there are two central vertices u and v with a common neighbor w outside D'. Notice that x_1x_2 is one of the edges uw or vw. Then without loss of generality, we may assume that x_1 is a central vertex in D' and x_2 is an isolated neighbor of a central vertex of D' distinct from x_1.

4 An Extension of Brooks Theorem

If $D \in \mathcal{D}(G)$, then a vertex of $I(D)$ could be a central and pick vertex of another diamond of $\mathcal{D}(G)$. Denote by $I_c(D)$ and $I_p(D)$ the subset of all such vertices of $I(D)$, respectively. By Lemma 2(b), sets $I_c(D)$ and $I_p(D)$ are disjoint. Finally, let $I_s(D)$ be the vertices of $I(D)$ that are neither in $I_c(D)$, nor in $I_p(D)$.

Lemma 7. *Let G be a K_{k+2}-free (k,s)-dart graph with given $\mathcal{D}(G) \neq \emptyset$ and $s \leq k$. Then, in a constant time we construct a K_{k+2}-free (k,s)-dart graph G^* together with $\mathcal{D}(G^*)$ such that*

(*a*) $|E(G^*)| < |E(G)|$;
(*b*) *From any $(k+1)$-coloring λ of G^* one can construct a $(k+1)$-coloring of G in a constant time.*

Proof. In the construction of G^* we use a bounded number of vertex/edge additions and deletions. Similarly, we obtain $\mathcal{D}(G^*)$ from $\mathcal{D}(G)$ in a finite number of steps. This will preserve that constructions are completed in a constant time. In the sequel consider the following cases:

Case 1. *There exists $v \in V(G)$ of degree $\leq k$.* Then v is not a central vertex. Thus, by Lemma 3(a), $G^* := G - v$ is a (k,s)-dart graph with $|E(G^*)| < |E(G)|$. By the same lemma, one can construct $\mathcal{D}(G^*)$ from $\mathcal{D}(G)$ in a constant time. Obviously, G^* is a K_{k+2}-free graph. A coloring of G^* can be easily extended to a coloring of G by assigning to v a color that miss in its neighborhood.

Case 2. *There exists $v \in C(D)$, $D \in \mathcal{D}(G)$, having no isolated neighbor.* By Lemma 3(b), $G^* := G - D$ is a (k,s)-dart graph and $\mathcal{D}(G^*)$ can be constructed from $\mathcal{D}(G)$ in a constant time. Obviously, G^* is a K_{k+2}-free graph and $|E(G^*)| < |E(G)|$. Let λ^* be a $(k+1)$-coloring of G^*. Since each pick vertex of D has at most one neighbor outside D and since $|P(D)| < k+1$, it follows that there exists a color that we can assign to all pick vertices. Since v has no isolated neighbor, we can apply Lemma 1 to extend λ^* to the central vertices of D.

Case 3. *There exists $D \in \mathcal{D}(G)$, such that $I_c(D) \cup I_s(D) \neq \emptyset$, or some two vertices of $I_p(D)$ do not belong to the same $D' \in \mathcal{D}(G)$.* We can assume that Case 2 does not hold, whence $|I_c(D)| + |I_p(D)| + |I_s(D)| = k$. Let $x_1, x_2 \in I(D)$ be two distinct vertices. And, let $a_i \in C(D)$ be the neighbor of x_i for $i = 1, 2$.

Now, consider the graph $G^* = G - x_1a_1 - x_2a_2 + x_1x_2$. If none of the exceptions of Lemmas 5 or 6 holds, then G^* is a K_{k+2}-free (k,s)-dart graph, and by Lemma 4, we can modify any coloring of G^* to a proper coloring of G in a constant time. Moreover, $|E(G^*)| < |E(G)|$ and $\mathcal{D}(G)$ can be determined in a constant time from $\mathcal{D}(G^*)$.

Assume that for any pair $x_1, x_2 \in I(D)$, exeptions of Lemmas 5 or 6 is satisfied. This implies immediately that $|I_c(D)| \leq 1$.

Let $x_1, x_2 \in I_s(D)$. Then, the exception Lemma 6(b) cannot occur. Thus, x_1 and x_2 are the pick vertices of a $(k,2)$-diamond D' and $D' \notin \mathcal{D}(G)$ (because $x_1, x_2 \notin I_p(D)$). Since $D' \notin \mathcal{D}(G)$ all vertices of $C(D')$ are of degree $k+1$, which means no vertex of $C(D')$ has isolated neighbor. Since $k \geq 3$, there exists $x \in I(D)$, $x \neq x_1, x_2$. Now, using x, x_1 instead of x_1, x_2, we do not obtain any of the exceptions of Lemmas 5 and 6. Hence $|I_s(D)| \leq 1$.

Thus $|I_p(D)| \geq 1$ (because $k \geq 3$). Then $x_1 \in I_s(D) \cup I_c(D)$ and $x_2 \in I_p(D)$ do not satisfy exeptions of Lemmas 5 or 6, whence $I_s(D) \cup I_c(D) = \emptyset$. Thus all vertices of $I(D)$ must be pick vertices of one diamond of $\mathcal{D}(G)$. This contradicts the assumptions of Case 3.

Case 4. *None of Cases 1.–3. occurs.* Thus, by Case 3., for each $D \in \mathcal{D}(G)$, $I_c(D) \cup I_s(D) = \emptyset$, and all vertices of $I_p(D)$ belongs to the same (k,k)-diamond $\varphi(D) \in \mathcal{D}(G)$. Furthermore, there exists a perfect matching between $C(D)$ and $P(\varphi(D))$.

Case 4.1. *There exists $D \in \mathcal{D}(G)$, such that $\varphi^2(D) = D$.* Then vertices of D and $\varphi(D)$ induce a component G' of G. Let $G^* := G - G'$. Obviously, G^* is a (k,k)-dart graph, $|E(G^*)| < |E(G)|$ and $\mathcal{D}(G^*) = \mathcal{D}(G) \setminus \{D, \varphi(D)\}$. Moreover, we can construct a $(k+1)$-coloring of G' in a constant time: just color all vertices of $P(D)$ and $P(\varphi(D))$ by the color $k+1$, and assign colors $1, \dots, k$ to the vertices of $C(D)$ and $C(\varphi(D))$.

Case 4.2. *For each $D \in \mathcal{D}(G)$, $\varphi^2(D) \neq D$.* By the assumptions of lemma, there exists $D \in \mathcal{D}(G) \neq \emptyset$. Let G^* be the graph, we obtain by removing the vertices of $\varphi(D)$ and inserting a perfect matching between $C(D)$ and $P(\varphi^2(D))$. Obviously G^* is a (k,k)-dart graph with less edges than G and $\mathcal{D}(G^*) = \mathcal{D}(G) \setminus \{\varphi(D)\}$. Let λ^* be a $(k+1)$-coloring of G^*. Then λ^* assigns the same color c to all vertices of $P(\varphi^2(D))$. Assign c also to all vertices of $P(\varphi(D))$ and to each of the vertices of $C(\varphi(D))$ an unique color from $\{1, \dots, k+1\} \setminus \{c\}$. This gives a required coloring of G, completing the proof.

Now we are ready to prove the main result.

Theorem 1. *Let G be a (k,s)-dart graph with $k \geq 3$ and $k \geq s$. Then G is $(k+1)$-colorable if and only if it has no component isomorphic to K_{k+2}. Furthermore, if G is $(k+1)$-colorable, then a $(k+1)$-coloring of G can be constructed in a linear time.*

Proof. The necessity of the first part of the theorem is trivial. To see the sufficiency, observe that a (k,s)-dart graph is K_{k+2}-free if and only if it has no component isomorphic to a K_{k+2}. The same is true if G is a graph with vertex degree at most $k+1$. Therefore, the sufficiency follows from Lemma 7 and Brooks' Theorem [1].

We can check whether a dart graph G is K_{k+2}-free in linear time. Analogously, we can find the set $\mathcal{D}(G)$ in linear time. Consequently, by means of Lemma 7 we can create in linear time a K_{k+2}-free graph G' without vertices of degree more than $k+1$ such that any $(k+1)$-coloring of G' can be transformed into a $(k+1)$-coloring of G in linear time. By [7] (see also [9,6]), a $(k+1)$-coloring of G' can be found in linear time, which proves the statement.

5 NP-Completeness

In this section we show that Theorem 1 cannot be extended for (k,s)-dart graphs where $s > k \geq 2$ unless P = NP.

We need some more notation. Take n vertex disjoint copies of $(k,k+1)$-diamonds $D_1, \dots, D_n$, $k, n \geq 2$. For $i = 1, \dots, n$, denote by $v_{i,1}, \dots, v_{i,k}$ and $u_{i,1}, \dots, u_{i,k+1}$ the central and pick vertices of D_i, respectively. Add nk new edges $v_{i,j}u_{i+1,j}$, $i = 1 \ldots, n$, $j = 1, \ldots, k$ (considering the sum $i+1$ mod n). Then the resulting graph is called a $(n, k+1)$-*bracelet* and vertices $u_{1,k+1}, \ldots, u_{n,k+1}$ are called its *connectors*.

We study complexity of the following problem.

DART-(k, s)-$(k+1)$-COL
Instance: A (k, s)-dart graph G, $2 \leq k, s$.
Question: Is G $(k+1)$-colorable?

Theorem 2. *The problem* DART-(k, s)-$(k+1)$-COL, $k \geq 2$, *is*

(a) NP*-complete for* $s > k$,
(b) *solvable in linear time for* $s \leq k$.

Proof. Claim (b) holds true by Theorem 1 (for $k \geq 3$) and by [6, Theorem 4.3] (for $k = 2$). We prove (a). Let G be a graph. Replace each vertex v of G of degree ≥ 2 by a $(d_G(v), k+1)$-bracelet H_v. Let H_v be an isolated vertex if $d_G(v) = 1$. Each edge uv of G replace by an edge joining a connector of H_v with a connector of H_u so that each connector is attached to at most one new edge. Denote the resulting graph by G'. Clearly, G' is a $(k, k+1)$-dart graph. By any $(k+1)$-coloring of H_v, $v \in V(G)$, all connectors of H_v must be colored by the same color. Hence G' is $(k+1)$-colorable iff G is so. Thus the problem from item (a) can be polynomially reduced to the problem of $(k+1)$-coloring. This problem is NP-complete for every fixed $k \geq 2$ by Garey and Johnson [3, GT4].

References

1. Brooks, R.L.: On coloring the nodes of a network. Proc. Cambridge Phil. Soc. 37, 194–197 (1941)
2. Bryant, V.: A characterisation of some 2-connected graphs and a comment on an algorithmic proof of Brooks' theorem. Discrete Math. 158, 279–281 (1996)
3. Garey, M.R., Johnson, D.S.: Computers and Intractability. W.H. Freeman, San Francisco (1979)
4. Garey, M.R., Johnson, D.S., Stockmeyer, L.: Some simplified NP-complete graph problems. Theor. Comput. Sci. 1, 237–267 (1976)
5. Diestel, R.: Graph Theory, 3rd edn. Springer, Heidelberg (2005)
6. Kochol, M., Lozin, V., Randerath, B.: The 3-colorability problem on graphs with maximum degree four. SIAM J. Comput. 32, 1128–1139 (2003)
7. Lovász, L.: Three short proofs in graph theory. J. Combin. Theory Ser. B 19, 269–271 (1975)
8. Randerath, B., Schiermeyer, I.: A note on Brooks theorem for triangle-free graphs, Australasian. J. Combin. 26, 3–10 (2002)
9. Skulrattanakulchai, S.: Δ-list vertex coloring in linear time. In: Penttonen, M., Schmidt, E.M. (eds.) SWAT 2002. LNCS, vol. 2368, pp. 240–248. Springer, Heidelberg (2002)

Worst Case Efficient Single and Multiple String Matching in the RAM Model

Djamal Belazzougui

LIAFA, Univ. Paris Diderot - Paris 7, 75205 Paris Cedex 13, France
dbelaz@liafa.jussieu.fr

Abstract. In this paper, we explore worst-case solutions for the problems of pattern and multi-pattern matching on strings in the RAM model with word length w. In the first problem, we have a pattern p of length m over an alphabet of size σ, and given any text T of length n, where each character is encoded using $\log \sigma$ bit, we wish to find all occurrences of p. For the multi-pattern matching problem we have a set S of d patterns of total length m and a query on a text T consists in finding all the occurrences in T of the patterns in S (in the following we refer by *occ* to the number of reported occurrences). As each character of the text is encoded using $\log \sigma$ bits and we can read w bits in constant time in the RAM model, the best query time for the two problems which can only possibly be achieved by reading $\Theta(w/\log \sigma)$ consecutive characters, is $O(n\frac{\log \sigma}{w} + occ)$. In this paper, we present two results. The first result is that using $O(m)$ words of space, single pattern matching queries can be answered in time $O(n(\frac{\log m}{m} + \frac{\log \sigma}{w}) + occ)$, and multiple pattern matching queries answered in time $O(n(\frac{\log d + \log y + \log \log m}{y} + \frac{\log \sigma}{w}) + occ)$, where y is the length of the shortest pattern. Our second result is a variant of the first result which uses the four Russian technique to remove the dependence on the shortest pattern length at the expense of using an additional space t. It answers to multi-pattern matching queries in time $O(n\frac{\log d + \log \log_\sigma t + \log \log m}{\log_\sigma t} + occ)$ using $O(m + t)$ words of space.

1 Introduction

The problems of string pattern matching and multiple string pattern matching are classical algorithmic problems in the area of pattern matching. In the multiple string matching problem, we have to preprocess a dictionary of d strings of total length m characters over an alphabet of size σ so that we can answer to the following query: given any text of length n, find all occurrences in the text of any of the d strings. In the case of single string matching, we simply have $d = 1$.

The textbook solutions for the two problems are the Knuth-Morris-Pratt [15] (KMP for short) automaton for the single string matching problem and the Aho-Corasick [1] automaton (AC for short) for the multiple string matching problem. The AC automaton is actually a generalization of the KMP algorithm. Both algorithms achieve $O(n + occ)$ query time (where *occ* denotes the number of reported occurrences) using $O(m \log m)$ bits of space. The query time

C.S. Iliopoulos and W.F. Smyth (Eds.): IWOCA 2010, LNCS 6460, pp. 90–102, 2011.

of both algorithms is in fact optimal if the matching is restricted to read all the characters of the text one by one. However it was noticed that in many cases, it is actually possible to avoid reading all the characters of the text and hence achieve a better performance. This stems from the fact that by reading some characters at certain positions in the text, one could conclude whether a match is possible or not. This has led to various algorithms with so-called sublinear query time assuming that the characters of the patterns and/or the text are drawn from some random distribution. The first algorithm which exploited that fact was the the Boyer-Moore algorithm [5]. Later algorithms with provably average-optimal performance were devised. Most notably the BDM and BNDM for single string matching and the multi-BDM [9,8] and multi-BNDM [18] for multiple string matching. Those algorithms achieve $O(n\frac{\log m}{m\log\sigma}+occ)$ time for single string matching (which is optimal according to the lower bound in [24]) and $O(n\frac{\log d+\log y}{y\log\sigma}+occ)$ time for multiple string matching, where y is the length of the shortest string in the set. Still in the worst case those algorithms may have to read all the text characters and thus have $\Omega(n+occ)$ query time (actually many of those algorithms have an even worse query time in the worst-case, namely $\Omega(nm+occ)$).

A general trend has appeared in the last two decades when many papers have appeared trying to exploit the power of the word RAM model to speed-up and/or reduce the space requirement of classical algorithms and data structures. In this model, the computer operates on words of length w and usual arithmetic and logic operations on the words all take one unit of time.

In this paper we focus on the worst-case bounds in the RAM model with word length w. That is we try to improve on the KMP and AC in the RAM model assuming that we have to read all the characters of the text which are assumed to be stored in a contiguous area in memory using $\log\sigma$ bits per characters. That means that it is possible to read $\Theta(w/\log\sigma)$ consecutive characters of the text in $O(1)$ time. Thus given a text of length n characters, an optimal algorithm should spend $O(n\frac{\log\sigma}{w}+occ)$ time to report all the occurrence of matching patterns in the text. The main result of this paper is a worst case efficient algorithm whose performance is essentially the addition of a term similar to the average optimal time presented above plus the time necessary to read all the characters of the text in the RAM model. Unlike many other papers, we only assume that $w=\Omega(\log(n+m))$, and not necessarily that $w=\Theta(\log(n+m))$. That is we only assume that a pointer to the manipulated data (the text and the patterns), fit in a memory word but the word length w can be arbitrarily larger than $\log m$ or $\log n$. This assumption makes it possible to state time bounds which are independent of m and n, implying larger speedups for small values of m and n.

In his paper Fredriksson presents a general approach [13] which can be applied to speed-up many pattern matching algorithms. This approach which is based on the notion of super-alphabet relies on the use of tabulation (four russian technique). If this approach is applied to our problems of single and multiple string matching queries, given an available precomputed space t, we can get a $\log_\sigma(t/m)$ factor speedup. In his paper [4], Bille presented a more space efficient method for single string matching queries which accelerates the KMP algorithm to answer to queries in time $O(\frac{n}{\log_\sigma n} + occ)$ using $O(n^\varepsilon + m)$ words of space for any constant ε such that $0 < \varepsilon < 1$. More generally, the algorithm can be tuned to use an additional amount t of tabulation space in order to provide a $\log_\sigma t$ factor speedup.

At the end of his paper, Bille asked two questions: the first one was whether it is possible to get an acceleration proportional to the machine word length w (instead of $\log n$ or $\log t$) using $O(m)$ words of space only. The second one was whether it is possible to obtain similar results for the multiple string matching problem. We give partial answers to both questions. Namely, we prove the following two results:

1. Our first result implies that for d strings of minimal length y, we can construct an index which occupies $O(m)$ words of space and answers to queries in time $O(n(\frac{\log d + \log y + \log\log m}{y} + \frac{\log\sigma}{w}) + occ)$. This result implies that we can get a speedup factor $\frac{w}{(\log d + \log w)\log\sigma}$ if $y \geq \frac{w}{\log\sigma}$ and get the optimal speedup factor $\frac{w}{\log\sigma}$ if $y \geq (\log d + \log w)\frac{w}{\log\sigma}$.
2. Our second result implies that for d patterns of arbitrary lengths and an additional amount of memory t, we can obtain a factor $\frac{\log_\sigma t}{\log d + \log\log_\sigma t + \log\log m}$ speedup using $O(m + t)$ words of memory.

Our first result compares favorably to Bille's and Fredriksson approaches as it does not use any additional tabulation space. In order to obtain any significant speedup, the algorithms of Bille and Fredriksson require a substantial amount of space t which is not guaranteed to be available. Even if such an amount of space was available, the algorithm could run much slower in case $m \ll t$ as modern hardware is made of memory hierarchies, where random access to large tables which do not fit in the fast levels of the hierarchy might be much slower than access to small data which fit in faster levels of the hierarchy.

Our second result is useful in case the shortest string is very short and thus, the first result do not provide any speedup. The result is slightly less efficient than that of Bille for single string matching, being a factor $\log\log_\sigma t + \log\log m$ slower (compared to the $\log_\sigma t$ speedup of Bille's algorithm). However, our second result efficiently extends to multiple string matching queries, while Bille's algorithms seems not to be easily extensible to multiple string matching queries.

In a recent work [3], we have tried to use the power of the RAM model to improve the space used by the AC representation to the optimal (up to a constant factor) $O(m\log\sigma)$ bits instead of $O(m\log m)$ bits of the original representation, while maintaining the same query time. In this paper, we attempt to do the converse. That is, we try to use the power of the RAM model to improve the query time of the AC automaton while using the same space as the original representation.

We emphasize that our results are mostly theoretical in nature. The constants in space usage and query time of our data structures seem rather large. Moreover, in practice average efficient algorithms which have been tuned for years are likely to behave much better than any worst-case efficient algorithm. For example, for

DNA matching, it was noted that DNA sequences encountered in practice are rather random and hence average-efficient algorithms tend to perform extremely well for matching in DNA sequences (see [20] for example).

2 Outline of the Results

2.1 Problem Definition, Notation and Preliminaries

In this paper, we aim at addressing two problems: the single string pattern matching and the multiple string pattern matching problems. In the single string pattern matching problem we have to build a data structure on a single pattern (string) of length m over an alphabet of size $\sigma \leq m$. In the multiple string pattern matching problem, we have a set S of d patterns of total length m characters where each character is drawn from an alphabet of size $\sigma \leq m$. In the first problem, we have to identify all occurrences of the pattern in a text T of length n. In the second problem, we have to identify all occurrences of any of the d patterns.

In this paper, we assume a unit-cost RAM model with word length w, and assume that $w = \Omega(\log m + \log n)$. However w could be arbitrarily larger than $\log m$ or $\log n$. We assume that the patterns and the text are drawn from an alphabet of size $\sigma \leq m$. We assume that all usual RAM operations (multiplications, additions, divisions, shifts, etc...) take one unit of time.

For any string x we denote by $x[i,j]$ (or $x[i..j]$) the substring of x which begins at position i and ends at position j in the string x. For any integer m we note by $\log m$ the integer number $\lceil \log_2 m \rceil$.

In the paper we make use of two kinds of ordering on the strings: the prefix lexicographic order which is the standard lexicographic ordering (strings are compared right-to-left) and the suffix-lexicographic order which is defined in the same way as prefix lexicographic, but in which string are compared left-to-right instead of right-to-left. The second ordering can be thought as if we write the strings in reverse before comparing them. Unless otherwise stated, string lengths are expressed in terms of number of characters. We make use of the fixed integer bit concatenation operator $(\cdot)$ which operates on fixed length integers, where $z = x \cdot y$ means that z is the integer whose bit representation consists in the concatenation of the bits of the integers x as most significant bits followed by the bits of the integer y as least significant bits. We define the function $sucount_X(s)$, which returns the number of elements of a set X which have a string s as a suffix. Likewise we define the function $prcount_X(s)$, which returns the number of elements of a set X which have a string s as a prefix. We also define two other functions $surank_X(s)$ and $prrank_X(s)$ as the functions which return the number of elements of a set X which precede the string s in suffix and prefix lexicographic orders respectively.

2.2 Results

The results of this paper are summarized by the following two theorems:

Theorem 1. *Given a set S of d strings of total length m, where the shortest string is of length y, we can build a data structure of size $O(m \log m)$ bits such that given any text T of length n, we can find all occurrences of strings of S in T in time $O(n(\frac{\log d+\log y+\log\log m}{y} + \frac{\log \sigma}{w}) + occ)$.*

The theorem give us the following interesting corollaries.

Corollary 1. *Given a string p of length m, we can build a data structure of size $O(m \log m)$ bits of space such that given any text T of length n, we can find all occ occurrences of p in T in time $O(n(\frac{\log m}{m} + \frac{\log \sigma}{w}) + occ)$.*

For multiple string matching, we have the following two corollaries:

Corollary 2. *Given a set S of d strings of total length m where each string is of length at least $\frac{w}{\log \sigma}$ characters, we can build a data structure of size $O(m \log m)$ bits of space such that given any text T of length n, we can find all occurrences of strings of S in T in time $O(n\frac{(\log d+\log w)\log \sigma}{w} + occ)$.*

For the case of even larger minimal length, we can get the following corollary which improves on corollary 2 by giving an optimal query time in case of sufficiently long strings:

Corollary 3. *Given a set S of d strings of total length m where each string is of length at least $(\log d + \log w)\frac{w}{\log \sigma}$ characters, we can build a data structure occupying $O(m \log m)$ bits of space such that given any text T of length n, we can find all occurrences of strings of S in T in the optimal $O(n\frac{\log \sigma}{w} + occ)$ time.*

The dependence of the bounds in theorem 1 and its corollaries on minimal patterns lengths is not unusual. This dependence exists also in average-optimal algorithms like BDM, BNDM and their multiple patterns variants [9,8,18]. Those algorithms achieve a $\frac{y \log \sigma}{\log d+\log y}$ speedup factor on average requiring that the strings are of minimal length y. Our query time is the addition of a term which represents the time necessary to read all the characters of text in the RAM model and a term which is similar to the query time of the average optimal algorithms.

We also show the following theorem which is mostly useful in case the minimal length is too short:

Theorem 2. *Given a set S of d strings of total length m and a parameter s, we can build a data structure occupying $O(m \log m + \sigma^s \log^2 s \log m)$ bits of space such that given any text T of length n, we can find all occ occurrences of strings of S in T in time $O(n\frac{\log d+\log s+\log\log m}{s} + occ)$.*

The theorem could be interpreted in the following way: having some additional amount t of available memory space, we can achieve a speedup factor $\frac{s}{\log d+\log s}$ for $s = \log_\sigma t$ using a data structure which occupies $O(m \log m + t)$ bits of space.

The theorem gives us two interesting corollaries which depend on the relation between m and n. In the case where $n \geq m$, by setting $t = n^\varepsilon$ for any $0 < \varepsilon < 1$, we get the following corollary:

Corollary 4. *Given a set S of d strings of total length m, we can build a data structure occupying $O(m \log m + n^{\varepsilon})$ bits of space such that given any text T of length n, we can find all occurrences of strings of S in T in time $O(n\frac{\log d+\log\log_\sigma n+\log\log m}{\log_\sigma n} + occ)$, where ε is any constant such that $0 < \varepsilon < 1$.*

In the case $m \geq n$ (which is only possible in the case of multiple string matching, we can get a better speedup by setting $t = m$:

Corollary 5. *Given a set S of d strings of total length m, we can build a data structure occupying $O(m \log m)$ bits of space such that given any text T of length n, we can find all occurrences of strings of S in time $O(n\frac{\log d+\log\log m}{\log_\sigma m} + occ)$.*

We note that in the case $d = 1$, the result of theorem 2 is worse by a factor $\log\log_\sigma n + \log\log m$ than that of Bille which achieves a query time of $O(\frac{n}{\log_\sigma n} + occ)$. However the result of Bille does not extend naturally to $d \geq 1$. The straightforward way of extending Bille's algorithm is to build d data structures and to match the text against all the data structures in parallel. This however would give a running time of $O(n\frac{d}{\log_\sigma n} + occ)$ which is worse than our running time $O(n\frac{\log d+\log\log_\sigma n+\log\log m}{\log_\sigma n} + occ)$ which is linear in $\log d$ rather than d.

As of the technique of Fredriksson, in order to obtain query time $O(\frac{n}{s} + occ)$, it needs to use at least space $\Omega(m\sigma^s)$ which can be too much in case s is too large.

3 Multiple String Matching without Tabulation

In this section we give a proof of the first theorem. However, before that we present the results about the components which are used in our construction.

3.1 Components

For our results, we use the following lemmata:

Lemma 1. *[23] Given a collection of n intervals over universe U where for any two intervals s_1 and s_2 we have either $s_1 \cap s_2 = s_1$, $s_1 \cap s_2 = s_2$ or $s_1 \cap s_2 = \emptyset$ (for any two intervals either one is included in the other or the two intervals are disjoint). We can build a data structure which uses $O(n \log n)$ bits of space such that for any point x, we can determine the interval which most tightly encloses x in $O(n \log\log n)$ time (the smallest interval which encloses x).*

For implementing the lemma, we store the set of interval endpoints in a predecessor data structure, namely the Willard's y-fast trie [23] which is a linear space version of the Van Emde Boas tree [22]. Then those points divide the universe of size U into $2n + 1$ segments and each segment will point to the interval which most tightly encloses the segment. Then a predecessor query will point to the segment which in turn points to the relevant interval. This problem can be thought as a restricted $1D$ stabbing problem.

Lemma 2. *Given a collection S of n strings of arbitrary lengths and a function f from S into $[0, m-1]$, we can build a data structure which uses $O(n \log m)$ bits which computes $f(x)$ for any $x \in S$ in time $O(|x|/w)$ (where $|x|$ is the length of x in bits). When queried for any $y \notin S$ the function returns any value from the set $f(S)$.*

This result can easily be obtained using minimal perfect hashing [12,14]. Though perfect hashing is usually defined for fixed $O(w)$ bits integers, a standard string hash function [10] can be used to first reduce the strings to integers before constructing the minimal perfect hashing on the generated integers.

Lemma 3. *[7, Theorem 1] Given a collection S of n strings of variable lengths occupying a memory area of m characters (the strings can possibly overlap), we can build an index which uses $O(n \log m)$ bits so that given any string x, we can find the string $s \in S$ which is the longest among all the strings of S which are prefix of x in time $O(|x|/w + \log n)$ (where $|x|$ is the length of x in bits). More precisely, the data structure returns $prrank_S(s)$. Moreover the data structure is able to tell whether $x = s$.*

This result which is obtained using a string B-tree [11] combined with an LCP array and a compacted trie [16] built on the set of strings, and setting the block size of the string B-tree to $O(1)$. The following lemma is symmetric of the previous one.

Lemma 4. *Given a collection S of n strings of variable lengths occupying a memory area of m characters of space (the strings can possibly overlap), we can build an index which uses $O(n \log m)$ bits so that given any string x, we can find the string $s \in S$ which is the longest among all the strings of S which are suffix of x in time $O(|x|/w + \log n)$ (where $|x|$ is the length of x in bits). More precisely, the data structure returns $surank_S(s)$. Moreover the data structure is able to tell whether $x = s$.*

Lemma 5. *[6] Given a set of n rectangles in the plane, we can build a data structure which uses $O(n \log n)$ bits of space so that given any point $[v, z]$, we can report all the k occurrences of rectangles which enclose that point in time $O(\log n + k)$.*

The problem solved by lemma 5 is called the $2D$ stabbing problem or sometimes called the point enclosure. The lemma uses the best linear space solution to the problem which is due to Chazelle [6] (which is optimal according to the lower bound in [19]).

3.2 Overview

The goal of this paper is to simulate the running of the AC automaton [1], by processing the characters of the scanned text in blocks of b characters. The central idea of the main result relies on a reduction of the problem of dictionary matching to the $1D$ and $2D$ stabbing problems, in addition to the use of standard string data structures namely, string B-trees, suffix arrays and minimal perfect hashing on strings. At each step, we first read b characters of the text, find the matching patterns which end at one of those characters and finally jump to the state which would have been reached after reading the b characters by the AC automaton (thereby simulating all *next* and *fail* transitions which would have been traversed by the standard AC automaton for the b characters). Finding the matching patterns is reduced to the $2D$ stabbing problems, while jumping to the next state is reduced to $1D$ stabbing problem. The geometric approach has already been used for dictionary matching problem and for text pattern matching algorithms in general. For example, it has been recently used in order to devise compressed indexes for substring matching [17,7]. Even more recently the authors of [21] have presented a compressed index for dictionary matching which uses a reduction to $2D$ stabbing problem.

3.3 The Data Structure

We now describe the data structure in more detail. Given the set S of d patterns, we note by P the set of the prefixes of the patterns in S (note that $|P| \leq m+1$). It is a well-known fact that there is a bijective relation between the set P and the set of states of the AC automaton. We use the same state representation as the one used in [3]. That is we first sort the states of the automaton in the suffix-lexicographic order of the prefixes to which they correspond, attributing increasing numbers to the states from the interval $[0, m]$. Thus the state corresponding to the empty string gets the number 0, while the state corresponding to the greatest element of P (in suffix-lexicographic order) gets the largest number which is at most m. We define *state*(p) as the state corresponding to the prefix $p \in P$.

Now, the characters of the scanned text, are to be scanned in blocks of b characters. For finding occurrences of the patterns in a text T, we do $\lceil n/b \rceil$ steps. At each step $i \in [0, \lceil n/b \rceil - 1]$ we do three actions:

- Read b characters of the text, $T[ib, (i+1)b-1]$ (or $n - ib \leq b$ characters of the text, $T[ib, n)$ in the last step).
- Identify all the occurrence of patterns which end at a position j of the text such that $j \in [ib, (i+1)b)$ ($j \in [ib, n)$ in the last step).
- If not in the last step go to the next state corresponding to the longest element of P which is a suffix of $T[0, (i+1)b]$.

The details of the implementation of each of the last two actions is given in the full version.

Our AC automaton representation has the following components:

1. An array A which contains the concatenation of all of the patterns. This array clearly uses mb bits of space.
2. Let $P_{0<i\leq b}$ be the set of prefixes of S of lengths in $[1, b]$. We use an instance of lemma 3, which we denote by B_1 and in which we store the set $P_{0<i\leq b}$

(by means of pointers into the array A). Clearly B_1 uses $O(db \log m) = O(m \log m)$ bits of space (we have db elements stored in B_1 and pointers into A take $\log m$ bits). We additionally store a vector of $|P_{0<i\leq b}| \leq db$ elements which we denote by T_1 and which associates an integer in $[0, m)$ with each element stored in B_1. The table T_1 uses $O(db \log m) = O(m \log m)$.

3. We use an instance of lemma 2, which we denote by B_2 and in which we store all the suffixes of strings in P (or equivalently all factors of the strings in S) of length b and for each suffix, store a pointer to its ending position in the array A (if the same factor occurs multiple times in the S we store it only once). As we have at most m elements in P and each pointer (in the array A) to each factor can be encoded using $O(\log m)$ bits, we conclude that B_2 uses at most $O(m \log m)$ bits of space.
4. We use an instance of lemma 4 which we denote by B_3 and in which we store all the suffixes of strings of S of lengths in $[1, b]$ (We note that set by $U_{0<i\leq b}$). It can easily be seen that B_3 also uses $O(db \log m) = O(m \log m)$ bits of space.
5. We use a $1D$ stabbing data structure (lemma 1) in which we store m segments where each segment corresponds to a state of the automaton. This data structure which uses $O(m \log m)$ bits of space is used in order to simulate the transitions in the AC automaton. We also store a vector of integers of size m which we denote by T_2 and which associates an integer with each interval stored in the $1D$ stabbing data structure. The table T_2 uses $O(m \log m)$ bits of space.
6. We use a $2D$ stabbing data structure (lemma 5) in which we store up to db rectangles. The space used by this data structure is $O(db \log(db)) = O(m \log m)$ bits. We also use a table T_3 which stores triplets of integers associated with each rectangle. The table T_3 will also use $O(db \log m) = O(m \log m)$ bits.

We deffer the details about the contents of each component to the full version which uses also to the full version. Central to the working of our data structure is the following technical lemma:

Lemma 6. *Given a set of strings X. We have that for any two strings $x \in X$ and $y \in X$:*

- *$prrank_X(y) \in [prrank_X(x), prrank_X(x) + prcount_X(x) - 1]$ iff x is a prefix of y.*
- *$surank_X(y) \in [surank_X(x), surank_X(x) + sucount_X(x) - 1]$ iff x is a suffix of y.*

The proof of the lemma is omitted.

3.4 Simulating Transitions

We will use the representation of states similar to the one used in [3]. That is each state of the automaton corresponds to a prefix $p \in P$ and is represented as an integer $state(p) = surank_P(p)$. The main idea for accelerating transitions is to read the text into blocks of size b characters and then find the next destination state attained after reading those b characters using B_1, T_1, B_2, T_2 and the $1D$ stabbing data structure. More precisely being at a state $state(p)$ and after reading next b characters of the text which form a string q, we have to find next state which is the state $state(x)$ such that $x \in P$ is the longest element of P which is suffix of pq. For that purpose the $1D$ stabbing data structure is used in combination with B_1 (which is queried on string q) in order to find $state(x)$ in case $|x| \geq b$. Otherwise if no such x is found the data structure B_2 will be used to find $state(x)$, where $|x| < b$. The following lemma summarizes the time and the space of the data structures needed to simulate a transition.

Lemma 7. *We can build a data structure occupying $O(m \log m)$ bits of space such that if the automaton is in a state t_i, the state t_{i+b} reached after doing all the transitions on b characters, can be computed in $O(\log d+\log b+\log\log m+\frac{b\log\sigma}{w})$ time.*

The proof of the lemma is deferred to the full version.

3.5 Identifying Matching Occurrences

In order to identify matching patterns the $2D$ stabbing data structure is used in combination with B_1. The details are deferred to the full version.

Lemma 8. *Given a parameter b and a set S of variable length strings of total length m characters over an alphabet of size σ, we can build a data structure occupying space $O(m \log m)$ bits, such that if the automaton is at a state t_i after reading i characters of a text T, all the occ_i matching occurrences of T which end at any position in $T[i, i+b]$ (or $T[i, |T|-1]$ if $i+b \geq |T|$) and begin at any position in $T[0, i]$ can be computed in $O(\log d + \log b + \frac{b\log\sigma}{w} + occ_i)$ time.*

Theorem 1 is obtained by combining lemma 7 with lemma 8. Namely by setting $b = y$, where y is the shortest pattern in S in both lemmata we can simulate the running of the automaton in $\lceil n/y \rceil$ steps at each step i, spending $O(\log d + \log b + \log\log m + \frac{y\log\sigma}{w}) + occ_i)$ to find the occ_i matching occurrences (through lemma 8) and $O(\log d+\log b+\log\log m+\frac{b\log\sigma}{w})$ time to simulate the transitions (through lemma 7). Summing up over all the $\lceil n/y \rceil$ steps, we get the query time stated in the theorem.

4 Tabulation Based Solution

We now prove theorem 2. A shortcoming of theorem 1 is that it gives no speedup in case the length of the shortest string in S is too short. In this case we resort to tabulation in order to accelerate matching of short patterns. More specifically, in case, we have a specified quantity t of available memory space (where $t < 2^w$ as obviously we can not address more than 2^w words of memory), we can precompute lookup tables using a standard technique known as the four russian technique [2] so that we can handle queries in time $O(n\frac{\log d+\log\log_\sigma t+\log\log m}{\log_\sigma t} + occ)$. In theorem 1 our algorithm reads the text in blocks of size $b = y$, where y

is the length of the shortest pattern. In reality we can not afford to read more than y characters at the each step, because by doing so we may miss a substring of the block of length y. Thus in order to be able to choose a larger block size b, we must be able to efficiently identify all substrings of any block of (at most)b characters which belong to S. The idea is then to use tabulation to answer to such queries in constant times (or rather in time linear in the number of reported occurrences). More in detail, for each possible block of $u \leq b$ characters, we have a total of $(u-1)(u-2)/2$ substrings which could begin at all but the first position of the block. For each possible block of u characters, we could store a list of all substrings belonging to S and each list takes at most $(u-1)(u-2)/2 = O(u^2)$ pointers of length $\log m$ bits. As we have a total of σ^u possible characters, we can use a precomputed table of total size $t = O((\sigma^u)u^2 \log m)$ bits.

Lemma 9. *For a parameter $u \leq \varepsilon w / \log \sigma$ (where ε is any constant such that $0 < \varepsilon < 1$) and a set S of patterns where each pattern is of length at most u, we can build a data structure occupying $O(\sigma^u \log^2 u \log m)$ bits of space such that given any string T of length u, we can report all the occ occurrences of patterns of S in T in $O(occ)$ time.*

Theorem 2 is obtained by combining lemmata 7, 8 and 9. Suppose we are given the parameter s: for implementing transitions, we can just use lemma 7 in which we set $b = s$, where the transitions are built on the set containing all the patterns. Now in order to report all the matching strings, we build an instance of lemma 8 on the set S and in which we set $b = s$ and also build $s-1$ instances of lemma 9 for every u such that $1 \leq u < s$. More precisely let $S_{\leq u}$ be the subset of strings in S of length at most u, then the instance number u will be built on the set $S_{\leq u}$ using parameter u and will thus for all possible strings of length u, store all matching patterns in S of length at most u.

A query on a text of T will work in the following way: we begin at step $I = 0$ and the automaton is at state 0 which corresponds to the empty string. Recognizing the patterns will consist in the following actions done at each step I:

1. Read the substring $T[i, j]$, where $i = Ib$ and $j = (I+1)b - 1$ (or $j = n - 1$ if $n > (I+1)b - 1$).
2. Recognize all the pattern occurrences which start at any position $i' \leq i$ and which terminate at any position $j' \in [i, j]$ using lemma 8.
3. Recognize all the matching strings of lengths at most b which are substrings of $T[i+1, j]$ using the instance number $j - i$ of lemma 9.
4. Increment step I by setting $I = I + 1$. Then if $Ib > n$, stop the algorithm immediately.
5. Do a transition using lemma 7 and return to action 1.

5 Conclusion

In this paper, we have proposed two solutions to the problems of single and multiple pattern matching on strings in the RAM model. In this model we assume that we can read $\Theta(w/\log \sigma)$ consecutive characters of any string in $O(1)$ time.

The first solution has a query time which depends on the length of the shortest pattern (or the length of the only pattern in case of single string matching) in a way similar to that of the previous algorithms which aimed at average-optimal expected performance (not worst-case performance as in our case). The first solution achieves an optimal query time if the shortest pattern is sufficiently long. The second solution has no dependence on the length of the shortest pattern but uses additional precomputed space. The second result is an interesting alternative to the previous tabulation approaches by Bille [4] and Fredriksson [13].

This paper gives rise to two interesting open problems:

- In order to obtain any speedup we either rely on the length of the shortest pattern being long enough (theorem 1) or have to use additional precomputed space (theorem 2). An important open question is whether it is possible to obtain any speedup without relying on any of the two assumptions.
- The space usage of both solutions is $\Omega(m \log m)$ bits, but the patterns themselves occupy only $O(m \log(\sigma))$ bits only. The space used is thus at least a factor $\Omega(\log_\sigma m)$ larger than the space occupied by the patterns. An interesting open problem is whether it is possible to obtain an acceleration compared to the standard AC automaton while using only $O(m \log \sigma)$ bits of space.

Acknowledgements

The author wishes tò thank Mathieu Raffinot for his many helpful comments and suggestions and two anonymous reviewers for their helpful remarks and corrections.

References

1. Aho, A.V., Corasick, M.J.: Efficient string matching: An aid to bibliographic search. ACM Commun. 18(6), 333–340 (1975)
2. Arlazarov, V.L., Dinic, E.A., Kronrod, M.A., Faradzev, I.A.: On economical construction of the transitive closure of a directed graph. Soviet Mathematics Doklady 11(5), 1209–1210 (1970)
3. Belazzougui, D.: Succinct dictionary matching with no slowdown. In: Amir, A., Parida, L. (eds.) CPM 2010. LNCS, vol. 6129, pp. 88–100. Springer, Heidelberg (2010)
4. Bille, P.: Fast searching in packed strings. In: Kucherov, G., Ukkonen, E. (eds.) CPM 2009. LNCS, vol. 5577, pp. 116–126. Springer, Heidelberg (2009)
5. Boyer, R.S., Moore, J.S.: A fast string searching algorithm. ACM Commun. 20(10), 762–772 (1977)
6. Chazelle, B.: Filtering search: A new approach to query-answering. SIAM J. Comput. 15(3), 703–724 (1986)
7. Chien, Y.-F., Hon, W.-K., Shah, R., Vitter, J.S.: Geometric Burrows-Wheeler transform: Linking range searching and text indexing. In: DCC, pp. 252–261 (2008)
8. Crochemore, M., Czumaj, A., Gasieniec, L., Jarominek, S., Lecroq, T., Plandowski, W., Rytter, W.: Speeding up two string-matching algorithms. Algorithmica 12(4/5), 247–267 (1994)

9. Crochemore, M., Rytter, W.: Text Algorithms. Oxford University Press, Oxford (1994)
10. Dietzfelbinger, M., Gil, J., Matias, Y., Pippenger, N.: Polynomial hash functions are reliable (extended abstract). In: ICALP, pp. 235–246 (1992)
11. Ferragina, P., Grossi, R.: The string b-tree: A new data structure for string search in external memory and its applications. J. ACM 46(2), 236–280 (1999)
12. Fredman, M.L., Komlós, J., Szemerédi, E.: Storing a sparse table with 0(1) worst case access time. J. ACM 31(3), 538–544 (1984)
13. Fredriksson, K.: Faster string matching with super-alphabets. In: Laender, A.H.F., Oliveira, A.L. (eds.) SPIRE 2002. LNCS, vol. 2476, pp. 44–57. Springer, Heidelberg (2002)
14. Hagerup, T., Tholey, T.: Efficient minimal perfect hashing in nearly minimal space. In: Ferreira, A., Reichel, H. (eds.) STACS 2001. LNCS, vol. 2010, pp. 317–326. Springer, Heidelberg (2001)
15. Knuth, D.E., Morris Jr., J.H., Pratt, V.R.: Fast pattern matching in strings. SIAM J. Comput. 6(2), 323–350 (1977)
16. Manber, U., Myers, E.W.: Suffix arrays: A new method for on-line string searches. SIAM J. Comput. 22(5), 935–948 (1993)
17. Navarro, G.: Indexing text using the ziv-lempel trie. J. Discrete Algorithms 2(1), 87–114 (2004)
18. Navarro, G., Raffinot, M.: A bit-parallel approach to suffix automata: Fast extended string matching. In: Farach-Colton, M. (ed.) CPM 1998. LNCS, vol. 1448, pp. 14–33. Springer, Heidelberg (1998)
19. Patrascu, M.: (data) structures. In: FOCS, pp. 434–443 (2008)
20. Rivals, E., Salmela, L., Kiiskinen, P., Kalsi, P., Tarhio, J.: mpscan: Fast localisation of multiple reads in genomes. In: Salzberg, S.L., Warnow, T. (eds.) WABI 2009. LNCS, vol. 5724, pp. 246–260. Springer, Heidelberg (2009)
21. Tam, A., Wu, E., Lam, T.W., Yiu, S.-M.: Succinct text indexing with wildcards. In: Karlgren, J., Tarhio, J., Hyyrö, H. (eds.) SPIRE 2009. LNCS, vol. 5721, pp. 39–50. Springer, Heidelberg (2009)
22. van Emde Boas, P., Kaas, R., Zijlstra, E.: Design and implementation of an efficient priority queue. Mathematical Systems Theory 10, 99–127 (1977)
23. Willard, D.E.: Log-logarithmic worst-case range queries are possible in space theta(n). Inf. Process. Lett. 17(2), 81–84 (1983)
24. Yao, A.C.-C.: The complexity of pattern matching for a random string. SIAM J. Comput. 8(3), 368–387 (1979)

The (2,1)-Total Labeling Number of Outerplanar Graphs Is at Most $\Delta + 2$

Toru Hasunuma[1], Toshimasa Ishii[2], Hirotaka Ono[3], and Yushi Uno[4]

[1] Department of Mathematical and Natural Sciences, The University of Tokushima, Tokushima 770–8502 Japan
hasunuma@ias.tokushima-u.ac.jp
[2] Department of Information and Management Science, Otaru University of Commerce, Otaru 047-8501, Japan
ishii@res.otaru-uc.ac.jp
[3] Department of Economic Engineering, Kyushu University, Fukuoka 812-8581, Japan
hirotaka@en.kyushu-u.ac.jp
[4] Department of Mathematics and Information Sciences, Graduate School of Science, Osaka Prefecture University, Sakai 599-8531, Japan
uno@mi.s.osakafu-u.ac.jp

Abstract. A (2, 1)-total labeling of a graph G is an assignment f from the vertex set $V(G)$ and the edge set $E(G)$ to the set $\{0, 1, \ldots, k\}$ of nonnegative integers such that $|f(x) - f(y)| \geq 2$ if x is a vertex and y is an edge incident to x, and $|f(x) - f(y)| \geq 1$ if x and y are a pair of adjacent vertices or a pair of adjacent edges, for all x and y in $V(G) \cup E(G)$. The (2, 1)-total labeling number $\lambda_2^T(G)$ of G is defined as the minimum k among all possible assignments. In [D. Chen and W. Wang. (2,1)-Total labelling of outerplanar graphs. Discr. Appl. Math. 155 (2007)], it was conjectured that all outerplanar graphs G satisfy $\lambda_2^T(G) \leq \Delta(G)+2$, where $\Delta(G)$ is the maximum degree of G, while they also showed that it is true for G with $\Delta(G) \geq 5$. In this paper, we solve their conjecture completely, by proving that $\lambda_2^T(G) \leq \Delta(G) + 2$ even in the case of $\Delta(G) \leq 4$.

1 Introduction

In the channel assignment problems, we need to assign different frequencies to 'close' transmitters so that they can avoid interference. The $L(p, q)$-labelings of a graph have been extensively studied as one of important graph theoretical models of this problem, where an *$L(p, q)$-labeling* of a graph G is an assignment f from the vertex set $V(G)$ to the set $\{0, 1, \ldots, k\}$ of nonnegative integers such that $|f(x) - f(y)| \geq p$ if x and y are adjacent and $|f(x) - f(y)| \geq q$ if x and y are at distance 2, for all x and y in $V(G)$. The *$L(p, q)$-labeling number* is defined as the minimum k among all possible assignments and is denoted by $\lambda_{p,q}(G)$. Notice that we can use $k+1$ different labels when $\lambda_{p,q}(G) = k$ since we can use 0 as a label for conventional reasons. We can find related results on $L(p, q)$-labelings in comprehensive surveys by Calamoneri [3] and by Yeh [14].

In [13], Whittlesey et al. studied the $L(2, 1)$-labeling number of incidence graphs, where the *incidence graph* of a graph G is the graph obtained from G by replacing each edge with a path with length two. Observe that an $L(p, 1)$-labeling of the incidence

C.S. Iliopoulos and W.F. Smyth (Eds.): IWOCA 2010, LNCS 6460, pp. 103–106, 2011.

graph of a given graph G can be regarded as an assignment f from $V(G) \cup E(G)$ to the set $\{0, 1, \ldots, \ell\}$ of nonnegative integers such that $|f(x) - f(y)| \geq p$ if x is a vertex and y is an edge incident to x, and $|f(x) - f(y)| \geq 1$ if x and y are a pair of adjacent vertices or a pair of adjacent edges, for all x and y in $V(G) \cup E(G)$. Such a labeling of G is called a $(p, 1)$-*total labeling* of G, introduced by Havet and Yu [7,8]. The $(p, 1)$-*total labeling number* is defined as the minimum value ℓ among all possible $(p, 1)$-total labelings of G, and denoted by $\lambda_p^T(G)$.

We can see that a $(1, 1)$-total labeling of G is equivalent to a total coloring of G. Generalizing the Total Coloring Conjecture [2,11], Havet and Yu [7,8] conjectured that

$$\lambda_p^T(G) \leq \Delta(G) + 2p - 1 \tag{1}$$

holds for any graph G, where $\Delta(G)$ denotes the maximum degree of a vertex in G. They also showed that (i) $\lambda_p^T(G) \leq \min\{2\Delta(G) + p - 1, \chi(G) + \chi'(G) + p - 2\}$ for any graph G where $\chi(G)$ and $\chi'(G)$ denote the chromatic number and the chromatic index of G, respectively, (ii) $\lambda_2^T(G) \leq 2\Delta(G)$ if $p = 2$ and $\Delta(G) \geq 2$, (iii) $\lambda_2^T(G) \leq 2\Delta(G) - 1$ if $p = 2$ and $\Delta(G)$ is an odd ≥ 5, and (iv) $\lambda_p^T(G) \leq n + 2p - 2$ if G is the complete graph where $n = |V(G)|$; the conjecture (1) is true if (a) $p \geq \Delta(G)$, (b) $p = 2$ and $\Delta(G) \leq 3$, or (c) G is the complete graph. By (i), it follows that $\lambda_p^T(G) \leq \Delta(G) + p$ for any bipartite graph [1,4,7,8] (by $\chi(G) \leq 2$ and König's theorem). Also, Bazzaro et al. [1] investigated that $\lambda_p^T(G) \leq \Delta(G) + p + s$ for any s-degenerated graph (by $\chi(G) \leq s + 1$ and $\chi'(G) \leq \Delta(G) + 1$), where an *s-degenerated graph* G is a graph which can be reduced to a trivial graph by successive removal of vertices with degree at most s, and that $\lambda_p^T(G) \leq \Delta(G) + p + 3$ for any planar graph (by the Four-Color Theorem). They also showed sufficient conditions about $\Delta(G)$ and girth for which the conjecture (1) holds. In [10], Montassier and Raspaud proved that $\lambda_p^T(G) \leq \Delta(G) + 2p - 2$ when $p \geq 2$ and $\Delta(G)$ and the maximum average degree of G satisfy some conditions. In [9], Lih et al. showed that $\lambda_p^T(G) \leq \lfloor 3\Delta(G)/2 \rfloor + 4p - 3$ for any graph G. They also investigated $\lambda_p^T(K_{m,n})$ of the complete bipartite graphs $K_{m,n}$.

2 The Main Theorem

Let G be an outerplanar graph. In [1], Bazzaro et al. pointed out that $\lambda_p^T(G) \leq \Delta(G) + p + 1$ for any outerplanar graph other than an odd cycle. This improves the bound (1) for $p > 2$ (it is the same for $p = 2$). Moreover, Chen and Wang [4] conjectured that in the case of $p = 2$, $\lambda_2^T(G) \leq \Delta(G) + 2$. They also proved that this conjecture is true if (i) $\Delta(G) \geq 5$, (ii) $\Delta(G) = 3$ and G is 2-connected, or (iii) $\Delta(G) = 4$ and every two faces consisting of three vertices have no vertex in common. In the case of $\Delta(G) = 2$ (i.e., G is a path or a cycle), we can easily see that $\lambda_2^T(G) \leq 4$, since the incidence graph of a path (resp., a cycle) is also a path (resp., a cycle), and the $L(2, 1)$-labeling number $\lambda_{2,1}(C_n)$ for a cycle C_n with n vertices is at most 4 [5]. The cases of $\Delta(G) \in \{0, 1\}$ are trivial. However, the general cases of $\Delta(G) \in \{3, 4\}$ were left open. In this paper, we solve Chen and Wang's conjecture completely, by showing that it holds for the remaining cases of $\Delta(G) \in \{3, 4\}$; namely, we show the following theorem.

Theorem 1. *If* $G = (V, E)$ *is an outerplanar graph with* $\Delta(G) \leq 4$, *then* $\lambda_2^T(G) \leq \Delta(G) + 2$.

On the other hand, the bound $\Delta(G) + 2$ is tight, since there exist infinitely many outerplanar graphs G such that $\lambda_2^T(G) = \Delta(G) + 2$ if $\Delta(G) \geq 2$, as investigated in [4].

Here we define some notations. A graph G is an ordered set of its vertex set $V(G)$ and edge set $E(G)$ and is denoted by $G = (V(G), E(G))$. Throughout the paper, assume that $G = (V, E)$ is undirected and simple. An edge with end vertices u and v is denoted by (u, v). For a vertex set $V' \subseteq V(G)$, let $G - V'$ be the subgraph of G induced by $V(G) - V'$. The *degree* of a vertex v in G is denoted by $d_G(v)$. A vertex v is called a *cut vertex* (*of* G) if $G - v$ is disconnected. A subgraph G' of G is called a *2-connected component* if no two vertices u and u' in G' are disconnected by any cut vertex v of G with $v \notin \{u, u'\}$ and G' is maximal to this property. In particular, if $G' = G$ and $|V(G)| \geq 3$, then G is called *2-connected*.

A graph is called *planar* if it can be drawn in the plane without generating a crossing by two edges, and a *plane graph* is a particular drawing of a planar graph. A plane graph divides the plane into regions, and a *face* of a plane graph is the maximal region of the plane that contains no vertex in that region. A face whose vertex set is $\{u_1, u_2, \ldots, u_k\}$ with $(u_i, u_{i+1}) \in E(G)$, $i = 1, 2, \ldots, k$ (where $u_{k+1} = u_1$) is denoted by $[u_1 u_2 \cdots u_k]$. We call a face consisting of k vertices a *k-face*. A planar graph G is called *outerplanar* if it can be drawn in the plane so that all vertices lie on the boundary of some face called the *outer face*. Such a drawing is referred to as an *outerplane graph*. An edge not belonging to the boundary of the outer face is called an *inner edge*.

3 Proof Sketch of Theorem 1

We give a proof sketch of Theorem 1 (see [6] for its complete proof).

Let G be an outerplane graph. We prove this by induction on $k = |V(G)| + |E(G)|$. The theorem clearly holds if $k = 1$. Consider the case of $k \geq 2$ and assume that for each $k' < k$, this theorem holds. If $\Delta(G) \leq 2$, $\lambda_2^T(G) \leq 4$ holds as mentioned in Section 1. We also assume that G is connected, since otherwise we can treat each component separately. Thus, $1 \leq \delta(G) \leq 2$, where $\delta(G)$ denotes the minimum degree of G.

Consider the case where $\delta(G) = 1$. Let u_1 be a vertex with $d_G(u_1) = 1$. By the induction hypothesis, $G - u_1$ has a (2,1)-total labeling $f : V(G - u_1) \cup E(G - u_1) \to \mathcal{L}_{\Delta(G)+2}$, where $\mathcal{L}_k = \{0, 1, \ldots, k\}$. Let u_2 be the neighbor of u_1 in G and $e_1, \ldots, e_{\Delta(G)-1}$ be edges incident to u_2 in $G - u_1$ where $e_i = e_j$ may occur. Hence we can extend f to the edge (u_1, u_2) and the vertex u_1 as follows: assign a label $a \in \mathcal{L}_{\Delta(G)+2} - \{f(u_2)-1, f(u_2), f(u_2)+1, f(e_1), \ldots, f(e_{\Delta(G)-1})\}$ to (u_1, u_2), and then a label in $\mathcal{L}_{\Delta(G)+2} - \{f(u_2), a-1, a, a+1\}$ to u_1.

Consider the case of $\delta(G) = 2$. There are the following two possible cases: (Case-I) $\Delta(G) = 3$ and (Case-II) $\Delta(G) = 4$.

(Case-I) In [4], Chen and Wang showed that if G is 2-connected, then $\lambda_2^T(G) \leq 5$; they gave an algorithm (CW algorithm) for finding a feasible labeling $g : V(G) \cup E(G) \to \mathcal{L}_5$ of a 2-connected graph G. Consider the case where G has a cut vertex. We can observe that G has a 2-connected component G_1 which has exactly one cut vertex v_c of G. By $\delta(G) = 2$, $|V(G_1)| \geq 3$ and $d_{G_1}(v_c) \geq 2$. By $\Delta(G) = 3$, it follows that $d_{G_1}(v_c) = 2$ holds and G_1 and $G - V(G_1)$ are connected by one edge. By the induction hypothesis, $H = G - (V(G_1) - \{v_c\})$ has a (2,1)-total labeling $f : V(H) \cup E(H) \to \mathcal{L}_5$. Then, we modify CW algorithm so that it can provide a more flexible labeling, and extend f to a (2,1)-total labeling $f' : V(G) \cup E(G) \to \mathcal{L}_5$ of G by using the modified algorithm.

(Case-II) It is known that any outerplane graph G' with $\delta(G') = 2$ contains one of the following configurations (C1)–(C3): (C1) two adjacent vertices u_1 and u_2 with $d_{G'}(u_1) = d_{G'}(u_2) = 2$, (C2) a 3-face $[u_1u_2u_3]$ with $d_{G'}(u_1) = 2$ and $d_{G'}(u_2) = 3$, and (C3) two 3-faces $[u_1u_2u_3]$ and $[u_3u_4u_5]$ such that $d_G(u_3) = 4$ and $d_{G'}(u_2) = d_{G'}(u_4) = 2$ [12]. In [4], Chen and Wang proved that if G contains (C1) or (C2) as its subgraph H, then a (2,1)-total labeling of the proper subgraph $G - H'$ of G for some subgraph H' of H can be extended to a (2,1)-total labeling of G. On the other hand, in the case where G contains neither (C1) nor (C2), such a property does not hold; for example, for some graph G which contains (C3) as its subgraph H, there exists a labeling of $G - H$ which we cannot extend to any feasible labeling in G. To overcome this, we derive the following new structural property, and show that a (2,1)-total labeling of the proper subgraph $G - H$ of G, where H is a subgraph of G corresponding to (C4), can be extended to a (2,1)-total labeling of G.

Lemma 1. *If G is an outerplane graph with $\Delta(G) = 4$ and $\delta(G) = 2$ which contains neither* (C1) *nor* (C2), *then G has the following configuration* (C4): (C4) *a family $\{[u_1u_2u_3], [u_3u_4u_5], \ldots, [u_{2t-1}u_{2t}u_{2t+1}]\}$ of 3-faces such that (u_1, u_{2t+1}) is an inner edge.*

References

1. Bazzaro, F., Montassier, M., Raspaud, A.: $(d,1)$-total labelling of planar graphs with large girth and high maximum degree. Discr. Math. 307, 2141–2151 (2007)
2. Behzad, M.: Graphs and their chromatic numbers. Ph.D. Thesis, Michigan State University (1965)
3. Calamoneri, T.: The $L(h,k)$-labelling problem: A survey and annotated bibliography. The Computer Journal 49, 585–608 (2006) (Updated version) (ver. October 19, 2009), `http://www.dsi.uniroma1.it/~calamo/PDF-FILES/survey.pdf`
4. Chen, D., Wang, W.: (2,1)-Total labelling of outerplanar graphs. Discr. Appl. Math. 155, 2585–2593 (2007)
5. Griggs, J.R., Yeh, R.K.: Labelling graphs with a condition at distance 2. SIAM J. Disc. Math. 5, 586–595 (1992)
6. Hasunuma, T., Ishii, T., Ono, H., Uno, Y.: A tight upper bound on the (2,1)-total labeling number of outerplanar graphs. CoRR abs/0911.4590 (2009)
7. Havet, F., Yu, M.-L.: $(d,1)$-Total labelling of graphs. Technical Report 4650, INRIA (2002)
8. Havet, F., Yu, M.-L.: $(p,1)$-Total labelling of graphs. Discr. Math. 308, 496–513 (2008)
9. Lih, K.-W., Liu, D.D.-F., Wang, W.: On $(d,1)$-total numbers of graphs. Discr. Math. 309, 3767–3773 (2009)
10. Montassier, M., Raspaud, A.: $(d,1)$-total labeling of graphs with a given maximum average degree. J. Graph Theory 51, 93–109 (2006)
11. Vizing, V.G.: Some unsolved problems in graph theory. Russian Mathematical Surveys 23, 125–141 (1968)
12. Wang, W., Zhang, K.: Δ-Matchings and edge-face chromatic numbers. Acta. Math. Appl. Sinica 22, 236–242 (1999)
13. Whittlesey, M.A., Georges, J.P., Mauro, D.W.: On the λ- number of Q_n and related graphs. SIAM J. Discr. Math. 8, 499–506 (1995)
14. Yeh, R.K.: A survey on labeling graphs with a condition at distance two. Discr. Math. 306, 1217–1231 (2006)

Upper and Lower I/O Bounds for Pebbling r-Pyramids

Desh Ranjan[1], John Savage[2], and Mohammad Zubair[3]

[1] Old Dominion University, Norfolk, Virginia 23529
[2] Brown University, Providence, Rhode Island 02912
[3] Old Dominion University, Norfolk, Virginia 23529

Abstract. Modern computers have several levels of memory hierarchy. To obtain good performance on these processors it is necessary to design algorithms that minimize I/O traffic to slower memories in the hierarchy. In this paper, we present I/O efficient algorithms to pebble r-pyramids and derive lower bounds on the number of I/O steps to do so. The r-pyramid graph models financial applications which are of practical interest and where minimizing memory traffic can have a significant impact on cost saving.

Keywords: Memory hierarchy, I/O, Lower bounds.

1 Introduction

Modern computers have several levels of memory hierarchy. To obtain good performance on these computers it is necessary to design algorithms that minimize I/O traffic to slower memories in the hierarchy [1,2]. The cache blocking technique is used to reduce memory traffic to slower memories in the hierarchy [1]. Cache blocking partitions a given computation such that the data required for a partition fits in a processor cache. For computations, where data is reused many times, this technique reduces memory traffic to slower memories in the hierarchy [1]. The memory traffic reduction that can be obtained using this technique depends on the application, memory hierarchy architecture, and the effectiveness of the blocking algorithm.

In this paper, we present I/O efficient algorithms to compute the values at vertices ("pebble" the vertices) of a computation graph that is an r-pyramid and derive lower bounds on its memory traffic complexity. A formal definition of memory traffic complexity is given later in the paper. For simplicity, in this paper we will only consider two levels of memory hierarchy. The results for two-levels can be extended to multiple-levels of memory hierarchy using the multiple-level memory hierarchy model outlined in [3]. (See also [4, Chapter 11].) This model is an extension of the red-blue model introduced by [5], a game played on directed acyclic graphs with red and blue pebbles.

The paper is motivated by a very practical financial application - that of computing option prices. An option contract is a financial instrument that gives the

C.S. Iliopoulos and W.F. Smyth (Eds.): IWOCA 2010, LNCS 6460, pp. 107–120, 2011.

right to its holder to buy or sell a financial asset at a specified price referred to as strike price, on or before the expiration date. The current asset price, volatility of the asset, strike price, expiration time, and prevailing risk-free interest rate determine the value of an option. Binomial and trinomial option valuation are two popular approaches that value an option using a discrete time model [6,7]. The binomial option pricing computation is modelled by the directed acyclic pyramid graph $G_{biop}^{(n)}$ with height n and $n+1$ leaves shown in Figure 1. Here the expiration time is divided into n intervals (defined by $n+1$ endpoints), the root is at the present time, and the leaves are at expiration times. We use $G_{biop}^{(n)}$ to determine the price of an option at the root vertex iteratively, starting from the leaf vertices.

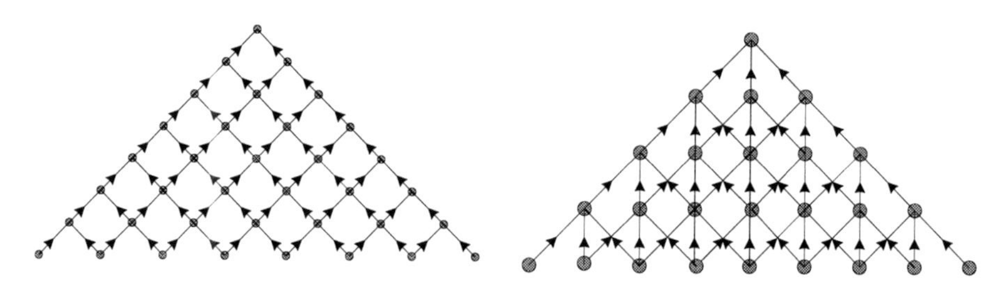

Fig. 1. A 2-pyramid representing binomial computation, and a 3-pyramid representing trinomial computation

The trinomial model improves over the binomial model in terms of accuracy and reliability [6]. The trinomial option pricing computation is represented using the directed acyclic graph with in-degree 3 denoted $G_{triop}^{(n)}$ of height n on $2n+1$ leaves shown in Figure 1. As in the binomial model, we divide the time to expiration into n intervals and let the root be at the present time and the leaves be at expiration times. As in the binomial model, we use $G_{triop}^{(n)}$ to determine the price of an option at the root vertex iteratively, starting from the leaf vertices. The trinomial model assumes that the price of an asset can go three ways: up, down, and remain unchanged. This is in contrast to the binomial model where the price can only go two ways: up and down.

In [8] the authors derived lower bounds for memory traffic at different levels of memory hierarchy for $G_{biop}^{(n)}$ and $G_{triop}^{(n)}$. The technique used in the paper is based on the concept of a S-span of the DAG [3]. The S-span intuitively represents the maximum amount of computation that can be done after loading data in a cache at some level without accessing higher levels (those further away from the CPU) memories.

In this paper we first define a general family of graphs called r-pyramids. $G_{biop}^{(n)}$ and $G_{triop}^{(n)}$ are sub families of this family. We then provide an algorithm to pebble $r-pyramids$ using S pebbles that requires roughly half the I/O needed by previously described algorithms [8]. We also provide a lower bound that is twice the previous best known lower bound for the same problem [8]. With these improvements, one can prove that the pebbling scheme presented here does no more than twice the I/O required by an optimal pebbling scheme.

Strengthening the lower bound by a constant factor, besides being of theoretical interest, is important for practical reasons. Deriving these strong bounds gives insight into deriving better algorithms, which are a factor of four to eight times better than the existing algorithms. These factors may look small but are significant in terms of cost saving for applications with real time constraints, such as financial application.

The rest of the paper is organized as follows. The required definitions and the memory hierarchy model that helps in developing memory complexity is discussed in Section 2. In Section 3 we present an efficient algorithm, in terms of memory I/O, for pebbling r-pyramid. Section 4 gives improved lower bounds for the r-pyramid graph. Finally, in Section 5 we present some open problems.

2 Background

2.1 Computation Graphs, Structures and Memory Traffic Complexity

We define here formally what we mean by a computation graph, a computation structure and memory traffic complexity of a computation structure. A *computation graph* is a directed acyclic graph $G = (V, E)$. The vertices of G with in-degree zero are called the *input vertices* and the vertices with out-degree zero are called the *output vertices*. The idea here is that we wish to compute the values at the output vertices given the values at the input vertices. The value at a vertex can be computed if and only if the value at all its predecessor vertices have been computed and are available. We say that the computation on G is complete if the values at all its output vertices have been computed. A *computation structure* is a parametric description of computation graphs. Formally, a computation structure is a function $\tilde{G} : N^k \to \{G \mid G \text{ is a computation graph}\}$.

Given a computation graph G, the computation on G can be carried out in many different ways. A *computation scheme* for a computation structure $\tilde{G}$ is an algorithm that completely specifies how to carry out the computation for each $\tilde{G}(t)$ where $t \in N^k$. An input in a 2-level memory hierarchy refers to a read from secondary memory, and an output refers to a write to the secondary memory. We now define the memory traffic complexity for a single processor with 2-levels of memory hierarchy with $\hat{\sigma} = \langle \sigma_0, \sigma_1 \rangle$ where σ_0 is the primary memory size, and σ_1 is the secondary memory size. Let $\tilde{G} : N^k \to \{G \mid G \text{ is a computation graph}\}$ be a computation structure. Let $T_1(\hat{\sigma}, \tilde{G})(t)$ be the minimum I/O required by any computation scheme for $\tilde{G}$ on input $\tilde{G}(t)$ where $t \in N^k$. The function $T_1(\hat{\sigma}, \tilde{G}) : N^k \to N$ as defined above is called the *memory traffic complexity* of $\tilde{G}$. A computation scheme that matches the memory traffic complexity for $\tilde{G}$ is called a *memory traffic optimal scheme* for $\tilde{G}$.

2.2 The Reb-Blue Pebble Game

The red-blue pebble game models data movement between adjacent levels of a two-level memory hierarchy. In the red-blue game, red pebbles identify values held in a fast primary memory whereas blue pebbles identify values held in a secondary memory. Recall, that an input refers to a read from the secondary memory, and an output refers to a write to a secondary memory. Since the red-blue pebble game is used to study the number of I/O operations necessary for a problem, the number of red pebbles is assumed limited and the number of blue pebbles is assumed unlimited. Before the game starts, blue pebbles reside on all input vertices. The goal is to place a blue pebble on each output vertex, that is, to compute the values associated with these vertices and place them in long-term storage. These assumptions capture the idea that data resides initially in the most remote memory unit and the results must be deposited there.

Red-Blue Pebble Game Rules

- (Initialization) A blue pebble can be placed on an input vertex at any time.
- (Computation Step) A red pebble can be placed on (or moved to) a vertex if all its immediate predecessors carry red pebbles.
- (Pebble Deletion) A pebble can be deleted from any vertex at any time.
- (Goal) A blue pebble must reside on each output vertex at the end of the game.
- (Input from Blue Level) A red pebble can be placed on any vertex carrying a blue pebble.
- (Output to Blue Level) A blue pebble can be placed on any vertex carrying a red pebble.

A pebbling strategy $\mathcal{P}$ is the execution of the rules of the pebble game on the vertices of a computation graph. We assign a step to each placement of a pebble, ignoring steps on which pebbles are removed. The I/O time of $\mathcal{P}$ on the graph G is the number of input and output (I/O) steps used by $\mathcal{P}$.

3 An Efficient Algorithm for Pebbling $P_r(n)$

3.1 An r-Pyramid

A directed graph $G = (V, E)$ is called a layered graph with n levels if V can be written as a disjoint union of n non-empty sets $V_1, V_2, \ldots, V_n$ such that $\forall\ e = (u, v) \in E, \exists\ i$ such that $u \in V_i$ and $v \in V_{i+1}$.

Definition 1. *An r-pyramid of height n, $P_r(n)$, is a graph $(V_r(n), E_r(n))$ with the following properties (see Figure 2):*

1. *$P_r(n) = (V_r(n), E_r(n))$ is a layered graph with height n. Here $V_r(n) = V_1 \cup V_2 \ldots \cup V_{n+1}$, V_i is the set of vertices on level i, and $E_r(n)$ are the edges.*
2. *V_i has $n_r(i) = (r-1)*(i-1)+1$ vertices labeled $v(i, 1), v(i, 2), \ldots, v(i, n_r(i))$*
3. *Vertex $v(i, j)$ has r incoming edges from vertices $v(i + 1, j), v(i + 1, j + 1), \ldots, v(i + 1, j + r - 1)$.*
4. *There are no other edges in $P_r(n)$.*

With this definition it is easy to see that $G_{biop}^{(n)}$ is a 2-pyramid of height n (or $P_2(n)$) and $G_{triop}^{(n)}$ is a 3-pyramid of height n (or $P_3(n)$). Also, note that an $P_r(n)$ has $|V_r(n)| = (n+1)((r-1)n+2)/2$ vertices. We note the nice recursive structure of r-pyramid. For any vertex v in the r-pyramid, the *subgraph rooted at* v is a smaller r-pyramid itself.

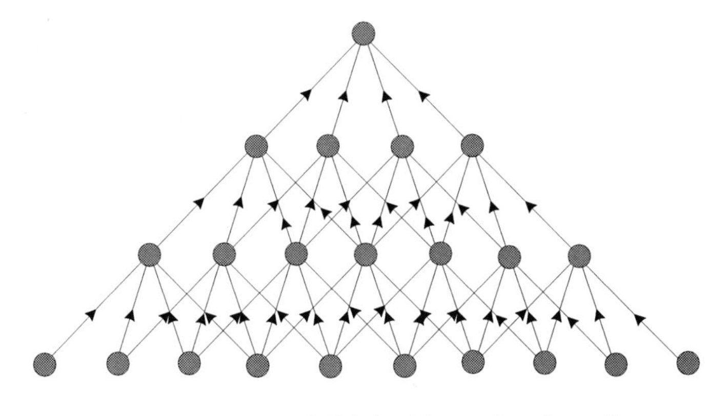

Fig. 2. r-pyramid $P_r(n)$ with $r = 4$ and $n = 3$

3.2 Algorithm

Let, $S = (r-1)m + 1$. We give an algorithm that we can pebble an r-pyramid $P_r(n) = (V_r(n), E_r(n))$ of height n with S red pebbles using no more than $2|V_r(n)|(r-1)/(S-1)$ I/O operations. Note that if $n \leq m$ then $P_r(n)$ can be pebbled without any intermediate I/O. Recall that we are assuming an unlimited supply of blue pebbles.

Let $D_{i,j}^k$ denote the "diagonal" shown in Figure 3 consisting of the k vertices $\{(i,j), (i+1, j+r-1), \ldots, (i+(k-1), j+(k-1)(r-1))\}$ that originate at the vertex (i,j).

The algorithm starts with the pebbling of the r-pyramid, $P_{n-m,1}^m$, of height m rooted at vertex $(n-m, 1)$. This pyramid shares inputs with the inputs to the full pyramid. This is done in a such way that it leaves S red pebbles on S vertices of $P_{n-m,1}^m$ one of which is $(n-m,1)$. The other vertices are those in $P_{n-m,1}^m$ that are required to compute $D_{n-m,2}^m$. More precisely, this is a collection of $(r-1)$ vertices at each of the lower $m-1$ levels. These vertices are

$$\begin{array}{r}
(n-m+1, 2), (n-m+1, 3), \ldots, (n-m+1, (r-1)) + 1) \\
(n-m+2, r), (n-m+2, 2), \ldots, (n-m+2, 2(r-1)+1) \\
\vdots \\
(n, (m-1)(r-1)), (n, (m-1)(r-1)+1), \ldots, (n, m(r-1))
\end{array}$$

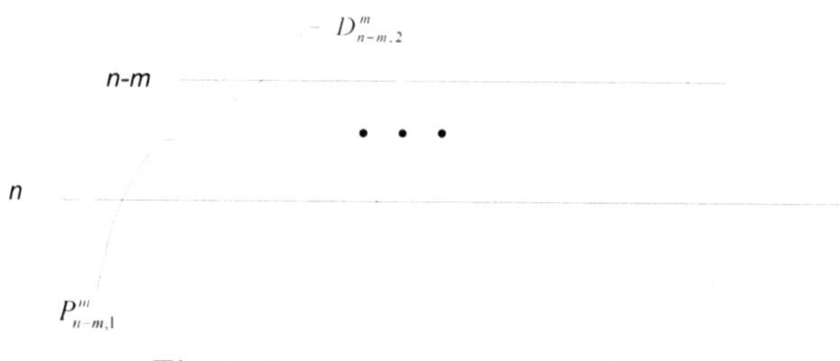

Fig. 3. Processing of r-pyramid at level k

Procedure *PebbleSubPyramid* given in Algorithm 1 explains how this is done.

Procedure *PebbleSubPyramid*(n)
if $n \leq m$ **then**
  Pebble the whole subpyramid using $(r-1)*n+1$ red pebbles ;
else
  $t \leftarrow S$;
  for $i = 1$ *to* t **do**
    Place a red pebble at vetex (n, i);
  end
  for $j = 0$ *to* $m-1$ **do**
    $t \leftarrow t - (r-1)$;
    for $k = 1$ *to* t **do**
      move pebble at $(n-j, k)$ to $(n-j-1, k)$;
    end
  end
end

Algorithm 1. An algorithm for pebbling an r-subpyramid of height m at position $(n-m, 1)$ using $S = (r-1)m+1$ red pebbles leaving the red pebbles at the vertices needed for future pebbling.

Next we repeatedly pebble the diagonals $D^m_{n-m,i}$ starting with $i = 2$ and progressing incrementally all the way to $D^m_{n-m,(n-m-1)(r-1)+1}$. Observe that pebbling of $D^m_{n-m,2}$ requires the red pebbles on exactly $S-1$ vertices from the pyramid $P^m_{n-m,1}$ that was pebbled earlier (using *PebbleSubpyramid*) and a red pebble on vertex $(n, s+1)$. We place a blue pebble at $(n-m, 1)$ move the red pebble at $(n-m, 1)$ left by *PebbleSubpyramid* to $(n, s+1)$.

It is now easy to verify that all the red pebbles are in exactly the needed locations to compute $D^m_{n-m,2}$. Moreover, we can maintain this property while pebbling consecutive diagonals. That is, after pebbling $D^m_{n-m,2}$ we leave S red pebbles on the vertices that are required for the processing of the next diagonal $D^m_{n-m,3}$ etc. Observe that in general, processing of diagonal $D^m_{n-m,j}$

requires input from vertices on diagonals $D^m_{n-m,j-1}, D^m_{n-m,j-2}, \ldots, D^m_{n-m,j-r+1}$. This way we continue processing diagonals until we process the last diagonal $D_{n-m,(r-1)(n-m-1)+1}$.

Also, observe that while processing these diagonals we only need to preserve vertices at $(n-m, 1), (n-m, 2), \ldots, (n-m, (r-1)(n-m-1)+1)$ for future processing. The basic idea is that with S pebbles we can pebble all vertices at the lower m levels blue pebbling only the vertices at level m. We then repeat this process for the r-pyramid of height $n-m$. The complete algorithm to process $P_r(n)$ is presented in Algorithm 2 and illustrated in Figure 3.

```
Procedure PebblePyramid(n)
PebbleSubPyramid(n);
for j = 2 to (r − 1)(n − m − 1) + 1 do
    place a blue pebble on (n − m, j − 1);
    move the red pebble on (n − m, j − 1) to (n, j + s − 1);
    for i = 0 to m − 1 do
        move the red pebble on (n − i − 1, (j + s − 1 − (r − 1)i) to
        (n − i, (j + s − 1 − (r − 1)i) ;
    end
end
PebblePyramid(n − m) ;
```

Algorithm 2. An algorithm to pebble an r-pyramid of height n

Notice that this pebbling scheme does not "re-pebble" any vertex, that is, a vertex is never pebbled red using the *computation step* rule (Section 2.2) more than once. Additionally, it uses a blue pebbled vertex exactly once for input. It is obviously an optimal scheme in terms of computation. It is natural to ask the question if this is also an I/O optimal scheme. We conjecture that this is indeed the case. To prove this, we need to establish lower bounds on pebbling schemes for pebbling an r-pyramid. We do so in the following section.

4 Lower Bounds for Pebbling an r-Pyramid

Lower bounds for pebbling an r-pyramid can be obtained by using S-span arguments [8].

4.1 A Lower Bound Based on the S-Span of a Graph

In this approach, to derive lower bounds for a given DAG, we first compute its S-span. This is a measure that intuitively represents the maximum amount of computation that can be done after loading data in a cache at some level without accessing higher level memories (those further away from the CPU).

Definition 2. *The S-span of a DAG G, $\rho(S, G)$, is the maximum number of vertices of G that can be pebbled starting with any initial placement of S red pebbles and using no blue pebbles.*

The S-span is a measure of how many vertices can be pebbled without doing any I/O. S pebbles are placed on the most fortuitous vertices of a graph and the maximum number of vertices that can be pebbled without doing I/O is the value of the S-span. Clearly, the measure is most useful for graphs that have a fairly regular structure. It has provided good lower bounds on communication traffic for matrix multiplication, the Fast Fourier Transform, the binomial graph and other graphs. This definition applies even if G is not a connected graph.

The following theorem [9] relates the S-span of the graph to its memory traffic complexity.

Theorem 1. *Let $\tilde{G}$ be a computation structure. Consider a pebbling of the DAG $\tilde{G}(t)$ in an 2-level memory hierarchy game. Let $\rho(S, \tilde{G}(t))$ be the S-span of $\tilde{G}(t)$ and $|V_t^*|$ be the number of vertices in $\tilde{G}(t)$ other than the inputs. Assume that $\rho(S, \tilde{G}(t))/S$ is a non-decreasing function of S.*

Then the memory traffic complexity for $\tilde{G}$, $T_1(\hat{\sigma}, \tilde{G})$, satisfies the following lower bound.

$$T_1(\hat{\sigma}, \tilde{G})(t) \geq \frac{\sigma_0 |V_t^*|}{\rho(2\sigma_0, \tilde{G}(t))}$$

Lemma 1. *For a given path π from a leaf vertex x_1 to the output vertex x_{p+1} in $P_r(p)$ consisting of vertices $x_1, x_2, x_3, \ldots, x_{p+1}$ there is a total of $(r-1)p$ distinct paths from leaf vertices to the x_i's for $i > 1$.*

Proof. We use induction on p to prove this result. The lemma holds for the base case $P_r(1)$. Assume the lemma is true for $P_r(p)$ rooted at x_{p+1}. Then for a given path π of length p in $P_r(p)$ consisting of vertices $x_1, x_2, \ldots, x_{p+1}$, we have $(r-1)p$ distinct paths from leaf vertices of $P_r(p)$ to x_i's for $i > 1$. Observe that the leaf vertices corresponding to these paths along with x_1 are the total number of leaf vertices in $P_r(p)$, which is $(r-1)p+1$. We now consider $P_r(p+1)$ rooted at x_{p+2}. $P_r(p+1)$ has $(r-1)(p+1)+1$ leaf vertices. Observe that $P_r(p)$ is a sub-graph of $P_r(p+1)$ and the vertex x_1 has r edges coming from the leaf vertices of $P_r(p+1)$, see Figure 4. Let one of these leaf vertices in $P_r(p+1)$ be x_0. Additionally, for every other leaf vertex of $P_r(p)$, we can identify a distinct leaf vertex in $P_r(p+1)$, which it is connected to, see Figure 4. This demonstrate that for a path in $P_r(p+1)$ consisting of vertices $x_0, x_1, x_2, \ldots, x_{p+2}$, there are a total of $(r-1)(p+1)$ distinct paths from leaf vertices to vertices on this path. This completes the proof.

Lemma 2. *$P_r(p)$ requires a minimum of $S = (r-1)p+1$ pebbles to place a pebble on the root vertex. The graph can be pebbled completely with S pebbles without repebbling any vertices.*

Proof. The proof uses an argument analogous to the last path argument used in [10]. We say that a path π from a leaf vertex $x_1 \in P_r(p)$ to the root vertex x_{p+1} is **blocked** (at some time instance t) if at least one vertex on the path has a pebble (at time t). Consider the time instance when the root vertex, x_{p+1}, of $P_r(p)$ was pebbled. At this time instance, all paths from all the leaf vertices of

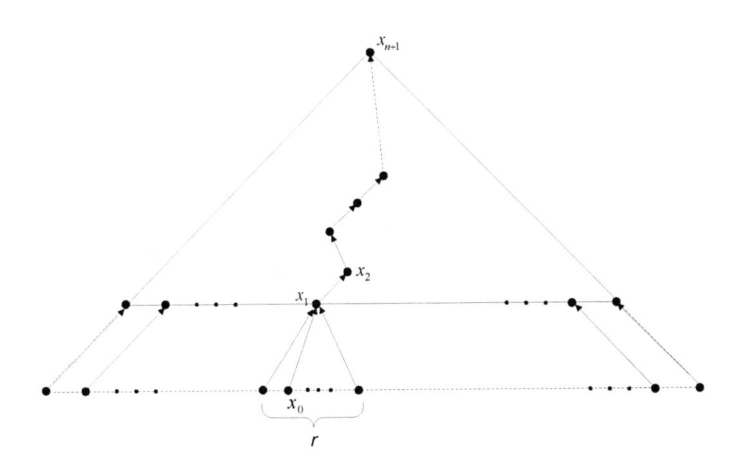

Fig. 4. A r-pyramid with a path π

$P_r(p)$ to x_{p+1} are blocked. Now let us consider the first time instance t' when all paths from all the leaf vertices to x_{p+1} were blocked. Then at time instance $t'-1$, there must have been an open path from one of the bottom level vertices to x_{p+1}. This implies that all vertices on this path did not have pebbles on them and that at time t' by placing a pebble at the leaf vertex all paths were blocked. Observe that when a pebble is placed on the leaf vertex to block π, the graph already had pebbles on each of the $(r-1)p$ distinct paths leading to each of the p other vertices on π (Lemma 1). Thus, when the input to π is pebbled, the graph has at least $(r-1)p+1$ pebbles on it.

To show that the graph can be pebbled completely without re-pebbling any vertices, place all $(r-1)p+1$ pebbles on the inputs. Then one can slide the leftmost pebble up one level and then proceed to slide (r-1)(p-1) more pebbles up one level to pebble the leaves of the subgraph $P_r(p-1)$ with $(r-1)(p-1)+1$ leaves. The rest follows by induction. Procedure PebbleSubpyramid provided earlier formally describes this process.

Theorem 2. *The S-span of an r-pyramid is*

$$\tfrac{1}{2}(\lfloor S/(r-1)\rfloor+1)(2S-(r-1)\lfloor S/(r-1)\rfloor).$$

We present a complete proof for $r=2$ in Appendix A. The proof for general r is analogous and is omitted because of space limitations.
Applying Theorem 1 we have the following result.

Theorem 3. *Let $\sigma_0 = S$. The memory traffic complexity of P_r on a 2-level memory hierarchy system, $T_1(\hat{\sigma}, P_r)$, satisfies*

$$T_1(\hat{\sigma}, P_r)(n) \geq \frac{Sn((r-1)(n-1)+2)}{(\lfloor 2S/(r-1)\rfloor+1)(4S-(r-1)\lfloor 2S/(r-1)\rfloor)}.$$

4.2 The Blue Pebble Strategy for Proving Pebbling Lower Bounds

The above results leave a gap of a factor of 4 between the bounds achieved by the scheme provided and the lower bounds obtained. We improve this by strengthening the lower bound. To do so, we develop a new technique for proving lower bounds on I/O in pebbling schemes. We start by making a simple observation.

Observation: Let $\mathcal{P}$ be any I/O optimal scheme for pebbling $P_r(n)$. Suppose $\mathcal{P}$ uses $f(n)$ blue pebbles. Then $In(P_r(n)) + 2f(n)$ is a lower bound on the number of I/Os for any I/O-optimal scheme for pebbling $P_r(n)$ where $In(P_r(n))$ is the number of input vertices in $V_r(n)$.

This is straightforward because in any I/O optimal pebbling scheme if a blue pebble is placed on a vertex then later a red pebble must be placed on this vertex using the rule that a red pebble can be placed on a blue pebble. If this is not the case, placing the blue pebble is redundant and we have a better pebbling scheme that simply does not place the blue pebble.

The Blue Pebble strategy for proving lower bounds in pebbling a graph G simply establishes a lower bound on the number of blue pebbles placed in any I/O optimal pebbling scheme. The overall lower bound for G is obtained through lower bounds for smaller subgraphs (not necessarily disjoint) and combining these lower bounds.

Theorem 4. *Let $G = (V, E)$ be any layered graph. Suppose that we have q subgraphs $H_1, H_2, \ldots H_q$ of $V^* = G - In(G)$ with the following properties:*

(i) In any complete pebbling of G, each H_i must have at least b blue pebbled vertices

(ii) No $v \in V$ belongs to more than l different H_i's.

Then, in any complete pebbling of G at least $q * b/l$ vertices of $\bigcup_i H_i$ are pebbled with blue pebbles.

Proof. Let S_i denote the set of blue pebbled vertices in the subgraph H_i. Then the set of blue vertices in $\bigcup_i H_i$ is $S = \bigcup_i S_i$. By assumption $\forall\, i\, |S_i| \geq b$. Consider the set $A = \{(v, i)\,|\, v \in S_i, 1 \leq i \leq q\}$. Then $|A| \geq q \times b$. For a vertex u denote by A_u the subset of A of pairs where the first component is u, that is, $A_u = \{(u, i) | 1 \leq i \leq q\}$. Then if $u \neq u'$, A_u and $A_{u'}$ are trivially disjoint. Also, by assumption (ii) for each u, $|A_u| \leq l$. Noticing that $A = \bigcup_{u \in S} A_u$, it then follows that $|S| \geq |A|/l = qb/l$.

To make use of the Blue Pebble strategy, one needs to identify an appropriate family of sub-graphs and establish a lower bound on number of blue pebbles on each of these sub-graphs. Naturally, the choice of the subgraphs can be driven by the ability to establish a lower bound on number of blue pebbled vertices in these subgraphs.

4.3 A Lower Bound for Pebbling $P_r(n)$

To obtain a lower bound on number of blue pebbles in a complete pebbling of $P_r(n)$ we first establish the following lemmas:

Lemma 3. *Consider any complete pebbling of $P_r(n)$ with S red pebbles and let $P_r(h)$ be any r-pyramid of height h in $P_r(n)$. Then $P_r(h)$ has at least $(r-1)h+1-S$ blue pebbled vertices.*

Proof. Using the argument of Lemma 2, we have at least $(r-1)h+1$ pebbles on the graph $P_r(n)$ when the last path from the leaf vertex of $P_r(n)$ to the root is blocked. Since there are only S red pebbles in total, it follows that at least $(r-1)h+1-S$ of the vertices in $P_r(h)$ have blue pebbles at this time.

We now use the Blue Pebble strategy to establish a lower bound for pebbling $P_r(n)$ with S red pebbles and unlimited number of blue pebbles. We choose for our subgraphs H_i, all r-pyramids of height h in $P_r(n)$. There is one such pyramid with root at each of the vertices at level $n-h$ and above. Hence there are $q=(r-1)(n-h+1)(n-h)/2+(n-h+1)$ such r-pyramids. From Lemma 1 it follows that in any complete pebbling of $P_r(n)$, each such r-pyramid of height h must have at least $b=(r-1)h+1-S$ blue pebbles. Notice that no vertex in $P_r(n)$ is shared by more than $l=|H_i|=(r-1)(h+1)h/2+(h+1)$ different subgraphs. It then follows that the number of blue pebbles in complete pebbling of $P_r(n)$ is at least $qb/l=q*[(r-1)h-(S-1)]/[(r-1)(h+1)h/2+(h+1)]$. Choosing, $(r-1)h=2(S-1)$, this gives us $qb/l=q*(S-1)/S*(h+1)=q*(S-1)/[S*(2(S-1)/(r-1)+1)]$. This is roughly $q(r-1)/2S$ which is roughly $|V|(r-1)/2S$ if $n>>S$. Hence the total number of I/O operations is bounded below by roughly $|V|(r-1)/S$.

5 Remarks and Conclusion

We presented an I/O efficient and computation optimal scheme for pebbling an r-pyramid. We also presented a new technique for proving lower bounds in pebbling and used it to prove improved lower bounds on I/O for pebbling r-pyramids. There is a gap of a factor of (roughly) 2 between the upper bound and lower bound presented for pebbling the r-pyramids. It will be nice to close this gap one way or the other. The pebbling scheme presented here does not use any "re-pebbling". We conjecture, that this is an I/O and (obviously simultaneously) computation optimal scheme for $P_r(n)$. For pebbling schemes that do not use re-pebbling, a better lower bound on the number of I/Os needed to pebble a 2-pyramid of height n has been established by the authors (manuscript available upon request). However, the technique used there does not immediately help to improve lower bounds on the number of I/Os for pebbling r-pyramids for $r>2$ even when re-pebbling is not allowed. Finally, it is worth noting that for general layered graphs re-pebbling can reduce the number of I/Os. However, our conjecture also implies that this is not the case for r-pyramids.

References

1. Hennessy, J.L., Patterson, D.A.: Computer Architecture: A Quantitative Approach. Morgan Kaufmann, San Francisco (2007)
2. Kumar, V., Sameh, A., Grama, A., Karypis, G.: Architecture, algorithms and applications for future generation supercomputers. In: Proceedings of the 6th Symposium on the Frontiers of Massively Parallel Computation, FRONTIERS 1996, Washington, DC, USA, p. 346. IEEE Computer Society, Los Alamitos (1996)
3. Savage, J.E.: Extending the Hong-Kung model to memory hierarchies. In: Li, M., Du, D.-Z. (eds.) COCOON 1995. LNCS, vol. 959, pp. 270–281. Springer, Heidelberg (1995)
4. Savage, J.E.: Models of Computation: Exploring the Power of Computing. Addison Wesley, Reading (1998)
5. Hong, J.W., Kung, H.T.: I/O complexity: The red-blue pebble game. In: Proc. 13th Ann. ACM Symp. on Theory of Computing, pp. 326–333 (1981)
6. Kwok, Y.: Mathematical Models of Financial Derivatives. Springer, Singapore (1998)
7. Cox, J.C., Ross, S.A., Rubinstein, M.: Option pricing: A simplified approach. Journal of Financial Economics 7(3), 229–263 (1979)
8. Savage, J.E., Zubair, M.: Cache-optimal algorithms for option pricing. ACM Trans. Math. Softw. 37(1), 1–30 (2010)
9. Savage, J.E., Zubair, M.: Evaluating multicore algorithms on the unified memory model. Scientific Programming 17(4), 295–308 (2009)
10. Cook, S.A.: An observation on storage-time trade off. J. Comp. Systems Sci. 9, 308–316 (1974)

Appendix

The S-Span of a 2-Pyramid

The basic intuition is that the S-span is obtained by placing the S pebbles on contiguous nodes at the same level and then pebbling all possible nodes from this placement. The number of such nodes is $S + (S-1) + \dots 1$ or $S(S+1)/2$ (this includes S nodes where the pebbles were originally placed). We provide a proof that this intuition is indeed correct.

Lemma 4. *The S-span of a 2-pyramid is at least $S(S+1)/2$.*

Proof. We can place all S pebbles contiguously on a single level and pebble $S(S+1)/2$ nodes by moving the pebbles up by one level from left to right (discarding the rightmost pebble) and then repeating this at the next level. This scheme pebbles $S(S+1)/2$ nodes. Hence the S-span for the 2-pyramid is at least $S(S+1)/2$.

We will next establish that for any placement X of S pebbles on the 2-pyramid, no more than a total of $S(S+1)/2$ nodes can be pebbled. We do so by first defining a function, $pp(X)$, that upper bounds the maximum number of nodes that can be possibly pebbled from a placement X of S pebbles. We then show that $pp(X) \leq S(S+1)/2$ for any placement X with S pebbles. The basic idea behind the definition is that if the maximum number of nodes that can be possibly pebbled at a level i is k_i then the maximum number of new nodes that can be possibly pebbled level $i+1$ is at most $(k_i - 1)$ (except when k_i is zero in which case this is zero).

Definition 3. *Let X be any placement of S pebbles. Let l denote the lowest level on which there is at least one pebble in X. Let h be the highest such level. Let $m = h-l+1$ and let $s_i \geq 0$ denote the number of pebbles on the ith level starting from level l (i.e. s_1 is the number of pebbles on level l, s_2 on level $l+1$ ... s_m on level $l+m-1 = h$). Then, $pp(X) = \Sigma_{i=1}^{i=m} max_i + (max_m - 1)(max_m)/2$ where max_i is defined recursively as below:*

$$max_1 = s_1$$
$$max_i = s_i + max_{i-1} - 1 \text{ if } 1 < i \leq m \text{ and } max_{i-1} > 0$$
$$max_i = s_i \text{ if } 1 < i \leq m \text{ and } max_{i-1} = 0$$

It is easy to observe that $pp(X)$ is an upper bound on the number of nodes that can be possibly pebbled by *any* pebbling scheme starting with placement X.

Lemma 5. *For any placement X of S pebbles $pp(X) \leq S(S+1)/2$.*

Proof. We first consider the case where all the S pebbles are placed on a single level (say level 1). Then no more than $S-1$ nodes can be possibly pebbled at level 2, consequently, no more than $S-2$ nodes at level 3 and in general no more than $S-i$ at level $i+1$. It then follows that $pp(X) \leq S+(S-1)+\ldots 1 = S(S+1)/2$.

If the maximum value of $pp(X)$ is obtained by placing all the pebbles on one level we have nothing further to prove. Else, let us consider a placement X of pebbles that maximizes $pp(X)$. By our assumption, X places at least one pebble on more than one levels. Among all placements that maximize $pp(X)$, let us consider the one that has the minimum number of levels between the lowest and the highest levels with non-zero pebbles.

As in Definition 3 let m denote the number of levels between the lowest and highest levels (both included) with non-zero pebbles. Let us label the levels as $1, 2 \ldots m$ with 1 being the lowest level. Let s_i denote the number of pebbles on the ith level in the placement S. Note that, while $s_1, s_m > 0$, some of the other s_is can be zero and also that $\Sigma_i s_i = s$. Let us now consider the value $pp(X) = \Sigma_{i=1}^{i=m} max_i + (max_m - 1)max_m/2$. We contend that by choice of X, none of the max_is is zero and hence for all $1 < i \leq m$ $max_i = s_i + max_{i-1} - 1$. If this is not true then consider the lowest j where $max_j = 0$. Then $s_j = 0$ and $s_{j-1} = 1$. Consider a new placement X' of S pebbles which is identical to X except that all the pebbles below level j are moved one level up. Then $pp(X') = pp(X)$ but X' has fewer levels contradicting our assumption. We now show that $pp(X) \leq S(S+1)/2$.

Expanding out the definition of max_i we get,

$$
\begin{aligned}
&max_1 = s_1 \\
&max_2 = s_2 + s_1 - 1 \\
&max_3 = s_3 + s_2 + s_1 - 2 \\
&\vdots \\
&max_m = s_m + s_2 + \ldots s_1 - (m-1) = (s - (m-1))
\end{aligned}
$$

Hence

$$
\begin{aligned}
&pp(X) = \Sigma_{i=1}^{i=m} max_i + (max_m - 1)max_m/2 \\
&= ms_1 + (m-1)s_2 + \ldots 1.s_m - m(m-1)/2 + (S-m)(S-(m-1))/2 \\
&= m(\Sigma_{i=1}^{i=m} s_i) - \Sigma_{i=2}^{i=m}(i-1)s_i - m(m-1)/2 + (S-m)(S-(m-1))/2 \\
&\leq mS - m(m-1)/2 + (S-m)(S-(m-1))/2 \\
&= ms - m(m-1)/2 + (S^2 - (2m-1)S + m(m-1))/2 = S(S+1)/2.
\end{aligned}
$$

Theorem 5. *The S-span of a 2-pyramid is $S(S+1)/2$.*

Single Parameter FPT-Algorithms for Non-trivial Games

Vladimir Estivill-Castro and Mahdi Parsa

School of ICT, Griffith University, Queensland, Australia 4111
{v.estivill-castro,m.parsa}@griffith.edu.au,

Abstract. We know that k-UNIFORM NASH is $W[2]$-Complete when we consider imitation symmetric win-lose games (with k as the parameter) even when we have two players. However, this paper provides positive results regarding Nash equilibria. We show that consideration of sparse games or limitations of the support result in fixed-parameter algorithms with respect to one parameter only for the k-UNIFORM NASH problem. That is, we show that a sample uniform Nash equilibrium in r-sparse imitation symmetric win-lose games is not as hard because it can be found in FPT time (i.e polynomial in the size of the game, but maybe exponential in r). Moreover, we show that, although **NP**-Complete, the problem of BEST NASH EQUILIBRIUM is also fix-parameter tractable.

Keywords: Algorithmic Game Theory, Computational Complexity.

1 Introduction

Game theory analyzes interactions between self-interested agents, with the recent interest in artificial intelligence, multi-agents systems, and automatic decision making it has received much study. The first complexity results for computing Nash equilibria used classic notions of complexity theory [11]. Later, several researchers have introduced different types of equilibria and games. These **NP**-hardness results have been extended to the other games and solution concepts [1,2,5,6].

We study the fixed-parameter tractability of **NP**-Hard problems for the computation of Nash equilibria. One of the most recently studied class of games are *win-lose games* [2,5]. In these games, all payoff values are 0 or 1. We study the parameterized complexity of finding uniform Nash equilibria in imitation win-lose games because:

- The computation complexity of a Nash equilibrium in win-lose games is as hard as for general bi-matrix games [1].
- There is a corresponding one-to-one relation between Nash equilibria of two-player games and Nash strategies for the row player in an imitation game [5].
- "A uniform mixed strategy is the simplest way of mixing pure strategies", but deciding the existence of uniform Nash equilibria in win-lose games is **NP**-Complete [2] and it is $W[2]$-Hard [8] in bi-matrix games.

C.S. Iliopoulos and W.F. Smyth (Eds.): IWOCA 2010, LNCS 6460, pp. 121–124, 2011.

- Deciding whether an imitation symmetric win-lose game has a uniform Nash equilibrium with support of size k is $W[2]$-Complete [7].
- It has been observed [12] that the lower bounds of Chen et al [3] and the $W[2]$-hardness results imply that unless FPT=$W[1]$, there is no $n^{o(k)}$ time algorithm for computing a Nash equilibrium with support size at most k in a bi-matrix game.

In contrast, if the support is known, the equilibrium can be found in polynomial time. Thus, there is much interest in studying the complexity of the support size or if the support is included in a set of strategies. We show that restrictions of the support result in fixed-parameter algorithms.

2 FPT Results for Win-Lose Games

A win-lose game $\mathcal{G}=(A,B)$ is called *r-sparse* if there are at most r nonzero entries in each row and each column of the matrices A and B. The first natural step to parameterize the computation of a sample Nash equilibrium is to consider the r as a parameter in r-sparse games. But, Chen, Deng and Teng [4] showed that it is unlikely to find an ϵ-approximate equilibrium for a 10-sparse game in time polynomial both in ϵ and n (the size of game). Therefore, it is unlikely to find an FPT-time algorithm that just considers r as the parameter. We have proved the parameterized tractability of Nash equilibria in a subclass of r-sparse games.

Definition 1. *Let* ***ISWLG*** *be the class of all* Imitation Symmetric Win-Lose Games *$(I_{n\times n},M_{n\times n})$ where the matrix M is a symmetric matrix, and has diagonal equal to zero.*

If a game (I,M) is in **ISWLG**, then this game represents a simple undirected graph $G=(V,E)$ where the matrix M corresponds with G's adjacency matrix. We have shown that any maximal clique in the graph representation of game $\mathcal{G}=(I,M)$ corresponds to a uniform Nash equilibrium, but the reverse is not true.

Lemma 2. *Let G be the graph for game $\mathcal{G}=(I,M)$ in* ***ISWLG*** *and $G_{\boldsymbol{x}}$ be a maximal clique of size k of G. Then the mixed strategy $(\boldsymbol{x},\boldsymbol{x})$ is a uniform Nash equilibrium of $\mathcal{G}$ where* $x_i = \begin{cases} 1/k, & \text{if } i \text{ is vertex of } G_{\boldsymbol{x}}; \\ 0, & \text{otherwise.} \end{cases}$

We study the effect of sparsity. Existence of uniform Nash equilibria is not an issue since every graph has a maximal clique. By Lemma 2, every game in **ISWLG** has a uniform Nash equilibrium.

Theorem 3. *Finding a uniform Nash equilibrium for a r-sparse game in* ***ISWLG*** *is polynomial in the size of the game but exponential in r.*

We used the link between graph theory and Nash equilibria to show our FPT results. Now, we can provide many results regarding families of graphs where finding a maximal clique is in FPT. For example we show that a sample uniform Nash equilibrium can be found in FPT-time where the treewidth of graph is considered as the parameter.

Theorem 4. *Let $\mathcal{G}$=(I,M) be an imitation symmetric win-lose game with graph G. If G has bounded tree-width ω, then a uniform Nash equilibrium of the $\mathcal{G}$ can be found in $O(2^\omega \cdot \omega \cdot |I|)$ time.*

3 FPT Results When Searching Nash Equilibrium on a Set

We can obtain a result for general bi-matrix games. The following problem has been shown to be **NP**-Complete.

NASH EQUILIBRIUM IN A SUBSET
 Instance : A game $\mathcal{G}$=(A,B).
 Parameter: A subset of strategies $E_i \subseteq \{1, \ldots, n\}$ for each player i.
 Question : Does there exists a Nash equilibrium of $\mathcal{G}$ in which all strategies not included in E_i are played with probability zero?

There is a *Feasibility Program* [13], which is a linear program, and, if the support of a Nash equilibrium is known, then the computation of corresponding Nash equilibrium can be done in polynomial time. We use this Feasibility Program to proof following theorem.

Theorem 5. NASH EQUILIBRIUM IN A SUBSET *is in FPT* .

4 FPT Results for Congestion Games

In congestion games (also *routing games*), players choose several links, one link to route their traffic [10, and references].

Definition 6. *A* routing game $\mathcal{G}$ *consists of:*

- *a set of m parallel links from a source node s to a terminal node t and a capacity c^j for each link $j \in \{1, 2, \ldots, m\}$,*
- *a set $N = \{1, 2, \ldots, n\}$, of n users,*
- *traffic weights, w_1,w_2,..., w_n, where the i-th user has traffic $w_i > 0$.*

A pure strategy for a user i is a link j in $\{1, 2, \ldots, m\}$. Analogously, a pure strategy profile is an n-tuples $(l_1, l_2, \ldots l_n)$, when user i chooses link l_i in $\{1, 2, \ldots, m\}$. The *cost* for a user i, when users choose a pure strategy profile $P = (l_1, l_2, \ldots, l_n)$ is $C_i(P) = \sum_{k:l_k=l_i} w_k/c^{l_i}$. Every routing game admits at least one pure Nash equilibrium [9]. However, the individual (non-cooperative) optimization of utility does not lead to a social optimal outcome. Therefore, the *price of stability* is a measure inefficiency of equilibria. It differentiates between games that all Nash equilibria are inefficient or some of them are inefficient. Formally, the price of stability of a game is the ratio between the best objective function value of a Nash equilibrium of the game and the optimal outcome. We consider the makespan as the social objective function. The makespan of a strategy profile $P = (l_1, l_2, \ldots, l_n)$ is defined as: $C_{max}(P) = \max_{i\in\{1,2,\ldots,n\}} C_i(P)$.

BEST NASH EQUILIBRIUM
Instance : A routing game $\mathcal{G}$ with identical links.
Parameter : k a positive integer.
Question : Is there a pure Nash equilibrium P with $C_{max}(P) \leq k$?

BEST NASH EQUILIBRIUM on identical links is a **NP**-Hard problem [9], but we showed it is fix parameter tractable. with a parameterized reduction to INTEGER LINEAR PROGRAMMING. The INTEGER LINEAR PROGRAMMING problem (with a bounded number of variables) is FPT.

Theorem 7. BEST NASH EQUILIBRIUM *is in FPT.*

References

1. Abbott, T., Kane, D., Valiant, P.: On the complexity of two-player win-lose games. In: 46th Annual IEEE Symp. on Foundations of Computer Science, FOCS 2005, pp. 113–122. IEEE Computer Society Press, Los Alamitos (2005)
2. Bonifaci, V., Iorio, U.D., Laura, L.: The complexity of uniform Nash equilibria and related regular subgraph problems. Theoretical Computer Science 401(1-3), 144–152 (2008)
3. Chen, J., Chor, B., Fellows, M., Huang, X., Juedes, D., Kanj, I., Xia, X.: Tight lower bounds for certain parameterized NP-hard problems. Information and Computation 201(2), 216–231 (2005)
4. Chen, X., Deng, X., Teng, S.-H.: Sparse games are hard. In: Spirakis, P.G., Mavronicolas, M., Kontogiannis, S.C. (eds.) WINE 2006. LNCS, vol. 4286, pp. 262–273. Springer, Heidelberg (2006)
5. Codenotti, B., Stefankovic, D.: On the computational complexity of Nash equilibria for (0,1) bimatrix games. Information Processing Letters 94, 145–150 (2005)
6. Conitzer, V., Sandholm, T.: Complexity results about Nash equilibria. In: Gottlob, G., Walsh, T. (eds.) 18th Int. Joint Conf. on Artificial Intelligence, IJCAI 2003, pp. 765–771. Morgan Kaufmann, Acapulco (2003)
7. Estivill-Castro, V., Parsa, M.: The parameterized complexity of uniform Nash equilibria in win-lose games (submitted for publication)
8. Estivill-Castro, V., Parsa, M.: Computing Nash equilibria gets harder— new results show hardness even for parameterized complexity. In: Downey, R., Manyem, P. (eds.) The Australasian Theory Symposium (CATS 2009). CRPIT, vol. 94. Australian Computer Society, Inc., Wellington (2009)
9. Fotakis, D., Kontogiannis, S., Koutsoupias, E., Mavronicolas, M., Spirakis, P.: The structure and complexity of Nash equilibria for a selfish routing game. Theoretical Computer Science 410(36), 3305–3326 (2009)
10. Gassner, E., Hatzl, J., Krumke, S., Sperber, H., Woeginger, G.: How hard is it to find extreme Nash equilibria in network congestion games?. Theoretical Computer Science 410(47-49), 4989–4999 (2009)
11. Gilboa, I., Zemel, E.: Nash and correlated equilibria: Some complexity considerations. Games and Economic Behavior 1(1), 80–93 (1989)
12. Hermelin, D., Huang, C.C., Kratsch, S., Wahlstrom, M.: Parameterized two-player Nash equilibrium, personal communication
13. von Stengel, B.: Computing equilibria for two-person games. In: Aumann, R.J., Hart, S. (eds.) Handbook of Game Theory. ch.45, vol. 3, pp. 1723–1759. Elsevier, Amsterdam (2002)

The Complexity Status of Problems Related to Sparsest Cuts

Paul Bonsma[1], Hajo Broersma[2], Viresh Patel[2], and Artem Pyatkin[2,⋆]

[1] Humboldt Universität zu Berlin, Computer Science Department,
Unter den Linden 6, 10099 Berlin
bonsma@informatik.hu-berlin.de
[2] School of Engineering and Computing Sciences,
Durham University, Science Laboratories,
South Road, Durham DH1 3LE, U.K.
{hajo.broersma,viresh.patel,artem.pyatkin}@dur.ac.uk

Abstract. Given an undirected graph $G = (V, E)$ with a capacity function $w : E \longrightarrow \mathbb{Z}^+$ on the edges, the sparsest cut problem is to find a vertex subset $S \subset V$ minimizing $\sum_{e \in E(S,V \setminus S)} w(e)/(|S||V \setminus S|)$. This problem is NP-hard. The proof can be found in [16]. In the case of unit capacities (i. e. if $w(e) = 1$ for every $e \in E$) the problem is to minimize $|E(S, V \setminus S)|/(|S||V \setminus S|)$ over all subsets $S \subset V$. While this variant of the sparsest cut problem is often assumed to be NP-hard, this note contains the first proof of this fact. We also prove that the problem is polynomially solvable for graphs of bounded treewidth.

Keywords: NP-hardness, sparsest cut, densest cut, MSSC, bounded treewidth.

1 Introduction

Motivation

It is fair to say that the two results of this paper (Theorem 1 and Theorem 2 below) are not very surprising. In fact, the former of the two results – an NP-completeness result – has been assumed to be true by many authors during the last two decades, but no proof is known to us. One of the reasons for this assumption might be that these authors either directly or indirectly refer to a paper by Matula and Shahrokhi [16] in which NP-completeness of a more general weighted version of the problem has been established. Here we provide a solid basis for all the papers that build on the assumption that the unweighted version is also NP-complete. This assumption might seem reasonable and easy

⋆ The first author is supported by DFG grant BO 3391/1-1. The last three authors are supported by EPSRC Grant EP/F064551/1. The fourth author is also supported by RFBR (projects 08-01-00516 and 09-01-00032).

C.S. Iliopoulos and W.F. Smyth (Eds.): IWOCA 2010, LNCS 6460, pp. 125–135, 2011.

to justify, but we know of cases where the unweighted version of an NP-hard optimization problem is polynomially solvable. We therefore think it is useful for the graph theory and computational complexity community to disseminate this NP-completeness proof via this contribution.

Our second result shows that both the weighted and the unweighted version of the considered problem can be solved in polynomial time on graphs of bounded treewidth. This result, although perhaps unsurprising, is useful, and seems not to have been previously observed. We use a reasonably straightforward dynamic programming approach, but it seems the result cannot be deduced by formulating the problem in monadic second order logic.

Background

Some problems in theoretical computer science have a strange status: many people assume that they are NP-hard but there is no proof of their NP-hardness. One instance of this concerns the Minimum Sum of Squares Clustering (MSSC) problem.

Given a set $V = \{v_1, v_2, \ldots, v_N\}$ of Euclidean vectors, a positive integer $k > 1$, and a positive real K, the *MSSC problem* is to determine whether there exists a partition of V into nonempty subsets (clusters) $C_1, C_2, \ldots, C_k$ such that

$$\sum_{i=1}^{k} \sum_{v \in C_i} \|v - w_i\|^2 \leq K,$$

where $w_j = \sum_{v \in C_j} v/|C_j|$ is the center of the cluster C_j for $j = 1, 2, \ldots, k$.

This problem is assumed to be NP-complete by more than 20 authors. Some authors do not provide any references at all, some authors cite the standard reference book by Garey and Johnson [13], and some authors cite other papers in which clustering problems with other criteria are shown to be NP-hard (see [2] for further details). The first supposed proof of the NP-completeness of the MSSC problem appeared in [12]; however, it was shown in [2] that this proof contained an error. Two years later Aloise et al. [1] provided a correct NP-completeness reduction for the MSSC problem; however, the problem they reduced from – the unit-capacity densest cut problem – has not been shown to be NP-complete. Let us define this problem.

For disjoint $S, T \subset V(G)$, we write $E_G(S,T)$ for the set of all edges having one end in S and the other in T. For $G = (V, E)$ a graph and S a nonempty strict subset of V, we write $\bar{S} = V \setminus S$. Any set of edges of the form $E_G(S, \bar{S})$ with $S \neq \emptyset$, $S \neq V$ is called a *cut* of G. Given a positive weighting $w_G : E \rightarrow \mathbb{Z}^+$ of the edges, we define the weight of $E_G(S,T)$ to be

$$w_G(S,T) = \sum_{e \in E_G(S,T)} w_G(e).$$

Define the *density* of the cut $E_G(S, \bar{S})$ to be

$$d_G(S, \bar{S}) = \frac{w_G(S, \bar{S})}{|S||\bar{S}|}.$$

We will omit subscripts when the graph is clear from the context. A cut $E(S,\bar{S})$ that minimizes (maximizes) d_G over all cuts is called a *sparsest* (*densest*) cut of G.

The *sparsest* (resp. *densest*) *cut problem* is the following. Given an undirected graph $G = (V, E)$, a weighting $w : E \rightarrow \mathbb{Z}^+$ of the edges, and a positive rational D, determine whether there exists a subset $S \subset V$ such that $d(S,\bar{S}) \leq D$ (resp. $\geq D$). The same problem, but where $w(e) = 1$ for all $e \in E$, is referred to as the *unit capacity sparsest* (resp. *unit capacity densest*) *cut problem.* Note that $E_G(S,\bar{S})$ is a densest cut of G if and only if $E_{\bar{G}}(S,\bar{S})$ is a sparsest cut of the complementary graph $\bar{G}$. (The complement of an edge weighted graph or *weighted complement* is obtained by first introducing edges with zero weight between all non-adjacent vertex pairs, and subsequently changing every edge weight $w(e)$ to $M - w(e)$, where M is the maximum edge weight.) So, the problems of finding sparsest and densest cuts are equivalent. We remark that in the literature this problem is also called the *uniform* sparsest cut problem (see e.g. [15]), to distinguish it from the more general problem where, in addition, an edge weighted *demand graph* H with $V(H) = V(G)$ is given, and the objective is to minimize $w_G(S,\bar{S})/w_H(S,\bar{S})$. We will call this more general problem the *non-uniform sparsest cut problem.* (The sparsest cut problem corresponds to the case where H is the complete graph with unit weights.) The sparsest cut problem plays an important role in theoretical computer science. In particular approximation algorithms have received a lot of study; see e.g. [4,5,15], which have many other algorithmic applications [15].

In [16], Matula and Shahrokhi proved the NP-completeness of the sparsest cut problem, and in the process, proved the NP-completeness of the densest cut problem. The unit capacity versions of these problems, however, were not shown to be NP-complete in [16]; nonetheless some authors claiming the NP-completeness of the unit capacity versions [1,8,15] refer to [16]. We remedy this situation by giving a proof of the NP-completeness of the unweighted problems. While our reduction follows that of Matula and Shahrokhi [16], it is not a completely trivial adaptation of their reduction. Here is our first main theorem, which we prove in the next section.

Theorem 1. *The unit capacity densest cut problem is NP-complete.*

We remark that recently it has also been shown that the unit capacity sparsest cut problem admits no polynomial time approximation scheme (PTAS), unless NP-complete problems can be solved in randomized subexponential time [3]. Since this is a stronger assumption than P$\neq$NP, this does not imply Theorem 1 however.

In the second part of the paper we prove that the sparsest cut problem is polynomially solvable for graphs of bounded treewidth.

Theorem 2. *Let G be a graph on n vertices, for which a tree decomposition of width k is given. In time $O^*(n^3 2^k)$, a sparsest cut of G can be found.*[1]

[1] The O^* notation omits polynomial factors, provided that exponential factors in the same variable are present.

For detailed definitions related to tree decompositions, see Section 3. For graphs of treewidth at most k (fixed), a tree decomposition of width at most k can be found in linear time [6]. Thus, Theorem 2 shows that the sparsest cut problem can be solved in polynomial time for such graphs. Examples of graph classes with bounded treewidth include series-parallel graphs, outerplanar graphs and Halin graphs, which have treewidth at most 2, 2 and 3 respectively.

Note that many graph problems can be shown to be linear-time solvable on bounded-treewidth graphs by expressing the problem in monadic second order logic [11]. However, it seems that the sparsest cut problem cannot be expressed in this way. Theorem 2 uses a fairly standard dynamic programming approach [7], but some slightly unusual choices are made, which leads to the cubic complexity bound.

There have been a few other positive results on the sparsest cut problem when restricted to certain graph classes. Matula and Shahrokhi [16] show that the non-uniform sparsest cut problem can be solved in polynomial time on trees. They also show that the non-uniform sparsest cut problem can be solved in polynomial time on 3-connected planar graphs G when the demand graph H only contains edges between vertices that lie on the outer face of G. In [9] it is shown that sparsest cuts can be computed in polynomial time for unit interval graphs, and the sparsest cuts of complete bipartite graphs are characterized. In [8] it is shown that sparsest cuts of cartesian product graphs $G \times H$ can be obtained from sparsest cuts of G and H, which gives polynomial time algorithms for various graph classes.

2 NP-Completeness of the Densest Cut Problem

We shall reduce the max cut problem to the unit capacity densest cut problem. Given a graph $G = (V, E)$ and a positive integer k, the *max cut* problem is to determine whether there exists a cut $E(S, \bar{S})$ such that $|E(S, \bar{S})| \geq k$. It is known that the max cut problem is NP-complete [14].

For the sake of completeness we present the original proof from [16] of the NP-completeness of the densest cut problem. Our proof for the case of unit capacities uses a similar type of reduction, but the proof in our case requires more calculations.

Theorem 3. *[16] The densest cut problem is NP-complete.*

Proof. Given an instance of max cut, that is, a graph $G = (V, E)$ and a positive integer k, construct a weighted graph H in the following way. Take two copies $G_1 = (V_1, E_1)$ and $G_2 = (V_2, E_2)$ of the graph G and connect each vertex in V_1 with its copy in V_2 by an edge of capacity M. Set the capacities of all other edges to be 1. We show that, for M large enough, G has a cut of cardinality at least k if and only if H has a cut of density at least $(nM + 2k)/n^2$, where $n = |V|$. Indeed, let $E_H(S, \bar{S})$ be a densest cut in H. For M large enough, S must contain exactly one copy of each vertex from G; otherwise $d_H(V_1, V_2) > d_H(S, \bar{S})$. Let $T_1 = S \cap V_1$ and $\bar{T_1} = \bar{S} \cap V_1$. It is easy to see that

$$d_H(S, \bar{S}) = \frac{nM + 2|E_{G_1}(T_1, \bar{T_1})|}{n^2}.$$

Thus, for $E_H(S, \bar{S})$ a densest cut of H, we have $d(S, \bar{S}) \geq (nM + 2k)/n^2$ if and only if $|E_{G_1}(T_1, \bar{T_1})| \geq k$. □

To prove Theorem 1 we need two easy facts.

Proposition 1. *The maximum cut of the complete bipartite graph $K_{n,n}$ with parts A and B is $E(A, B)$ with cardinality n^2. All other cuts have cardinality at most $n^2 - n$.*

Proof. The first part is trivial. For the second part, let $S \subset A \cup B$ such that $S \neq A$ and $S \neq B$. Let $a = |S \cap A|$ and $b = |S \cap B|$. Then

$$|E(S, \bar{S})| = a(n - b) + b(n - a) = (a + b)n - 2ab.$$

If $|S| < |\bar{S}|$, then $(a + b) \leq n - 1$ and $|E(S, \bar{S})| \leq n^2 - n$. If $|S| = |\bar{S}|$, then $a + b = n$, but $a \neq n$ and $b \neq n$, so that $ab \geq n - 1$. Then we have $|E(S, \bar{S})| \leq n^2 - 2(n - 1) \leq n^2 - n$ for $n \geq 2$. The case $n = 1$ is trivial. □

Proposition 2. *If n, m, M are positive integers such that $n > 1$ and $M \geq 2m + 1$, then for every $t \in [1, nM - 1]$ the inequality $t/n > 2m/(nM - t)$ holds.*

Proof. Fix $n \geq 1$ and $M \geq 2m + 1$. The value of $t \in [1, nM - 1]$ that minimizes $f(t) = t(nM - t)$ must occur at one of the end points of the interval since f is a concave function. Since $f(1) = f(nM - 1) = nM - 1$, then for $t \in [1, nM - 1]$ we have

$$t(nM - t) \geq nM - 1 \geq n(2m + 1) - 1 > 2nm.$$

Dividing by $n(nM - t)$ gives the desired inequality. □

Now we are ready to prove Theorem 1.

Proof (of Theorem 1). Let the graph $G = (V, E)$ and the positive integer k be an instance of max cut. Let $V = \{v_1, \ldots, v_n\}$ and let m be the number of edges in G. Construct the graph H in the following way.

For each $v \in V$ we have two sets I_v and I'_v of vertices in H, each of size $M = 2m+1$. Thus, H has $2nM$ vertices. For each $v \in V$, connect each vertex in I_v to each vertex in I'_v. Pick one distinguished vertex from each I_v to form a set A of n vertices, and pick one distinguished vertex from each I'_v to form a set A' of n vertices. Insert edges in A and A' to create two copies of G. The resulting graph is H (see Fig. 1). Note that the degree of every vertex in H is equal to M plus possibly the degree of corresponding vertex in G.

We show that G has a cut of cardinality at least k if and only if H has a cut of density at least $(nM^2 + 2k)/(Mn)^2$.

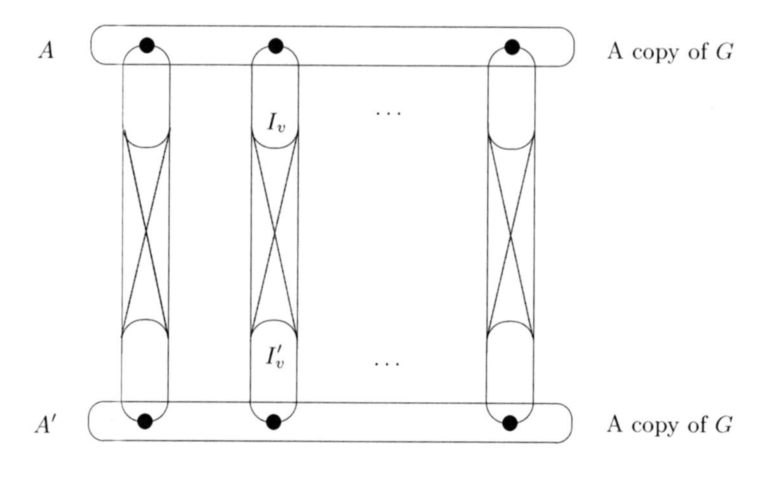

Fig. 1. Graph H

First of all we show that if $E_H(S, \bar{S})$ is a densest cut of H with $|S| \le |S'|$ then $|S| = Mn$. Indeed, assume that $|S| = Mn - t$ where $t \in [1, Mn-1]$. Writing

$$X = \bigcup_{v \in V} I_v \quad \text{and} \quad X' = \bigcup_{v \in V} I'_v,$$

note that $d_H(X, X') = nM^2/(Mn)^2 = 1/n$. Since $E_H(S, \bar{S})$ is a densest cut, the inequality $d_H(S, \bar{S}) \ge 1/n$ must hold. But since the cut has at most $2m$ edges within X and X' and at most $M|S|$ edges between X and X', we have

$$d_H(S, \bar{S}) \le \frac{M|S| + 2m}{|S||\bar{S}|} = \frac{M}{Mn+t} + \frac{2m}{M^2n^2 - t^2} =$$

$$\frac{1}{n} - \frac{t}{Mn^2 + nt} + \frac{2m}{M^2n^2 - t^2} = \frac{1}{n} - \frac{1}{Mn+t}\left(\frac{t}{n} - \frac{2m}{Mn-t}\right) < \frac{1}{n}$$

by Proposition 2, giving a contradiction. So, $t = 0$ and $|S| = Mn$.

Now assume that for some $v \in V(G)$ both $S \cap I_v$ and $S \cap I'_v$ are nonempty. Then the cut has at most $2m$ edges within X and X', at most $M^2(n-1)$ edges between $X \setminus I_v$ and $X' \setminus I'_v$ and at most $M^2 - M$ edges between I_v and I'_v (by Proposition 1). By the choice of M we have

$$d_H(S, \bar{S}) \le \frac{2m + M^2(n-1) + M^2 - M}{(Mn)^2} = \frac{M^2n - 1}{(Mn)^2} < \frac{1}{n}.$$

So, for every $v \in V(G)$, either $S \cap (I_v \cup I'_v) = I_v$ or $S \cap (I_v \cup I'_v) = I'_v$. Let $T = \{v \in V \mid I_v \subseteq S\}$. Then clearly

$$d_H(S, \bar{S}) = \frac{M^2n + 2|E_G(T, \bar{T})|}{(Mn)^2}$$

and so $d_H(S, \bar{S}) \ge (nM^2 + 2k)/(Mn)^2$ if and only if $|E_G(T, \bar{T})| \ge k$. Thus Theorem 1 is proved. $\square$

We sketch an alternative NP-completeness proof for Theorem 1, the ingredients of which might be useful to some of the readers. It is based on the observation that in the proof of Theorem 3 it is sufficient, 'for M large enough', to take $M = n^2$. This shows that in fact the densest cut problem is NP-complete for instances G with positive integer edge weights at most $M \leq n^2$, where $n = |V(G)|$, and n is even. By taking the weighted complement of the graph, this statement then also holds for the sparsest cut problem. Because the weights are polynomially bounded, this allows the following polynomial transformation to the unweighted problem. Let G be a sparsest cut instance on n vertices, with edge weights at most n^2. Construct a unit capacity sparsest cut instance H as follows: For every vertex v of G, introduce a clique K_v on n^4 vertices. For every edge uv in G with capacity $c(uv)$, introduce $c(uv)$ edges between K_u and K_v in an arbitrary way. (Since $c(uv) \leq n^2$, no multiedges are needed). This yields the (simple) unit capacity graph H on n^5 vertices (see Fig. 2).

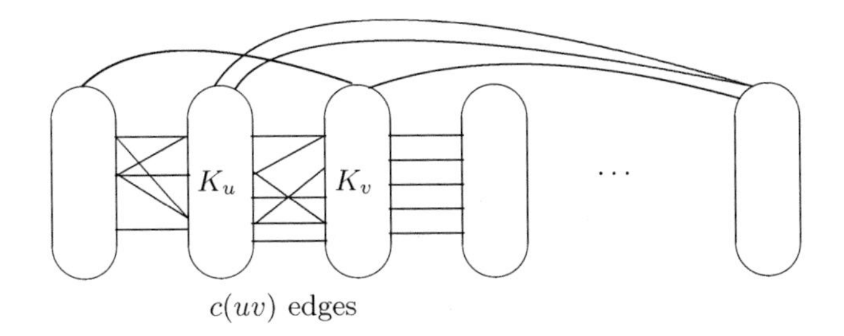

Fig. 2. Graph H for the alternative proof

Partitions of vertices in G correspond to partitions of cliques in H. In this way, cuts $E_G(S, \overline{S})$ of G correspond to cuts $E_H(S', \bar{S}')$ in H with proportional density; to be precise $d_H(S', \bar{S}') = d_G(S, \bar{S})/n^8$. We only need to verify that the sparsest cut $E_H(S', \bar{S}')$ of H does not separate any of the cliques, that is, for every $v \in V(G)$, either $V(K_v) \subseteq S'$ or $V(K_v) \subseteq \bar{S}'$. Any cut that separates a clique contains at least $n^4 - 1$ edges, and therefore has density at least

$$d_H(S', \bar{S}') \geq \frac{n^4 - 1}{(n^5/2)^2} = 4/n^6 - 4/n^{10} > 3/n^6.$$

On the other hand, by taking an arbitrary cut $E_H(S, \bar{S})$ in H that contains $n/2$ cliques in S and $n/2$ cliques in $\bar{S}$ (recall that n is even) we find a cut with lower density, at most

$$d_H(S, \bar{S}) \leq \frac{M(n/2)^2}{(n^5/2)^2} = n^4/n^{10} = 1/n^6.$$

So, a sparsest cut of H corresponds to a sparsest cut of G, and therefore G has a cut of density D if and only if H has a cut of density D/n^8.

3 The Case of Bounded Treewidth

A tuple (X, T) is a *tree decomposition* of a graph $G = (V, E)$ if T is a tree, and $X = \{X_v : v \in V(T)\}$ is a family of subsets of V such that:

- $\cup_{v \in V(T)} X_v = V$,
- for all $xy \in E$, there exists a $v \in V(T)$ with $x, y \in X_v$, and
- for every $x \in V$, the subgraph $T[\{v \in V(T) : x \in X_v\}]$ is connected.

The *width* of a tree decomposition (X, T) is $\max_{v \in V(T)} |X_v| - 1$. The *treewidth* of a graph G is the minimum width over all tree decompositions of G. To distinguish between vertices of G and vertices of T, the latter will be called *nodes*. If T is a rooted tree, (X, T) is called a *rooted* tree decomposition. A rooted tree decomposition (X, T) of G is *nice* [7] if every node of T is of one of the following types:

- *leaf nodes* u are leaves of T and have $|X_u| = 1$.
- *introduce nodes* u have one child v with $X_u = X_v \cup \{x\}$ for some $x \in V(G)$.
- *forget nodes* u have one child v with $X_u = X_v \backslash \{x\}$ for some $x \in X_v$.
- *join nodes* u have two children v and z, with $X_u = X_v = X_z$.

For fixed k, it can be decided in linear time if a given graph has treewidth at most k, and in that case, a tree decomposition of width at most k can be found [6]. In fact, it can be checked that in polynomial time this can be made into a *nice* tree decomposition (X, T) of width at most k, with $|V(T)| \in O(kn)$, where $|V(G)| = n$. For two vertices u and v of a rooted tree T, we write $v \succeq u$ if u is a predecessor of v, or $u = v$. For a rooted tree decomposition (X, T) of G and a node $v \in V(T)$, we define the subgraph $G(v) = G[\cup_{z \succeq v} X_z]$.

Let (X, T) be a rooted tree decomposition of a graph G on n vertices with edge weights w. Denote the root node by r. For $u \in V(T)$, $S' \subseteq X_u$, $i \leq n$, we define $w(u, S', i)$ to be the minimum cut weight over all cuts $E_{G(u)}(S, \bar{S})$ of $G(u)$ that satisfy $S \cap X_u = S'$ and $|S| = i$, if such a cut exists. If $i = 0$ then $w(u, S', i) = 0$, provided that $S' = \emptyset$. If $i = |V(G(u))|$ then $w(u, S', i) = 0$, provided that $S' = X_u$. In all other cases, we define $w(u, S', i) = \infty$. Since $G(r) = G$, the following proposition follows immediately from the above definition.

Proposition 3. *Let (X, T) be a tree decomposition of G with root r. The density of the sparsest cut of G equals the minimum of $\frac{w(r,S',i)}{i(n-i)}$ taken over all $S' \subseteq X_r$ and $1 \leq i \leq n - 1$.*

So to compute the density of a sparsest cut, we only need to compute the values of $w(r, S', i)$ for all $S' \subseteq X_r$ (possibly empty) and $i \in \{0, \ldots, n\}$.

Lemma 1. *Let (X, T) be a nice tree decomposition of width k, of a graph G on n vertices. In time $O^*(n^3 2^k)$, the values $w(u, S', i)$ can be calculated for all combinations of $u \in V(T)$, $S' \subseteq X_u$ and $i \in \{0, \ldots, n\}$.*

Proof. We show how $w(u, S', i)$ can be computed, when all values $w(v, S'', j)$ are known for all children v of u.

If u is a leaf node of T, then $i \in \{0,1\}$. Let $w(u,S',i) = 0$ if $|S'| = i$ and $w(u,S',i) = \infty$ otherwise. Suppose u is an introduce node with a child v, and let $X_u \backslash X_v = \{x\}$. For all $S' \subseteq X_u$:

$$\begin{array}{ll} w(u,S',i) = w(v,S' \setminus \{x\}, i-1) + w_G(\{x\}, X_u \backslash S') & \text{if } x \in S', \\ w(u,S',i) = w(v,S',i) + w_G(\{x\}, S') & \text{if } x \notin S'. \end{array}$$

Suppose u is a forget node with child v, and let $X_v \backslash X_u = \{x\}$. Then

$$w(u,S',i) = \min \{w(v,S',i), w(v,S' \cup \{x\}, i)\}.$$

Finally, suppose u is a join node with children v and z. By the third property in the definition of tree decomposition, we know that $V(G(v)) \cap V(G(z)) = X_u$, so for the cut $E_{G(u)}(S,\bar{S})$ of $G(u)$ that determines $w(u,S',i)$ the set S contains j vertices of $G(v)$ and $i-j+|S'|$ vertices of $G(z)$, for some $j \in \{|S'|,\ldots,i\}$. Therefore,

$$w(u,S',i) = \min_{j:|S'| \leq j \leq i} w(v,S',j) + w(z,S',i+|S'|-j) - w_G(S', X_u \backslash S').$$

So all values $w(u,S',i)$ can be computed using the above expressions, if the nodes of T are treated in the proper order. Now we consider the time complexity. As a first step, we build an adjacency matrix for G in time $O(n^2)$, which also contains the weights of the edges. This allows us to determine the existence and weight of a possible edge between two vertices in constant time. For every node, at most $n2^{k+1}$ values need to be computed. In the case of leaf, introduce or forget nodes, computing a value using the above expressions takes time $k^{O(1)}$. (Note that for introduce nodes, the weight of some cut in $G[X_u]$ needs to be computed. To bound the time this takes by a function of k, we have to use the adjacency matrix.) In the case of join nodes, the computation requires time $nk^{O(1)}$. So for every node, the complexity is bounded by $n^2 O^*(2^k)$. Since we can ensure that $|V(T)| \in O(kn)$, computing all values for all nodes of T then requires time $O^*(n^3 2^k)$. □

Lemma 1 and Proposition 3 together prove Theorem 2. Note that we can not only compute the density of a sparsest cut, but also construct one with the same time complexity. Furthermore, by inspecting the $w(r,S',i)$ values, other important cut problems can be solved as well: a cut $E_G(S,\bar{S})$ is an *α-balanced cut* if $\min\{|S|,|\bar{S}|\} \geq \alpha|V(G)|$, for $0 \leq \alpha \leq \frac{1}{2}$. In particular, if $\alpha = \frac{1}{2}$, it is a *bisection*.

Theorem 4. *Let G be a graph on n vertices, for which a tree decomposition of width k is given. For any $\alpha \leq \frac{1}{2}$, in time $O^*(n^3 2^k)$, a minimum α-balanced cut of G can be found.*

We remark that in the above proof, we first built an adjacency matrix in time $O(n^2)$. One might wonder how the many *linear time* dynamic programming algorithms over tree decompositions manage to avoid this step. Indeed, although it is rarely mentioned explicitly, all such algorithms that are known to us require (usually for introduce nodes) that checking whether two vertices are adjacent can be done in time bounded by some function of k (or preferably constant time), which cannot be guaranteed with adjacency lists or similar small encodings of the graph. Note that this small encoding assumption also implies that for every tree node u, we only know the vertex lists X_u, not the induced subgraphs $G[X_u]$. This problem can be solved however by generating and storing an adjacency matrix using a *lazy array* data structure [17], which takes time $O(|E(G)|) \subseteq O(kn)$, but $O(n^2)$ space. In short, the adjacency matrix is only initialized on the non-zero entries, and in addition a pointer is stored here to the corresponding edge in the adjacency lists or edge list of G. This enables checking in constant time whether an entry of the matrix is correct. This improvement may also be applied for the above algorithm.

4 Conclusions

We gave an NP-completeness proof for the unit capacity densest (sparsest) cut problem. We also showed that the sparsest cut problem can be solved in polynomial time for graphs of bounded treewidth. One may ask how far this can be generalized to the non-uniform sparsest cut problem. The algorithm from Section 3 can easily be generalized to give a pseudopolynomial time algorithm in the case where demands are determined by vertex weights $x(v)$ in the following way: the weight of edge uv in the demand graph equals $x(u)x(v)$. It seems however impossible to generalize this approach further, which leads to the following question:

Is the non-uniform sparsest cut problem NP-hard for graphs of bounded treewidth? This question has very recently been answered affirmatively in [10], where the authors use a reduction from the max cut problem to show that the non-uniform sparsest cut problem is NP-hard on graphs with pathwidth 2. In the same preprint, the authors claim a constant-factor approximation algorithm for the non-uniform sparsest cut problem for graphs of bounded treewidth, using linear programming relaxation techniques.

We complete this contribution with two open complexity problems related to graphs of bounded treewidth:

(1) the case where the input graph G has bounded treewidth and unit weights, and the demand graph H has unit weights, and
(2) the case where both G and H have bounded treewidth (but possibly both are weighted).

References

1. Aloise, D., Deshpande, A., Hansen, P., Popat, P.: NP-hardness of Euclidean sum-of-squares clustering. Journal of Machine Learning 75, 245–248 (2009)
2. Aloise, D., Hansen, P.: On the complexity of minimum sum-of-squares clustering. Cahiers du GERAD, G–2007–50 (2007), `http://www.gerad.ca`

3. Ambühl, C., Mastrolilli, M., Svensson, O.: Inapproximability Results for Sparsest Cut, Optimal Linear Arrangement, and Precedence Constrained Scheduling. In: FOCS, pp. 329–337 (2007)
4. Arora, S., Hazan, E., Kale, S.: $O(\sqrt{\log n})$ Approximation to SPARSEST CUT in $\tilde{O}(n^2)$ Time. SIAM J. Comput. 39(5), 1748–1771 (2010)
5. Arora, S., Rao, S., Vazirani, U.V.: Expander flows, geometric embeddings and graph partitioning. In: STOC, pp. 222–231 (2004)
6. Bodlaender, H.L.: A linear-time algorithm for finding tree-decompositions of small treewidth. SIAM J. Comput. 25, 1305–1317 (1996)
7. Bodlaender, H.L.: Treewidth: algorithmic techniques and results. In: Privara, I., Ružička, P. (eds.) MFCS 1997. LNCS, vol. 1295, pp. 19–36. Springer, Heidelberg (1997)
8. Bonsma, P.: Sparsest cuts and concurrent flows in product graphs. Discrete Applied Mathematics 136(2-3), 173–182 (2004)
9. Bonsma, P.: Linear time algorithms for finding sparsest cuts in various graph classes. In: CS 2006, Prague. Electronic Notes in Discrete Mathematics, vol. 28, pp. 265–272 (2007)
10. Chlamtac, E., Krauthgamer, R., Raghavendra, P.: Approximating sparsest cuts in graphs of bounded treewidth. arXiv:1006.3970v2 [cs.Ds] (June 23, 2010)
11. Courcelle, B.: Graph rewriting: an algebraic and logic approach. In: Handbook of Theoretical Computer Science, vol. B, pp. 193–242. Elsevier, Amsterdam (1990)
12. Drineas, P., Frieze, A., Kannan, R., Vempala, S., Vinay, V.: Clustering large graphs via the singular value decomposition. Journal of Machine Learning 56, 9–33 (2004)
13. Garey, M.R., Johnson, D.S.: Computers and Intractability, A Guide to the Theory of NP-Completeness. W.H. Freeman and Company, New York (1979)
14. Garey, M.R., Johnson, D.S., Stockmeyer, L.: Some Simplified NP-Complete Graph Problems. Theoretical Computer Science 1, 237–267 (1976)
15. Leighton, F.T., Rao, S.: Multicommodity max-flow min-cut theorems and their use in designing approximation algorithms. J. ACM 46(6), 787–832 (1999)
16. Matula, D.W., Shahrokhi, F.: Sparsest cuts and bottlenecks in graphs. Discrete Applied Mathematics 27, 113–123 (1990)
17. Moret, B., Shapiro, H.: Algorithms from P to NP. Benjamin/Cummings (1990)

On Approximation Complexity of Metric Dimension Problem*

Mathias Hauptmann, Richard Schmied**, and Claus Viehmann***

Dept. of Computer Science, University of Bonn

Abstract. We study the approximation complexity of the *Metric Dimension problem* in bounded degree, dense as well as in general graphs. For the general case, we prove that the Metric Dimension problem is not approximable within $(1-\epsilon)\ln n$ for any $\epsilon > 0$, unless $NP \subseteq DTIME(n^{\log\log n})$, and we give an approximation algorithm which matches the lower bound. Even for bounded degree instances it is APX-hard to determine (compute) the exact value of the metric dimension which we prove by constructing an approximation preserving reduction from the bounded degree Vertex Cover problem.

The special case, in which the underlying graph is superdense turns out to be APX-complete. In particular, we present a greedy constant factor approximation algorithm for these kind of instances and construct a approximation preserving reduction from the bounded degree Dominating Set problem. We also provide first explicit approximation lower bounds for the Metric Dimension problem restricted to dense and bounded degree graphs.

Keywords: Metric Dimension, Bounded Degree Instances, Dense Instances, Approximation Algorithms, Approximation Lower Bounds.

1 Introduction

In a connected graph $G = (V, E)$, a vertex $v \in V$ *resolves* or *distinguishes* a pair $u, w \in V$ if $d(v, u) \neq d(v, w)$, where $d(\cdot, \cdot)$ denotes the length of a shortest path between two vertices in G. A *resolving set* of G is a subset $V' \subseteq V$ such that for each pair $u, w \in V$ there exists some $v \in V'$ that distinguishes u and w. The minimum cardinality of a resolving set is called the *metric dimension* of G, denoted by $dim(G)$. The *Metric Dimension* problem asks to find a resolving set of minimum cardinality. We call here a graph $G = (V, E)$ k-superdense if the degree of every vertex is at least $|V| - k$ where k is a constant. Throughout the paper, we will use the notation $n := |V|$.

2 Related Work

The notion of resolving sets were introduced independently by Harary and Melter [12] and Slater [18]. Applications of resolving sets arise in various areas including

* The full version under `http://theory.cs.uni-bonn.de/~schmied`

** Work supported by Hausdorff Doctoral Fellowship.

*** Work partially supported by Hausdorff Center for Mathematics, Bonn.

C.S. Iliopoulos and W.F. Smyth (Eds.): IWOCA 2010, LNCS 6460, pp. 136–139, 2011.

coin weighing problems [20], drug discovery [6], robot navigation [16], network discovery and verification [1], connected joins in graphs [17], and strategies for the Mastermind game [8]. The Metric Dimension problem has been widely investigated from the graph theoretical point of view [19,6,9,3,13,21,5,4]. So far only a few papers discuss the computational complexity issues of this problem. The NP-hardness of the Metric Dimension problem was first mentioned in Gary and Johnson [10]. An explicit reduction from the 3-SAT problem was given by Khuller, Raghavachari, and Rosenfeld [16]. They also obtain for the Metric Dimension problem a $(2\ln(n) + \Theta(1))$-approximation algorithm based on the well-known greedy algorithm for the Set Cover problem and showed that the Metric Dimension problem is polynomial-time solvable for trees. Beerliova et al. [1] show that the Metric Dimension problem cannot be approximated within a factor of $o(\log(n))$ unless $P = NP$.

Berman, DasGupta, and Kao [2] study various *Test Set* problems and in particular give a $(1 + \ln(n))$-approximation algorithm for the *Test Set Collection* (TSC) problem. The Metric Dimension problem can be seen as a variant of the Test Set Collection problem where only certain combinations of tests (corresponding to the vertices of the input graph) are available.

Halldórsson, Halldórsson, and Ravi [11] study the Test Set Collection problem with bounded test size. They give a $(3 + 3\ln(k))$-approximation algorithm for the Test Set Collection problem with test of size at most k.

The approximation complexity of dense and superdense instances of various optimization problems was studied in [15,14].

3 Our Contributions

This work is the first, best to our knowledge, providing explicit approximation lower bounds for both bounded degree and dense instances of the Metric Dimension problem. Furthermore, we improve the upper bounds for general and dense instances as well as the lower bound for general instances. We also prove that the Metric Dimension problem restricted to point sets in $\mathbb{R}^d$ is polynomial-time solvable.

Theorem 1. *For each $d \in \mathbb{N}$, the Metric Dimension problem restricted to finite sets of points in $\mathbb{R}^d$ with the Euclidean distance is in PO.*

In contrast, we prove that the general problem is as hard to approximate as the Set Cover problem.

Theorem 2. *For any constant $\epsilon > 0$, solutions of the Metric Dimension problem cannot be approximated in polynomial time to within a factor of $(1-\epsilon)\ln(n)$, unless $NP \subset DTIME(n^{log(log(n))})$.*

Berman, DasGupta, and Kao [2] provided a simple greedy heuristic, which they called the Information Content Heuristic (ICH), for the TSC problem.

Here we use notations from [2] to define an entropy $H_{V'}$ for subsets $V' \subset V$ and to obtain an improved approximation algorithm for the Metric Dimension problem, that matches the lower bound of Theorem 2.

For each subset $V' \subseteq V$ we have an associated equivalence relation $\equiv_{V'}$ given by $u \equiv_{V'} w \iff [\forall v \in V' : d(v,u) = d(w,v)]$. Let $A_1, .., A_k$ be the equivalence classes of $\equiv_{V'}$, then the entropy of V' is defined as $H_{V'} = \log_2(\Pi_{i=1}^k |A_i|!)$. The information content of a vertex $v \in V$ with respect to V' is defined as $IC(v, V') = H_{V'} - H_{V' \cup \{v\}}$. Modified ICH is now ICH applied to the information content function $IC(v, V') := H_{V'} - H_{V' \cup \{v\}}$.

Modified ICH

$V' := \emptyset$
while $(H_{V'} \neq 0)$ do
select a $v \in argmax_{v \in V \setminus V'}(IC(v, V'))$
$V' := V' \cup \{v\}$
endwhile

Theorem 3. *Modified ICH is a polynomial-time approximation algorithm for the Metric Dimension problem with ratio* $1 + \ln(|V|) + \ln(\ln_2(|V|))$.

We also investigate the Metric Dimension problem in bounded degree graphs. In particular, we give the following explicit approximation lower bound.

Theorem 4. *The B-bounded Metric Dimension Problem is APX-hard for every* $B \geq 3$ *and it is NP-hard to approximate within any constant better than* $\frac{353}{352}$.

We show that even on k-superdense graphs the metric dimension is APX-hard $k \geq 6$. In particular, we obtain explicit approximation lower bounds.

Theorem 5. *It is NP-hard to approximate the Metric Dimension on k-superdense graphs to within any better than* $\frac{3511}{3510}$ *for* $k = 6$, $\frac{1090}{1089}$ *for* $k = 7$ *and* $\frac{677}{676}$ *for* $k = 8$.

We also provide a constant factor approximation algorithm with approximation ratio $(2 + 2\ln(k) + \ln(\log_2(k-1)) + o(1))$ for k-superdense instances. Previously, Halldórsson, Halldórsson, and Ravi [11] used a similar approach for the Test Set Collection with bounded test sizes, based on a twofold application of the greedy k-set cover algorithm. Here, we apply first the greedy k-set cover algorithm and afterwards use the Modified ICH to generate a resolving set. Since in a k-superdense graph we have $d(v,w) \in \{0,1,2\}$, only three equivalence classes occur. For every $v \in V$, let A_0^v, A_1^v and A_2^v be the equivalence classes under $\equiv_v$. We present the algorithm Pre-ICH:

1. Apply the greedy algorithm for the Min k-Set Cover problem to instance $SC(G) := (V, \{A_0^v \cup A_2^v \mid v \in V\})$ with solution $\{A_0^v \cup A_2^v \mid v \in V''\}$.
2. Apply Modified ICH with initial set $V' := V''$.

Theorem 6. *Pre-ICH is a* $(2 + 2\ln(k) + \ln(\log_2(k-1)) + o(1))$*-approximation algorithm for the Metric Dimension problem on k-superdense graphs.*

Acknowledgment

We thank Marek Karpinski for a number of interesting discussions and for his support.

References

1. Beerliova, Z., Eberhard, F., Erlebach, T., Hall, A., Hoffmann, M., Mihalák, M., Ram, L.: Network Discovery and Verification. IEEE J. on Selected Areas in Communications 24, 2168–2181 (2006)
2. Berman, P., DasGupta, B., Kao, M.: Tight approximability results for test set problems in bioinformatics. J. Comput. Syst. Sci. 71, 145–162 (2005)
3. Cáceres, J., Hernando, C., Mora, M., Pelayo, I., Puertas, M., Seara, C., Wood, D.: On the Metric Dimension of Cartesian Products of Graphs. SIAM J. Disc. Math. 21, 423–441 (2007)
4. Cáceres, J., Hernando, C., Mora, M., Pelayo, I., Puertas, M.: On the Metric Dimension of Infinite Graphs. Electr. Notes in Disc. Math. 35, 15–20 (2009)
5. Chappell, G., Gimbel, J., Hartman, C.: Bounds on the metric and partition dimensions of a graph. Ars Combinatoria 88, 349–366 (2008)
6. Chartrand, G., Eroh, L., Johnson, M., Oellermann, O.: Resolvability in graphs and the metric dimension of a graph. Disc. Appl. Math. 105, 99–133 (2000)
7. Chlebík, M., Chlebíková, J.: Approximation hardness of dominating set problems in bounded degree graphs. Inf. Comput. 206, 1264–1275 (2008)
8. Chvátal, V.: Mastermind. Combinatorica 3, 325–329 (1983)
9. Fehr, M., Gosselin, S., Oellermann, O.: The metric dimension of Cayley digraphs. Disc. Math. 306, 31–41 (2006)
10. Garey, M., Johnson, D.: Computers and Intractability: A Guide to the Theory on NP-Completeness. Freeman, New York (1979)
11. Halldórsson, B., Halldórsson, M., Ravi, R.: On the Approximability of the Minimum Test Collection Problem. In: Meyer auf der Heide, F. (ed.) ESA 2001. LNCS, vol. 2161, pp. 158–169. Springer, Heidelberg (2001)
12. Harary, F., Melter, R.: On the metric dimension of a graph. Ars Combinatoria 2, 191–195 (1976)
13. Hernando, C., Mora, M., Pelayo, I., Seara, C., Wood, D.: Extremal Graph Theory for Metric Dimension and Diameter. Electr. Notes in Disc. Math. 29, 339–343 (2007)
14. Karpinski, M., Schudy, W.: Linear time approximation schemes for the Gale-Berlekamp game and related minimization problems. In: Proc. 41st ACM STOC, pp. 313–322 (2009)
15. Karpinski, M., Zelikovsky, A.: Approximating Dense Cases of Covering Problems. In: Proc. DIMACS Workshop on Network Design: Connectivity and Facilities Location 1997, pp. 169–178 (1997); also published in ECCC TR97-004
16. Khuller, S., Raghavachari, B., Rosenfeld, A.: Landmarks in graphs. Disc. Appl. Math. 70, 217–229 (1996)
17. Sebö, A., Tannier, E.: On Metric Generators of Graphs. Math. Oper. Res. 29, 383–393 (2004)
18. Slater, P.: Leaves of trees. In: Hoffman, et al. (eds.) Southeastern Conf. on Combin., Graph Theory, and Comp., Congressus Numerantium, vol. 14, pp. 549–559. Utilitas Mathematica, Winnipeg (1975)
19. Slater, P.: Dominating and reference sets in graphs. J. Math. Phys. Sci. 22, 445–455 (1988)
20. Shapiro, H., Söderberg, S.: A combinatory detection problem. Amer. Math. Monthly 70, 1066–1070 (1963)
21. Tomescu, I.: Discrepancies between metric dimension and partition dimension of a connected graph. Disc. Math. 308, 5026–5031 (2008)

Collision-Free Routing in Sink-Centric Sensor Networks with Coarse-Grain Coordinates*

Alfredo Navarra and Cristina M. Pinotti

Dipartimento di Matematica e Informatica, Università degli Studi di Perugia,
Via Vanvitelli 1, 06123 Perugia, Italy
{navarra,pinotti}@dmi.unipg.it

Abstract. When the environment does not allow to access directly to disseminated data, a sensor network could be one of the most appropriate solution to retrieve the map of interesting areas. Based on existing approaches, we start our study from the standard random deployment of a sensor network and then we consider a coarse-grain localization algorithm which associates sensors with coordinates related to a central node, called *sink*. Once each sensor is related to an estimated position, it starts to send data to the sink according to a provided scheduling of communications which takes care of energy consumption, collisions and time. We propose a scheduling of communications based on distributed and fast coloring algorithms which require $O(1)$ computational time. As the localization is referred to coarse-grain coordinates, it happens that more than one sensor is associated with the same coordinates, hence leader-election mechanism is considered.

1 Introduction

A duty-cycle wireless sensor and sink network (DC-WSN) consists of many randomly deployed tiny low-cost *sensors* which follow a sleep-awake cycle, and a few powerful entities, called *sinks*. Clearly DC-WSNs are an extension of wireless sensor networks (WSNs) as we address uncertainty about the existence of a wireless link originating from the random sleep-awake schedules.

Specifically, we consider a dense DC-WSN where each sink is mobile and, upon reaching a specific location, remains there to collect data from the sensors in the surrounding area, called *sink-region*. Sensors are randomly deployed and are employed in applications where they remain unattended in a vast, possibly hostile, geographical area for long period of times (e.g., environment monitoring and intruder tracking). Sensors sense the physical world in their proximity, while sinks, equipped with much better processing capabilities, higher transmission power, and longer battery life, move around the area to collect, aggregate, and transmit to the external world the sensed data collected by the sensors [1, 8, 15]. When a sink reaches an area of interest in the network, sensors in its vicinity must be organized into a short-lived and mission-oriented subnetwork called *sink-centric network*.

* The research was partially funded by "Fondazione Cassa di Risparmio di Perugia" (Italy), under the project "Ricerca di base 2009".

C.S. Iliopoulos and W.F. Smyth (Eds.): IWOCA 2010, LNCS 6460, pp. 140–153, 2011.

In the rest of this paper, we will focus on the sink-centric network. We consider localization algorithms in order to provide a *virtual infrastructure* surrounding the sink which will be used for routing purposes. Such localization protocols (also referred to as *training* protocols) impose a *discrete coordinate system* on the *sink-region*. Sensors that acquire identical coordinates form a *cluster* of indistinguishable nodes. This means that the information sent from a cluster to the sink will be always the same, regardless the sending sensor. This suggests the usage of leader election mechanisms within each cluster in order to save energy.

Once sensors in the sink-region are localized, sensory data are relayed to the sink based on a geographical routing protocol. Latency, energy efficiency, and collision avoidance are addressed in the design of the routing protocol. We assume that a collision occurs when a sensor receives more than one message at the same time. Therefore, to avoid message collisions, communication schedules have to be designed. The main contribution of the paper is the design of a communication scheduling based on fast and distributed coloring algorithms. The proposed coloring algorithms are then applied in order to accomplish collision-free leader election and routing tasks.

1.1 Outline

The next section introduces the model assumptions under which the routing of sensory data must be performed, and defines the virtual infrastructure commonly used to organize DC-WSNs with respect to a central sink. Section 2.1 describes the first contribution of the paper. In particular, the virtual infrastructure is modified in favor of a uniform usage of the involved sensors. Section 3 introduces and optimally solves a coloring problem arising from the requirement of scheduling the communications from the sensors toward the sink without collisions. Section 4 describes how the proposed coloring can be used for both leader election and routing purposes. Finally, Section 5 provides concluding remarks, and points out possible directions for further investigations.

2 The Model

Time is assumed to be divided into slots. The sensors and the sink use equally long, in-phase slots, but they do not necessarily start counting time from the same slot. A sensor possesses three basic capabilities: sensing, computation, and wireless communication; and operates subject to the following constraints:

a. Each sensor alternates between *sleep* periods and *awake* periods – sleep-awake cycle has a total length of L time slots, out of which the sensor is in sleep mode for $L-d$ slots and in awake mode for d slots;
b. Each sensor is *asynchronous* – it wakes up for the first time according to its internal clock and it is not engaged in an explicit synchronization protocol with either the sink or other sensors. Sensors that wake up simultaneously at time slot x are said of *type* x or equivalently, they belong to time-zone x;

c. Individual sensors are *unattended* – once deployed it is neither feasible or practical to devote attention to individual sensors;
d. No sensor has global information about the network topology, but each one can hear transmissions from the sink;
e. The sensors are *anonymous* – they are not associated with unique IDs;
f. Each sensor has a modest non-renewable energy budget and a limited transmission range r;
g. Sensors can transmit and receive on multiple frequency channels.

Concerning the training protocol which will be further discussed later, it imposes a virtual coordinate system onto the sensor network by establishing:

1. *Coronas*: The sink-region area is divided into k coronas $C_0, C_1, \ldots, C_{k-1}$ each of fixed width $\rho > 0$. The coronas are centered at the sink and determined by k concentric circles whose radii are $\rho, 2\rho, \cdots, k\rho$, respectively;
2. *Sectors*: The sink-region is divided into h equiangular sectors $S_0, S_1, \ldots, S_{h-1}$, originated at the sink, each having a width of $\frac{2\pi}{h}$ radians.

In particular, a cluster is the intersection between a corona and a sector where all sensors acquire the same coordinates. Once the training protocol has terminated, we assume a data logging application, where the sensors are required to send their sensory data to the sink. When sensors transmit, if an awake sensor receives more than one message concurrently on the same frequency channel, we assume that it hears noise, i.e., a *collision* occurs.

2.1 Localization

Many research papers have provided different approaches to make anonymous sensors aware of their coarse-grain positions [2–6, 10–14]. In order to perform the training, two main procedure are usually executed. In the first, the sink makes use of its isotropic antenna for the corona training. In the second, the sink makes use of the directional antenna for the sector training. Our interest is in the final virtual infrastructure that a training protocol delivers. Our little modification to the previous approaches is to maintain the area of each cluster roughly the same among the whole network. In this way we better guarantee a uniform usage of the disseminated sensors in favor of better performances, and of an extended network lifespan. In order to obtain the desired configuration, let ℓ be the number of sectors imposed in corona 1. Considering $\rho = 1$, the number of sectors will be doubled at each corona $c = 2^p$, $0 < p \leq \lfloor \log_2(k-1) \rfloor$. In fact, corona $c = 2^p$ has area $\pi(2^{p+1}+1)$ which is almost the double than the area of corona $c = 2^{p-1}$. In doing so, we obtain that the proposed subdivision guarantees the following result:

Lemma 1. *The ratio given by the area spanned by two generic clusters is at most* 2.

Proof. Let (c, s), $c > 1$, be a generic cluster of the imposed virtual infrastructure, and let $p = \lfloor \log_2 c \rfloor$ which implies $2^p \leq c < 2^{p+1}$. The area spanned by corona 1 is 3π and it is divided into ℓ sectors. The area spanned by the generic corona c is $((c+1)^2 - c^2)\pi$ and it is divided by construction into $2^p\ell$ sectors. Hence, the area of one cluster in corona 1 is equal to $A_1 = 3\frac{\pi}{\ell}$, while the area of one cluster in corona c is equal to $A_c = \frac{(2c+1)\pi}{2^p\ell}$. The ratio gives:

$$\frac{A_1}{A_c} = 3\frac{\pi}{\ell} \times \frac{2^p\ell}{(2c+1)\pi} \leq \frac{3 \cdot 2^p}{2 \cdot 2^p} = \frac{3}{2},$$

and

$$\frac{A_1}{A_c} = 3\frac{\pi}{\ell} \times \frac{2^p\ell}{(2c+1)\pi} \geq \frac{3 \cdot 2^p}{2((2^{p+1}-1)+1)} \geq \frac{3 \cdot 2^p}{4 \cdot 2^p} = \frac{3}{4}.$$

Hence, the biggest ratio between the area of two generic clusters of the imposed virtual infrastructure gives:

$$\frac{3}{2}A_1 \times \frac{4}{3}\frac{1}{A_1} = 2. \qquad \square$$

Figure 1 illustrates the virtual infrastructure when $\ell = 4$. The sectors in corona c are numbered from 0 to $h_c - 1$ starting to count from the sector above the x-axis. Noting that the outmost corona $c = k - 1$ will be divided into $h = \ell 2^{\lfloor \log_2(k-1) \rfloor}$ sectors, the virtual infrastructure can be obtained as an ordinary coordinate system with k coronas and h sectors, in which the inner coronas just ignore further subdivisions of their coronas if not required with respect to the defined virtual infrastructure.

3 Coloring

Once that sensors are placed and localized, we need to schedule their communications toward the sink in order to deliver the sensory data. Communications should take care of required time, energy efficiency and collisions.

To this aim, we introduce a frequency channel assignment to schedule the communications on the adjacency graph associated with the virtual infrastructure imposed by the localization algorithm.

Recalling that ℓ is the number of clusters in corona 1 of the virtual infrastructure, the adjacency graph G_ℓ has one node for each cluster in corona $c \geq 1$ and one edge for each pair of nodes corresponding to adjacent clusters. Formally:

Definition 1. *The* adjacency graph G_ℓ *has one node* (c, s)*, with* $1 \leq c \leq k-1$ *and* $\ell \leq s \leq h_c$*, for each cluster in corona* $c \geq 1$ *of the virtual infrastructure. Two nodes* (c, s) *and* (c', s')*, with* $c \geq c'$*, are* adjacent *if*

1. $c = c'$ *and* $|s - s'| = 1$*, or*
2. $c = c' + 1$ *and for some* $x \in \mathbb{N}$*,* $2^{x-1} \leq c' < c < 2^x$ *and* $s = s'$ *or*
3. *for some* $x \in \mathbb{N}$*,* $c = c' + 1 = 2^x$ *and* $s' = \lfloor \frac{s}{2} \rfloor$*.*

Figure 1 shows the virtual infrastructure when $\ell = 4$ and the associated adjacency graph G_ℓ. For the rest of our discussion, we do not take into consideration corona 0, as the scheduling of communications in there (included forwarding communications from outer coronas) is not necessary due to the proximity of the sensors with the sink which can retrieve the information by itself. It is like assuming that if a transmission reaches corona 0 then it reaches the sink.

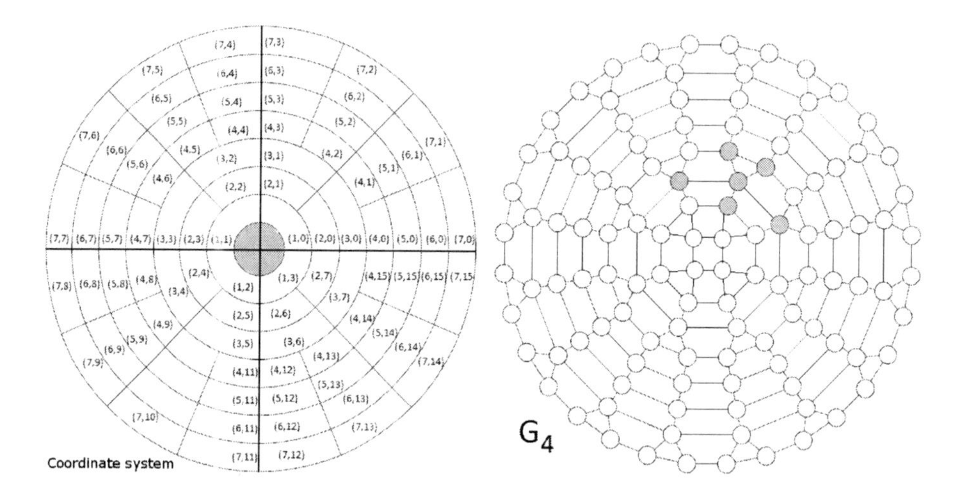

Fig. 1. On the left, the virtual infrastructure divided into clusters uniquely identified when $\ell = 4$. On the right, the corresponding adjacency graph G_4. The shadowed nodes represent a maximal subset of nodes at pairwise distance at most 2 in the graph, i.e. each pair of nodes in the subset is at distance at most 2.

In the rest of this section we focus on colorings of the graph G_ℓ. Formally:

Definition 2. *A distance-two coloring* (or frequency channel assignment)*, is a function that assigns to each node of G_ℓ a color such that the same color cannot be assigned to two nodes at pairwise distance smaller than or equal to 2.*

In the following, we will refer to *distance-two coloring* simply as *coloring* algorithm. We will postpone to Section 4 how such a coloring (or, scheduling) can be used for leader election and/or for routing purposes.

First of all, observe that by construction, the largest subset of pairwise nodes at distance at most 2 has size 6 in both G_4 and G_3 (see Figures 1 and 2 for an illustration). Thus:

Lemma 2. *Any coloring of G_ℓ, $\ell > 0$, which satisfies distance-two constraint requires at least 6 colors.*

Proof. A generic node $x \in G_\ell$ corresponds to a cluster (c, s) in the imposed coordinate system. The neighbors of x corresponding to other clusters in c are at most 2. By construction, x admits only one neighbor corresponding to corona

$c-1$ while at most two neighbors corresponding to corona $c+1$. In fact, clusters are doubled at every corona labelled by a power of two. Moreover, the intersection established at some corona, is maintained along all the coronas with bigger labels. Thus, (c, s) admits only one neighbor $(c-1, s)$ in corona $c-1$ and at most two neighbors in corona $c+1$ (this happens when $c+1$ is labelled by a power of two). All these neighbors along with x form then the biggest set of nodes in G_ℓ at pairwise distance at most 2, hence 6 colors are required by any coloring satisfying the distance-two constraint. □

In the following, we propose two coloring algorithms, called *Col*4 and *Col*3, which color, respectively, the adjacency graphs G_4 and G_3. Algorithm *Col*4 uses 8 colors, while *Col*3 is optimal because it uses exactly 6 colors. Both algorithms provide a very useful property, that is, each cluster can be colored in constant time, although *Col*4 is simpler than *Col*3. The sensors compute their colors in constant time once they know the coordinates of the cluster where they reside.

3.1 Algorithm *Col*4

Algorithm *Col*4 is based on Table 1 that can be used by each sensor in order to acquire the corresponding color of the cluster where it resides. The table has four rows and four columns. Let $|i|_j$ denotes the *modulo* operation, that is the nonnegative remainder of the division of i by j. Cluster (c, s), $0 < c < k$ and $0 \leq s < h$, will get the color according to entry $[|c-1|_4\,,\ |s|_4]$, We have to show

Table 1. Algorithm *Col*4: cluster (c, s), $0 < c < k$ and $0 \leq s < h$, will get the color according to entry $[|c-1|_4\,,\ |s|_4]$

	0	1	2	3
0	RED	GREEN	ORANGE	YELLOW
1	BLUE	WHITE	CYAN	PINK
2	GREEN	YELLOW	RED	ORANGE
3	WHITE	PINK	BLUE	CYAN

that such a coloring respects the imposed distance two constraint. First of all, we point out that two clusters belonging to two different adjacent coronas necessarily acquire two different colors. In fact, from Table 1 we have two different subsets of colors used for even and odd rows, respectively. Another simple observation is that if two clusters of the same color belong to the same corona, then they are at a distance which is a multiple of 4. The next lemma shows the remaining cases that must be addressed for a correct coloring (see Figure 3 for a visualization).

Lemma 3. *Consider two clusters (c, s) and (c', s'). If $c = c'+2$ and (a) $s = s'$, or (b) $s = 2s'-1$ or (c) $s = 2s'$, then $Col4(c, s) \neq Col4(c', s')$.*

Proof. Case (a) can be simply derived by observing Table 1, since at the same column no colors are repeated. For case (b), if $|s'|_4$ equals 0 then $|s|_4$ equals 3; if $|s'|_4$ equals 1 then $|s|_4$ equals 1; if $|s'|_4$ equals 2 then $|s|_4$ equals 3; if $|s'|_4$ equals

3 then $|s|_4$ equals 1. For case (c), if $|s'|_4$ equals 0 then $|s|_4$ equals 0; if $|s'|_4$ equals 1 then $|s|_4$ equals 2; if $|s'|_4$ equals 2 then $|s|_4$ equals 0; if $|s'|_4$ equals 3 then $|s|_4$ equals 2. In all the cases the corresponding entries associate two different colors to the considered clusters. □

Corollary 1. *Algorithm Col4 assigns colors to clusters satisfying the distance-two constraint.*

A natural question is whether there exists an optimal algorithm which makes use of six colors in constant time, or at least an algorithm that uses less than eight colors. This is an open problem that remains to be further investigated.

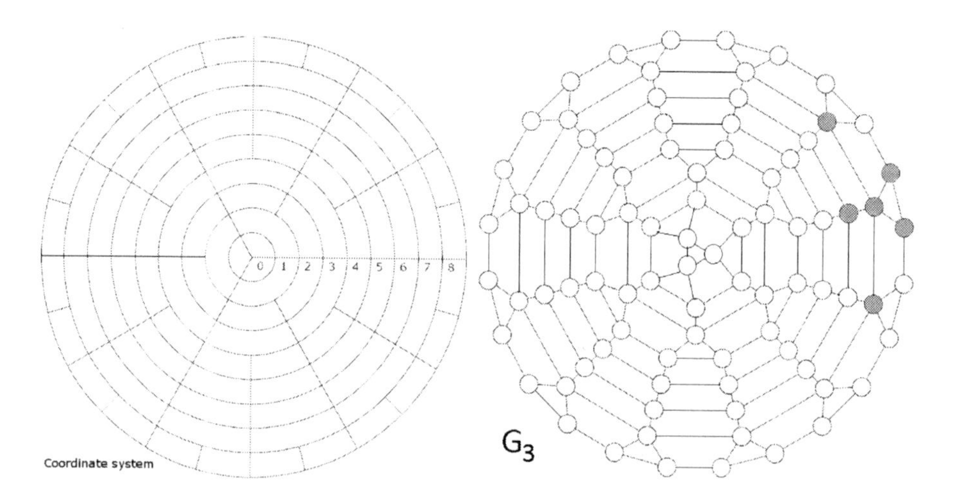

Fig. 2. On the left, the virtual infrastructure obtained by starting with 3 sectors. On the right, the corresponding adjacency graph G when corona 1 is divided into 3 sectors. The shadowed nodes represent a maximal subset of nodes at pairwise distance at most 2 in the graph.

3.2 Algorithm *Col*3

An optimal coloring *Col*3 can instead be found for the virtual infrastructure which partitions the first corona in 3 sectors (Figure 2) and whose adjacency graph is denoted as G_3. Algorithm *Col*3 is based again on two subsets of colors as for *Col*4 but each of three colors $\{RED; BLUE; GREEN\}$, $\{WHITE; PINK; CYAN\}$. The first set is used for odd coronas, the second for even coronas. This again realizes the property for which two clusters at two adjacent different coronas cannot get the same color. Moreover, for each corona a sequence of the colors is properly selected and repeated for coloring all the sectors in anti-clockwise order. Thus, two clusters associated with the same color at the same corona are at distance which is a multiple of 3. Starting from corona 1, we color the clusters using the sequence of colors $\{RED; BLUE; GREEN\}$ in an anti-clockwise order. Then corona 2 will be colored at the same way but using the sequence $\{WHITE; PINK; CYAN\}$ twice. In what follows, sometimes we refer to a sequence of colors by means of the sequence of their cardinal numbers in the sequence instead of the name of the colors. In doing so, any other cluster (c, s) is colored according to the following actions:

Shifting: Given a sequence of colors $\{0, 1, 2\}$, a shifting operation consists in summing $|-1|_3$ to each element of the sequence, hence obtaining $\{2, 0, 1\}$.

Swapping: Given a sequence of colors $\{0, 1, 2\}$, a swapping operation consists in exchanging the first element of the sequence with the third one, hence obtaining the sequence $\{2, 1, 0\}$.

Any cluster (c, s), $c > 2$, is colored in the following way: if the number of clusters in corona c is the same as in corona $c - 2$, then corona c is colored with the sequence obtained from the sequence used in corona $c-2$ by applying a shifting operation. If the number of clusters in corona c is doubled with respect to corona $c - 2$, then corona c is colored with the sequence obtained from the sequence used in corona $c - 2$ by applying a swapping operation.

Lemma 4. *Algorithm Col3 assigns colors to clusters satisfying the distance-two constraint.*

Proof. It has been already pointed out how different colors are assigned to clusters at distance one. Moreover, if two clusters of the same color belong to the same corona, then they are at a distance which is a multiple of 3. Therefore, the proof only needs to show the correctness of the coloring for clusters at distance two in two different coronas. Let (c, s) and (c', s') be two clusters at distance two, with $c > c'$.

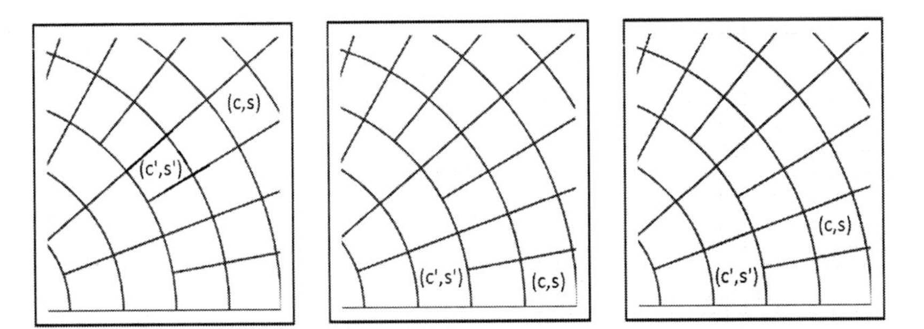

Fig. 3. The three possible configurations for two clusters at distance 2 when $c > c'$

Figure 3 shows the possible configurations. If the number of sectors in c is the same as in c', then s must be equal to s'. In this case, the sequence of colors used to color c is obtained form the sequence used in the c' after a shifting. This implies that colors assigned to (c, s) and (c', s') are different. Another configuration occurs when the number of sectors in c is doubled with respect to c', then s is equal either to $2s'$ or to $2s' + 1$. Since in this case the sequence of colors used to color c has been obtained from the one used in c' after a swapping operation, by construction (c, s) may assume any color in the sequence but the one associated to (c', s'). In fact, let $\{0, 1, 2\}$ be the sequence of colors used in c', then the sequence $\{2, 1, 0\}$ is used in c. Hence, if $Col3((c', s'), \{0, 1, 2\}) = 0$ (1, 2, resp.), then $Col3((c, 2s'), \{2, 1, 0\}) = 2$ (0, 1, resp.). Similarly, if $Col3((c', s'), \{0, 1, 2\}) = 0$ (1, 2, resp.), then $Col3((c, 2s' + 1), \{2, 1, 0\}) = 1$ (2, 0, resp.). □

We now show that each color assigned by $Col3$ to a generic cluster (c, s) can be evaluated in a constant number of steps with the only assumption of knowing the sequences of colors used in corona 1 and 2. For the ease of the analysis purposes, from now on we focus only on one set of three colors for coloring all the coronas instead of presenting two specular arguments for odd and even coronas, respectively.

Assuming we know the sequence of colors used at a generic corona c', the sequence used to color any $c > c'$ that has the same number of clusters of c' can be easily evaluated. In fact, it is sufficient to apply the shifting operation $c - c'$ times. As such operation is associative, the result does not change if we decrease by $|-(c - c')|_3$ each single element of the sequence. However, when the number of sectors is doubled with respect to c' then we need a more careful computation. Actually, we evaluate the first and the third colors of the sequence independently. The second then comes as a consequence. The next technical lemma provides a first contribution to the evaluation of the required sequence of colors.

Lemma 5. *Let $\{0, 1, 2\}$ be the sequence of colors used for corona 1, $c = 2^p$ for some $p > 0$, and $\{X', Y', Z'\}$ be the sequence of colors used for corona $c' = 2^{p-1}$, then the sequence of colors $\{X, Y, Z\}$ used for corona c can be evaluated as follows:*

(a) if $|p|_2 = 0$ then $X = X'$, $Z = |Z' + 1|_3$ and $Y = \{X', Y', Z'\} \setminus \{X, Z\}$;
(b) if $|p|_2 = 1$ then $X = |X' - 1|_3$, $Z = |Z' + 1|_3$ and $Y = \{X', Y', Z'\} \setminus \{X, Z\}$.

Proof. We prove the lemma by induction on p. The base of the induction is given for the two cases $p = 1$, and $p = 2$. In the first case, corona $c = 2^p = 2$ is colored by using the sequence $\{2, 1, 0\}$ obtained from the one of corona 1 by applying a swapping operation, hence obtaining $X = 2 = |X' - 1|_3$, $Z = 0 = |Z' + 1|_3$ and $Y = 1$. In the second case, corona $c = 2^p = 4$ is colored by using the sequence $\{2, 0, 1\}$ obtained from the one of corona 2 by first applying a shifting operation and then a swapping one, hence obtaining $X = 2 = X'$, $Z = 1 = |Z' + 1|_3$ and $Y = 0$. We assume the claim as true for any $p - 1 \leq 2$ and we prove it for p. Corona $c = 2^p$ is colored by using the sequence $\{X, Y, Z\}$ obtained from the sequence $\{X', Y', Z'\}$ used in corona $c' = 2^{p-1}$ after applying $2^{p-1} - 1$ shifting operations and one swapping operation. Hence, $X = |Z' - (2^{p-1} - 1)|_3$, $Y = |Y' - (2^{p-1} - 1)|_3$ and $Z = |X' - (2^{p-1} - 1)|_3$. This leads to $X = |Z' - (2^{p-1} - 1)|_3 = |Z' - (2^{\lfloor \frac{p-1}{2} \rfloor 2 + |p-1|_2} - 1)|_3 = |Z' - 4^{\lfloor \frac{p-1}{2} \rfloor} 2^{|p-1|_2} + 1|_3 = |Z' - 2^{|p-1|_2} + 1|_3$; $Y = |Y' - 2^{|p-1|_2} + 1|_3$; $Z = |X' - 2^{|p-1|_2} + 1|_3$. If $|p|_2 = 0$ then $X = |Z' - 1|_3$, $|Y' - 1|_3$ and $Z = |X' - 1|_3$. If $|p|_2 = 1$ then $X = Z'$, $Y = Y'$ and $Z = X'$. This implies that Y is always different from X and Z, as it is obtained from Y' different from X' and Z' by applying the same rules.

By induction, the sequence $\{X', Y', Z'\}$ used in corona c' is obtained by applying the claim to the sequence $\{X'', Y'', Z''\}$ used in corona $c'' = 2^{p-2}$, i.e. $X' = |X'' - 1|_3$, $Z' = |Z'' + 1|_3$ and Y' is the remaining available color. By the above calculations, if $|p|_2 = 0$, $X' = Z''$, $Y' = Y''$ and $Z' = X''$, and we obtain $X = |Z' - 1|_3 = |X'' - 1|_3$, $Y' = Y''$, $Z = |X' - 1|_3 = |Z'' - 1|_3 = |Z'' + 2|_3$ that is equivalent to apply first rule (b) and then rule (a) from $\{X'', Y'', Z''\}$.

If $|p|_2 = 1$, $X' = |Z'' - 1|_3$, $Y' = |Y'' - 1|_3$ and $Z' = |X'' - 1|_3$, and we obtain $X = Z' = |X'' - 1|_3$, $Y' = |Y'' - 1|_3$, $Z = X' = |Z'' - 1|_3 = |Z'' + 2|_3$ that is equivalent to apply first rule (a) and then rule (b) from $\{X'', Y'', Z''\}$. □

In other words, Lemma 5 provides the tool for evaluating the coloring in a distributed way by each sensor in constant time with respect to the network size. In fact, as shown by the next theorem, a sensor requires only calculations involving values c and s defining the cluster where it resides. Such quantities are negligible with respect to the size n of the network.

Theorem 1. *Let $\{X, Y, Z\}$ be the sequence of colors used to color corona 1, then $Col3(c, s)$ can be evaluated in constant time independently of the other clusters.*

Proof. Starting from the sequence of colors[1] used in corona 1, in order to guess the sequence of colors used at a generic corona c, it suffices to evaluate $p = \lfloor \log_2 c \rfloor$. Then, by Lemma 5, we can find the sequence of colors $\{X', Y', Z'\}$ used at corona $c' = 2^p$ by applying the following rules. Decrease X by $|\lceil \frac{p}{2} \rceil|_3$, increase Z by $|p|_3$, and choose for Y the remaining available color. Finally, by applying $c - c'$ shifting operations, i.e., by decreasing each element of the sequence evaluated for c' by $|c - c'|_3$, we obtain the sequence for c. More formally:

$$X' = X - \left| \left\lceil \frac{p}{2} \right\rceil \right|_3 - |c - c'|_3$$
$$Z' = Z + |p|_3 - |c - c'|_3$$
$$Y' \neq X' \neq Z' \text{ and } Y' \in \{0, 1, 2\}$$

Once the sequence of colors $\{X', Y', Z'\}$ used to color corona c is known, the corona will be colored in anti-clockwise order from sector 0. Precisely:

$$Col3(c, s) = \begin{cases} X' \text{ if } |s|_3 = 0 \\ Y' \text{ if } |s|_3 = 1 \\ Z' \text{ if } |s|_3 = 2 \end{cases}$$

Thus obtaining $Col3(c, s)$ takes constant time. □

Figure 4 shows the correct coloring obtained for both the odd and the even coronas by applying the described $Col3$ algorithm. The initial step is constituted by starting with the coloring of corona 1 with the sequences $\{0, 1, 2\}$ and $\{5, 4, 3\}$ for odd and even coronas, respectively.

---

[1] Note that the sequence of colors used for corona 1 may refer, without distinction, to the set of three colors used for odd coronas or even coronas.

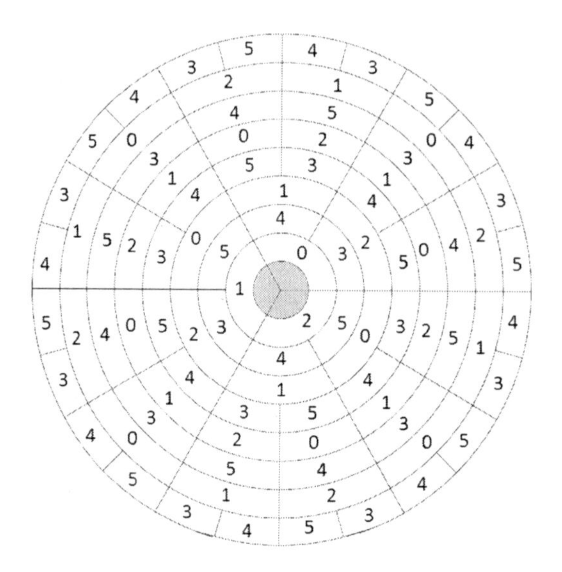

Fig. 4. The coloring obtained by applying algorithm *Col*3 on the virtual infrastructure with three sectors at corona 1, and considering the sequences $\{0, 1, 2\}$ and $\{5, 4, 3\}$ for corona 1 for coloring the odd and the even coronas, respectively

Once the suitable coloring has been performed among sensors, it is used to schedule communications. Different colors specify different communication frequency channels. This implies that adjacent clusters can perform in parallel their communications without incurring in collisions. However, before showing how the routing of sensory data can be performed over the virtual infrastructure, we provide a further step in the set-up of the network by electing inside each cluster one leader for each type (*time-zone*). In this way, we avoid redundant communications among sensors belonging to the same cluster (hence saving energy) while we ensure at least one active sensor at any time. Actually, we could schedule the repetition of the leader election procedure in order to rotate among sensors, hence prolonging the network lifespan.

4 Leader Election and Routing

In this section, we describe how the routing and the leader election can be performed in the sensor network without collisions by means of the coloring algorithms presented in the previous section.

Our routing algorithm requires that, in any cluster, there is a sensor ready to forward the message going toward the sink at any time slot t. Such a sensor will be the leader of the sensors that wake up at time t. From now on, we assume that there is at least one leader awake and ready to forward the message at any time in any cluster. Specifically, during the routing process, we assume that sensors transmit during the second time slot of their awake period, while they are listening during their first one. A message that originates at time t in cluster

(c, s), will be transmitted by the leader of time-zone t in cluster c at time $|t+1|_L$. Such a message will be then received and handled by the awake leader of time-zone $|t+1|_L$ in the cluster destination that receives the message at time $|t+1|_L$ and forward it toward the sink at time $|t+2|_L$. Note that, the destination cluster is $(c-1, s)$ if c is not a power of two, and cluster $(c-1, \lfloor \frac{s}{2} \rfloor)$ otherwise. In this way, a message originated in corona c can be potentially routed in c hops to the sink. To this aim, observe that a leader transmission reaches the cluster destination as well as the other adjacent clusters because, during the routing protocol, sensors broadcast with a radius equal to the corona width. Therefore, to avoid that a cluster is simultaneously reached by two different leader transmissions on the same frequency channel, two leaders that use the same frequency channel must reside in two clusters which are at least at distance three. Thus, any frequency channel assignment (or, coloring) suitable for routing without collisions must satisfy the distance-two constraint discussed earlier.

It is worthy to note that a weaker constraint on the distance of leaders transmitting on the same frequency channel is sufficient for the leader election protocol. Indeed, as it will be explained below, during such a protocol, a message that originates in cluster (c, s) has for destination the cluster itself. Thus, it is sufficient that two leaders that transmit on the same frequency channel reside in two clusters at distance two to avoid collisions. Thus, any coloring suitable for our routing algorithm is also suitable for the leader election.

A brief description of the routing and leader election protocols follow. Once the coloring of the virtual infrastructure has been performed, each sensor residing in a specific cluster is aware of its color. We consider one different frequency channel for each used color. Hence, each sensor will be aware of the frequency channel it has to use for transmission tasks. Our first goal is to elect inside each cluster one leader for each time-zone $0 \leq x \leq L-1$. To this aim, we make use of the well-known *uniform leader election for radio networks* protocol presented in [9]. In particular, we can consider the so called *Scenario 2* in which an upper bound to the number of sensors competing for the leader election inside each cluster and for each time-zone is known. In fact, by exploiting the arguments presented in [6, 7, 10] such an upper bound is $u \leq \frac{4}{3}A_1\Lambda = \frac{4\pi}{\ell}\Lambda$, where Λ is an estimation of the density of sensors related to one specific time-zone. From [9], the sensors require on average $\ln\ln u + o(\ln\ln u)$ transmissions. In practice, the protocol works by assigning a probability of transmission to each sensor. A sensor is elected as leader when it is the only one transmitting during the time-slot. If more than one sensor transmit or no one transmits, than the probability to transmit at the subsequent appropriate time-slot decreases or increases, respectively.

In our setting, we perform L leader elections, one for each time-zone, distributed over $O(\ln\ln u)$ subsequent sleep-awake periods. Only the sensors with the same time-zone are involved in one election. For each time-zone, each sensor performs one step of its leader election during every period. At the i-th time slot of the j-th awake period, the sensors of time-zone i perform the j-th step of their leader election. Each sensor transmits only during the first time-slot of its awake period. In doing so, we obtain the required leader election for all time-zones and all clusters. In fact, the protocol is performed in parallel in all clusters, each cluster transmitting on the frequency channel assigned to it by the coloring protocol. Finally, the routing earlier described can start.

Since in each cluster we have elected one leader for each time-zone, there will always be one leader, in the destination cluster, awake and ready to forward the message. Moreover, since each communication is performed according to the frequency channels that satisfy the distance-two constraint, in each time-slot the message will decrease by one hop its distance from the sink. Thus, using a multi-hop technique, a message originated in corona c reaches the sink in c time slots.

5 Conclusions

We investigate a virtual organization of a sink-centric subnetwork in a dense DC-WSN, which imposes a generalized coordinate system. Such a system provides a coarse-grained location to the sensors and allows a naive geographic routing algorithm. All the sensors that acquire the same coordinates form a cluster. For routing purposes, we assume that the sensors can transmit using different frequency channels. Following a multi-hop approach along the cluster-sink path, sensors in the outer coronas of the virtual infrastructure transmit their messages to the sink through intermediate coronas. The message stream can continuously proceed if there is, at any time, a relaying sensor awake and ready to transmit and no collisions arise on the frequency channels. To avoid collisions, a frequency channel assignment (or, coloring) that satisfies a distance-two constraint is provided for the graph G_ℓ that represents the cluster adjacencies of the virtual infrastructure that has ℓ clusters in corona 1. An optimal coloring for G_3 has been provided. Such coloring is fully distributed and requires constant time. Moreover, to avoid redundant messages during the routing protocol, we elect leaders in each cluster which act as relaying sensors. To this aim, we adapt a known uniform leader election protocol to our scenario. In the future, we intend to implement our routing algorithm in both simulated and real settings. Moreover, the study of optimal colorings for adjacency graphs G_ℓ with an arbitrary number ℓ of clusters in corona 1 is an interesting open problem.

References

1. Akyildiz, I.F., Kasimoglu, I.H.: Wireless sensor and actor networks: research challenges. Ad Hoc Networks 2(4), 351–367 (2004)
2. Barsi, F., Bertossi, A., Lavault, C., Navarra, A., Olariu, S., Pinotti, M.C., Ravelomanana, V.: Efficient binary search for training heterogeneous sensor and actor networks. In: Proceedings of the 1st ACM Workshop on Heterogeneous Sensor and Actor Networks (HeterSANET), pp. 17–24 (2008)
3. Barsi, F., Bertossi, A.A., Sorbelli, F.B., Ciotti, R., Olariu, S., Pinotti, M.C.: Asynchronous corona training protocols in wireless sensor and actor networks. IEEE Trans. on Parallel and Distributed Systems 20(8), 1216–1230 (2009)

4. Barsi, F., Bertossi, A.A., Lavault, C., Navarra, A., Pinotti, C.M., Olariu, S., Ravelomanana, V.: Efficient location training protocols for heterogeneous sensor and actor networks. IEEE Transactions on Mobile Computing (to appear)
5. Bertossi, A.A., Olariu, S., Pinotti, C.M.: Efficient corona training protocols for sensor networks. Theor. Comput. Sci. 402(1), 2–15 (2008)
6. Sorbelli, F.B., Ciotti, R., Navarra, A., Pinotti, M.C., Ravelomanana, V.: Cooperative training in wireless sensor and actor networks. In: Proceedings of the 6th International ICST Conference on Heterogeneous Networking for Quality, Reliability, Security and Robustness (QShine). Lecture Notes of the Institute for Computer Science, Social-Informatics and Telecommunications Engineering, vol. 22, pp. 569–583 (2009)
7. Ghidini, G., Pinotti, C.M., Das, S.K.: A Semi-Distributed Localization Protocol for Wireless Sensor and Actor Networks. In: 6th IEEE International Workshop on Sensor Networks and Systems for Pervasive Computing, PerSeNS (2010)
8. Melodia, T., Pompili, D., Akyildiz, I.F.: A communication architecture for mobile wireless sensor and actor networks. In: 3rd Annual IEEE Communcations Society Conference on Sensor, Mesh and Ad Hoc Communications and Networks (SECON), pp. 109–118 (2006)
9. Nakano, K., Olariu, S.: Uniform Leader Election Protocols for Radio Networks. IEEE Trans. on Parallel and Distributed Systems 13(5), 516–525 (2002)
10. Navarra, A., Pinotti, C.M., Ravelomanana, V., Sorbelli, F.B., Ciotti, R.: Cooperative Training for High Density Sensor and Actor Networks. IEEE Journal of Selected Areas in Communications 28(5), 753–763 (2010)
11. Olariu, S., Wada, A., Wilson, L., Eltoweissy, M.: Wireless sensor networks: leveraging the virtual infrastructure. IEEE Network 18(4), 51–56 (2004)
12. Olariu, S., Xu, Q., Wada, A., Stojmenovic, I.: A Virtual Infrastructure for Wireless Sensor Networks. John Wiley & Sons, Chichester (2005)
13. Sohrabi, K., Gao, J., Ailawadhi, V., Pottie, G.J.: Protocols for self-organization of a wireless sensor network. IEEE Personal Communications 7, 16–27 (2000)
14. Wadaa, A., Olariu, S., Wilson, L., Eltoweissy, M., Jones, K.: Training a wireless sensor network. Mob. Netw. Appl. 10(1-2), 151–168 (2005)
15. Yick, J., Mukherjee, B., Ghosal, D.: Wireless sensor network survey. Computer Networks 52(12), 2292–2330 (2008)

Complexity of Most Vital Nodes for Independent Set in Graphs Related to Tree Structures

Cristina Bazgan[1], Sonia Toubaline[1], and Zsolt Tuza[2,*]

[1] Université Paris-Dauphine, LAMSADE, France
{bazgan,toubaline}@lamsade.dauphine.fr
[2] Computer and Automation Institute, Hungarian Academy of Sciences, Budapest and Department of Computer Science and Systems Technology, University of Veszprém, Hungary
tuza@sztaki.hu

Abstract. Given an undirected graph with weights on its vertices, the k most vital nodes independent set problem consists of determining a set of k vertices whose removal results in the greatest decrease in the maximum weight of independent sets. We also consider the complementary problem, minimum node blocker independent set that consists of removing a subset of vertices of minimum size such that the maximum weight of independent sets in the remaining graph is at most a specified value. We show that these problems are NP-hard on bipartite graphs but polynomial-time solvable on unweighted bipartite graphs. Furthermore, these problems are polynomial also on graphs of bounded treewidth and cographs. A result on the non-existence of a ptas is presented, too.

Keywords: most vital nodes, independent set, complexity, NP-hard, ptas, bipartite graph, bounded treewidth, cograph.

Mathematics Subject Classification: 05C85, 05C69.

1 Introduction

In many applications involving the use of communication or transportation networks, we often need to identify vulnerable or critical infrastructures. By critical infrastructure we mean a set of nodes/lines whose damage causes the largest increase in the cost within the network. Modeling the network by a weighted graph, identifying a vulnerable infrastructure amounts to finding a subset of vertices/edges of a given size whose removal from the graph causes the largest inconvenience to a particular property of the graph in question. In the literature this problem is referred to as the *k most vital nodes/edges* problem. A complementary problem consists of determining a set of vertices/edges of minimum size whose removal involves that the cost within the network is at most a given value. In the literature this problem is referred to as the *min node/edge blocker* problem.

* Research supported in part by the Hungarian Scientific Research Fund, OTKA grant 81493.

C.S. Iliopoulos and W.F. Smyth (Eds.): IWOCA 2010, LNCS 6460, pp. 154–166, 2011.

The problems of k most vital nodes/edges and min node/edge blocker have been studied for various problems, including shortest path, spanning tree, maximum flow, assignment, and maximum matching. The k most vital edges problem with respect to shortest path was proved NP-hard [1]. Later, k most vital edges/nodes shortest path (and min node/edge blocker shortest path) were proved not 2-approximable (not 1.36-approximable, respectively) if P$\neq$ NP [8]. For spanning tree, k most vital edges is NP-hard [5] and $O(\log k)$-approximable [5] and randomized 2-approximable [15]. In [17] it is proved that k most vital edges maximum flow is NP-hard. Also k most vital edges and min edge blocker assignment are proved NP-hard and not 2-approximable (not 1.36-approximable, respectively) if P$\neq$NP [2]. For maximum matching, min edge blocker is NP-hard even for bipartite graphs [16], but polynomial for grids and trees [14].

In this paper, we are interested in determining a subset of k *vertices* of the graph whose deletion causes the largest decrease in the maximum weight of an independent set. This problem is referred to as k MOST VITAL NODES INDEPENDENT SET. We also consider the complementary version of this problem, where given a threshold, we have to determine a subset of vertices of minimum cardinality that has to be removed such that in the resulting graph the maximum weight of an independent set is at most this threshold. This problem is referred to as MIN NODE BLOCKER INDEPENDENT SET.

In Section 3 we consider bipartite graphs. It turns out that a substantial jump in complexity occurs between unweighted and weighted graphs for these problems. More precisely we show that the unweighted versions are polynomial while the weighted versions are NP-hard and the most vital nodes problem even has no ptas, unless P=NP. In Section 4 we deal with graphs with weights on their vertices, which have either a tree-like structure or a representation associated with trees. These include trees themselves, cycles, more generally graphs of bounded treewidth, and cographs (graphs containing no induced P_4). For these classes we design polynomial-time algorithms for the problems mentioned above.

In fact, trees and cycles have treewidth 1 and 2, respectively, therefore our general algorithm for bounded treewidth works for the former classes, too. Nevertheless, the algorithms on trees and cycles are simpler and this is why we include them here. It should be noted further that for k fixed, there are only polynomially many subsets of k removable vertices, therefore k MOST VITAL NODES INDEPENDENT SET is solvable efficiently on every graph class where the largest independent set is tractable. On the other hand if $k \to \infty$ then a formula expressing the present problems in second-order monadic logic would have unbounded length. Consequently, the general approach to linear-time algorithms via MSOL is not applicable here. This fact is relevant for both treewidth and cliquewidth.

2 Preliminaries

Let $G = (V, E)$ be an undirected graph, with $V = \{v_1, \ldots, v_n\}$, where each vertex v_i has a weight w_i. For an edge $v_iv_j \in E$, we could write $v_i, v_j \in e$ and if $v_i, v_j \in V'$ then we consider that $e \subset V'$. When removing a set V' of vertices from G, let us denote the remaining graph by $G - V'$. If H is a subgraph of G then $V(H)$ denotes the vertex set of H. Moreover, for a subset V' of vertices from G, the subgraph induced by V' is denoted by $G[V']$. A maximum-weight independent set of G is a subset of vertices of maximum total weight where any two vertices are nonadjacent. A minimum-weight vertex cover of G is a subset of vertices of minimum total weight where every edge of G has at least one vertex in the set. We denote by $\alpha(G)$ the maximum weight of an independent set and by $\tau(G)$ the minimum weight of a vertex cover. Moreover, $\alpha(k)$ represents the minimum of $\alpha(G - V')$ after removing any set of vertices V' of size k. A matching is a set of mutually vertex-disjoint edges. The largest number of edges in a matching is denoted by $\nu(G)$.

In this paper we are interested in the complexity of the following problems.

k Most Vital Nodes Independent Set
Input: An undirected graph $G = (V, E)$ where each vertex v_i has a weight w_i, and an integer k.
Output: A subset $V' \subseteq V$ of size k such that the maximum weight $\alpha(G - V')$ of an independent set in $G - V'$ is minimum.

Min Node Blocker Independent Set
Input: An undirected graph $G = (V, E)$ where each vertex v_i has a weight w_i, and an integer U.
Output: A subset $V' \subseteq V$ of minimum cardinality such that the maximum weight $\alpha(G - V')$ of an independent set in $G - V'$ is at most U.

Remark 1. The exact versions of k Most Vital Nodes Independent Set and Min Node Blocker Independent Set are polynomial-time equivalent. Indeed, if an algorithm $\mathcal{A}_k$ solves k Most Vital Nodes Independent Set for all $1 \leq k \leq n$, then we can run $\mathcal{A}_k$ for $k = 1, \dots, n$ and choose the smallest k yielding optimum at most U. Conversely, if an algorithm $\mathcal{B}_U$ solves Min Node Blocker Independent Set with any bound U, we can apply binary search to locate the smallest U that requires the removal of at most k vertices.

Theorem 1. *If there exists an algorithm that solves the k most vital nodes version of an optimization problem $\mathcal{P}$ on graphs with n vertices in $O(t)$ time, then the min node blocker version of $\mathcal{P}$ can be solved in $O(t \log\log n)$ time.*

Proof. If the value of an optimum solution is at most U then the optimum size is 0. Otherwise, we combine the algorithm for the k most vital nodes version with an accelerated version of *approximate binary search.* On the size k of a min node blocker we maintain a lower bound ℓ and an upper bound u, initialized to $\ell_0 = 1$ and $u_0 = n$. Instead of using a standard binary search with $v = \frac{\ell+u}{2}$, we iteratively set $v = \sqrt{\ell u}$, as suggested in [7]. More precisely, although computing the exact value $\sqrt{\ell u}$ can be time consuming, it is shown in [7] that an approximate value of $\sqrt{\ell u}$ can be computed without affecting the time complexity. The number of tests for obtaining a lower bound ℓ and an upper bound u such that

$u = \ell + 1$ is $O(\log\log \frac{u_0}{\ell_0})$ (see [7] for more details), which means $O(\log\log n)$ iterations in our case. Since one iteration takes $O(t)$, finding the smallest k for which the solution has value at most U takes total running time $O(t \log\log n)$.□

For the proof concerning the non-existence of a ptas (polynomial-time approximation scheme) we shall use an approximation-preserving reduction, called L-reduction, which was introduced by Papadimitriou and Yannakakis in [12]. Let A and B be two optimization problems. Then A is said to be *L-reducible* to B if there are two constants $a, b > 0$ such that

1. there exists a function, computable in polynomial time, which transforms each instance x of A into an instance x' of B such that $opt_B(x') \leq a \cdot opt_A(x)$,
2. there exists a function, computable in polynomial time, which transforms each solution y' of x' into a solution y of x such that $|val(x,y) - opt_A(x)| \leq b \cdot |val(x',y') - opt_B(x')|$.

For us the important property of this reduction is that if A is L-reducible to B and A has no ptas then B has no ptas.

3 Complexity on Bipartite Graphs

Maximum-weight independent set is polynomial-time solvable on bipartite graphs. We show in this section that the k most vital nodes or min node blocker versions become NP-hard on bipartite graphs, and most vital nodes has no ptas. Nevertheless, these problems remain polynomial-time solvable in the unweighted case. We first prove this latter fact.

Theorem 2. k Most Vital Nodes Independent Set *and also its complementary problem* Min Node Blocker Independent Set *are polynomial for unweighted bipartite graphs. Moreover, if a largest matching and a smallest vertex cover are given with the input, these problems are solvable in linear time.*

Proof. Let $G = (V, E)$ be a bipartite input graph on n vertices. From Kőnig's theorem [10] we know that $\tau(G) = \nu(G)$ holds; let us denote here their common value by t. The classical proof of the equality $\tau = \nu$ is algorithmic and also yields a maximum matching $M = \{e_1, \ldots, e_t\}$ and a minimum vertex cover $X = \{v_1, \ldots, v_t\}$ in polynomial time. Moreover, we have $\alpha(G) = n - t$ (known as a Gallai-type identity) and $V \setminus X$ is a largest independent set in G. Let us introduce the further notation $R = V \setminus V(M)$ and $r = |R| = n - 2t$; i.e., the number and the set of vertices not contained in any of the matching edges in M.

We can show now that these problems are solvable in linear time, as follows.

k Most Vital Nodes Independent Set

If $k \leq |R|$, we remove any k vertices from R. Since the remaining graph (of order $n - k$) still contains the matching M of size t, the independence number cannot be larger than $n - k - t$. It is also clear that α cannot be decreased by more than k if we remove just k vertices, hence the solution obtained is optimal.

If $k > |R|$, we remove the entire R and the vertices of $\lfloor(k-r)/2\rfloor$ edges from M, and one further vertex if $k-r$ is odd. This decreases the size of M by $\lceil(k-r)/2\rceil$ and the independence number by $\lfloor(k+r)/2\rfloor$, and hence the new value is $\lceil(n-k)/2\rceil$ (originally we had $\alpha(G) = (n+r)/2$). This decrease is optimal, because after the removal of k vertices at least half of the remaining $n-k$ belong to the same vertex class.

Min Node Blocker Independent Set

If $U \geq n-t$, no vertices need to be removed. If $t \leq U < n-t$, we remove $n-t-U$ vertices of R. If $U = t-\ell$ where $1 \leq \ell \leq t$, we remove the entire R and the 2ℓ vertices of ℓ arbitrarily chosen edges from M. All these choices are optimal, as follows from the proof concerning most vital nodes. □

We show in the following that these problems become NP-hard in the weighted case. The following notion will be of essence.

Definition 1. *Let $G = (V, E)$ be an undirected graph. The* bipartite incidence graph *of G is the bipartite graph H whose vertex set is $V \cup E$ and there is an edge in H between $v \in V$ and $e \in E$ if and only if e is incident to v in G.*

Theorem 3. *k Most Vital Nodes Independent Set and Min Node Blocker Independent Set are strongly NP-hard even for bipartite graphs.*

Proof. We first prove hardness for k Most Vital Nodes Independent Set. Let $G = (V, E)$ be an instance of the decision problem associated to Independent Set with n vertices and m edges, and an integer ℓ; and let H denote the bipartite incidence graph of G. The construction of H from G requires linear time only. Each vertex of E in H has weight 1 and each vertex of V in H has weight n^2. Due to this rather unbalanced weighting, the unique maximum-weight independent set in H is V; i.e., $\alpha(H) = n^3$.

We show in the following that if there is an independent set of size at least ℓ in G then H contains a set S of ℓ vertices such that $\alpha(H-S) = (n-\ell)n^2$, and otherwise removing any subset S of ℓ vertices from H, we have $\alpha(H-S) \geq (n-\ell)n^2+1$. Since vertices from V have weight n^2 and those from E have weight 1, in order to have a maximum-weight independent set as small as possible after removing a set S of size ℓ, S has to be included in V.

If G contains an independent set S of size ℓ, then removing S from the vertex set of H, we obtain a graph whose maximum-weight independent set is $V \setminus S$. This set has weight $(n-\ell)n^2$.

If G contains no independent set of size ℓ, then any $S \subset V$ of size ℓ contains at least an edge $e \in E$ in G, and this e in H is nonadjacent to the entire $V \setminus S$. Thus, when we remove any set S of ℓ vertices from H, $\alpha(H-S) \geq (n-\ell)n^2+1$.

Due to Remark 1, Min Node Blocker Independent Set is also strongly NP-hard. □

We are going to prove an approximation hardness result, too. In the reduction, the following problem will be used.

MAX k VERTEX COVER
Input: A graph $G = (V, E)$ with $k \leq |V|$.
Output: The maximum number of edges in G that can be covered by a subset $V' \subseteq V$ of cardinality k.

MAX k VERTEX COVER-B is the version of MAX k VERTEX COVER where the maximum degree of the graph is at most B.

We shall apply the following version of some known results.

Lemma 1. *For appropriately chosen B, MAX $n/2$ VERTEX COVER-B has no ptas on graphs $G = (V, E)$ with $m = \Theta(n)$ and $\alpha(G) = \tau(G) = n/2$, where $n = |V|$ and $m = |E|$, unless P=NP.*

Proof. An approximation algorithm for VERTEX COVER on graphs with $\tau(G) \geq |V(G)|/2$ is also an approximation algorithm with the same ratio for general instances of VERTEX COVER [11]. Using the APX-hardness of VERTEX COVER-B [12] and the gap reduction from VERTEX COVER-B to MAX k VERTEX COVER-B [13] for $k \geq n/2$, we conclude that MAX k VERTEX COVER-B has no ptas on graphs with n vertices when $k \geq \tau(G) \geq n/2$. We can reduce this last problem to the same problem on instances with $k \geq \tau = n/2$ by inserting $2\tau - n$ isolated vertices. Moreover, these last instances are reducible to instances where $k = 2\tau = n/2$ by inserting $k - \tau$ isolated edges. □

We extract the key points of the reduction in the following lemma on independent sets.

Lemma 2. *Let $G = (V, E)$ be a graph with n vertices and m edges, and let H be the bipartite incidence graph of G. Then the following properties are valid.*

(a) Suppose that G has maximum degree at most B, and the weights in H are $w_v = B + 1$ for all $v \in V$ and $w_e = 1$ for all $e \in E$. Then, for any $V' \subset V$ and any independent set S disjoint from V' in H, there exists an independent set S' of H such that $w(S') \geq w(S)$ and $S' \cap V = V \setminus V'$. Thus, if S' is maximal, then

$$S' = (V \setminus V') \cup \{e \in E \mid e \subset V'\}$$

and, in particular, $\alpha(H - V') \geq (B + 1) \cdot (n - |V'|)$.

(b) Under the conditions of (a), for any $V' \subset V \cup E$ with $|V'| < |V|$ there exists a $V'' \subset V$ such that $|V''| = |V'|$ and the maximum weight of an independent set in $H - V''$ is not larger than that in $H - V'$. As a consequence,

$$\alpha(H - V') \geq \alpha(H - V'') = (B + 1) \cdot (n - |V'|) + |\{e \in E \mid e \subset V''\}|.$$

Moreover, the set V'' can be found efficiently.

Proof. (a) If S contains all vertices of $V \setminus V'$, then we have nothing to prove. Otherwise we modify S step by step, keeping it independent and not decreasing its value, until it contains the entire $V \setminus V'$. Hence, assume that $v \in V$ is a vertex such that $v \notin V' \cup S$. If v has no neighbor in $S \cap E$, then $S \cup \{v\}$ is a proper extension. Suppose that this is not the case; i.e., there is an edge $e \in E \cap S$ such that $v \in e$. We now modify S to $(S \setminus N_H(v)) \cup \{v\}$, where $N_H(v)$ denotes the set of vertices adjacent to v in H, that is the set of edges incident to v in G. In this way we have removed at most B neighbors of v from S, each of weight 1, and inserted v of weight $B+1$, hence the total weight of the modified set is at least $w(S)$. Moreover, the set remains independent because all neighbors of v have been removed. Thus, after $|(V \setminus V') \setminus S|$ steps, the required set S' is obtained.

(b) If $V' \subset V$, then $V'' = V'$ is a proper choice. Hence suppose $V' \cap E \neq \emptyset$. Let us introduce the notation $n' = |V' \cap V|$, $m' = |E(G[V' \cap V]) \setminus (V' \cap E)|$. By ($a$) we see that $\alpha(H - V') = (B+1) \cdot (n - n') + m'$ holds. Choose $e \in V' \cap E$ and $v \in V \setminus V'$, and modify V' to the set $(V' \setminus \{e\}) \cup \{v\}$. This keeps cardinality unchanged, while the first term $(B+1) \cdot (n - n')$ decreases by precisely $B+1$. Moreover, since G has maximum degree at most B, the second term m' can increase by at most B when we insert v into the set, and can further increase by at most 1 when we omit e. Thus, sum does not increase. Repeatedly eliminating all $e \in E$ from V', the required V'' is obtained. Then (a) implies that the independent set of maximum weight in $H - V''$ consists of all $v \notin V''$ and all $e \subset V''$. □

Theorem 4. k Most Vital Nodes Independent Set *has no ptas even for bipartite graphs if* $P \neq NP$.

Proof. We prove the non-existence of a ptas for $k = n/2$, constructing an L-reduction from Max $n/2$ Vertex Cover-B to $n/2$ Most Vital Nodes Independent Set, where instances of the former problem are restricted to graphs G of maximum degree at most B and also satisfying $\alpha(G) = \tau(G) = n/2$. In this case, let H denote the bipartite incidence graph of the input graph $G = (V, E)$, the latter having n vertices and m edges. The vertices of H have weight $w_v = B + 1$ for all $v \in V$ and $w_e = 1$ for all $e \in E$.

Consider first an optimum solution V' in G. As $\tau(G) = n/2$ has been assumed, $opt_1 = m$ holds and V' covers all edges of G. Then removing $V \setminus V'$ from the vertex set of H, we obtain a graph in which the maximum weight of an independent set is $((B+1)/2) \cdot n$, as implied by part (a) of Lemma 2. On the other hand, parts (a) and (b) together yield that after the removal of any $n/2$ vertices from H, there always remains an independent set of at least that large weight, thus

$$opt_2 = \frac{B+1}{2} \cdot n \leq (B+1) \cdot opt_1,$$

the upper bound being valid since $opt_1 \geq n/2$ surely holds by the assumption $\tau(G) = n/2$.

Consider now any subset V' of $n/2$ vertices in H, and denote $val_2 = \alpha(H - V')$. Now we apply part (b) of Lemma 2 to obtain an appropriate set V'' of $n/2$ vertices, which is a subset of V. We view $V \setminus V''$ as a solution in G and denote its value by val_1. In this way we obtain

$$val_2 - opt_2 \geq \alpha(H - V'') - opt_2 = ((B+1) \cdot (n - |V''|) + |E(G[V''])|) - \frac{B+1}{2} \cdot n$$

$$= |E(G[V''])| = opt_1 - val_1,$$

the last equation being valid because $opt_1 = m$ and $E(G[V''])$ is precisely the set of edges not covered by the vertices of $V \setminus V''$. This completes the proof of the theorem. □

4 Graph Classes Related to Tree Structures

In this section we consider graph classes representable over tree structures, and prove that they admit algorithms solving the considered problems in polynomial time. Efficient solvability for the graph classes in the first two subsections are implied by the results of the third subsection, too, but the methods for the former are simpler.

4.1 Trees

Theorem 5. k MOST VITAL NODES INDEPENDENT SET *is polynomial on trees. On trees of order n the algorithm runs in $O(nk^2)$ time, for any $k \geq 1$.*

Proof. Our general approach is to find not only a set of k most vital nodes but simultaneously also the value of a corresponding largest independent set. For this purpose we view the input as a *rooted* tree with an arbitrarily chosen root, and organize computation according to a postorder traversal.

Consider any tree T with vertices $v_1, \ldots, v_n$. Each vertex v_i can have three positions in a solution, that we shall denote by marks $+, -, 0$ as follows:

- '+' means that v_i is selected into an independent set;
- '−' means that v_i is selected for deletion;
- '0' means that v_i is none of the above two types.

In a solution exactly k marks '−' have to occur.

The subtree rooted in v_i is denoted by T_i. For each $i = 1, \ldots, n$, each $* \in \{+, -, 0\}$, and each $j = 0, 1, \ldots, k$, a value $z_i(j, *)$ will be computed. This $z_i(j, *)$ represents the minimum achievable weight of a largest independent set on T_i under the conditions that *exactly* j vertices are removed from T_i and v_i has mark $*$. For the recursive computation the children of v_i with degree d will be denoted by $v_{i_1}, \ldots, v_{i_d}$. We traverse T in postorder and apply dynamic programming.

Recursion. If v_i is marked '+', then all its children must have '−' or '0', since otherwise two vertices selected for the independent set would be adjacent. Moreover, $z_i(j, *)$ requires that the total number of vertices marked '−' should be exactly j in T_i. On the other hand, we have one and only one way to make the final result as small as possible: decide which of the vertices should be marked with '−'. Once this has been decided, the distribution of '+' and '0' positions aims at maximizing the total weight of '+'. This leads to the following general recursions:

$$z_i(j, +) = w_i + \min_{\substack{j_1, \ldots, j_d \geq 0 \\ j_1 + \ldots + j_d = j}} \sum_{\ell=1}^{d} \min\left(z_{i_\ell}(j_\ell, -), z_{i_\ell}(j_\ell, 0)\right),$$

$$z_i(j,-) = \min_{\substack{j_1,\ldots,j_d \geq 0 \\ j_1+\ldots+j_d=j-1}} \sum_{\ell=1}^{d} \min\left(z_{i_\ell}(j_\ell,-), \max\left(z_{i_\ell}(j_\ell,+), z_{i_\ell}(j_\ell,0)\right)\right),$$

$$z_i(j,0) = \min_{\substack{j_1,\ldots,j_d \geq 0 \\ j_1+\ldots+j_d=j}} \sum_{\ell=1}^{d} \min\left(z_{i_\ell}(j_\ell,-), \max\left(z_{i_\ell}(j_\ell,+), z_{i_\ell}(j_\ell,0)\right)\right),$$

For a leaf v_i we clearly have $z_i(0,+) = w_i$ and $z_i(1,-) = z_i(0,0) = 0$. Further, to indicate that all other combinations of $j \in \{0,1,\ldots,k\}$ and $* \in \{+,-,0\}$ are infeasible, we set a dummy symbol $z_i(j,*) = \mathsf{NIL}$ for them. In the recursive step, terms with value NIL on the right-hand side are neglected, except when all terms are the same, and in this case we define $z_i(j,*) = \mathsf{NIL}$, too.

Finding an optimal solution. Assuming that T has root v_{i_0}, after the removal of k properly chosen vertices the smallest possible value of α is

$$\min\left(z_{i_0}(k,-), \max\left(z_{i_0}(k,+), z_{i_0}(k,0)\right)\right).$$

In fact, inserting a new vertex v_0 with weight $w_0 = 0$ as new root and parent for v_{i_0} does not change the optimum, and then we would have $z_0(k,+) \leq opt = z_0(k,0) \leq z_0(k,-)$. A set of k most vital nodes can also be determined in $O(n)$ additional steps if we make a little more administration. At the recursive step for each $z_i(j,*)$ we register for each edge $v_i v_{i_\ell}$ the corresponding value of j_ℓ in the optimal distribution $(j_1,\ldots,j_d)$ for j and also the mark $* \in \{+,-,0\}$ of i_ℓ which gave the optimum for v_i. Once these data are available for all v_i and all pairs $(j,*)$, we can traverse T in preorder and select the vertices having '$-$' mark for the most vital set.

Efficient implementation. The key point is to find in polynomial time a best distribution $(j_1,\ldots,j_d)$ for the 'max' and 'min' functions acting on the sums. This can be done, despite that the number of possibilities can even be exponential if d is proportional to n.

If $d = 2$ then we have at most $j+1$ combinations of feasible pairs j_1, j_2. Hence, optimal choice can be made in $O(k)$ steps for any one particular j, and in $O(k^2)$ steps for all $0 \leq j \leq k$. If d is larger, we can split the children of v_i into two sets of (nearly) equal size, $\{v_\ell \mid 1 \leq \ell \leq \lfloor d/2 \rfloor\}$ and $\{v_\ell \mid \lfloor d/2 \rfloor + 1 \leq \ell \leq d\}$, make all computation in them separately, and then combine the results for v_i. (Splitting corresponds to inserting a 'supernode' above each of the two sets, which has weight zero and becomes a virtual child of v_i.) This requires $d-1$ rounds for v_i. Since T is a tree, those $d-1$ sum up to $n-2$, thus the overall running time is $O((k^2+1)n)$, and never exceeds $O(n^3)$. (Here '+1' is needed for $k = 0$.) Note that there are no 'hidden large constants' in the 'O' notation. □

Theorem 6. Min Node Blocker Independent Set *is polynomial on trees. On trees with n vertices the algorithm runs in $O(n^3 \log\log n)$ time.*

Proof. The above algorithm in one iteration for any $1 \leq v \leq n$ runs in $O(v^2 n) = O(n^3)$ time. Hence, using Theorem 1, finding the smallest k for which the solution has value at most U takes total running time $O(n^3 \log\log n)$. □

Remark 2. The algorithm proposed in Theorem 5 solves the k MOST VITAL NODES INDEPENDENT SET problem on *paths* in $O(kn)$ time. In fact, in the general time bound $O(nk^2)$ for trees, the factor k^2 occurs due to the presence of vertices with more than one child. This observation implies further that the algorithm proposed in Theorem 6 solves MIN NODE BLOCKER INDEPENDENT SET on paths in $O(n^2 \log \log n)$ time.

4.2 Cycles

Theorem 7. *k MOST VITAL NODES INDEPENDENT SET is polynomial on cycles. On cycles of order n the algorithm runs in $O(kn^2)$ time, for any $k \geq 1$.*

Proof. Let $S^* = \{v_1, \dots, v_r\} \subset V$ be a maximum-weight independent set of a given cycle $C = (V, E)$. An optimal solution $V' \subset V$ of k MOST VITAL NODES INDEPENDENT SET must contain at least one node of S^*, since otherwise $\alpha(C - V')$ is not smaller than $\alpha(C)$. Thus, for each $v_j \in S^*$, $j = 1, \dots, r$, we determine the $k-1$ further nodes to remove in the resulting path as follows. We delete v_j from C and determine a maximum-weight independent set in the resulting path $C - v_j$ by applying the algorithm given in Theorem 5 in order to find an optimal solution $R_j^* \subset V \setminus \{v_j\}$ of $k-1$ MOST VITAL NODES INDEPENDENT SET on the path $C - v_j$. Then, an optimal solution for k MOST VITAL NODES INDEPENDENT SET on C is $R_\ell^* \cup \{v_\ell\}$ such that $\alpha(C - v_\ell - R_\ell^*) = \min_{1 \leq j \leq r} \alpha(C - v_j - R_j^*)$. If the root is chosen to be an endpoint of the path, the complexity of the algorithm given in Theorem 5 for path $C - v_j$ is $O(kn)$. Since $|S^*| \leq n$, in this way k MOST VITAL NODES INDEPENDENT SET is solved in $O(kn^2)$. □

Theorem 8. MIN NODE BLOCKER INDEPENDENT SET *is polynomial on cycles. On cycles of order n the algorithm runs in $O(n^3 \log \log n)$ time.*

Proof. The theorem follows from Theorem 7 and Theorem 1. □

4.3 Graphs of Bounded Treewidth

A *tree decomposition* of a graph $G = (V, E)$ without isolated vertices is a pair $(T, \mathcal{X})$ where

- $T = (X, F)$ is a tree graph with a set $X = \{x_1, \dots, x_m\}$ of *nodes* and a set F of *lines*;
- $\mathcal{X} = \{X_1, \dots, X_m\}$ is a set system over V (i.e., over the vertex set of G), where each X_q is associated with node x_q of T;
- each edge $v_i v_j \in E$ of G is contained in at least one X_q for some $1 \leq q \leq m$;
- for any $v_i \in V$, if $v_i \in X_{q'}$ and $v_i \in X_{q''}$, then $v_i \in X_q$ for all q such that x_q lies on the $x_{q'}$–$x_{q''}$ path in T.

The width of $(T, \mathcal{X})$ is $\max_{1 \leq q \leq m} |X_q| - 1$, and the *treewidth* of G, denoted by $tw(G)$, is the smallest integer t for which G admits a tree decomposition of width t. For undefined details on tree decomposition we refer to [9].

Theorem 9. *k MOST VITAL NODES INDEPENDENT SET is polynomial on bounded treewidth graphs. On graphs of order n, the algorithm runs in $O(nk^2)$ time for any $k \geq 1$.*

Due to space limitation, the proof of this result is omitted and will appear in the extended version of the paper.

Theorem 10. *MIN NODE BLOCKER INDEPENDENT SET is polynomial on bounded treewidth graphs. On graphs of order n the algorithm runs in $O(n^3 \log\log n)$ time.*

Proof. The result follows from Theorem 9 and Theorem 1. □

4.4 Cographs

To each cograph G with n vertices, we can associate a rooted tree T, called the *cotree* of G. Leaves of T correspond to vertices of the graph G and internal nodes of T are labeled with either '$\cup$' (union-node) or '$\times$' (join-node). A subtree rooted at node '$\cup$' corresponds to the union of the subgraphs defined by the children of that node, and a subtree rooted at node '$\times$' corresponds to the join of the subgraphs defined by the children of that node; that is, we add an edge between every two vertices corresponding to leaves in different subtrees. Cographs can be recognized in linear time and the cotree representation can be obtained efficiently [4,6]. Moreover, this cotree can easily be transformed in linear time to a binary cotree with $O(n)$ nodes.

Theorem 11. *k MOST VITAL NODES INDEPENDENT SET is polynomial on cographs. On cographs of order n, the algorithm runs in $O(nk^2)$ time, for any $k \geq 1$.*

Proof. Consider a cograph G with n vertices $v_1, \ldots, v_n$. Given a binary cotree representation T of G, we show in the following how to solve the k MOST VITAL NODES INDEPENDENT SET using dynamic programming.

Let $x_1, \ldots, x_t$ be the nodes of T where x_r is its root and t is in $O(n)$. For $i = 1, \ldots, t$, denote by T_i the subtree rooted at x_i, G_i the subgraph induced by the vertices corresponding to the leaves of T_i, and V_i these vertices.

Recursion. We associate a $(k+1)$-vector to each node x_i of T, $i = 1, \ldots, t$. In the following, a $(k+1)$-*vector* is simply call a vector. For each i and each $j = 0, 1, \ldots, k$, we compute $z_i(j)$ that is the minimum weight of a maximum independent set on G_i where exactly j vertices are removed from G_i. These vectors are computed "bottom-up" in the cotree. So, we start by computing vectors of leaves and after that the vector of an internal node if the vectors of its two children are already computed.

Given a node x_i of the cotree, the corresponding vector is obtained as follows:

- If x_i is a union-node with two children x_ℓ and x_r, we have no edges between G_ℓ and G_r. Then the maximum independent set in G_i is the union of those in G_ℓ and G_r. Thus, since we want to find a maximum-weight independent set as small as possible, the best choice is given by $z_i(j) = \min_{j_1+j_2=j}(z_\ell(j_1) + z_r(j_2))$.

– If x_i is a join-node with two children x_ℓ and x_r, every vertex in V_ℓ is adjacent to every vertex in V_r. Then each independent set in G_i is entirely contained either in G_ℓ or in G_r. So, $z_i(j) = \min_{j_1+j_2=j}(\max(z_\ell(j_1), z_r(j_2)))$.
– If x_i is a leaf then $z_i(0) = w_i$, $z_i(1) = 0$, and $z_i(j) = \mathsf{NIL}$ for $j = 2, \ldots, k$ which means that the latter configurations are infeasible. In the recursive step, terms with value NIL on the right-hand side are neglected, except when all terms are the same, and in this case we define $z_i(j) = \mathsf{NIL}$, too.

Finding an optimal solution. An optimal solution is obtained at the root x_r of T and its weight is equal to $z_r(k)$. Moreover, an optimal set of k removed vertices can be computed step by step in the recursion. Indeed, let $S_i^-(j)$ be the subset of j removed vertices in G_i. For a leaf x_i we have $S_i^-(0) = \emptyset$, $S_i^-(1) = \{v_i\}$ and $S_i^-(j) = \emptyset$ for $j = 2, \ldots, k$. For a union-node or a join-node x_i with two children x_ℓ and x_r, recursion yields $S_i^-(j) = S_\ell^-(j_1^*) \cup S_r^-(j_2^*)$ where j_1^* and j_2^* are the indices that realize the minimum for $z_i(j)$.

Time analysis. For k MOST VITAL NODES INDEPENDENT SET, vectors are computed in $O(k)$ for each leaf and in $O(k^2)$ for each union-node and each join-node. Since $t = O(n)$, the algorithm runs in $O(nk^2)$. □

Theorem 12. MIN NODE BLOCKER INDEPENDENT SET *is polynomial on cographs. On cographs of order n, the algorithm runs in $O(n^3 \log\log n)$ time.*

Proof. The theorem follows from Theorem 11 and Theorem 1. □

5 Conclusion

In this paper we studied the complexity of the k most vital nodes and min node blocker versions of the maximum-weight independent set problem. While maximum-weight independent set is polynomial on bipartite graphs, the k most vital nodes and min node blocker versions become NP-hard. Nevertheless the unweighted versions remain polynomial on bipartite graphs. In a graph, a maximum-weight independent set is the complementary set of a minimum-weight vertex cover. In sharp contrast to this, however, concerning the k most vital nodes or min node blocker versions an optimum solution for maximum-weight independent set may be substantially different from an optimum solution for minimum-weight vertex cover. Our results on the latter will be included in an extended paper. We show in this paper that the k most vital nodes version has no ptas. An interesting open question would be to establish other positive and negative results concerning the approximability of these versions. In particular it remains open to decide weather min node blocker on bipartite graphs has a ptas.

Another interesting perspective is to study the complexity of the k most vital nodes and min node blocker versions of the maximum-weight independent set problem for graphs of bounded cliquewidth and graphs of bounded NLC-width, that generalize cographs. Moreover, the study of the complexity and the approximation of these versions for further classes of graphs for which maximum-weight independent set and minimum-weight vertex cover are polynomial is also of interest.

Note added in Proof. The time bound in Theorem 1 and its applications should read $O(t(\log\log n + \log 1/\epsilon))$ and is meant for finding a $(1+\epsilon)$-approximation.

References

1. Bar-Noy, A., Khuller, S., Schieber, B.: The complexity of finding most vital arcs and nodes, Technical Report CS-TR-3539, Department of Computer Science, University of Maryland (1995)
2. Bazgan, C., Toubaline, S., VanderPoten, D.: Détermination des éléments les plus vitaux pour le problème d'affectation. In: Actes de la 10ème Conférence de la Société Française de Recherche Opérationelle et d'Aide à la Décision, ROADEF 2010 (2010)
3. Bodlaender, H.L.: A linear time algorithm for finding tree-decompositions of small treewidth. SIAM Journal on Computing 25, 1305–1317 (1996)
4. Corneil, D.G., Perl, Y., Stewart, L.K.: A linear recognition algorithm for cographs. SIAM Journal on Computing 14(4), 926–934 (1985)
5. Frederickson, G.N., Solis-Oba, R.: Increasing the weight of minimum spanning trees. In: Proceedings of the 7th Annual ACM-SIAM Symposium on Discrete Algorithms (SODA 1996), pp. 539–546 (1996)
6. Habib, M., Paul, C.: A simple linear time algorithm for cograph recognition. Discrete Applied Mathematics 145(2), 183–197 (2005)
7. Hassin, R.: Approximation schemes for the restricted shortest path. Mathematics of Operations Research 17(1), 36–42 (1992)
8. Khachiyan, L., Boros, E., Borys, K., Elbassioni, K., Gurvich, V., Rudolf, G., Zhao, J.: On short paths interdiction problems: total and node-wise limited interdiction. Theory of Computing Systems 43(2), 204–233 (2008)
9. Kloks, T.: Treewidth, computations and approximations. LNCS, vol. 842. Springer, Heidelberg (1994)
10. Kőnig, D.: Graphs and matrices. Math. Fiz. Lapok 38, 116–119 (1931) (in Hungarian)
11. Nemhauser, G.L., Trotter, L.E.: Vertex packing: structural properties and algorithms. Mathematical Programming 8, 232–248 (1975)
12. Papadimitriou, C., Yannakakis, M.: Optimization, approximation and complexity classes. Journal of Computer and System Science 43(3), 425–440 (1991)
13. Petrank, E.: The hardness of approximation: gap location. Computational Complexity 4, 133–157 (1994)
14. Ries, B., Bentz, C., Picouleau, C., de Werra, D., Costa, M., Zenklusen, R.: Blockers and Transversals in some subclasses of bipartite graphs: When caterpillars are dancing on a grid. Discrete Mathematics 310(1), 132–146 (2010)
15. Shen, H.: Finding the k most vital edges with respect to minimum spanning tree. Acta Informatica 36, 405–424 (1999)
16. Zenklusen, R., Ries, B., Picouleau, C., de Werra, D., Costa, M., Bentz, C.: Blockers and Transversals. Discrete Mathematics 309(13), 4306–4314 (2009)
17. Wood, R.K.: Deterministic network interdiction. Mathematical and Computer Modeling 17(2), 1–18 (1993)

Computing Role Assignments of Proper Interval Graphs in Polynomial Time*

Pinar Heggernes[1], Pim van 't Hof[2], and Daniël Paulusma[2]

[1] Department of Informatics, University of Bergen,
P.O. Box 7803, N-5020 Bergen, Norway
pinar.heggernes@ii.uib.no

[2] School of Engineering and Computing Sciences, Durham University,
Science Laboratories, South Road, Durham DH1 3LE, England
pimvanthof@gmail.com, daniel.paulusma@durham.ac.uk

Abstract. A homomorphism from a graph G to a graph R is locally surjective if its restriction to the neighborhood of each vertex of G is surjective. Such a homomorphism is also called an R-role assignment of G. Role assignments have applications in distributed computing, social network theory, and topological graph theory. The Role Assignment problem has as input a pair of graphs (G, R) and asks whether G has an R-role assignment. This problem is NP-complete already on input pairs (G, R) where R is a path on three vertices. So far, the only known non-trivial tractable case consists of input pairs (G, R) where G is a tree. We present a polynomial time algorithm that solves Role Assignment on all input pairs (G, R) where G is a proper interval graph. Thus we identify the first graph class other than trees on which the problem is tractable. As a complementary result, we show that the problem is Graph Isomorphism-hard on chordal graphs, a superclass of proper interval graphs and trees.

1 Introduction

Graph homomorphisms form a natural generalization of graph colorings: there is a homomorphism from a graph G to the complete graph on k vertices if and only if G is k-colorable. A *homomorphism* from a graph $G = (V_G, E_G)$ to a graph $R = (V_R, E_R)$ is a mapping $r : V_G \to V_R$ that maps adjacent vertices of G to adjacent vertices of R, i.e., $r(u)r(v) \in E_R$ whenever $uv \in E_G$. A homomorphism r from G to R is *locally surjective* if the following is true for every vertex u of G: for every neighbor y of $r(u)$ in R, there is a neighbor v of u in G with $r(v) = y$. We also call such an r an *R-role assignment*. See Figure 1 for an example.

Role assignments originate in the theory of social behavior [7,19]. A role graph R models roles and their relationships, and for a given society we can ask if its individuals can be assigned roles such that relationships are preserved: each person playing a particular role has exactly the roles prescribed by the model among

* This work has been supported by EPSRC (EP/D053633/1 and EP/G043434/1) and by the Research Council of Norway.

C.S. Iliopoulos and W.F. Smyth (Eds.): IWOCA 2010, LNCS 6460, pp. 167–180, 2011.

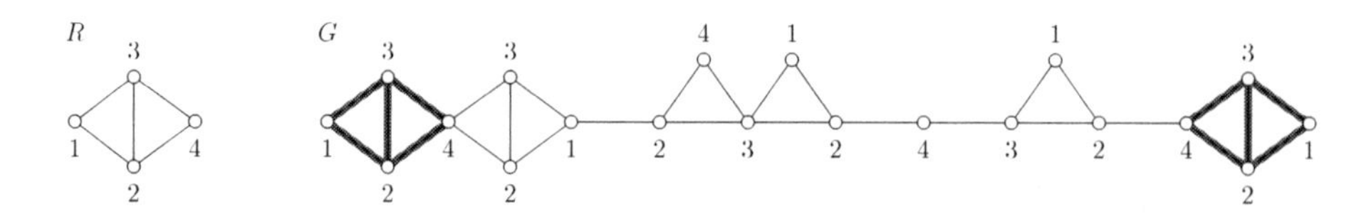

Fig. 1. A graph R and a proper interval graph G with an R-role assignment

its neighbors. In this way, a large network of individuals can be compressed into a smaller network that still gives some description of the large network. Role assignments are also useful in the area of distributed computing, in which one of the fundamental problems is to arrive at a final configuration where all processors have been assigned unique identities. Chalopin et al. [4] show that, under a particular communication model, this problem can be solved on a graph G representing the distributed system if and only if G has no R-role assignment for a graph R with fewer vertices than G. Role assignments are useful in topological graph theory as well, where a main question is which graphs G allow role assignments to planar graphs R [21].

The Role Assignment problem has as input a pair of graphs (G, R) and asks whether G has an R-role assignment. It is NP-complete on arbitrary graphs G, even when R is any fixed connected bipartite graph on at least three vertices [10]. Hence, for polynomial time solvability, our only hope is to put restrictions on G. So far, the only known non-trivial graph class that gives tractability is the class of trees: Role Assignment is polynomial time solvable on input pairs (G, R) where G is a tree and R is arbitrary [11]. Are there other graph classes on which Role Assignment can be solved in polynomial time?

We show that Role Assignment can be solved in polynomial time on input pairs (G, R) where G is a proper interval graph and R is arbitrary. Our work is motivated by the above question and continues the research direction of Sheng [23], who characterizes proper interval graphs that have an R-role assignment for some fixed role graphs R with a small number of vertices. Proper interval graphs, also known as unit interval graphs or indifference graphs, are widely known due to their many theoretical and practical applications [3,14,22]. By our result, they form the first graph class other than trees on which Role Assignment is shown to be polynomial time solvable. To obtain our algorithm we prove structural properties of clique paths of proper interval graphs related to role assignments. This enables us to give an additional result, namely a polynomial time algorithm for the problem of deciding whether there exists a graph R with fewer vertices than a given proper interval graph G such that G has an R-role assignment. Recall that this problem stems from the area of distributed computing [4]. It is co-NP-complete in general [5]. Finally, to indicate that Role Assignment might remain hard on larger graph classes, we show that it is Graph Isomorphism-hard for input pairs (G, R) where G belongs to the class of chordal graphs, a superclass of both proper interval graphs and trees.

2 Preliminaries

All graphs considered in this paper are undirected, finite and simple, i.e., without loops or multiple edges. A graph is denoted $G = (V_G, E_G)$, where V_G is the set of vertices and E_G is the set of edges. We will use the convention that $n = |V_G|$ and $m = |E_G|$. For a vertex u of G, $N_G(u) = \{v \mid uv \in E_G\}$ denotes the set of *neighbors* of u, also called the *neighborhood* of u. The *degree* of a vertex u is $\deg_G(u) = |N_G(u)|$. A graph $H = (V_H, E_H)$ is a *subgraph* of G if $V_H \subseteq V_G$ and $E_H \subseteq E_G$. For $U \subseteq V_G$, the graph $G[U] = (U, \{uv \in E_G \mid u, v \in U\})$ is called the subgraph of G *induced* by U. A graph is *complete* if it has an edge between every pair of vertices. A set of vertices $A \subseteq V_G$ is a *clique* if $G[A]$ is complete. A clique is *maximal* if it is not a proper subset of any other clique.

An *isomorphism* from a graph G to a graph H is a bijective mapping $f : V_G \to V_H$ such that for any two vertices $u, v \in E_G$, we have $uv \in E_G$ if and only if $f(u)f(v) \in E_H$. We say that G is *isomorphic* to H and write $G \simeq H$.

Let u and v be two vertices of a graph G. Then a *path* between u and v is a sequence of distinct vertices $P = u_1u_2\cdots u_p$ starting at $u_1 = u$ and ending at $u_p = v$, where each pair of consecutive vertices u_i, u_{i+1} forms an edge of G. If uv is an edge as well we obtain a *cycle*. Sometimes we fix an orientation of P. In that case we write $u_i\overrightarrow{P}u_j = u_iu_{i+1}\cdots u_j$ and $u_j\overleftarrow{P}u_i = u_ju_{j-1}\cdots u_i$ to denote the subpath from u_i to u_j, or from u_j to u_i, respectively. The *length* of a path or cycle is its number of edges. The set of vertices of a path or cycle P is denoted by V_P. A graph is *connected* if there is a path between every pair of vertices. A *connected component* of G is a maximal connected subgraph of G.

Let $A_1, \ldots, A_p$ be a sequence of sets. For $i = 1, \ldots, p$, we use shorthand notation $A_{\leq i} = A_1 \cup \cdots \cup A_i$ and $A_{\geq i} = A_i \cup \cdots \cup A_p$.

2.1 Chordal, Interval, and Proper Interval Graphs

A graph isomorphic to the graph $K_{1,3} = (\{a, b_1, b_2, b_3\}, \{ab_1, ab_2, ab_3\})$ is called a *claw* with *center* a and *leaves* b_1, b_2, b_3. A graph is called *claw-free* if it does not have a claw as an induced subgraph. An *asteroidal triple (AT)* in a graph G is a set of three mutually nonadjacent vertices u_1, u_2, u_3 such that G contains a path P_{ij} from u_i to u_j with $P_{ij} \cap N_G(u_k) = \emptyset$ for all distinct $i, j, k \in \{1, 2, 3\}$. A graph is called *AT-free* if it does not have an AT.

A graph is *chordal* if it contains no induced cycle of length at least 4. A graph is an *interval graph* if intervals of the real line can be associated with its vertices such that two vertices are adjacent if and only if their corresponding intervals overlap. Interval graphs are a subclass of chordal graphs: a chordal graph is an interval graph if and only if it is AT-free [17].

The following characterization of interval graphs is also well known. Let G be a connected graph with maximal cliques $K_1, \ldots, K_p$ and let $\mathcal{K}_v$ denote the set of maximal cliques in G containing vertex $v \in V_G$. Then G is an interval graph if and only if G has a path decomposition that is a *clique path* [12], i.e., a path $P = K_1 \cdots K_p$ such that for each $v \in V_G$ the set $\mathcal{K}_v$ induces a connected subpath in P. We say that the maximal cliques of G are the *bags* of P. A bag K_i

introduces a vertex u of G if $u \in K_i$ for $i = 1$ or $u \in K_i \setminus K_{i-1}$ for some $i \geq 2$. In that case, by the definition of a clique path, u is not in a bag K_h with $h \leq i-1$. If $u \in K_i$ for $i = p$ or $u \in K_i \setminus K_{i+1}$ for some $i \leq p-1$, then we say that K_i *forgets* u. Note that every bag introduces at least one vertex, and forgets at least one vertex. Because G is connected, we also observe that each bag, except K_1, contains at least one vertex from a previous bag. We denote the index of the bag in P that introduces a vertex u (the *first* bag in which u appears) by $f_P(u)$ and the index of the bag that forgets u (the *last* bag in which u appears) by $l_P(u)$. We say that u *transcends* a vertex v in P if $f_P(u) < f_P(v)$ and $l_P(v) < l_P(u)$. A clique path has at most n bags, and can be constructed in linear time (see e.g. [12]).

An interval graph is *proper interval* if it has an interval representation in which no interval is properly contained in any other interval. An interval graph is a proper interval graph if and only if it is claw-free [22]. Equivalently, a chordal graph is a proper interval graph if and only if it is AT-free and claw-free. Chordal graphs, interval graphs, and proper interval graphs can all be recognized in linear time, and have at most n maximal cliques (see e.g. [3,14]). The following theorem will be used heavily in our proofs.

Theorem 1 ([15]). *A connected chordal graph is a proper interval graph if and only if it has a unique clique path in which no vertex transcends any other vertex.*

Two adjacent vertices u and v of a graph G are *twins* if $N_G(u) \cup \{u\} = N_G(v) \cup \{v\}$. Let G be a connected proper interval graph with clique path $P = K_1 \cdots K_p$. Note that two vertices u and v of G are twins if and only if $f_P(u) = f_P(v)$ and $l_P(u) = l_P(v)$. We partition V_G into sets of twins. A vertex that has no twin appears in its twin set alone. We order the twin sets with respect to P, and label them $T_1, \ldots, T_S$, in such a way that $i < j$ if and only if for all $u \in T_i, v \in T_j$, either $f_P(u) < f_P(v)$, or $f_P(u) = f_P(v)$ and $l_P(u) < l_P(v)$. We call $T_1, \ldots, T_s$ the *ordered twin sets* of G. The following observation immediately follows from this definition and the definition of a clique path. Hence, this observation is even valid for interval graphs that are not proper.

Observation 1. *Let G be a connected proper interval graph with clique path $P = K_1 \ldots K_p$ and ordered twin sets $T_1, \ldots, T_s$. Then for $h = 1, \ldots, s-1$, there exists a bag that contains twin sets T_h and T_{h+1}. Furthermore, if a bag contains twin sets T_b and T_c with $b < c$ then it contains twin sets $T_{b+1}, \ldots, T_{c-1}$ as well.*

2.2 Role Assignments

If r is a homomorphism from G to R and $U \subseteq V_G$, then we write $r(U) = \bigcup_{u \in U} r(u)$. Recall that r is an R-role assignment of G if $r(N_G(u)) = N_R(r(u))$ for every vertex u of G. Graph R is called a *role graph* and its vertices are called *roles*. Throughout the paper, we use n and m to refer to the number of vertices and edges of G. We frequently make use of the following two known results.

Observation 2 ([10]). *Let G be a graph and let R be a connected graph such that G has an R-role assignment. Then each vertex $x \in V_R$ appears as a role of some vertex $u \in V_G$, i.e., $r(u) = x$. Furthermore, if $|V_G| = |V_R|$ then $G \simeq R$.*

Lemma 1 ([10]). *Let G and R be two graphs such that G has an R-role assignment r, and let $x, y \in V_R$ be roles connected by a path $z_1 \cdots z_\ell$ in R, with $x = z_1$ and $y = z_\ell$. Then for each $u \in V_G$ with $r(u) = x$ there exists a vertex $v \in V_G$ and a path $t_1 \cdots t_\ell$ in G, with $u = t_1$ and $v = t_\ell$, such that $r(t_i) = z_i$ for $i = 1, \ldots, \ell$.*

Our first result, given in Theorem 2, shows that chordal graphs, interval graphs, and proper interval graphs are closed under role assignments, and it is needed in Section 3. We postpone its proof to the journal version of this paper. Note that, for each of the three statements in Theorem 2, the reverse implication is not valid. In order to see this let G be the 6-cycle and R be the 3-cycle.

Theorem 2. *Let G be a graph and let R be a connected graph such that G has an R-role assignment.*
(i) If G is a chordal graph then R is a chordal graph.
(ii) If G is an interval graph then R is an interval graph.
(iii) If G is a proper interval graph then R is a proper interval graph.

3 Role Assignments on Proper Interval Graphs

We start with the following key result. Note that this result is easy to verify for paths.

Theorem 3. *Let G and R be two connected proper interval graphs such that G has an R-role assignment r. Let P and P' be the clique paths of G and R, respectively. Then the bags of P and P' can be ordered such that $P = K_1 \cdots K_p$ and $P' = L_1 \cdots L_q$, with $q \leq p$, and $r(K_i) = L_i$, for $i = 1, \ldots, q$.*

Proof. By the definition of a role assignment, $|V_G| \geq |V_R|$ holds. Assume first that $|V_G| = |V_R|$. Then, as a result of Observation 2, G and R are isomorphic. By Theorem 1 the clique paths of G and R are unique. Hence the ordering of the bags in each path is unique up to reversal. We can try each direction for one of the paths, and the statement of the theorem holds.

For the rest of the proof, assume that $|V_G| > |V_R|$. Then at least one vertex of R is the role of more than one vertex of G. Let x be such a role. Then there exist vertices u and u' in G with $r(u) = r(u') = x$. Assume $l_P(u) = h$ and $f_P(u') = i$, where we may assume that $h < i$ because K_h and K_i are cliques, and vertices with the same role can not be adjacent. Let x be chosen in such a way that every vertex in $K_{\leq i-1}$ has a unique role, i.e., $|r(K_{\leq i-1})| = |K_{\leq i-1}|$.

Claim 1. Every vertex of R occurs as a (unique) role of a vertex of $K_{\leq i-1}$.

We prove this claim by contradiction. Suppose there is a role y that does not occur as a role of a vertex in $K_{\leq i-1}$. As a result of Observation 2, there exists a vertex v in G with $r(v) = y$. Let $f_P(v) = j$. Since y does not appear as a role on $K_{\leq i-1}$, we find that $j \geq i$. We may choose v such that there is no vertex in $K_{\leq j-1}$ with role y. Because K_j is a clique, we find that v is the only vertex of K_j with role y.

Let $Q' = z_1 \cdots z_\ell$, with $x = z_1$ and $y = z_\ell$, be a shortest path between x and y in R. By Lemma 1 we find that G contains a path $Q_1 = t_1 \cdots t_\ell$ with $u = t_1$, such that $r(t_i) = z_i$ for $i = 1, \ldots, \ell$. Since Q' is a shortest path from x to y in R, and there is no other vertex in K_j with role y, our choice of v implies that we may assume that $v = t_\ell$.

By the same reasoning we find a path $Q_2 = t'_1 \cdots t'_\ell$, with $u' = t'_1$ and $v = t'_\ell$, such that $r(t'_i) = z_i$ for $i = 1, \ldots, \ell$. Hence Q_1 and Q_2 are two paths with $r(V_{Q_1}) = r(V_{Q_2}) = V_{Q'}$ and $|V_{Q_1}| = |V_{Q_2}| = |V_{Q'}|$. Consequently, u is not on Q_2 and u' is not on Q_1. However, since $l_P(u) = h < f_P(u') = i \leq f_P(v) = j$ and K_i, K_j are cliques, we find that Q_1 contains a neighbor w of u'.

Suppose $i = j$. Then u' and v are neighbors in G, and consequently, xy is an edge of R. This means that u and v are neighbors in G. Hence, there is a bag in P containing both of them. This means that $h = l_P(u) \geq f_P(v) = j$. However, this is not possible since $h < i \leq j$.

Suppose $i < j$. Then $w = t_2$ as otherwise r maps the path $u'w\overrightarrow{Q_1}t_\ell$ to a path from x to y in R that is shorter than Q'. By the same reasoning, we find that w is the only neighbor of u' on Q_1. Since Q_1 is a shortest path and $uu' \notin E_G$, this means that G contains an induced claw with center t_2 and leaves u, u', t_3, which contradicts the assumption that G is a proper interval graph. This completes the proof of Claim 1.

By Claim 1 we find that $r(K_{\leq i-1}) = V_R$, and consequently, as $|r(K_{\leq i-1})| = |K_{\leq i-1}|$, we obtain $|K_{\leq i-1}| = |V_R|$. Let r' be the restriction of r to $K_{\leq i-1}$.

Claim 2. r' is an R-role assignment of $G[K_{\leq i-1}]$.

We prove Claim 2 as follows. Suppose r' is not an R-role assignment of $G[K_{\leq i-1}]$. Because r is a homomorphism from G to R, we find that r' is an homomorphism from $G[K_{\leq i-1}]$ to R. Hence, there must exist a vertex $t \in K_{\leq i-1}$ and vertices $z, z' \in V_R$ with $r'(t) = r(t) = z$, $zz' \in E_R$ and $z' \notin r'(N_G(t))$. Since r is an R-role assignment of G, we find that $z' \in r(N_G(t))$. Hence $l_P(t) \geq i + 1$. Consequently, as $t \in K_{\leq i-1}$, we find that t belongs to K_i. We proceed as follows. Since $r(K_{\leq i-1}) = V_R$, there exists a vertex $t' \in K_{\leq i-1}$ with $r'(t') = r(t') = z'$. By definition of r, we find that t' has a neighbor s in G with $r(s) = z$. Because t has no neighbor with role z', we find that t and t' are not adjacent in G. Hence $s \neq t$ holds. Since every vertex of $K_{\leq i-1}$ has a unique role and vertex $t \in K_{\leq i-1}$ already has role z, we find that $s \notin K_{\leq i-1}$. This means that K_i does not only contain t but also contains t'. However, since K_i is a clique, t and t' must be adjacent. With this contradiction we have completed the proof of Claim 2.

Due to Claim 2 and the aforementioned observation that $|K_{\leq i-1}| = |V_R|$, we may apply Observation 2 and obtain that $G[K_{\leq i-1}]$ is isomorphic to R. By Theorem 1, the clique paths of $G[K_{\leq i-1}]$ and R are unique. Hence, $i = q + 1$, and the statement of the theorem follows. □

Note that Theorem 3 is not valid for interval graphs, which can be seen with the following example. Let G be the path $u_1u_2u_3u_4$ to which we add a vertex u_5 with edge u_2u_5 and a vertex u_6 with edge u_3u_6. Let $P = K_1 \cdots K_5$ be a clique path of G with $K_1 = \{u_1, u_2\}$, $K_2 = \{u_2, u_5\}$, $K_3 = \{u_2, u_3\}$, $K_4 = \{u_3, u_6\}$ and $K_5 = \{u_3, u_4\}$. Let R be the 4-vertex path 1234. The unique clique path of R is $P' = L_1L_2L_3$ with $L_1 = \{1, 2\}$, $L_2 = \{2, 3\}$ and $L_3 = \{3, 4\}$. However, we find that G has an R-role assignment r with $r(u_1) = r(u_5) = 1$, $r(u_2) = 2$, $r(u_3) = 3$, and $r(u_4) = r(u_6) = 4$.

Also note that we can apply Theorem 3 twice depending on the way the bags in the clique path of the proper interval graph G are ordered. This leads to a rather surprising corollary that might be of independent interest.

Corollary 1. *Let G be a connected proper interval graph with clique path $P = K_1 \cdots K_p$. If G has an R-role assignment and R is connected, then $R \simeq G[K_{\leq i}]$ and $R \simeq G[K_{\geq p-i+1}]$, for some $1 \leq i \leq p$.*

As an illustration of Corollary 1 we have indicated the two copies of R in G with bold edges in Figure 1. Due to Theorem 2 we do not need to restrict R to be a proper interval graph in the statement of the above corollary. Hence for any two connected graphs G and R, where G is proper interval with $|V_G| > |V_R|$, if G has an R-role assignment then G contains two (not necessarily vertex-disjoint) induced subgraphs isomorphic to R.

Theorem 3 only shows what an R-role assignment r of a proper interval graph G looks like at the beginning and end of the clique path of G. To derive our algorithm, we need to know the behavior of r in the middle bags as well. We therefore give the following result, which is valid when R has at least three maximal cliques and the number of maximal cliques in G is not too small. Its proof is postponed to the journal version of this paper. The special cases when R has just one or two maximal cliques or G has few maximal cliques will be dealt with separately in the proof of Theorem 4.

Lemma 2. *Let G be a connected proper interval graph with clique path $P = K_1 \cdots K_p$. Let R be a connected proper interval graph with clique path $P' = L_1 \cdots L_q$ and ordered twin sets $X_1, \ldots, X_t$. Let r be an R-role assignment of G with $r(K_q) = L_q$. Let T be the subset of K_q that consists of all vertices with roles in X_t. Then the following holds if $q \geq 3$ and $p \geq 2q + 1$.*

(i) If there is a vertex in T not in K_{q+1}, then there exists an index $i \geq q+1$ such that $K_{\geq q+1} \setminus K_{\leq q} \subseteq K_{\geq i}$ and the restriction of r to $K_{\geq i}$ is an R-role assignment of $G[K_{\geq i}]$ with $r(K_i) = L_q$. Furthermore, if $i > q+1$ then $r(K_h) \subseteq X_t$ for $h = q + 1, \ldots, i - 1$.

(ii) If all vertices in T are in K_{q+1}, then there exists an index $i \geq q + 1$ such that $T = K_{\leq i-1} \cap K_i$ and $T \cap K_{i+1} = \emptyset$, and the restriction of r to $K_{\geq i}$ is an R-role assignment of $G[K_{\geq i}]$ with $r(K_i) = L_q$.

Let G and R be two connected proper interval graphs with clique paths $P = K_1 \cdots K_p$ and $P' = L_1 \cdots L_q$, respectively. A mapping $r : K_{\leq i} \to V_R$ for some

$1 \leq i \leq p$ is a *starting R-role assignment* of $G[K_{\leq i}]$ if for all $u \in K_{\leq i} \setminus K_{i+1}$ we have that $r(N_G(u)) = N_R(r(u))$, and for all $u \in K_{\leq i} \cap K_{i+1}$ we have that $r(N_G(u)) \subseteq N_R(r(u))$. Note that a starting R-role assignment of $G[K_{\leq i}]$ is an R-role assignment of G if and only if $i = p$.

Let $1 \leq i \leq p$, and let r be a starting R-role assignment of $G[K_{\leq i}]$. We say that $v \in K_{\leq i} \cap K_{i+1}$ is *missing* role $x \in V_R$ if x is a neighbor of $r(v)$, and x is not a role of a neighbor of v in $K_{\leq i}$. Let $X_1, \ldots, X_t$ be the ordered twin sets of R. We denote the set of missing roles of v that are in X_c by $M_c(v)$. We say that r can be *finished by* r^* if r^* is an R-role assignment of G with $r^*(u) = r(u)$ for all $u \in K_{\leq i}$.

The following lemma is important for our algorithm.

Lemma 3. *Let G and R be two connected proper interval graphs. Let G have clique path $P = K_1 \cdots K_p$, and let R have ordered twin sets $X_1, \ldots, X_t$. Let $r : K_{\leq i} \to V_R$ be a starting R-role assignment of $G[K_{\leq i}]$ for some $1 \leq i \leq p$. Then $K_{\leq i} \cap K_{i+1}$ does not contain two vertices u, v such that $M_c(u) \setminus M_c(v) \neq \emptyset$ and $M_c(v) \setminus M_c(u) \neq \emptyset$ for some $1 \leq c \leq t$.*

Proof. In order to derive a contradiction, assume that such vertices u and v exist. Note that u and v are adjacent, because both of them belong to bag K_{i+1}. Let $x \in M_c(u) \setminus M_c(v)$ and $y \in M_c(v) \setminus M_c(u)$. Because u misses x and $x \in X_c$, we find that $r(u)$ is adjacent to all roles in $X_c \setminus \{r(u)\}$. Hence $r(u)$ is adjacent to $y \in X_c$, unless $r(u) = y$. However, the latter case is not possible, because in that case v, being adjacent to u, would not miss y. So, indeed $r(u)$ and y are adjacent. From $y \in M_c(v) \setminus M_c(u)$ we then deduce that u already has a neighbor $w \in K_{\leq i}$ with role $r(w) = y$. Since v misses y and R contains no self-loop, we find that $r(v) \neq y$, and consequently $w \neq v$. Since v misses y, the edge uw must be in a bag before v got introduced. Hence, we obtain $f_P(u) < f_P(v)$. Analogously, we get $f_P(v) < f_P(u)$. This is not possible, and we have proven Lemma 3. □

We are now ready to present our main result.

Theorem 4. Role Assignment *can be solved in polynomial time on input pairs (G, R) where G is a proper interval graph and R is an arbitrary graph.*

Proof. First we give an algorithm with running time $\mathcal{O}(n^3)$ that takes as input a *connected* proper interval graph G and a *connected* graph R, and decides whether G has an R-role assignment.

If $|V_R| > n$ or R is not a proper interval graph, then we know by respectively Observation 2 and Theorem 2 that the answer is `NO`. These conditions can be checked in linear time, as explained in the preliminaries. Thus we assume that $|V_R| \leq n$ and R is a proper interval graph.

Let G have clique path $P = K_1 \cdots K_p$. Recall that P can be constructed in linear time. Let R have clique path $P' = L_1 \cdots L_q$ and ordered twin sets $X_1, \ldots, X_t$. Because $|V_R| \leq n$, we find that $q \leq p$ and that we can compute P' and the ordered twin sets in $\mathcal{O}(|V_R| + |E_R|) = \mathcal{O}(n^2)$ time. Since Lemma 2 applies only when $q \geq 3$, we distinguish between the cases where $q = 1$, $q = 2$, and $q \geq 3$.

Case 1. $q = 1$. Then R is a complete graph. By Theorem 3, we find that $|K_1| = |L_1|$ must hold, and we give each vertex in K_1 a different role. This yields a starting R-role assignment r of $G[K_1]$.

Suppose $i \geq 1$ and that we have extended r to a starting R-role assignment of $G[K_{\leq i}]$. By Lemma 3 we can order the vertices in $K_i \cap K_{i+1}$ as $u_1, \ldots, u_b$ such that $M_1(u_a) \subseteq M_1(u_{a+1})$ for $a = 1, \ldots, b-1$. We assign different roles to the vertices of $K_{i+1} \setminus K_i$, where we first use the roles of $M_1(u_a)$ before using any roles of $M_1(u_{a+1})$ for $a = 1, \ldots, b-1$. If we have used all the roles and there are still vertices in K_{i+1} with no role yet, we output NO. Otherwise we must verify if the resulting mapping is a starting R-role assignment of $G[K_{\leq i+1}]$ by checking if all vertices in $K_{i+1} \backslash K_{i+2}$ have neighbors with all the required roles. If this is not the case, we output NO, because any R-role assignment is a starting role assignment of $G[K_{\leq i+1}]$. If this is the case, we stop if $i+1 = p$, because a starting R-role assignment of $G[K_{\leq p}] = G$ is an R-role assignment of G; otherwise we repeat the above procedure with $i := i+1$.

It is clear that this algorithm is correct. It runs in $\mathcal{O}(n^3)$ time, because ordering the vertices in $K_i \cap K_{i+1}$ takes $\mathcal{O}(n^2)$ time and there are $\mathcal{O}(n)$ bags.

Case 2. $q = 2$. The algorithm for this case uses similar arguments as above (but in a more advanced way). Due to space restrictions we postpone its proof.

Case 3. $q \geq 3$. First suppose $p \leq 2q$. By Theorem 3, both $G[K_{\leq q}]$ and $G[K_{\geq p-q+1}]$ must be isomorphic to R and have an R-role assignment, in case G has an R-role assignment. Because $p \leq 2q$, every vertex of G is in $K_{\leq q} \cup K_{\geq p-q+1}$. Hence, there are just four possibilities of assigning roles to vertices of G, namely two possibilities for $K_{\leq q}$ combined with two possibilities for $K_{\geq p-q+1}$. We check if one of them leads to an R-role assignment of G. Verifying whether a mapping $V_G \rightarrow V_R$ is an R-role assignment of G can be done in $\mathcal{O}(n^3)$ time by considering each vertex and checking if it has the desired roles occurring in its neighborhood.

Suppose $p \geq 2q+1$. We first check if $G[K_{\leq q}]$ is isomorphic to R. This can be done in linear time [18]. If $G[K_{\leq q}]$ is not isomorphic to R then we output NO due to Theorem 3. Suppose $G[K_{\leq q}] \simeq R$ and that without loss of generality we have a starting R-role assignment r of $G[K_{\leq q}]$ with $r(K_i) = L_i$ for $i = 1, \ldots, q$. We now check whether we are in situation (i) or (ii) of Lemma 2. Then in both situations we can determine in $\mathcal{O}(n)$ time the desired index i and afterwards we continue with the graph $G[K_{\geq i}]$ unless we found no starting R-role assignment of $G[K_{\leq i}]$; in that case we output NO. The total running time of this procedure is $\mathcal{O}(n^3)$.

We have thus presented and proved the correctness of an algorithm with running time $\mathcal{O}(n^3)$ for testing whether a connected proper interval graph G has an R-role assignment for a connected graph R. If G is disconnected then we run the algorithm on each connected component separately. The total running time is still $\mathcal{O}(n^3)$. It remains to study the case when R is disconnected. In this case we cannot assume that $|V_R| \leq |V_G|$. Let c_R be the number of connected components of R. By the definition of a role assignment, G has an R-role assignment if and only if each connected component of G has an R'-role assignment for some connected component R' of R. Hence we can run our algorithm on every pair of connected components of G and R. This gives a total running time $\mathcal{O}(n^3 \cdot c_R)$, which is clearly polynomial. □

Recall that the problem of testing if a graph G has an R-role assignment for some smaller graph R is co-NP-complete in general [5]. Theorem 4 together with Corollary 1 has the following consequence.

Corollary 2. *There exists a polynomial time algorithm that has as input a proper interval graph G and that tests whether there exists a graph R with $|V_R| < |V_G|$ such that G has an R-role assignment.*

Proof. Let G be a proper interval graph on n vertices. First assume that G is connected. Let $P = K_1 \dots K_p$ be the clique path of G. Recall that $p \leq n$. By Corollary 1 we find that G only has an R-role assignment if $R \simeq G[K_{\leq i}]$ for some $1 \leq i \leq p$. This means that we need to apply the $\mathcal{O}(n^3)$ time algorithm for connected proper interval graphs of Theorem 4 at most $p \leq n$ times. Hence we find that testing whether G has an R-role assignment for some graph R with $|V_R| < |V_G|$ takes $\mathcal{O}(n^4)$ time.

Now assume that G is disconnected. Let $G_1, \dots, G_a$ with $a \geq 2$ be the connected components of G. For $j = 1, \dots, a$ we define $n_j = |V_{G_j}|$. As long as $j \leq a - 1$ we do as follows. We consider G_j and check if G_j has an R_j-role assignment for some role graph R_j with $|V_{R_j}| \leq n_j$. If so, then we replace connected component G_j by connected component R_j in G, i.e., we output $R = G_1 \oplus \dots G_{j-1} \oplus R_j \oplus G_j \oplus \dots \oplus G_a$, where $\oplus$ denotes the disjoint union operation on graphs. Suppose not. Then we consider G_{j+1}. If $j = a$ and we did not find a suitable role graph R in this way, then we output NO. Because we need $\mathcal{O}(n_j^4)$ time for each G_j and $n = n_1 + \dots + n_a$, the total running time of this algorithm is $\mathcal{O}(n^4)$, which is polynomial, as desired. □

As a consequence, we have in fact a stronger result: given a proper interval graph G, we can list in polynomial time all graphs R (up to isomorphism) with $|V_R| < n$ such that G has an R-role assignment.

4 Complementary Results and an Open Question

A homomorphism r from a graph G to a graph R is *locally injective* if $|r(N_G(u))| = |N_G(u))|$ for every $u \in V_G$, and r is *locally bijective* if $r(N_G(u)) = N_R(r(u))$ and $|r(N_G(u))| = |N_G(u))|$ for every $u \in V_G$. Locally injective homomorphisms, also called *partial coverings*, have applications in frequency assignment [8] and telecommunication [9]. Locally bijective homomorphisms are also called *coverings* and have applications in topological graph theory [20] and distributed computing [1,2]. The corresponding decision problems, called Partial Cover and Cover respectively, are NP-complete for arbitrary G even when R is fixed to be the complete graph on four vertices [9,16].

In this section, to give a complete picture, we study the computational complexity of all three locally constrained homomorphisms on chordal, interval, and proper interval graphs. Our findings can be summarized in the table below, where the three problems have input (G, R) and the left column indicates the graph class that G belongs to. In the table, R is assumed to be an arbitrary graph.

	PARTIAL COVER	COVER	ROLE ASSIGNMENT
Chordal	NP-complete	GI-complete	GI-hard
Interval	NP-complete	Polynomial	?
Proper Interval	NP-complete	Polynomial	Polynomial

We start with the following result, which allows us to conclude several of the entries in the above table, and which can be viewed as interesting on its own.

Theorem 5. *Let G be a chordal graph and let R be a connected graph. Then there exists a locally bijective homomorphism from G to R if and only if every connected component of G is isomorphic to R.*

Proof. If G is disconnected then we consider each connected component of G separately. Assume that G is connected. If G is isomorphic to R, then the identity mapping from G to R is our desired locally bijective homomorphism.

For the reverse implication, suppose that there exists a locally bijective homomorphism r from G to R. Because any locally bijective homomorphism is also locally surjective, we can apply Theorem 2 in order to find that R is chordal. For the same reason we can apply Observation 2 in order to find that each vertex in R appears as a role of at least one vertex in G. We claim that each vertex in R appears as a role of exactly one vertex in G. In order to derive a contradiction, suppose there exists a vertex $x \in V_R$ such that $r^{-1}(x)$ has size at least two.

Let v and v' be two different vertices of G belonging to $r^{-1}(x)$. Let P be a shortest path from v to v' in G. Because P is shortest, P is an induced path. From the definition of a locally bijective homomorphism we deduce the following two statements. Firstly, because two vertices with the same role cannot be adjacent, we find that $|V_P| \neq 2$. Secondly, because a vertex has no two neighbors with the same role, we find that $|V_P| \neq 3$. Hence, P is an induced path with $|V_P| \geq 4$. This, together with $r(v) = r(v') = x$, means that $r(P)$ forms an induced cycle D in R with $|V_D| = |V_P| - 1$. Because R is chordal, D must consist of three vertices, say $D = xyzx$. Consequently, $|V_P| = 4$ holds.

Let C be the connected component of $G[r^{-1}(x) \cup r^{-1}(y) \cup r^{-1}(z)]$ that contains v and v'. By definition of a locally bijective homomorphism, every vertex is of degree two in D. This means that D is an induced cycle in G. Because every vertex of P belongs to D, and $|V_P| = 4$, we find that $|V_D| \geq 4$. This contradicts our assumption that G is chordal. We conclude that indeed each vertex in R appears as a role of exactly one vertex in G. This means that r is an isomorphism between G and R, and we find that $G \simeq R$, as desired. □

It is known that GRAPH ISOMORPHISM is GRAPH ISOMORPHISM-complete even for pairs (G, R) where G and R are chordal graphs [18]. This implies together with Theorem 5 that COVER is GRAPH ISOMORPHISM-complete for pairs (G, R) where G and R are chordal graphs. On the other hand, COVER is polynomial time solvable on interval graphs, and hence also on proper interval graphs, since isomorphism between two interval graphs can be checked in polynomial time [18]. Because every locally bijective homomorphism is locally surjective, we can use Theorem 2 to deduce that these three results stay valid for input pairs (G, R) where only G is required to be chordal and R may be an arbitrary graph. This explains the three corresponding entries in the table.

Unfortunately, as indicated in the table, the problem PARTIAL COVER remains NP-complete even on pairs (G, R) where G is a proper interval graph (and R is an arbitrary graph). To see this, observe that a complete graph G allows a locally injective homomorphism to an arbitrary graph R if and only if R contains G as a subgraph. This gives a reduction from the well-known NP-complete problem CLIQUE (cf. [13]).

We present one more complexity result on the ROLE ASSIGNMENT problem. This result explains a corresponding entry in the table after applying Theorem 2. It shows that, unless GRAPH ISOMORPHISM is polynomial time solvable, we do not have hope of solving ROLE ASSIGNMENT in polynomial time on chordal graphs.

Theorem 6. ROLE ASSIGNMENT *is* GRAPH ISOMORPHISM*-hard on input pairs* (G, R) *where* G *and* R *are chordal graphs.*

Proof. As we argued above, COVER is GRAPH ISOMORPHISM-complete on input pairs (G, R) where both G and R are chordal graphs. It is not hard to see that we may also assume that G and R are connected and have the same number of vertices. We give a polynomial time reduction from COVER to ROLE ASSIGNMENT. Let G and R be two connected role graphs with $|V_G| = |V_R|$. We claim that G allows a locally bijective homomorphism to R if and only if G allows a locally surjective homomorphism to R.

Suppose G allows a locally bijective homomorphism r to R. Because any locally bijective homomorphism is locally surjective by definition, r is a locally surjective homomorphism from G to R. To prove the reverse implication, suppose G allows a locally surjective homomorphism to R. Recall that $|V_G| = |V_R|$. Then we use Observation 2 to deduce that $G \simeq R$. Hence, G allows a locally bijective homomorphism to R, namely the identity mapping. This completes the reduction and the proof. □

Just as for ROLE ASSIGNMENT, we denote the problems COVER and PARTIAL COVER as R-COVER and R-PARTIAL COVER, respectively, if R is fixed, i.e., not a part of the input. In that case we obtain the following result.

Proposition 1. *For any fixed* R*, the problems* R-ROLE ASSIGNMENT, R-COVER, *and* R-PARTIAL COVER *can be solved in linear time on chordal graphs.*

Proof. We first observe that a homomorphism from G to R maps the vertices in a clique of G to different vertices of R. Hence, in order to get a YES answer, a largest clique in G can have at most $|V_R|$ vertices. We compute the number of vertices in a largest clique of G in linear time. If this number is greater than $|V_R|$, we output NO. Otherwise, because the treewidth of a chordal graph is equal to the number of vertices in a largest clique minus 1, we find that G has treewidth bounded by $|V_R|$, which is a constant, as R is fixed. Since all three problems are expressible in monadic second order logic, linear time solvability follows from a well-known result of Courcelle [6]. □

We conclude with the following two open questions resulting from the table.

1. Is Role Assignment NP-complete on input pairs (G, R) when G is a chordal graph?
2. What is the computational complexity of Role Assignment on input pairs (G, R) when G is an interval graph?

References

1. Angluin, D.: Local and global properties in networks of processors. In: Proceedings of STOC 1980, pp. 82–93. ACM, New York (1980)
2. Bodlaender, H.L.: The classification of coverings of processor networks. Journal of Parallel and Distributed Computing 6, 166–182 (1989)
3. Brandstädt, A., Le, V.B., Spinrad, J.: Graph Classes: A Survey. SIAM, Philadelphia (1999)
4. Chalopin, J., Métivier, Y., Zielonka, W.: Local computations in graphs: the case of cellular edge local computations. Fundamenta Informaticae 74, 85–114 (2006)
5. Chalopin, J., Paulusma, D.: Graph labelings derived from models in distributed computing. In: Fomin, F.V. (ed.) WG 2006. LNCS, vol. 4271, pp. 301–312. Springer, Heidelberg (2006)
6. Courcelle, B.: The Monadic Second-Order Logic of Graphs. I. Recognizable Sets of Finite Graphs. Information and Computation 85, 12–75 (1990)
7. Everett, M.G., Borgatti, S.: Role colouring a graph. Mathematical Social Sciences 21, 183–188 (1991)
8. Fiala, J., Kratochvíl, J., Kloks, T.: Fixed-parameter complexity of λ-labelings. Discrete Applied Mathematics 113, 59–72 (2001)
9. Fiala, J., Kratochvíl, J.: Partial covers of graphs. Discussiones Mathematicae Graph Theory 22, 89–99 (2002)
10. Fiala, J., Paulusma, D.: A complete complexity classification of the role assignment problem. Theoretical Computer Science 349, 67–81 (2005)
11. Fiala, J., Paulusma, D.: Comparing universal covers in polynomial time. Theory of Computing Systems 46, 620–635 (2010)
12. Fulkerson, D., Gross, O.: Incidence matrices and interval graphs. Pacific Journal of Mathematics 15, 835–855 (1965)
13. Garey, M.R., Johnson, D.S.: Computers and Intractability. W. H. Freeman and Co., New York (1979)
14. Golumbic, M.C.: Algorithmic Graph Theory and Perfect Graphs. In: Annals of Discrete Mathematics, vol. 57. Elsevier B.V., Amsterdam (2004)
15. Ibarra, L.: The clique-separator graph for chordal graphs. Discrete Applied Mathematics 157, 1737–1749 (2009)
16. Kratochvíl, J., Proskurowski, A., Telle, J.A.: Covering regular graphs. Journal of Combinatorial Theory, Series B 71, 1–16 (1997)

17. Lekkerkerker, C., Boland, D.: Representation of finite graphs by a set of intervals on the real line. Fundamenta Mathematicae 51, 45–64 (1962)
18. Lueker, G.S., Booth, K.S.: A linear time algorithm for deciding interval graph isomorphism. Journal of the ACM 26, 183–195 (1979)
19. Pekec, A., Roberts, F.S.: The role assignment model nearly fits most social networks. Mathematical Social Sciences 41, 275–293 (2001)
20. Reidemeister, K.: Einführung in die kombinatorische Topologie. Braunschweig: Friedr. Vieweg. Sohn A.-G. XII, 209 S (1932)
21. Rieck, Y., Yamashita, Y.: Finite planar emulators for $K_{4,5} - 4K_2$ and $K_{1,2,2,2}$ and Fellows' conjecture. European Journal of Combinatorics 31, 903–907 (2010)
22. Roberts, F.S.: Indifference Graphs. In: Proof Techniques in Graph Theory, pp. 139–146. Academic Press, New York (1969)
23. Sheng, L.: 2-Role assignments on triangulated graphs. Theoretical Computer Science 304, 201–214 (2003)

Efficient Connectivity Testing of Hypercubic Networks with Faults

Tomáš Dvořák[1], Jiří Fink[1,⋆], Petr Gregor[1,⋆⋆],
Václav Koubek[1,⋆], and Tomasz Radzik[2]

[1] Faculty of Mathematics and Physics, Charles University, Prague, Czech Republic
{dvorak@ksvi,fink@kam,gregor@ktiml,koubek@ktiml}.mff.cuni.cz
[2] Department of Computer Science, King's College London, United Kingdom
Tomasz.Radzik@kcl.ac.uk

Abstract. Given a connected graph G and a set F of faulty vertices of G, let $G - F$ be the graph obtained from G by deletion of all vertices of F and edges incident with them. Is there an algorithm, whose running time may be bounded by a polynomial function of $|F|$ and $\log |V(G)|$, which decides whether $G - F$ is still connected? Even though the answer to this question is negative in general, we describe an algorithm which resolves this problem for the n-dimensional hypercube in time $O(|F|n^3)$. Furthermore, we sketch a more general algorithm that is efficient for graph classes with good vertex expansion properties.

1 Introduction

A study of interconnection networks, originally initiated by particular applications in telephone and computer networks, has become fairly pervasive in many different areas in the recent decade. In a whole avenue of problems that arise in the course of network design, a good deal of attention has been paid to the aspect of reliability: If some nodes of the network become overloaded or unavailable, can the network still preserve its functionality?

If interconnection networks are modeled as simple undirected graphs, our problem may be formulated as follows: Suppose we are given a class $\mathcal{G}$ of graphs such that each graph $G \in \mathcal{G}$ has a property $\mathcal{P}$. Note that to describe an arbitrary vertex of G, we need a string of length $\Omega(\log |V(G)|)$ in the worst case. Let $n_G : V(G) \to 2^{V(G)}$ be an oracle which for a given vertex $v \in V(G)$ returns the set $N(v)$ of all neighbors of v in G in time $O(|N(v)| \cdot \log |V(G)|)$.

Problem 1.1. Is there an algorithm which

- given an oracle n_G for a graph $G \in \mathcal{G}$ and a set F of faulty vertices of G,
- decides whether the graph $G - F$, obtained from G by deletion of all vertices of F and edges incident with them, still possesses the property $\mathcal{P}$,
- whose running time is bounded by a polynomial function of $|F|$ and $\log |V(G)|$?

⋆ The Institute for Theoretical Computer Science (ITI) is supported by project 1M0545 of the Czech Ministry of Education.

⋆⋆ Partially supported by the Czech Science Foundation Grant 201/08/P298.

C.S. Iliopoulos and W.F. Smyth (Eds.): IWOCA 2010, LNCS 6460, pp. 181–191, 2011.

The requirement on the time complexity is motivated by practical considerations. Recall that to describe an arbitrary input F, a string of length $|F|\Omega(\log|V(G)|)$ is needed. It is plausible to presume that the number of nodes which may become faulty at the same time would be just a fraction of the total number of nodes of the network. A typical instance of this problem may be a network topology modeled by the graph of the n-dimensional hypercube with 2^n vertices, while the number of faults is bounded by $O(n^k)$ for some natural number k [5,7]. In this case it would be useful to design an algorithm for Problem 1.1, running in time proportional to the length of the input F, possibly searching only some local neighborhood of F in G, rather than exploring the whole graph $G - F$.

A natural requirement imposed on each reasonable interconnection network is its connectivity. In this paper we therefore study an instance of Problem 1.1 where property $\mathcal{P}$ equals connectivity. To the best of our knowledge, no results on this problem have been reported previously.

It should be noted that although graph connectivity is a textbook example of an algorithmic problem that may be solved in linear time [3], our variant is more involved. In particular, we claim that if G may be an arbitrary connected graph, an algorithm testing the connectivity of $G - F$ in time $|F|^k$ for some natural number k does not exist. Indeed, suppose that $G_{x,y}$ and $G_{x,z}$ are two connected graphs containing distinct vertices $x \neq y$ and $x \neq z$, respectively. Let G_1 be the graph obtained from $G_{x,y}$ and $G_{x,z}$ by gluing together vertex x of $G_{x,y}$ with vertex x of $G_{x,z}$, and G_2 be the graph obtained from G_1 by adding edge yz. In order to verify the connectivity of $G_1 - \{x\}$ or $G_2 - \{x\}$, it is necessary to check the presence of edge yz, since only this edge distinguishes the connected graph $G_2 - \{x\}$ from disconnected $G_1 - \{x\}$. Since the choice of y and z was quite arbitrary, it follows that any algorithm that correctly decides on the connectivity of $G - F$ must necessarily read all edges of this graph. It follows that its running time is bounded from below by the size of the input graph, which need not be necessarily a polynomial in $|F|$ and $\log|V(G)|$.

This argument shows that it is necessary to restrict class $\mathcal{G}$ to some proper subclass of connected graphs. In this paper, we resolve our problem for the class of hypercubes, which has served for decades as a popular topology of interconnection networks for parallel or distributed computing [10]

It is worth mentioning that a fairly special instance of Problem 1.1 for $\mathcal{G}$ being the class of hypercubes and property $\mathcal{P}$ being the existence of

(i) Hamiltonian cycles and paths [4],
(ii) long cycles and paths [5,7],

has been studied previously. There are positive results for a special case when the number of faults is bounded by a certain linear (i) or quadratic (ii) function of n. On the other hand, when the number of faults is not limited, the problems are NP-hard [1,6].

The main results of this paper is an algorithm which verifies the connectivity of the n-dimensional hypercube with f faults in time $O(fn^3)$. We also describe a more general algorithm based on vertex-expansion properties that for the class of hypercubes works in time $O(f^2n^{3.5})$. The rest of the paper is laid out as follows.

After introducing some necessary concepts and notations, we start with vertex-expansion approach in Section 3. In Section 4 we study walk transformations. This is our main technical tool, applied in Section 5 to derive a theorem relating connectivity of faulty hypercube with that of certain local neighborhood of the set of faults. Based on these theoretical results, in Section 6 we describe an algorithm for connectivity testing and analyze its time complexity. The paper is concluded with some open problems and directions for further research.

2 Preliminaries

The concepts used in this paper but undefined below may be found e. g. in [3]. In the rest of this text, n always denotes a positive integer while $[n]$ stands for the set $\{1, 2, \ldots, n\}$.

Vertex and edge sets of a graph G are denoted by $V(G)$ and $E(G)$, respectively. Given a set $V \subseteq V(G)$ let $G[V]$ denote the subgraph of G induced by V while $G - V$ stands for $G[V(G) \setminus V]$. The *distance* between vertices u, v in G is denoted by $d_G(u, v)$, the subscript being omitted if no ambiguity may arise. A *square* of the graph G, denoted by G^2, is the graph on vertices of G and edges between every two distinct vertices that are at distance at most two in G. Given a vertex u, an edge vw and sets $S, T \subseteq V(G)$, we define

$$\begin{aligned} d(u, S) &= \min\{d(u, v) \mid v \in S\}, \\ d(S, T) &= \min\{d(u, T) \mid u \in S\}, \\ d(u, vw) &= d(u, \{v, w\}), \\ N(u) &= \{v \in V(G) \mid d(u, v) = 1\}, \\ N(S) &= \{v \in V(G) \mid d(v, S) = 1\}. \end{aligned}$$

The *n-dimensional hypercube* Q_n is a graph with all binary vectors of length n as vertices, an edge joining two vertices whenever they differ in a single coordinate. For two vertices u, v of Q_n let $u \triangle v$ be the set of coordinates in which u and v differ. Note that $|u \triangle v| = d(u, v)$. The *direction* of an edge uv of Q_n is the integer $i \in [n]$ in which u and v differ; that is, $u \triangle v = \{i\}$.

3 Expansion Approach

In this section we describe an algorithm for testing vertex-deleted connectivity which works efficiently for graph classes with good vertex expansion.

The set $N(S)$ contaning neighbors of vertices from $S \subseteq V(S)$ that are not in S is called the *boundary* of S. The graph G is said to have *vertex expansion* ε if $|N(S)| \geq \varepsilon \cdot |S|$ for every $S \subseteq V(G)$ with $|S| \leq |V(G)|/2$. Note that nonzero expansion implies connectedness; otherwise, a component of at most half of the vertices would have empty boundary.

Theorem 3.1. *Let $(G_n)_{n \in \mathbb{N}}$ be a sequence of graphs G_n with vertex expansion $\varepsilon_n > 0$ and maximal degree Δ_n. There is an algorithm that for input $n \in \mathbb{N}$, $F \subseteq V(G_n)$, $|V(G_n)|$, and $\varepsilon_n > 0$ tests the connectivity of $G_n - F$ in time*

$$O\left(\frac{|F|^2 \cdot \Delta_n^2 \cdot \log(|V(G_n)|)}{\varepsilon_n}\right).$$

Proof (A sketch.). A component of $G_n - F$ induced by vertices $S \subseteq V(G_n) \setminus F$ is said to be

- *major* if $|S| > |V(G_n)|/2$;
- *small* if $|S| \leq |F|/\varepsilon_n$.

Obviously, there is at most one major component. If $|F| > |V(G_n)|\varepsilon_n/2$, we can afford to run a standard search algorithm in $G_n - F$. Otherwise, it follows from vertex expansion of G_n that every component of $G_n - F$ is either major or small.

The key idea is that we can afford searching through small components completely. Furthermore, if we find more than $|F|/\varepsilon_n$ vertices in the same component, we know that we are in the major component (and thus we can stop our search).

Hence, the algorithm works as follows. We start searching $G_n - F$ from each (non-faulty) neighbor v of a faulty vertex u. If we find a component of more than $|F|/\varepsilon_n$ vertices, we stop the search from v with a remark that there exists a major component. If we have found a complete small component, we report that the graph $G_n - F$ is disconnected. Otherwise, we continue the search until we check all non-faulty neighbors v of all faulty vertices u. In this case, we report that the graph $G_n - F$ is connected.

The time complexity is obtained as follows. There are at most $|F| \cdot \Delta_n$ non-faulty neighbors of faulty vertices. From each of them we search for at most $|F|/\varepsilon_n$ vertices. For every vertex found we ask oracle for its neighbors, and each query takes $O(\Delta_n \cdot \log(|V(G_n)|))$ time. □

It follows from classical results of Harper [8] on isoperimetric problems that the hypercube Q_n has a vertex expansion $\frac{c}{\sqrt{n}}$ for some constant c.

Corollary 3.1. *There is an algorithm for testing connectivity of $Q_n - F$ that runs in $O(|F|^2 \cdot n^{3.5})$ time.*

4 Transformations of Walks in Hypercubes

In this section we introduce a useful concept of transformations of one walk to another walk of the hypercube. This is where the structure of the hypercube plays its role.

A *walk* in a simple graph G is a sequence $W = (v_0, v_1, \ldots, v_k)$ of vertices in G such that v_i and v_{i+1} are adjacent for all $0 \leq i < k$. If W starts with the vertex u and ends with the vertex v, we say that W is a *uv-walk*.

Let $W = (v_0, v_1, \ldots, v_k)$ be a walk in Q_n. Let d_i be the direction of the edge between v_{i-1} and v_i for every $i \in [k]$. Then the sequence $(d_1, d_2, \ldots, d_k)$ is called the *transitional sequence* of the walk W. For a sequence τ over $[n]$ and $i \in [n]$ let $\#(\tau, i)$ be the number of occurrences of i in τ. It is easy to see that a sequence τ over $[n]$ is a transitional sequence of some uv-walk in Q_n if and only if

$$u \bigtriangleup v = \{i \in [n];\ \#(\tau, i) \text{ is odd}\}. \tag{4.1}$$

Thus, we may identify uv-walks in Q_n with sequences over $[n]$ satisfying (4.1). We use both representations of a uv-walk as a sequence of vertices and as its transitional sequence, depending on what is more convenient.

Let τ be a transitional sequence of a uv-walk W. Consider the following three operations on τ:

$$\begin{aligned}
&\mathsf{swap}(\tau_1, i, j, \tau_2) = (\tau_1, j, i, \tau_2) && \text{for } \tau = (\tau_1, i, j, \tau_2),\\
&\mathsf{insert}_i(\tau_1, \tau_2) = (\tau_1, i, i, \tau_2) && \text{for } \tau = (\tau_1, \tau_2),\\
&\mathsf{delete}(\tau_1, i, i, \tau_2) = (\tau_1, \tau_2) && \text{for } \tau = (\tau_1, i, i, \tau_2),
\end{aligned}$$

where τ_1, τ_2 are contiguous subsequences of τ and $i, j \in [n]$. Since these operations preserve (4.1), their results are also transitional sequences of some uv-walk.

We say that two uv-walks σ and τ in Q_n are *equivalent* if $\#(\sigma, i) = \#(\tau, i)$ for all $i \in [n]$. Note that the operation swap transforms a uv-walk to an equivalent uv-walk. Conversely, the following proposition holds.

Proposition 4.1. *For every two equivalent uv-walks σ and τ in Q_n, there is a sequence of* swaps *that transforms σ into τ.*

Proof. Since σ and τ are equivalent, they have the same length k. Moreover, there is a permutation $f : [k] \to [k]$ such that $\sigma(i) = \tau(f(i))$ for all $i \in [k]$. An arbitrary decomposition of f into consecutive transpositions gives us a sequence of swaps that transforms σ into τ. □

Let $W = (v_0, v_1, \ldots, v_k)$ be a walk in Q_n with a transitional sequence $\tau = (d_1, d_2, \ldots, d_k)$. We say that $\mathsf{insert}_i(\tau_1, \tau_2)$ on τ is performed in the vertex v_i where $0 \leq i \leq k$ if $\tau_1 = (d_1, d_2, \ldots, d_i)$ and $\tau_2 = (d_{i+1}, d_{i+2}, \ldots, d_k)$.

Proposition 4.2. *For every two uv-walks σ and τ, there are two sequences of* inserts *that transform σ into σ' and τ into τ', respectively, such that σ' and τ' are equivalent. Moreover, these* inserts *can be performed in arbitrary vertices.*

Proof. For every direction $i \in [n]$ we perform the operations insert_i on σ if $\#(\sigma, i) < \#(\tau, i)$, or on τ if $\#(\sigma, i) > \#(\tau, i)$ until we obtain $\#(\sigma, i) = \#(\tau, i)$. These inserts can be performed on any position. □

Since delete is an inverse of insert, we obtain the following corollary.

Corollary 4.1. *For every two uv-walks σ and τ in Q_n there is a sequence of* inserts, swaps *and* deletes *(in this order) that turns σ into τ.*

5 Local Connectivity

For a given set F of vertices in Q_n we define a subgraph $G(F) = (A \cup B \cup F, E)$ of Q_n by

$$A = N(F), \quad B = N(A) \setminus F, \quad E = \{uv \in E(Q_n);\ u \in A \cup F\}.$$

That is, $G(F)$ is the subgraph of Q_n on all vertices at distance at most 2 from F and with all edges at distance at most 1 from F. Our aim in this section is to show that if $G(F)$ is connected and $G(F) - F$ is disconnected, then $Q_n - F$ is also disconnected. Note that if $Q_n^2[F]$ is connected, then $G(F)$ is connected as well.

Let W be a walk in Q_n and let u be a vertex on W. We say that u is a *port* on W if $u \in A$ and exactly one of his neighbors on W is in F. Note that if u is a port on W and not an endvertex, then his second neighbor on W is in $A \cup B$. Furthermore, since u may have several occurrences on the walk W, the notion of ports is defined with respect to a particular occurrence of u on W, and not the vertex u itself.

For a connected component C of $G(F) - F$ let $p(C, W)$ denote the number of ports on the walk W from the component C. First, we show that swap performed on W preserves the parity of $p(C, W)$.

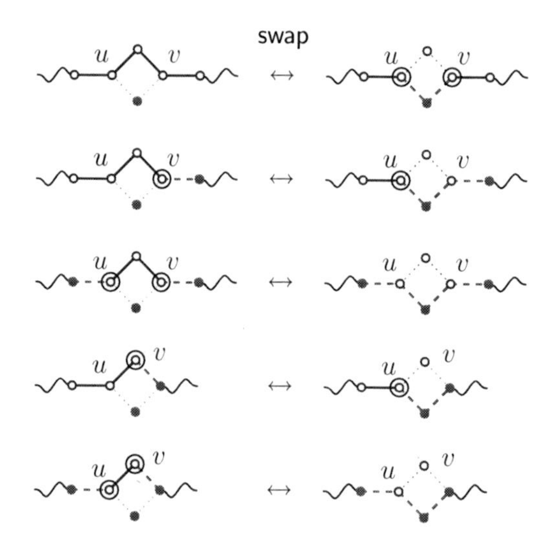

Fig. 1. All possible swaps that change ports

Lemma 5.1. *Let W_2 be a walk in Q_n obtained from a walk W_1 by a single* swap*, and let $F \subseteq V(Q_n)$. For every component C of $G(F) - F$, the numbers $p(C, W_1)$ and $p(C, W_2)$ differ by 0 or 2.*

Proof. See Figure 1 for all configurations of swaps that change ports. The vertices of F are full (red), the vertices of $A \cup B$ are empty (blue), the ports are circled.

The edges of walks W_1, W_2 that are incident to F are dashed (red), the edges of W from $G(F) - F$ are full (blue).

Note that in each case, the ports change on the vertices u and v. Since u and v are connected by edges of $G(F) - F$, they are in the same component C of $G(F) - F$. In the first, third and last case, the numbers of ports of C changes by 2, whereas in the second and fourth case, it remains unchanged. □

Corollary 5.1. *For every $F \subseteq V(Q_n)$, every component C of $G(F) - F$, and every equivalent walks W_1 and W_2 in Q_n, the parity of $p(C, W_1)$ and $p(C, W_2)$ is the same.*

Now we show that global connectivity implies local connectivity. That is, disconnected $Q_n - F$ can be recognized locally on $G(F) - F$.

Lemma 5.2. *Let $F \subseteq V(Q_n)$ be such that $G(F)$ is connected. If $Q_n - F$ is connected, then $G(F) - F$ is also connected.*

Proof. Suppose for a contradiction that there are vertices $u, v \in A \cup B = V(G(F) - F)$ that are connected in $Q_n - F$ by a walk P but are disconnected in $G(F) - F$. Clearly, the walk P contains some vertex x that is not from A; otherwise, P is in $G(F) - F$.

Let C_u and C_v denote the components of $G(F) - F$ containing the vertices u and v, respectively. Since $G(F)$ is connected, there is a uv-walk R in $G(F)$. As u and v are disconnected in $G(F) - F$, the walk R contains some vertex $y \in F$, and an odd number of ports from each component C_u and C_v.

By Proposition 4.2, the walks P and R can be transformed by inserts to walks P' and R' in Q_n, respectively, such that P' and R' are equivalent. Moreover, inserts on P and on R can be performed at the vertices x and y, respectively. It follows that the sets of ports on P and R do not change by these transformations. In particular, $p(C, P') = p(C, P)$ and $p(C, R') = p(C, R)$ for every component C of $G(F) - F$.

However, from Corollary 5.1 it follows that $p(C_u, P)$ and $p(C_v, P)$ have odd parity. Hence, the walk P contains some port, and consequently, some vertex of F. This is a contradiction with the assumption that P is a walk in $Q_n - F$. □

Lemma 5.3. *Let F be a set of vertices of Q_n such that $G(C) - C$ is connected for every component C of $Q_n^2[F]$. Then $Q_n - F$ is connected as well.*

Proof. Let $u, v \in V(Q_n) \setminus F$ and P be an arbitrary uv-walk in Q_n. If P contains no vertex from F, we are done. Otherwise it contains a subwalk $S = (x, y_1, \ldots, y_m, z)$ whose all vertices except x and z are in F. Then $y_1, \ldots, y_m$ belong to the same component C of $Q_n^2[F]$. By our assumption, $G(C) - C$ contains an xz-walk T. Replacing the subwalk S of P with T, we obtain a uv-walk which contains less vertices from F than P. Repeating this process for every subwalk of P of the described type, we finally obtain a uv-walk in $Q_n - F$, and the desired conclusion follows. □

Theorem 5.1. *Let $F \subseteq V(Q_n)$. The graph $Q_n - F$ is connected if and only if $G(C) - C$ is connected for every component C of $Q_n^2[F]$.*

Proof. Let $Q_n - F$ be connected and C be a component of $Q_n^2[F]$. By Lemma 5.2 it suffices to prove that $G(C)$ is connected in order to prove that $G(C) - C$ is connected.

Let u, v be vertices of $G(C)$ and our aim is to prove that u and v belong into the same component of $G(C)$. There exist vertices $u', v' \in C$ such that $d(u, u') \leq 2$ and $d(v, v') \leq 2$. Since C is a component of $Q_n^2[F]$, there exists a sequence $u' = w_1, w_2, \ldots, w_k = v'$ of vertices of C such that $d(w_i, w_{i+1}) \leq 2$ for every $1 \leq i < k$. Therefore, vertices $w_1, w_2, \ldots, w_k$ belong to the same component of $G(C)$. Consequently, u and v are in the same component of $G(C)$ as well.

The other implication follows from Lemma 5.3. □

6 Algorithm

In this section we apply Theorem 5.1 to design an algorithm for testing the connectivity of $Q_n - F$. To accomplish this task, we employ the following data structures.

List F of faulty vertices of Q_n.

Disjoint-set data structure D [3, Chapter 22] with operations

- MAKE(v, D) creates a singleton set $\{v\}$,
- FIND(v, D) returns a pointer to the set containing v,
- UNION(u, v, D) unites the sets containing u and v,

whose amortized time complexity may be loosely bounded by $O(\log m)$, provided that MAKE(v, D) was executed m times. We use D to detect the connectivity of $G(C)$ where C is a component of $Q_n^2[F]$.

Binary trie T [9, Section 6.3] which stores information about some vertices of Q_n. Each vertex of Q_n stored in T is represented by a leaf of T, which we denote by v_T. Moreover, v_T includes the following additional information:

- a pointer to v in the disjoin-set data structure D
- a boolean variable indicating whether v is *healthy* or *faulty*,
- a boolean variable *visited* indicating that v has been visited and $v \in N(F) \cup F$.

Note that we mark as visited only faulty vertices and their neighbors, even though our algorithm inspects also vertices at distance 2 from F. Given a vertex v of Q_n,

- INSERT(v, T) inserts v into T and returns v_T,
- RETRIEVE(v, T) returns v_T or NIL if it does not exist.

Both operations require $O(n)$ time.

Given a list F of faulty vertices of Q_n, Algorithm 6.1 finds all components of $Q_n^2[F]$ using a depth–first search (DFS), described as Procedure 6.2. For every $f \in F$, all vertices v at distance at most two from f are visited. If v is faulty, DFS is applied recursively on v, which ensures that the algorithm indeed finds the components of $Q_n^2[F]$. If v is healthy, then v is inserted into the disjoint-set data structure D. Furthermore, sets of D containing vertices u and v are united for every edge uv at distance one from f. In that way, after a call to DFS(f) (line 8 of Algorithm 6.1) is completed, disjoint sets of D represent components of the graph $G(C) - C$ for the component C of $Q_n^2[F]$ containing f. This verifies the condition of Theorem 5.1.

The trie T is used to store information about the vertices visited during the search. Note that due to the time and space constraints, T cannot contain each of 2^n vertices of Q_n. Faulty vertices are inserted into T during the initialization of Algorithm 6.1. Healthy vertices of $G(C)$ for a component C of $Q_n^2[F]$ are inserted into T during the DFS of C. When the whole component of $Q_n^2[F]$ is found, all healthy vertices are removed from T .

Algorithm 6.1. CONNECTIVITY(n, F)

Input: Positive integer n, a list F of faulty vertices of Q_n
Output: "$Q_n - F$ is connected" or "$Q_n - F$ is disconnected"

```
1  T ← empty trie
2  foreach f ∈ F do f_T ← INSERT(f, T) ; mark f_T as faulty and non-visited
3  foreach f ∈ F do
4  |   f_T ← RETRIEVE(f, T)
5  |   if f_T is not visited then
6  |   |   mark f_T as visited
7  |   |   D ← empty data structure for disjoint sets
8  |   |   DFS(f)          // DFS of the component C of Q_n^2[F] containing f
9  |   |   if D contains more than one set then return "Q_n − F is disconnected"
10 |   |   remove all healthy vertices from T and clean-up data structure D
11 return "Q_n − F is connected".
```

Removing all healthy vertices from trie T (line 10 of Algorithm 6.1) may be implemented using the depth-first search of T. Since the total number of calls to INSERT($\cdot, T$) is bounded by $O(|F|n^2)$, the total time complexity of this clean-up is $O(|F|n^3)$.

To analyze the time complexity of our algorithm, observe that DFS(f) is called exactly once for each faulty vertex $f \in F$. Next, considering the code of Procedure 6.2, the outer for-loop (line 1) is executed for every neighbor of f, while the inner for-loop (line 8) is executed for some vertices at distance two from f. Therefore, the total number of the inner loop executions is bounded by $|F|n^2$. The time critical operation are RETRIEVE($\cdot, T$) and INSERT($\cdot, T$), requiring $O(n)$ time for each call as noted above. Hence, the total running time of the algorithm is bounded by $O(|F|n^3)$.

Theorem 6.1. *Given an integer $n \geq 1$ and a set of vertices F of Q_n, the problem whether the graph $Q_n - F$ is connected can be decided in $O(n^3|F|)$ time.*

Procedure 6.2. DFS(f)

```
Input: Faulty vertex f of Q_n        // f belongs to a component C of Q_n^2[F]
Data: Binary trie T, disjoint-set data structure D
 1 foreach u ∈ N(f) do
 2     u_T ← RETRIEVE(u, T)
 3     if u_T = NIL then
 4         u_T ← INSERT(u, T); mark u_T as healthy and non-visited; MAKE(u, D)
 5     if u_T is not visited then
 6         mark u_T as visited
 7         if u_T is healthy then
 8             foreach v ∈ N(u) do
 9                 v_T ← RETRIEVE(v, T)
10                 if v_T = NIL then
11                     v_T ← INSERT(v, T); mark v_T as healthy and non-visited;
                       MAKE(v, D)
12                 if v_T is healthy then
13                     if FIND(u, D) ≠ FIND(v, D) then UNION(u, v, D)
14                                                 // edge uv belongs to G(C)
15                 else if v_T is not visited then mark v_T as visited; DFS(v)
16                                                 // faulty vertex v belongs to C
17         else DFS(u)                             // faulty vertex u belongs to C
```

7 Concluding Remarks

In this paper we have described two algorithms for testing the connectivity of the n-dimensional hypercube with f faulty vertices. The (more general) expansion algorithm runs in $O(f^2 n^{3.5})$ time, whereas the local connectivity algorithm runs in $O(fn^3)$ time.

It is worth pointing out the following corollary: If $|F| = O(n^k)$ for some $k \in \mathbb{N}$, the size of $Q_n - F$ is exponential in n, but our algorithm still tests the connectivity of $Q_n - F$ in time which is polynomial in n.

We believe that it would be interesting to find other classes of graphs for which the connectivity instance of Problem 1.1 has a positive solution. Natural candidates are other hypercubic networks [10] whose fault-tolerance has been investigated previously [2,11]. In some networks, transformations of walks are possible if we allow swaps on larger cycles (of bounded-size), e.g. hexagonal grids, n-dimensional torus C_d^n with fixed d, planar graphs with faces of bounded size. We think that the approach described in Section 5 works for such networks as well.

Another question is what other properties can be efficiently tested in vertex-deleted graphs. A biconnectivity can be defined such that a graph $G = (V, E)$ is biconnected if $G - \{x\}$ is connected for every vertex x. If F is a set of faulty vertices of a hypercube Q_n then every vertex x of distance at least 2 from F has a connected neighborhood of distance 2 in $Q_n - F$. Thus it suffices to verify connectedness for vertices from $A \cup B$. Hence there exists an algorithm deciding a biconnectivity for $Q_n - F$ which requires $O(|F|^2 \cdot n^5)$ time. An analogous idea can work for multidimensional meshes and also for multiconnectivity.

References

1. Chan, M.Y., Lee, S.-J.: On the existence of Hamiltonian circuits in faulty hypercubes. SIAM J. Discrete Math. 4, 511–527 (1991)
2. Chen, Y.-C., Huang, Y.-Z., Hsu, L.-H., Tan, J.J.M.: A family of Hamiltonian and Hamiltonian connected graphs with fault tolerance. J. Supercomput. 54, 229–238 (2010)
3. Cormen, T.H., Leiserson, C.E., Rivest, R.L., Stein, C.: Introduction to Algorithms. MIT Press, Cambridge (2001)
4. Dvořák, T., Gregor, P.: Partitions of faulty hypercubes into paths with prescribed endvertices. SIAM J. Discrete Math. 22, 1448–1461 (2008)
5. Dvořák, T., Koubek, V.: Long paths in hypercubes with a quadratic number of faults. Inf. Sci. 179, 3763–3771 (2009)
6. Dvořák, T., Koubek, V.: Computational complexity of long paths and cycles in faulty hypercubes. Theor. Comput. Sci. 411, 3774–3786 (2010)
7. Fink, J., Gregor, P.: Long paths and cycles in hypercubes with faulty vertices. Inf. Sci. 179, 3634–3644 (2009)
8. Harper, L.H.: Optimal Numberings and Isoperimetric Problems on Graphs. J. Comb. Theory 1, 385–393 (1966)
9. Knuth, D.E.: The Art of Computer Programming: Sorting and Searching, 2nd edn., vol. III. Addison-Wesley, Reading (1998)
10. Leighton, F.T.: Introduction to Parallel Algorithms and Architectures: Arrays, Trees, Hypercubes. Morgan Kaufmann, San Mateo (1992)
11. Park, J.-H., Kim, H.-C., Lim, H.-S.: Many-to-Many Disjoint Path Covers in the Presence of Faulty Elements. IEEE Trans. Comput. 58, 528–540 (2009)

Reductions of Matrices Associated with Nowhere-Zero Flows*

Martin Kochol[1], Naďa Krivoňáková[2], Silvia Smejová[2], and Katarína Šranková[2]

[1] MÚ SAV, Štefánikova 49, 814 73 Bratislava 1, Slovakia
kochol@mat.savba.sk
[2] FPV ŽU, Univerzitná 8215/1, 010 26 Žilina, Slovakia
{nada.krivonakova,silvia.smejova,katarina.srankova}@fpv.uniza.sk

Abstract. Recently we have developed a method excluding certain subgraphs from a smallest counterexample to the 5-flow conjecture. This is based on comparing ranks of two matrices of large size. The aim of this paper is to be more effective by applying these methods so that we reduce the size of matrices used in the computation.

1 Introduction

A graph admits a *nowhere-zero k-flow* if its edges can be oriented and assigned numbers $\pm 1, \ldots, \pm(k-1)$ so that for every vertex, the sum of the values on incoming edges equals the sum on the outgoing ones. It is well-known that a graph with a bridge (1-edge-cut) does not have a nowhere-zero k-flow for any $k \geq 2$ (see, e.g., [2,9]). The famous 5-*flow conjecture* of Tutte [7] is that every bridgeless graph has a nowhere-zero 5-flow.

Let $\overline{G}$ be a counterexample to the 5-flow conjecture of the smallest possible order. It is well-known (see cf. Jaeger [2]) that $\overline{G}$ must be a *snark* which is a cyclically 4-edge-connected cubic graph without a 3-edge-coloring and with girth (the length of the shortest cycle) at least 5. (Note that a graph is *cyclically k-edge-connected* if deleting fewer than k edges does not result in a graph having at least two components containing cycles.) In [4], we have proved that $\overline{G}$ must be cyclically 6-edge-connected applying a method using ideas from linear algebra. We further improve the method in [5], where we show that if a specified matrix $\overline{M}_k$ has the same rank as certain submatrix $\overline{M}'_k$, then $\overline{G}$ cannot have a circuit of order k.

In this paper we improve the methods from [4,5] and present an approach how to reduce the size of matrices $\overline{M}_k$ and $\overline{M}'_k$. For example, in [5] $\overline{M}_7$ and $\overline{M}'_7$ have size 819×162 and 483×162, respectively. In this paper we reduce the size of the matrices into 317×110 and 287×110, respectively.

2 Preliminaries

The graphs considered in this paper are all finite and unoriented. Multiple edges and loops are allowed. If G is a graph, then $V(G)$ and $E(G)$ denote the sets

* Supported by grants VEGA 2/0118/10 and RI 5/06 DFM.

C.S. Iliopoulos and W.F. Smyth (Eds.): IWOCA 2010, LNCS 6460, pp. 192–200, 2011.

of vertices and edges of G, respectively. By a *multi-terminal network*, briefly a *network*, we mean a pair (G, U) where G is a graph and $U = (u_1, \ldots, u_n)$ is an ordered set of pairwise distinct vertices of G. The vertices $u_1, \ldots, u_n$ are called the *outer* vertices of (G, U) and the others are called the *inner* vertices of (G, U).

To each edge connecting u and v (including loops) we associate two distinct (directed) arcs, one directed from u to v, the other directed from v to u. If one of these arcs is denoted x then the other is denoted x^{-1}. Let $D(G)$ denote the set of such arcs, so that $|D(G)| = 2|E(G)|$. If $v \in V(G)$, then $\omega_G(v)$ denotes the set of arcs of G directed from v to $V(G) \setminus \{v\}$.

If G is a graph and A is an additive Abelian group, then an *A-chain* in G is a mapping $\varphi : D(G) \to A$ such that $\varphi(x^{-1}) = -\varphi(x)$ for every $x \in D(G)$. Furthermore, the mapping $\partial\varphi : V(G) \to A$ such that $\partial\varphi(v) = \sum_{x \in \omega_G(v)} \varphi(x)$ $(v \in V(G))$ is called the *boundary* of φ. An A-chain φ in G is called *nowhere-zero* if $\varphi(x) \neq 0$ for every $x \in D(G)$. If (G, U) is a network, then an A-chain φ in G is called an *A-flow* in (G, U) if $\partial\varphi(v) = 0$ for every inner vertex v of (G, U).

By a (*nowhere-zero*) *A-flow* in a graph G we mean a (nowhere-zero) A-flow in the network $(G, \emptyset)$. Our concept of nowhere-zero flows in graphs coincides with the usual definition of nowhere-zero flows as presented in Jaeger [2]. By Tutte [7,8], a graph has a nowhere-zero k-flow if and only if it has a nowhere-zero A-flow for any Abelian group A of order k. Thus the study of nowhere-zero 5-flows is, in a certain sense, equivalent to the study of nowhere-zero $\mathbb{Z}_5$-flows. We use this fact and deal only with $\mathbb{Z}_5$-flows because they are easier to handle than integral flows.

A network (G, U), $U = (u_1, \ldots, u_n)$, is called *simple* if the vertices $u_1, \ldots, u_n$ have degree 1. If φ is a nowhere-zero $\mathbb{Z}_5$-flow in (G, U), then denote by $\partial\varphi(U)$ the n-tuple $(\partial\varphi(u_1), \ldots, \partial\varphi(u_n))$. By simple counting, we get $\sum_{i=1}^{n} \partial\varphi(u_i) = -\sum_{v \in V(G) \setminus U} \partial\varphi(v) = 0$ (see [3]). Furthermore, $\partial\varphi(u_i) \neq 0$ because u_i has degree 1 $(i = 1, \ldots, n)$. Thus $\partial\varphi(U)$ belongs to the set

$$S_n = \{(s_1, \ldots, s_n);\ s_1, \ldots, s_n \in \mathbb{Z}_5 - \{0\},\ s_1 + \ldots + s_n = 0\}.$$

For every $s \in S_n$, denote by $F_{G,U}(s)$ the number of nowhere-zero $\mathbb{Z}_5$-flows φ in (G, U) satisfying $\partial\varphi(U) = s$.

A partition $P = \{Q_1, \ldots, Q_r\}$ of the set $\{1, \ldots, n\}$, $n \geq 2$, is called *proper* if each of $Q_1, \ldots, Q_r$ has cardinality at least 2. Let $\mathcal{P}_n$ denote the set of proper partitions of $\{1, \ldots, n\}$ and let $p_n = |\mathcal{P}_n|$. If $s = (s_1, \ldots, s_n) \in S_n$, $P = \{Q_1, \ldots, Q_r\} \in \mathcal{P}_n$, and $\sum_{i \in Q_j} s_i = 0$ for $j = 1, \ldots, r$, then we say that P and s are *compatible*. (For example, $\{\{1,2\},\{3,4,5\}\} \in \mathcal{P}_5$ is compatible with $(1,4,1,2,2) \in S_5$.) In this paper, we consider $\mathcal{P}_n$ as an p_n-tuple $(P_{n,1}, \ldots, P_{n,p_n})$. For any $s \in S_n$, denote by $\chi_n(s)$ the integral vector $(c_{s,1}, \ldots, c_{s,p_n})$ so that $c_{s,i} = 1$ $(c_{s,i} = 0)$ if $P_{n,i}$ is (is not) compatible with s, $i = 1, \ldots, p_n$. In [4] is proved the following statement.

Lemma 1. *Let (G, U), $U = (u_1, \ldots, u_n)$, be a simple network. Then there exist integers $x_1, \ldots, x_{p_n}$ such that for every $s \in S_n$, $F_{G,U}(s) = \sum_{i=1}^{p_n} c_{s,i} x_i$ where $(c_{s,1}, \ldots, c_{s,p_n}) = \chi_n(s)$.*

3 Forbidden Networks

Let (H, U), $U = (u_1, \dots, u_n)$, be a simple network. (H, U) is called *quasicubic*, if every vertex of H has degree at most 3. By a *cubic order* of (H, U), denoted by $\nu_3(H, U)$, we mean the number of the vertices of H of degree 3. Denote by $S_{H,U} = \{s \in S_n; F_{H,U}(s) > 0\}$ and by $V_{H,U}$ the linear hull of $\{\chi_n(s); s \in S_{H,U}\}$ in $\mathbb{R}^{p_n}$.

We say that (H, U) is a *forbidden network* if H cannot be a subgraph of a graph homeomorphic to a smallest counterexample to the 5-flow conjecture. By [5, Lemma 2], if $V_{H,U}$ is equal to the linear hull of $\{\chi_n(s); s \in S_n\}$ in $\mathbb{R}^{p_n}$, then (H, U) is forbidden. In order to improve this result we need some more notations.

Assume that H is a subgraph of a graph G and (H', U'), $U' = (u'_1, \dots, u'_n)$, is a simple network. Let G' arises from G after deleting the vertices from $V(H) \setminus U$ and identifying u_i with u'_i for $i = 1, \dots, n$. We say that G' arises from G after *replacing* (H, U) by (H', U').

We say that (H, U) can be *regularly replaced* by (H', U') in a class of graphs $\mathcal{C}$, if for every graph G of $\mathcal{C}$, the graph G' arising from G after replacing (H, U) by (H', U') is always bridgeless.

Lemma 2. *Let (H, U), $U = (u_1, \dots, u_n)$, and (H', U'), $U' = (u'_1, \dots, u'_n)$, $n \geq 2$, be quasicubic networks such that $\nu_3(H, U) > \nu_3(H', U')$, $V_{H',U'} \subseteq V_{H,U}$, and (H, U) can be regularly replaced by (H', U') in the class of cyclically 6-edge connected quasicubic graphs. Then (H, U) is a forbidden network.*

Proof. Let G be a counterexample to the 5-flow conjecture of the smallest possible order. Then by [4], G is a cyclically 6-edge-connected cubic graph. Suppose that F is homeomorphic with G and H is a subgraph of F. Without abuse of generality we can assume that $u_1, \dots, u_n$ have all degree 2 in F. Let F' be the graph arising after replacing (H, U) by (H', U'). By assumptions, F' is bridgeless and homeomorphic with a cubic graph G'. Since $\nu_3(H, U) > \nu_3(H', U')$, the order of G' is smaller than the order of G, therefore G' and F' admit nowhere-zero 5-flows.

Let I (I') be the graph arising from F (F') after deleting the vertices from $V(H) \setminus U$ ($V(H') \setminus U'$). Then (I, U) and (I', U') are simple networks, and there is an isomorphism of I and I' which maps $u_1, \dots, u_n$ to $u'_1, \dots, u'_n$, respectively. Thus $F_{I,U}(s) = F_{I',U'}(s)$ for every $s \in S_n$.

If there exists $s \in S_n$ such that $F_{H,U}(s)$, $F_{I,U}(s) > 0$, then (H, U) and (I, U) have nowhere-zero $\mathbb{Z}_5$-flows φ_1 and φ_2, respectively, such that $\partial\varphi_1(U) = \partial\varphi_2(U) = s$ and the flows φ_1 and $-\varphi_2$ can be "pieced together" into a nowhere-zero $\mathbb{Z}_5$-flow in F, a contradiction. Thus $F_{H,U}(s)F_{I,U}(s) = 0$ for every $s \in S_n$. Since $S_{H,U} = \{s \in S_n; F_{H,U}(s) > 0\}$, we have $F_{I,U}(s) = 0$ for every $s \in S_{H,U}$.

By Lemma 1, there exist integers $x_1, \dots, x_{p_n}$ such that for every $s \in S_n$, $F_{I,U}(s) = \sum_{i=1}^{p_n} c_{s,i} x_i$ where $(c_{s,1}, \dots, c_{s,p_n}) = \chi_n(s)$. Choose n-tuples $t_1, \dots, t_r$ from $S_{H,U}$ so that $\chi_n(t_1), \dots, \chi_n(t_r)$ form a basis in $V_{H,U}$. Then for every $s \in V_{H,U}$, there are numbers $y_{s,1}, \dots, y_{s,r}$ such that $\chi_n(s) = \sum_{j=1}^{r} y_{s,j} \chi_n(t_j)$

and, therefore, $F_{I,U}(s) = \sum_{i=1}^{p_n} c_{s,i}x_i = \sum_{i=1}^{p_n}(\sum_{j=1}^{r} y_{s,j}c_{t_j,i})x_i = \sum_{j=1}^{r} y_{s,j}$ $(\sum_{i=1}^{p_n} c_{t_j,i}x_i) = \sum_{j=1}^{r} y_{s,j}F_{I,U}(t_j) = 0$ (because $t_1, \dots, t_r \in S_{H,U}$). Thus we have $F_{I,U}(s) = 0$ for every $s \in V_{H,U}$.

Since F' has a nowhere-zero $\mathbb{Z}_5$-flow, there exists $t \in S_{H',U'} \cap S_{I',U'}$, i.e., $F_{H',U'}(t), F_{I',U'}(t) > 0$. By assumptions, $V_{H',U'} \subseteq V_{H,U}$, whence $F_{I',U'}(t) = F_{I,U}(s) = 0$, which is a contradiction. This proves the statement.

Let C_n be the circuit of order n, i.e., the graph having vertices $v_1, \dots, v_n$ and edges $v_1v_2, v_2v_3, \dots, v_nv_1$. Let H_n arises from C_n after adding new vertices $u_1, \dots, u_n$ and edges $u_1v_1, \dots, u_nv_n$. Then (H_n, U_n), $U_n = (u_1, \dots, u_n)$, is a simple network. For $i = 1, \dots, n$, let x_i denote the arc of H_n directed from u_i to v_i and y_i denote the arc directed from v_i to v_{i+1} (considering the indices mod n).

Consider a graph H_{n-2} and change the notation of its vertices by adding primes, i.e, denote them by $v'_1, \dots, v'_{n-2}, u'_1, \dots, u'_{n-2}$. Similarly change the notation of the arcs. Add new vertices $v'_{n-1}, v'_n, u'_{n-1}, u'_n$ and edges $v'_{n-1}u'_{n-1}, v'_nu'_n$, $v'_{n-1}v'_n$. Furthermore, let x'_{n-1}, x'_n, and z'_n denote the arcs of H'_n directed from u'_{n-1} to v'_{n-1}, from u'_n to v'_n, and from v'_{n-1} to v'_n, respectively. Then (H'_n, U'_n), $U'_n = (u'_1, \dots, u'_n)$, is a simple network.

Lemma 3. *For $n \geq 6$, $\nu_3(H_n, U_n) > \nu_3(H'_n, U'_n)$ and (H_n, U_n) can be replaced by (H'_n, U'_n) regularly in the class of cyclically 6-edge connected cubic graphs.*

Proof. $\nu_3(H_n, U_n) = n > n - 2 = \nu_3(H'_n, U'_n)$. Let G' arises from G after replacing (H_n, U_n) by (H'_n, U'_n). If G' has a bridge, then G is not cyclically 6-edge-connected.

Thus to show that a smallest counterexample to the 5-flow conjecture has no circuit of length n, it suffices, by Lemmas 2 and 3, to prove that $V_{H'_n,U'_n} \subseteq V_{H_n,U_n}$.

4 Superproper Permutations

We say that $s = (s_1, \dots, s_n) \in S_n$ and $t = (t_1, \dots, t_n) \in S_n$ are θ_n-equivalent if $s_1 = t_1, \dots, s_{n-2} = t_{n-2}$ and $s_{n-1} + s_n = t_{n-1} + t_n$.

A proper partition $P = \{Q_1, \dots, Q_r\}$ of the set $\{1, \dots, n\}$, $n \geq 2$, is called *superproper* if n and $n-1$ are contained in the same set from $Q_1, \dots, Q_r$. Let $\mathcal{P}'_n$ denote the set of superproper partitions of $\{1, \dots, n\}$. (For example, $\{\{1,2\},\{3,4,5\}\} \in \mathcal{P}'_5$.) We consider $\mathcal{P}'_n$ as an p'_n-tuple $(P_{n,1}, \dots, P_{n,p'_n})$. For any $s \in S_n$, denote by $\chi'_n(s)$ the integral vector $(c_{s,1}, \dots, c_{s,p'_n})$ so that $c_{s,i} = 1$ $(c_{s,i} = 0)$ if $P_{n,i}$ is (is not) compatible with s, $i = 1, \dots, p'_n$. Vector $\chi'_n(s)$ contains the coordinates of $\chi_n(s)$ corresponding with the superproper partitions from $\mathcal{P}_n$.

Lemma 4. *If s and t are θ_n-equivalent elements of S_n, then $\chi'_n(s) = \chi'_n(t)$.*

Proof. Follows from the definitions of $\mathcal{P}'_n$ and χ'_n.

By [4,5], we know that for $n \geq 2$,

$$p_n = 1 + \sum_{i=2}^{n-2} \binom{n-1}{i-1} p_{n-i}. \tag{1}$$

Clearly, $p'_n = p_n = 1$ for $n = 2, 3$. If $n \geq 4$, then $\mathcal{P}'_n$ contains exactly p_{n-2} partitions such that n and $n-1$ are contained in a 2-element subset and exactly p_{n-1} partitions such that n and $n-1$ are contained in at least 3-element subset (in this case we can delete n and get all partitions from $\mathcal{P}_{n-1}$). Thus, for $n \geq 4$,

$$p'_n = p_{n-1} + p_{n-2}. \tag{2}$$

The main idea standing behind the reductions presented here, is that instead of vectors $\chi_n(s)$ we consider vectors of the form $\chi_n(s) - \chi_n(t)$ where $s \equiv t(\theta_n)$. By Lemma 4, vectors of this form have all coordinates corresponding to superproper partitions equal to 0. Thus instead of p_n dimensional vectors we deal with $p_n - p'_n = p_n - p_{n-1} - p_{n-2}$ dimensional vectors. For the case $n = 7$, we get reduction from $p_7 = 162$ to $p_7 - p_6 - p_5 = 162 - 41 - 11 = 110$.

5 θ_n-Classes

Lemma 5. *Let φ_1, φ_2 be nowhere-zero $\mathbb{Z}_5$-flows in (H_n, U_n) such that $\partial\varphi_1(U_n) = \partial\varphi_2(U_n)$. Then either $\varphi_1 = \varphi_2$ or $\varphi_1(y_i) \neq \varphi_2(y_i)$ for $i = 1, \dots, n$.*

Proof. Follows from the fact that for $i = 1, \dots, n$, $\varphi_1(y_1) - \varphi_1(y_i) = \sum_{j=2}^{r} \varphi_1(x_j) = \sum_{j=2}^{r} \varphi_2(x_j) = \varphi_2(y_1) - \varphi_2(y_i)$.

Let

$$\begin{aligned} C(n) &= \{s \in S_n;\ F_{H_n,U_n}(s) \neq \emptyset\}, \\ C_i(n) &= \{s \in C(n);\ F_{H_n,U_n}(s) = i\}, \\ C'(n) &= \{s \in S_n;\ F_{H'_n,U'_n}(s) \neq \emptyset\}, \\ C''(n) &= C'(n) \setminus C(n). \end{aligned}$$

By [6], $C(n) = C_1(n) \cup C_2(n) \cup C_3(n)$ for every $n \geq 2$.

Before formulating another lemma, we introduce some more technical notation. For every $s \in S_n$ and a proper network (G, U), denote by $\Phi_{G,U}(s)$ the set of nowhere-zero $\mathbb{Z}_5$-flows φ in (G, U) satisfying $\partial\varphi(U) = s$. Note that $|\Phi_{G,U}(s)| = F_{G,U}(s)$.

If φ is a nowhere-zero $\mathbb{Z}_5$-flow in (H_{n-2}, U_{n-2}) and $a \in \mathbb{Z}_5 \setminus \{0, -\varphi(y_{n-2}),\}$, then denote by $\varphi^{[a]}$ the nowhere-zero $\mathbb{Z}_5$-flow in (H_n, U_n) such that $\varphi^{[a]}(x_n) = -\varphi^{[a]}(x_{n-1}) = a$, $\varphi^{[a]}(y_n) = \varphi(y_{n-2})$, $\varphi^{[a]}(y_{n-1}) = \varphi(y_{n-2}) + a$, and $\varphi^{[a]}(x_i) = \varphi(x_i)$, $\varphi^{[a]}(y_i) = \varphi(y_i)$ for $i = 1, \dots, n-2$. (Note that writing $\varphi(x_i)$, $\varphi(y_i)$ we consider the arcs x_i, y_i to be from H_{n-2}, and writing $\varphi^{[a]}(x_i)$, $\varphi^{[a]}(y_i)$ we consider the arcs x_i, y_i to be from H_n.)

If φ is a nowhere-zero $\mathbb{Z}_5$-flow in (H_n, U_n) and $\varphi(x_{n-1}) + \varphi(x_n) = 0$ (resp. $\varphi(x_{n-1}) + \varphi(x_n) \neq 0$), then denote by $\overline{\varphi}$ the nowhere-zero $\mathbb{Z}_5$-flow in (H_{n-2}, U_{n-2})

(resp. (H_{n-1}, U_{n-1})) such that $\overline{\varphi}(x_i) = \varphi(x_i)$, $\overline{\varphi}(y_i) = \varphi(y_i)$ for $i = 1, \ldots, n-2$ (resp. $\overline{\varphi}(x_{n-1}) = \varphi(x_n) + \varphi(x_{n-1})$, $\overline{\varphi}(y_{n-1}) = \varphi(y_n)$, and $\overline{\varphi}(x_i) = \varphi(x_i)$, $\overline{\varphi}(y_i) = \varphi(y_i)$ for $i = 1, \ldots, n-2$).

If $(s_1, \ldots, s_n) \in S_n$ and $s_{n-1} + s_n = 0$ (resp. $s_{n-1} + s_n \neq 0$), then denote by $\overline{s} = (s_1, \ldots, s_{n-2}) \in S_{n-2}$ (resp. $\overline{s} = (s_1, \ldots, s_{n-2}, s_{n-1} + s_n) \in S_{n-1}$).

Lemma 6. *Let $s = (s_1, \ldots, s_n) \in S_n$. Then $s \in C''(n)$ if and only if $s_{n-1} + s_n = 0$, $\overline{s} \in C_1(n-2)$, and $s_n = -\varphi(y'_{n-2})$ where $\varphi \in \Phi_{H_{n-2}, U_{n-2}}(\overline{s})$.*

Proof. Suppose $\varphi' \in \Phi_{H'_n, U'_n}(s)$ and φ'' be the restriction of φ' to $D(H_{n-2})$. Then $s_n = \varphi'(x'_n) = \varphi'(z'_n) = -\varphi'(x'_{n-1}) = -s_{n-1}$ and $\varphi''(u'_1, \ldots, u'_{n-2}) = \overline{s}$. If $s_n \neq -\varphi''(y'_{n-2})$, then $\varphi''^{[s_n]} \in \Phi_{H_n, U_n}(s)$. Thus if $s \in C''(n)$, then s must satisfy the assumptions.

If s satisfies the assumptions, then $s \in C'(n)$ and we can choose $\varphi' \in \Phi_{H'_n, U'_n}(s)$. If also $s \in C(n)$, take $\varphi \in \Phi_{H_n, U_n}(s)$. Then $\overline{\varphi}$ must be the restriction of φ' to $D(H_{n-2})$ (because $\overline{s} \in C_1(n-2)$), whence $\varphi(x_n) = s_n = -\varphi(y_{n-2}) = -\varphi(y_n) = \varphi(y_n^{-1})$, which is not possible. Thus $s \in C''(n)$.

By the arithmetic in the group $\mathbb{Z}_5$, if $(s_1, \ldots, s_n) \in S_n$ and $s_{n-1} + s_n = 0$ ($s_{n-1} + s_n \neq 0$), then $|[s]\theta_n| = 4$ ($|[s]\theta_n| = 3$).

Lemma 7. *Let $s = (s_1, \ldots, s_n) \in S_n$.*

(1) If $s_{n-1} + s_n = 0$ and $\overline{s} \in C_1(n-2)$, then $|[s]\theta_n \cap C(n)| = 3$ and $|[s]\theta_n \cap C''(n)| = 1$.
(2) If $s_{n-1} + s_n = 0$ and $\overline{s} \in C_2(n-2) \cup C_3(n-2)$, then $|[s]\theta_n \cap C(n)| = 4$ and $|[s]\theta_n \cap C''(n)| = 0$.
(3) If $s_{n-1} + s_n \neq 0$ and $\overline{s} \in C_1(n-1)$, then $|[s]\theta_n \cap C(n)| = 2$ and $|[s]\theta_n \cap C''(n)| = 0$.
(4) If $s_{n-1} + s_n \neq 0$ and $\overline{s} \in C_2(n-1) \cup C_3(n-1)$, then $|[s]\theta_n \cap C(n)| = 3$ and $|[s]\theta_n \cap C''(n)| = 0$.
(5) If neither of the assumptions from (1)–(4) occurs, then $|[s]\theta_n \cap C(n)| = 0$ and $|[s]\theta_n \cap C''(n)| = 0$.

Proof. Let $s_{n-1} + s_n = 0$, $t \in [s]\theta_n \cap C(n)$ and $\varphi \in \Phi_{H_n, U_n}(t)$. Then $\overline{s} = \overline{t} \in S_{n-2}$ and $\overline{\varphi} \in \Phi_{H_{n-2}, U_{n-2}}(\overline{s})$, whence $\overline{s} \in C(n-2)$. If $\overline{s} \in C_1(n-2)$ ($\overline{s} \in C_2(n-2) \cup C_3(n-2)$), then by Lemma 6, $|[s]\theta_n \cap C''(n)| = 1$ ($|[s]\theta_n \cap C''(n)| = 0$). Using $\psi^{[a]}$ for $\psi \in \Phi_{H'_{n-2}, U_{n-2}}(\overline{s})$ and all $a \in \mathbb{Z}_5 \setminus \{0, -\psi(y_{n-2})\}$ we get that $|[s]\theta_n \cap C(n)| = 3$ ($|[s]\theta_n \cap C(n)| = 4$). This proves (1), (2), but also (5) for the case $s_{n-1} + s_n = 0$.

Let $s_{n-1} + s_n \neq 0$, $t \in [s]\theta_n \cap C(n)$ and $\varphi \in \Phi_{H_n, U_n}(t)$. Without loss of generality we can assume that $s_{n-1} + s_n = 1$. Then $\overline{s} = \overline{t} \in S_{n-1}$ and $\overline{\varphi} \in \Phi_{H_{n-1}, U_{n-1}}(\overline{s})$, whence $\overline{s} \in C(n-1)$.

Assume that $\overline{s} \in C_1(n-1)$ and $\varphi' \in \Phi_{H_{n-1}, U_{n-1}}(\overline{s})$. Then $\varphi'(y_{n-1})$, $\varphi'(y_{n-2}^{-1})$ can be either 1,3, or 3,1, or 2,2, respectively. In all three cases, there exist exactly two nowhere-zero $\mathbb{Z}_5$-flows φ_1, φ_2 in (H_n, U_n) such that $\varphi_1(x_n) \neq \varphi_2(x_n)$ and $\overline{\varphi_1} = \overline{\varphi_2} = \varphi'$. This proves (3).

Assume that $\overline{s} \in C_2(n-1) \cup C_3(n-1)$ and $\varphi_1', \varphi_2' \in \Phi_{H_{n-1},U_{n-1}}(\overline{s})$, $\varphi_1' \neq \varphi_2'$. Then we can choose the notation of φ_1', φ_2' so that $\varphi_1'(y_{n-1})$, $\varphi_1'(y_{n-2}^{-1})$, $\varphi_2'(y_{n-1})$, $\varphi_2'(y_{n-1}^{-1})$ are either 1,3,3,1, or 3,1,2,2, or 2,2,1,3, respectively. In all three cases, there exist nowhere-zero $\mathbb{Z}_5$-flows $\varphi_1, \varphi_2, \varphi_3$ in (H_n, U_n) such that $\varphi_1(x_n) \neq \varphi_2(x_n) \neq \varphi_3(x_n) \neq \varphi_1(x_n)$ and $\overline{\varphi_1}, \overline{\varphi_2}, \overline{\varphi_3} \in \{\varphi_1', \varphi_2'\}$. This proves (4).

We have also proved (5) for the case $s_{n-1} + s_n \neq 0$, because we have shown that if $|[s]\theta_n \cap C(n)| \neq 0$, then either (4) or (5) must hold.

6 Excluding Girth Seven

Lemma 8. *Let $\alpha_1, \ldots, \alpha_r, \beta_1, \ldots, \beta_{2s+r}$, $r < s$, be not necessarily different vectors from $\mathbb{R}^p$ and $\alpha_1 - \beta_1, \ldots, \alpha_r - \beta_r$ be contained in the linear hull of $\beta_{r+1} - \beta_{r+2}, \ldots, \beta_{2s+r-1} - \beta_{2s+r}$. Then $\alpha_1, \ldots, \alpha_r$ are contained in the linear hull of $\beta_1, \ldots, \beta_{2s+r}$.*

Proof. Follows immediately from properties of linear dependence in linear spaces.

Let $\mathcal{A}$ denote the automorphism group of $\mathbb{Z}_5$. The elements of $\mathcal{A}$ are $\alpha_0 = \mathrm{id}$, $\alpha_1 = (1,2,4,3)$, $\alpha_2 = (1,4)(2,3)$ and $\alpha_3 = (1,3,4,2)$. If $s = (s_1, \ldots, s_n) \in S_n$ and $\alpha \in \mathcal{A}$, then denote $\alpha(s) = (\alpha(s_1), \ldots, \alpha(s_n)) \in S_n$. We say that s and $\alpha(s)$ are σ_n-equivalent. Clearly, $\chi_n(s) = \chi_n(\alpha(s))$ and $F_{U,G}(s) = F_{U,G}(\alpha(s))$ for any simple network (G, U) with n outer vertices (φ is a nowhere-zero $\mathbb{Z}_5$-flow in (G, U) if and only if $\alpha(\varphi)$ is so). Therefore, we do not need to consider all elements from S_n, but only non σ_n-equivalent representatives of the σ_n-equivalence classes (each of them has exactly four elements). Thus, from now on, we consider only elements $s = (s_1, \ldots, s_n) \in S_n$ such that $s_1 = 1$. The same restriction we consider also for sets $C(n)$, $C'(n)$, $C''(n)$, and $C_i(n)$, $i = 1,2,3$. Let $c(n) = |C(n)|$, $c'(n) = |C'(n)|$, $c''(n) = |C''(n)|$, and $c_i(n) = |C_i(n)|$, $i = 1,2,3$.

Theorem 1. $V_{H_7', U_7'} \subseteq V_{H_7, U_7}$.

Proof. Following Lemma 7, we can denote the elements from $C(n) \cup C'(n)$ as $s^{[i,j,k]}$ where $1 \leq i \leq 4$, $1 \leq j \leq b_i(n)$, $1 \leq k \leq r_i(n)$,

$$\begin{aligned}
&b_1(n) = c_1(n-2),\ b_2(n) = c_2(n-2) + c_3(n-2), \\
&b_3(n) = c_1(n-1),\ b_4(n) = c_2(n-1) + c_3(n-1), \\
&r_1(n) = r_2(n) = 4\ r_3(n) = 2, \quad r_4(n) = 3,
\end{aligned}$$

assuming that $\{s^{[i,j,k]}; k = 1, \ldots, r_i(n)\}$, $i = 1, \ldots, 4$, $j = 1, \ldots, b_i(n)$, are pairwise θ_n-equivalent and satisfy the condition (i) from Lemma 7. If $i = 1$, then we also assume that $s^{[i,j,3]} \in C''(n)$ for $j = 1, \ldots, b_1(n)$.

Define

$$\begin{aligned}
A_n = \{&\alpha^{[i,j,k]} = \chi_n(s^{[i,j,k]}) - \chi_n(s^{[i,j,r_i(n)]}); \\
&1 \leq i, \leq 4,\ 1 \leq j \leq b_i(n),\ 1 \leq k < r_i(n)\}, \\
B_n = \{&\alpha^{[1,j,3]};\ 1 \leq j \leq b_1(n)\}.
\end{aligned}$$

Then $B_n \subseteq A_n$. Let W_n be the linear hull of $A_n \setminus B_n$ in $\mathbb{R}^{p_n}$. If $B_n \subseteq W_n$, then by Lemmas 8 and 7, $V_{H_n', U_n'} \subseteq V_{H_n, U_n}$. Using computers, we have verified that $B_7 \subseteq W_7$. Thus $V_{H_7', U_7'} \subseteq V_{H_7, U_7}$.

By Lemmas 2, 3 and Theorem 1, the smallest counterexample to the 5-flow conjecture cannot have a circuit of order 7.

7 Computations

Now we discuss the computations mentioned in the proof of Theorem 1. Let M_n be the matrix whose rows are the vectors from A_n. Furthermore, we assume that the first $|A_n \setminus B_n|$ rows correspond with the elements from $A_n \setminus B_n$, and denote the submatrix composed from these rows by M'_n. We can also assume that M_n does not contain the columns corresponding with the superproper permutations from $\mathcal{P}_n$ (because, all these entries are 0 by Lemma 4). Thus M_n has $q_n = p_n - p'_n$ columns and $a_n = 3b_1(n) + 3b_2(n) + b_3(n) + 2b_4(n)$ rows and M'_n has $a'_n = y_n - b_1(n)$ rows.

By [6], $c(1) = c_1(1) = c_2(1) = c_3(1) = c_1(2) = c_2(2) = 0$, $c(2) = c_3(2) = 1$, and for every $n \geq 3$,

$$\begin{aligned} c_1(n) &= 3c_1(n-2) + 2c_2(n-2) + 2c_1(n-1) + 2c_2(n-1), \\ c_2(n) &= 2c_2(n-2) + 3c_3(n-2) + \ c_2(n-1) + 3c_3(n-1), \\ c_3(n) &= \ c_3(n-2), \\ c(n) &= \ c_1(n) + c_2(n) + c_3(n). \end{aligned} \tag{3}$$

Let $v_n = |S_n|$. By [5], $v_2 = v_3 = 1$ and $v_n = 3v_{n-2} + 4v_{n-1}$ for $n \geq 4$. Using (1), (2) and (3), we can evaluate q_n, a_n and a'_n for $1 \leq n \leq 8$.

n	1	2	3	4	5	6	7	8
$c_1(n)$	0	0	0	6	30	120	420	1386
$c_2(n)$	0	0	3	6	15	30	63	126
$c_3(n)$	0	1	0	1	0	1	0	1
$c(n)$	0	1	3	13	45	151	483	1513
v_n	0	1	3	13	51	205	819	3277
p_n	0	1	1	4	11	41	162	715
q_n	0	0	0	2	6	26	110	512
a_n	0	0	2	9	29	99	317	999
a'_n	0	0	2	9	29	93	287	879

(4)

To check whether $B_n \subseteq W_n$ we need to apply Gauss elimination to a matrix M_n and check whether the nonzero entries are only in the rows of submatrix M'_n. We have applied this for $n = 7$, and we got that M'_7 and M_7 have the same rank. Note that M_7 and M'_7 have size 317×110 and 287×110, respectively. This is a significant reduction, because in [5] we had to check that a matrix of size 819×162 has the same rank as its submatrix of size 483×162.

Using computers we have applied the same approach also for $n = 8$. But in this case we get that $B_8 \not\subseteq W_8$. Thus our method cannot be applied for excluding girth eight.

References

1. Isaacs, R.: Infinite families of nontrivial trivalent graphs which are not Tait colorable. Amer. Math. Monthly 82, 221–239 (1975)
2. Jaeger, F.: Nowhere-zero flow problems. In: Beineke, L.W., Wilson, R.J. (eds.) Selected Topics in Graph Theory, vol. 3, pp. 71–95. Academic Press, London (1988)
3. Kochol, M.: Superposition and constructions of graphs without nowhere-zero k-flows. European J. Combin. 23, 281–306 (2002)
4. Kochol, M.: Reduction of the 5-flow conjecture to cyclically 6-edge-connected snarks. J. Combin. Theory Ser. B 90, 139–145 (2004)
5. Kochol, M.: Girth restrictions for the 5-flow conjecture. In: Proceedings of the 16th ACM-SIAM Symposium on Discrete Algorithms, SODA 2005, pp. 705–707 (2005)
6. Kochol, M., Krivoňáková, N., Smejová, S., Šranková, K.: Counting nowhere-zero flows on wheels. Discrete Math. 308, 2050–2053 (2008)
7. Tutte, W.T.: A contribution to the theory of chromatic polynomials. Canad. J. Math. 6, 80–91 (1954)
8. Tutte, W.T.: A class of Abelian groups. Canad. J. Math. 8, 13–28 (1956)
9. Zhang, C.-Q.: Integral Flows and Cycle Covers of Graphs. Dekker, New York (1997)

Blocks of Hypergraphs

Applied to Hypergraphs and Outerplanarity

Ulrik Brandes[1], Sabine Cornelsen[1], Barbara Pampel[1], and Arnaud Sallaberry[2]

[1] Fachbereich Informatik & Informationswissenschaft, Universität Konstanz
{Ulrik.Brandes,Sabine.Cornelsen,Barbara.Pampel}@uni-konstanz.de
[2] CNRS UMR 5800 LaBRI, INRIA Bordeaux - Sud Ouest, Pikko
arnaud.sallaberry@labri.fr

Abstract. A support of a hypergraph H is a graph with the same vertex set as H in which each hyperedge induces a connected subgraph. We show how to test in polynomial time whether a given hypergraph has a cactus support, i.e. a support that is a tree of edges and cycles. While it is $\mathcal{NP}$-complete to decide whether a hypergraph has a 2-outerplanar support, we show how to test in polynomial time whether a hypergraph that is closed under intersections and differences has an outerplanar or a planar support. In all cases our algorithms yield a construction of the required support if it exists. The algorithms are based on a new definition of biconnected components in hypergraphs.

1 Introduction

A *hypergraph* (see e.g. [2,28]) is a pair $H = (V, A)$ where V is a finite set and A is a (multi-)set of non-empty subsets of V. There are basically two different variants of drawing a hypergraph, the *edge-standard* (drawing each hyperedge $h \in A$ as a star or a tree whose leaves are the elements of h – see Fig. 1(a)) or the *subset standard* (drawing each hyperedge $h \in A$ as a simple closed region that contains exactly the vertices in h and no other vertices of V – see Fig. 2(b)). For drawings in the edge standard see, e.g., [7,11,18,20]. In this paper, we concentrate on the second variant which is also called the *Euler diagram* of the set of hyperedges. Simultaneous drawings of a graph and a hypergraph in the subset standard are called clustered graphs. Drawing graphs with overlapping clusters is discussed in [9,19]. There are different variants on when a hypergraph admits a nice drawing in the subset standard. Several of them are based on some graphs associated with the hypergraph.

A hypergraph $H = (V, E)$ is *Zykov-planar* [28] if and only if there is a plane multi-graph M with vertex set V such that each hyperedge equals the set of vertices of some face of M. The hypergraph H can be represented as a *bipartite graph* B_H with vertex set $V \cup A$ and an edge between a vertex $v \in V$ and $h \in A$ if and only if $v \in h$ (see Fig. 1(a)). A hypergraph is Zykov-planar if and only if its bipartite graph is planar [27]. Thus, Zykov-planarity can be tested in linear time [13].

C.S. Iliopoulos and W.F. Smyth (Eds.): IWOCA 2010, LNCS 6460, pp. 201–211, 2011.

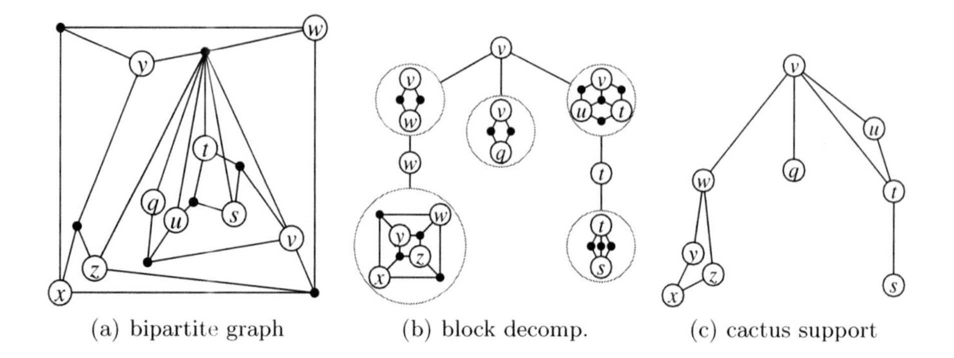

Fig. 1. Three representations of the hypergraph with hyperedges $\{s, t, v\}$, $\{s, t, u\}$, $\{q, u, v\}$, $\{w, x, z, v\}$, $\{x, y, z\}$, $\{w, x, y\}$, $\{q, s, t, u, v, w, z, y\}$

Some work on *Euler diagrams* and a definition on their well-formedness is summarized in [12]. The definition is associated with the *superdual* (or *combinatorial dual*) of H. Assuming that no two vertices of H are contained in the same set of hyperedges, the superdual is a graph on the vertex set V plus an artificial vertex that is not contained in any hyperedge. There is an edge between two vertices v and w if and only if the symmetric difference of the set of hyperedges containing v and the set of hyperedges containing w contains exactly one set h. Edge $\{v, w\}$ is then labeled h. Flower et al. [12] show that a hypergraph has a well-formed Euler diagram if and only if there is a plane subgraph of the super dual in which each hyperedge and its complement induces a connected subgraph and in which the labels around each face fulfill some condition. The superdual of the hypergraph H in Fig. 1 is highly non-connected and, hence, H has no well-formed Euler diagram. Verroust and Viaud [26] considered Euler diagrams for hypergraphs with at most 8 hyperedges. The complexity of Euler diagrams is discussed by Schaefer and Štefankovič [21]. Drawings of arbitray hypergraphs in an extended subset standart where the regions representing the hyperedges do not have to be connected are discussed by Simonetto and Auber [22,23].

A *support* [25,15] (or *host graph* [17]) of a hypergraph $H = (V, E)$ is a graph $G = (V, E)$ with the property that the subgraph of G induced by any hyperedge is connected. A hypergraph is *(vertex-)planar* [14] if it has a planar support. (The partial connectivity graphs of Chow [8] are planar supports of a dualized version of a hypergraph.) Planar hypergraphs are a generalization of both, Zykov-planar hypergraphs [25] and hypergraphs having a well-formed Euler-diagram [12]. It is $\mathcal{NP}$-complete to decide whether a hypergraph has a planar support [14] even if the set of hyperedges is closed under intersections and each hyperedge induces a path in the support. However, it can be decided in linear time whether a hypergraph has a support that is a tree [24], a path, or a cycle [6]. Tree supports with bounded degrees [6] and minimum weighted tree supports [16] can be constructed in polynomial time. Equivalent formulations for hypergraphs having a tree support can be found in [1].

To guarantee that each hyperedge can be drawn by a simple closed region, Kaufmann et al. [15] required *compact supports.* A support $G = (V, E)$ of a hypergraph is compact if G is planar, triangulated and no inner face of the subgraph of G induced by a hyperedge h contains a vertex not in h. It can be concluded from [14] that it is $\mathcal{NP}$-complete to decide whether a hypergraph has a compact support even if it is closed under intersections. However, a hypergraph has a compact support if it has an outerplanar support. So it would be interesting to know whether a hypergraph has an outerplanar support. So far the complexity of outerplanar supports is open. It is $\mathcal{NP}$-complete to decide whether a hypergraph has a 3-outerplanar support [6] or a 2-outerplanar support [5].

The *Hasse diagram* of a hypergraph $H = (V, A)$ is the directed acyclic graph with vertex set $A \cup V$ and there is an edge (h_1, h_2) (or (h_1, v) and $h_2 = \{v\}$) if and only if $h_2 \subsetneq h_1$ and there is no set $h \in A$ with $h_2 \subsetneq h \subsetneq h_1$. A hypergraph $H = (V, A)$ has an outerplanar support if its *based Hasse diagram*, i.e. the Hasse diagram of $A \cup \{V\}$ is planar [15].

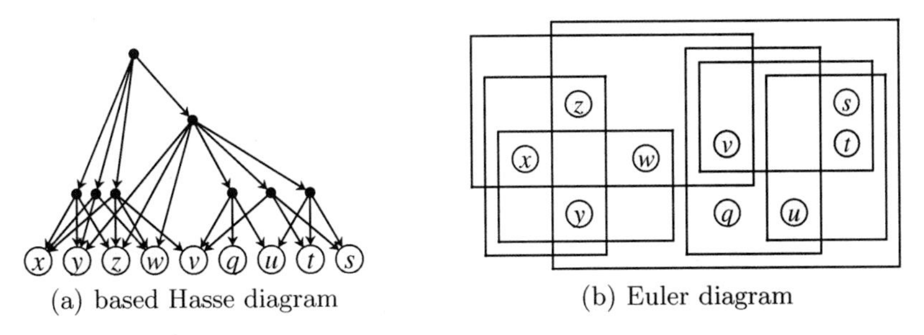

Fig. 2. Two more representations of the hypergraph with hyperedges $\{s, t, v\}$, $\{s, t, u\}$, $\{q, u, v\}$, $\{w, x, z, v\}$, $\{x, y, z\}$, $\{w, x, y\}$, $\{q, s, t, u, v, w, z, y\}$

In this paper, we consider special cases of outerplanar supports. A graph is a *cactus* if it is connected and each edge is contained in at most one cycle. A cactus can be used to represent the set of all minimum cuts of a graph [10]. Cactus supports also have applications in hypergraph coloring [17]. In Sect. 3, we show that a hypergraph has a cactus support if its based Hasse diagram is planar but the converse is not true. Further, we show how to decide in polynomial time whether a hypergraph has a cactus support. The construction is based on a new definition of biconnected components of a hypergraph introduced in Sect. 2 (see Fig. 1(b) for an illustration).

When drawing Euler diagrams it is desirable to visualize not only the hyperedges itself but also the intersection and the differences of two hyperedges. Motivated by this fact, we consider hypergraphs closed under intersections and differences (hcid) in Sect. 4. We show that it can be decided in polynomial time whether a hcid has an outerplanar or planar support.

In the remainder of the paper let $H = (V, A)$ be a hypergraph with $n = |V|$ vertices, $m = |A|$ hyperedges, and $N = \sum_{h \in A} |h|$ equals the sum of the sizes of all hyperedges. The *size of the hypergraph* is then $N + n + m$.

2 Biconnected Components

In this section, we show how to decompose a hypergraph into biconnected components that we will call blocks. This decomposition will be constructed in such a way that there is a support with the property that the blocks of the hypergraph correspond to the biconnected components of the support.

For a hypergraph $H = (V, A)$ and a subset $V' \subset V$ the hypergraph *induced* by V' is $H[V'] = (V', A[V'])$ with $A[V'] = \{h \cap V'; h \in A\} \setminus \{\emptyset, \{v\}; v \in V\}$. I.e., $A[V']$ contains from each hyperedge the part that is in V' omitting the empty set and the sets of size one to be consistent with the definition for ordinary graphs. Let $H|V' = (V', A|V')$ with $A|V' = \{h \in A;\ h \subseteq V'\}$. Note that $H[V']$ does not have to be planar if H is planar. However, $H|V'$ is planar if H is.

The sequence $p : v_0, h_1, v_1, \ldots, h_k, v_k$ is a v_0v_k*-path* in H if $h_1, \ldots, h_k \in A$, $v_0 \in h_1, v_k \in h_k$, and $v_i \in h_i \cap h_{i+1}, i = 1, \ldots k-1$. Vertices v_0 and v_k are the end vertices of p. Two vertices v, w of a hypergraph $H = (V, A)$ are *connected* if there is a vw-path in H. Connectivity is an equivalence relation on the set of vertices of a hypergraph and the hypergraphs induced by the equivalence classes are called *connected components* [28].

Let $v \in V$. The connected components of $H|(V \setminus \{v\})$ are the *parts* of v and v is an *articulation point* of H if v has more than one part. Note that v is an articulation point of H if and only if there is a support of H in which v is a cut vertex. E.g., vertex v is a cut vertex of the hypergraph in Fig. 1 and $\{w, x, y, z\}$, $\{q\}$, and $\{u, t, s\}$ are the parts of v.

A *decomposition into blocks* of a hypergraph $H = (V, A)$ is defined recursively. H is a block if and only if H is connected and does not contain an articulation point. If H is not connected then the blocks of H are the blocks of the connected components of H. If H is connected and contains an articulation point v, let $W_1, \ldots, W_k$ be the parts of v. Then the blocks of H are the blocks of $H[W_1 \cup \{v\}], \ldots, H[W_k \cup \{v\}]$.

Note that the blocks depend on the choices of the articulation points and are not uniquely defined. E.g., consider the hypergraph H in Fig. 1. Choosing the articulation points v, w, and t yields the subhypergraphs induced by the sets $\{v, w\}$, $\{w, x, y, z\}$, $\{v, q\}$, $\{t, u, v\}$, and $\{t, s\}$ as blocks. These are indicated within the circles of Fig. 1(b). Choosing s instead of t as an articulation point would yield the block $H[\{s, u, v\}]$ instead of $H[\{t, u, v\}]$.

Note that this definition of articulation points and blocks is related to but different from the definition given in [1]. Further note that the sum of the sizes of all blocks is at most three times the size of the hypergraph itself.

We will use the terminology analogously for the bipartite graph B_H on the vertex set $V \cup A$ representing the hypergraph $H = (V, A)$. The connected components of H correspond to the connected components of B_H. Vertex v is an articulation point of B_H if $B[V \setminus \{v\} \cup A \setminus \{h \in A; v \in h\}]$ contains more than one connected component which will again be called the *parts* of v. The blocks of B_H are the bipartite graphs representing the blocks of H. Then the blocks of B_H and, hence, of H can be constructed by determining n times the connected components of a subgraph of B_H.

Lemma 1. *The blocks of the hypergraph H can be found in $\mathcal{O}(nN + n + m)$ time.*

Proof. Since the connected components of B_H can be computed in $\mathcal{O}(N+n+m)$ time, we may assume that H is connected. Let $v_1, \ldots, v_n$ be any ordering of the vertices of H. The algorithm BLOCKFINDER(B, k) takes as argument a subgraph B of B_H and a $k = 0, \ldots, n$ such that $v_1, \ldots, v_k$ are not articulation points of B. It outputs a link to the list of blocks of B.

BLOCKFINDER(B, k)

- If there is no $k' > k$ such that $v_{k'}$ is contained in B return B
- Let $k' > k$ be minimal such that $v_{k'}$ is contained in B
- Remove $v_{k'}$ and all its adjacent vertices $h_1, \ldots, h_j$ from B and compute the connected components $B_1, \ldots, B_\ell$ of this bipartite graph.
- For $i = 1, \ldots, \ell$, add $v_{k'}$ and those hyperedges among $h_1, \ldots, h_j$ that contain some vertices of B_i with the corresponding edges to B_i.
- Return BLOCKFINDER(B_1, k'), ..., BLOCKFINDER(B_ℓ, k').

Then BLOCKFINDER(B_H, 0) finds a partition of H into blocks represented as bipartite graphs: Assume that BLOCKFINDER returns a subgraph B_i of B_H that contains an articulation point $v_{k'}$. Let P_1 and P_2 be two parts of $v_{k'}$ in B_i. Consider the subgraph B of B_H such that k' was chosen while proceeding BLOCKFINDER(B, k). Since in the end P_1 and P_2 are both in B_i there is a path p in B connecting P_1 and P_2 that does not contain $v_{k'}$. Let p have minimum length among all such paths. Then p is a path in B_i: Otherwise let $p : w_0, h_1, \ldots, h_\ell, w_\ell$ and assume that w_j is the first vertex of p not in B_i. Let $j' > j$ be the smallest index such that $w_{j'}$ is in B_i. Then there is an articulation point $v_\ell, \ell > k'$ of B_i with $v_\ell \in h_j \cap h_{j'}$. Hence, $w_0, h_1, \ldots, w_{j-1}, h_j, v_\ell, h_{j'}, w_{j'}, \ldots, h_\ell, w_\ell$ is a shorter path than p connecting P_1 and P_2. □

A decomposition of a hypergraph into blocks induces a "block-articulation-point tree" in the same way as block-cut-point trees for ordinary graphs: Let T be the bipartite graph that is constructed as follows. The vertices of T are the blocks of H and those vertices in V that are contained in more than one block. There is an edge between a vertex v and a block B if and only if v is contained in B. Then T is the *block-articulation-point tree* of the chosen decomposition of a hypergraph into blocks (see Fig. 1(b)).

Lemma 2. *A hypergraph has an (outer-)planar support if all its blocks have an (outer-)planar support.*

Proof. Let $B_1, \ldots, B_k$ be the blocks of a hypergraph $H = (V, A)$. Let $G_i = (V_i, E_i)$ be a support of B_i for $i = 1, \ldots, k$. Then $G = (V, E_1 \cup \ldots \cup E_k)$ is a support of H and $G_1, \ldots, G_k$ are the 2-connected components of G. Proceeding from the leaves of the block-articulation-point tree one can choose the embedding of the support of each block such that the articulation point with the parent block is on the outer face. Hence, if all G_i have an (outer-)planar support then so does G. □

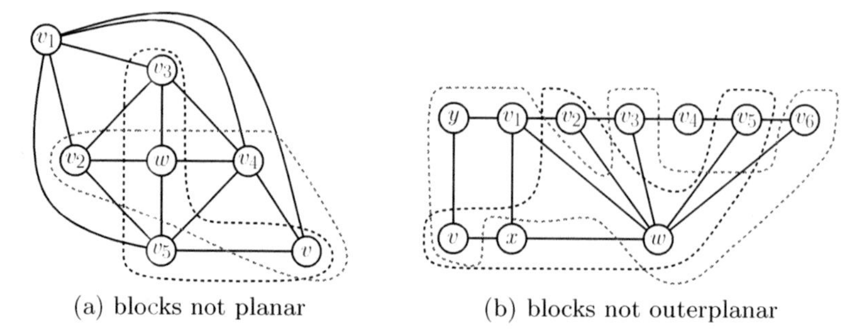

Fig. 3. Illustration of some examples. Solid edges indicate a support, dashed curves indicate hyperedges that contain more than two vertices.

The converse of Lemma 2 is not true. Let H be the hypergraph with hyperedges $\{v, v_1\}$, $\{v, v_4\}$, $\{v, v_5\}$, $\{v_2, v_4, v, w\}$, $\{v_3, v_5, v, w\}$, $\{v_1, v_2\}$, $\{v_1, v_3\}$, $\{v_1, v_4\}$, $\{v_1, v_5\}$, $\{v_2, v_3\}$, $\{v_3, v_4\}$, $\{v_4, v_5\}$, $\{v_2, v_5\}$. Then H is planar, v is an articulation point of H and $H[\{v_1, v_2, v_3, v_4, v_5, v\}]$ is a block of H that is not planar. See Fig. 3(a) for an illustration. In the outerplanar case consider the hyperedges $\{v_1, v_2\}$, $\{v_2, v_3\}$, $\{v_3, v_4\}$, $\{v_4, v_5\}$, $\{v_5, v_6\}$, $\{v, y\}$, $\{y, v_1\}$, $\{v, x\}$, $\{x, v_1\}$, $\{v, x, w, v_2, v_5\}$, and $\{v, y, v_1, w, v_3, v_6\}$ and the articulation point v. See Fig. 3(b) for an illustration. For hypergraphs closed under intersections, however, we have equivalence. A hypergraph $H = (V, A)$ is *closed under intersections* if $h_1 \cap h_2 \in A \cup \{\emptyset\} \cup \{\{v;\ v \in V\}\}$ for $h_1, h_2 \in A$.

Lemma 3. *A hypergraph that is closed under intersections has an (outer-) planar support if and only if each block has an (outer-) planar support.*

Proof. Let $H = (V, A)$ be a hypergraph that is closed under intersections and let $G = (V, E)$ be an (outer-)planar support of H. Let $v \in V$ and let W be a part of v. We show by induction on the number of vertices of $V \setminus W$ that $H[W \cup \{v\}]$ has an (outer-)planar support. There is nothing to show if $V = W \cup \{v\}$.

So let $w \in V \setminus (W \cup \{v\})$. We construct an (outer-)planar support G' of $H' = (V \setminus \{w\}, \{h' \in A; w \notin h'\} \cup \{h' \setminus \{w\}; v \in h' \in A\})$. If there is no hyperedge containing v and w let G' be the graph that results from G by deleting w and all its incident edges. Otherwise let h be the intersection of all hyperedges that contain v and w. Then there is a wv-path in $G[h]$. Let w' be the neighbor of w on this path. Then G' is constructed from G by merging w and w'. I.e., for each neighbor $u \neq w'$ of w add $\{u, w'\}$ to the edge set of G. Finally, remove w and all its incident edges from G.

If $V \setminus \{w\} = W \cup \{v\}$ then $H' = H[W \cup \{v\}]$. Otherwise v is an articulation point and W is a part of v in H'. Hence, by the inductive hypothesis $H'[W \cup \{v\}] = H[W \cup \{v\}]$ has an (outer-)planar support. □

3 Cactus Supports

A cactus is a connected graph that has an outerplanar embedding such that each edge is incident to the outer face. In this section, we relate cactus supports to planar based Hasse diagrams and we show how to utilize the decomposition into blocks to construct a cactus support if one exists.

It was shown by Kaufmann et al. [15] that a hypergraph $H = (V, A)$ has an outerplanar support if its based Hasse diagram is planar. In fact, in that case H has even a cactus support. In the construction of Kaufmann et al. [15] some unnecessary edges on the outer face have to be omitted. We briefly sketch their construction and our modification.

Theorem 1. *A hypergraph has a cactus support if its based Hasse diagram is planar.*

Proof. Let $H = (V, A)$ be a hypergraph, let $V \in A$, and let its Hasse diagram D be planar. Assume that a planar embedding of D is given. Let T be the DFS tree resulting from a directed left-first DFS and replace each non-tree arc $e = (h_1, h_2)$ in D by an arc (h_1, v) for some $v \in h_2$. According to Kaufmann et al. [15], this can be done by "sliding down" the arcs and thus maintaining planarity. Let D' be the thus constructed Hasse diagram and let A' be the set of vertices of D' that are not sinks. Let $H' = (V, \{\{v \in V;$ there is a directed hv-path in $D'\};\ h \in A'\}$. Then T remains a left-first DFS-tree of D' and any support of H' is a support of H.

Consider a simple closed curve C that visits the sequence $v_1, \ldots, v_n$ of leaves of T from left to right. We may assume that the vertex V of D is in the exterior of C, that C intersects no tree edges and that it intersects non-tree edges at most once. The *support sequence* $\sigma : w_1, \ldots, w_\ell$ is the sequence of vertices or targets of intersecting edges as they occur on C. Note that σ contains only vertices of V and that a vertex of V may occur several times in σ. As mentioned by Kaufmann et al. [15], each set $h \in A'$ corresponds then to a subsequence of σ.

Let now $w_{\ell+1} = w_1$. Then $G = (V, \{\{w_i, w_{i+1}\};\ i = 1, \ldots, \ell\})$ is a cactus support of H' and, hence, of H. In fact, the edges can be routed along C and the pieces of the arcs between C and $v_1, \ldots, v_n$. Then G has a planar embedding in which each edge is on the outer face. Further, each subsequence of W corresponds to a walk in G. Hence, G is a cactus support for H'. □

However, not only hypergraphs with a planar Hasse diagram have a cactus support. E.g., $A = \{\{i, i+1\}, i = 1, \ldots, 6; \{1, \ldots, 5\}, \{2, \ldots, 6\}, \{3, \ldots, 7\}\}$. In the following, we will show how to test efficiently whether any hypergraph has a cactus support and if so how to construct it in the same asymptotic run time.

Lemma 4. *A hypergraph has a support that is a cactus if and only if each block has a support that is a cycle or an edge.*

Proof. The if-part is analogous to Lemma 2. For the only-if-part let $H = (V, A)$ be a hypergraph and let $G = (V, E)$ be a cactus support of H. Let v be an articulation point and W a part of v. We show that $H[W \cup \{v\}]$ has a support that is a cactus.

We say that $u \in W$ is close to v if and only if there is a path in G from v to u not containing any edge of $G[W]$. Note that $G[W]$ is a connected subgraph of a cactus not containing v, hence there are at most two vertices in W that are

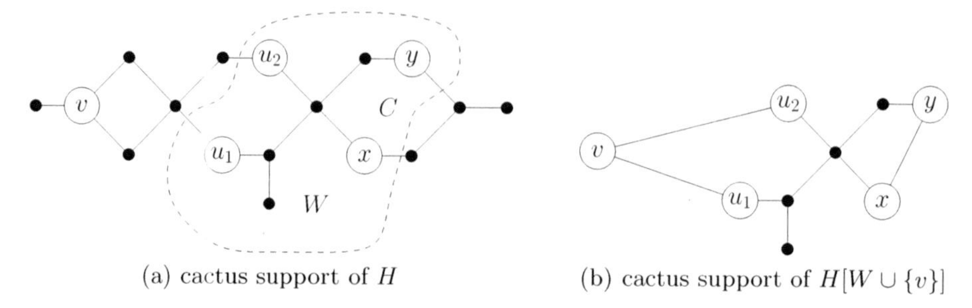

(a) cactus support of H (b) cactus support of $H[W \cup \{v\}]$

Fig. 4. Illustration of the proof of Lemma 4. Vertices inside the dashed curve are contained in a part W of v. Vertices u_1 and u_2 are close to v. Vertices x and y are end vertices of p_C.

close to v. A cactus support $G_W = (V_W, E_W)$ of $H[W \cup \{v\}]$ can be constructed as follows (see Fig. 4 for an illustration):

- Start with $G_W \leftarrow G[W \cup \{v\}]$
- For each $u \in W$ that is close to v, add $\{u, v\}$ to E_W
- For each cycle of G, let $C = \{e_1, e_2, \ldots, e_k\}$ be its set of edges . If $E[W] \cap C \neq \emptyset$ and $C \not\subseteq E[W]$ then $G[W \cap C]$ is a path p_C. If the end vertices x and y of p_C are not both close to v, add $\{x, y\}$ to E_W. □

A hypergraph $H = (V, A)$ has a support that is a cycle if and only if it has the *circular consecutive ones property*, i.e. if and only if there is an ordering $v_1, \ldots, v_n$ of the vertices such that for each hyperedge $h \in A$ there are $1 \leq j \leq k \leq n$ such that $h = \{v_j, \ldots, v_k\}$ or $V \backslash h = \{v_j, \ldots, v_k\}$. Summarizing, we have the following theorem.

Theorem 2. *It can be tested in $\mathcal{O}(nN + n + m)$ time whether a hypergraph has a support that is a cactus.*

Proof. Compute all blocks in $\mathcal{O}(nN + n + m)$ time. Test all blocks in linear time for the circular consecutive ones property [4]. □

4 Hypergraphs Closed under Intersections and Differences

Two hyperedges h_1, h_2 *overlap* if $h_1 \cap h_2 \neq \emptyset$, $h_1 \setminus h_2 \neq \emptyset$, and $h_2 \setminus h_1 \neq \emptyset$. An Euler diagram of two overlapping hyperedges is usually drawn such that the intersection of the two regions representing the two hyperedges is connected and such that the part of one of the regions that is not contained in the other is also connected. See Fig. 5 for an illustration. This motivates the following definition. A hypergraph $H = (V, A)$ is *closed under intersections and differences* if $h_1 \cap h_2 \in A \cup \{\{v\};\ v \in V\}$ and $h_1 \setminus h_2 \in A \cup \{\{v\};\ v \in V\}$ for two overlapping hyperedges $h_1, h_2 \in A$. In the remainder of this section we show that it is easy to decide whether a hypergraph closed under intersections and differences has a planar or an outerplanar support.

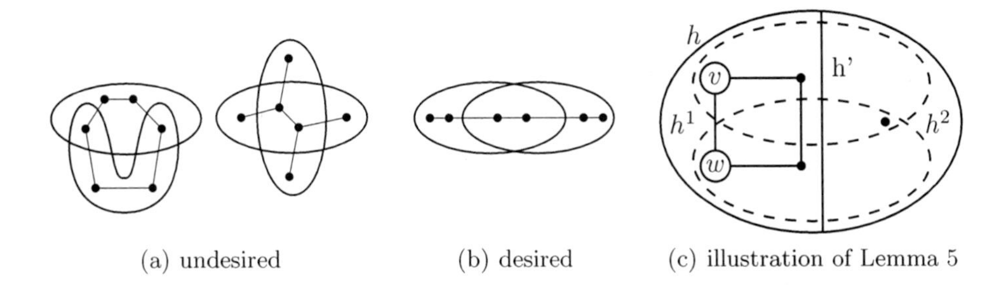

Fig. 5. (a) Undesired and (b) desired drawings of two overlapping hyperedges and (c) an illustration of the proof of Lemma 5. In (a) the intersection or the difference of two hyperedges is not connected, while in (b) it is.

For a hypergraph $H = (V, A)$ let $H_2 = (V, \{h \in A;\ |h| = 2\})$ be the graph of all hyperedges of H that contain exactly two vertices. We will show that H_2 is a support of H if H is a block.

Lemma 5. *If the hypergraph H is closed under intersections and differences and does not contain an articulation point then the hypergraph H_2 induced by all hyperedges of size two is a support of H.*

Proof. Let $H = (V, A)$ be a hypergraph that is closed under intersections and differences and assume that H does not contain an articulation point. Let h by a hyperedge of H. By induction on the size of h, we show that $H_2[h]$ is connected. There is nothing to show if $|h| \leq 2$. So assume that $|h| > 2$.

We first assume that $h \neq V$. Since H does not contain any articulation point there are at least two hyperedges h_1, h_2 with $h_1 \cap h \neq h_2 \cap h$ that overlap with h. We have $h \cap h_i, h \setminus h_i \in A \cup \{\{v\};\ v \in V\}, i = 1, 2$. By the inductive hypothesis, $H_2[h \cap h_i]$ and $H_2[h \setminus h_i], i = 1, 2$ are all four connected. If $h \cap h_1 \neq h \setminus h_2$ then it follows that $H_2[h]$ is connected.

So assume that for all pairs h_1, h_2 of hyperedges with $h \cap h_1 \neq h \cap h_2$ that overlap with h it holds that $h \cap h_1 = h \setminus h_2$. Hence there is a bisection h^1, h^2 of h such that for all hyperedges h_1 that overlap with h it holds that $h \cap h_1 = h^1$ or $h \cap h_1 = h^2$. See Fig. 5 for an illustration of this part of the proof. Note again that by the inductive hypothesis $H_2[h^i], i = 1, 2$ are both connected. Since h contains more than two vertices, we may assume without loss of generality that h^1 contains at least two vertices. If $|h^2| = 1$ there has to be a hyperedge $h' \subset h$ that overlaps h^1 and contains h^2. Otherwise every vertex in h^1 would be an articulation vertex. Similarly, if $|h^2| > 1$ there has to be a hyperedge h' that overlaps both, h^1 and h^2. Let h' be the smallest hyperedge with this property. Assume that $|h' \cap h^i| > 1$ for $i = 1$ or $i = 2$. Since $H_2[h^i]$ is connected there have to be vertices $v \in h^i \cap h', w \in h^i \setminus h'$ such that $\{v, w\}$ is a hyperedge. But then $h' \setminus \{v, w\} \in A$ is a smaller hyperedge than h' with the required property – a contradiction. It follows that $|h'| = 2$. Hence, $H_2[h]$ contains the connected subgraphs $H_2[h^i], i = 1, 2$ and the edge h' connecting them. Thus, $H_2[h]$ is connected.

Assume finally that $h = V$. If H contains more than two vertices then the hypergraph $(V, A \setminus \{V\})$ has to be connected. Otherwise all but at most one vertex of H would be articulation points. Since $H_2[h']$ is connected for all hyperedges $h' \neq V$ it thus follows that also $H_2[V]$ is connected. □

Note that the hyperedges of size two have to be contained in every support of a hypergraph. So we have the following corollary.

Corollary 1. *It can be decided in $\mathcal{O}(nN + n + m)$ time whether a hypergraph closed under intersections and differences has a planar or outerplanar support.*

Proof. First, decompose the hypergraph into blocks. Then test for each block whether the graph induced by the hyperedges of size two is planar or outerplanar, respectively (Lemma 3). □

5 Conclusions

In this paper, we newly defined a decomposition of a hypergraph into blocks. For any such decomposition there is a support with the property that the blocks of the hypergraph correspond to the biconnected components of the support. We then give two applications of the decomposition into blocks. A hypergraph has a cactus support if and only if each block has the cyclic consecutive one's property. A hypergraph that is closed under intersections and differences has an (outer-)planar support if and only if for each block the graph induced by the hyperedges of size two is (outer-)planar.

As a future work, we want to improve the run time of the decomposition into blocks and to solve the problem of testing whether an outerplanar support exists in more general cases.

References

1. Beeri, C., Fagin, R., Maier, D., Yannakakis, M.: On the desirability of acyclic database schemes. Journal of the Association for Computing Mashinery 30(4), 479–513 (1983)
2. Berge, C.: Graphs and Hypergraphs. North-Holland, Amsterdam (1973)
3. Blackwell, A.F., Marriott, K., Shimojima, A. (eds.): Diagrams 2004. LNCS (LNAI), vol. 2980. Springer, Heidelberg (2004)
4. Booth, K.S., Lueker, G.S.: Testing for the consecutives ones property, interval graphs, and graph planarity using PQ-tree algorithms. Journal of Computer and System Sciences 13, 335–379 (1976)
5. Buchin, K., van Kreveld, M., Meijer, H., Speckmann, B., Verbeek, K.: On planar supports for hypergraphs. Technical Report UU-CS-2009-035, Department of Information and Computing Sciences, Utrecht University (2009)
6. Buchin, K., van Kreveld, M., Meijer, H., Speckmann, B., Verbeek, K.: On planar supports for hypergraphs. In: Eppstein, D., Gansner, E.R. (eds.) GD 2009. LNCS, vol. 5849, pp. 345–356. Springer, Heidelberg (2010)
7. Chimani, M., Gutwenger, C.: Algorithms for the hypergraph and the minor crossing number problems. In: Tokuyama, T. (ed.) ISAAC 2007. LNCS, vol. 4835, pp. 184–195. Springer, Heidelberg (2007)

8. Chow, S.C.: Generating and drawing area-proportional Euler and Venn diagrams. PhD thesis, University of Victoria, British Columbia Canada (2007)
9. Didimo, W., Giordano, F., Liotta, G.: Overlapping cluster planarity. Journal on Graph Algorithms and Applications 12(3), 267–291 (2008)
10. Dinitz, Y., Karzanov, A.V., Lomonosov, M.: On the structure of a family of minimal weighted cuts in a graph. In: Fridman, A. (ed.) Studies in Discrete Optimization, pp. 290–306. Nauka (1976) (in Russian)
11. Eschbach, T., Günther, W., Becker, B.: Orthogonal hypergraph drawing for improved visibility. Journal on Graph Algorithms and Applications 10(2), 141–157 (2006)
12. Flower, J., Fish, A., Howse, J.: Euler diagram generation. Journal on Visual Languages and Computing 19(6), 675–694 (2008)
13. Hopcroft, J.E., Tarjan, R.E.: Efficient planarity testing. Journal of the Association for Computing Mashinery 21, 549–568 (1974)
14. Johnson, D.S., Pollak, H.O.: Hypergraph planarity and the complexity of drawing Venn diagrams. Journal of Graph Theory 11(3), 309–325 (1987)
15. Kaufmann, M., van Kreveld, M., Speckmann, B.: Subdivision drawings of hypergraphs. In: Tollis, I.G., Patrignani, M. (eds.) GD 2008. LNCS, vol. 5417, pp. 396–407. Springer, Heidelberg (2009)
16. Korach, E., Stern, M.: The clustering matroid and the optimal clustering tree. Mathematical Programming, Series B 98, 385–414 (2003)
17. Král', D., Kratochvíl, J., Voss, H.-J.: Mixed hypercacti. Discrete Mathematics 286, 99–113 (2004)
18. Mäkinen, E.: How to draw a hypergraph. International Journal of Computer Mathematics 34, 177–185 (1990)
19. Mutton, P., Rodgers, P., Flower, J.: Drawing graphs in Euler diagrams. In: Blackwell, et al. (eds.) [13], pp. 66–81
20. Sander, G.: Layout of directed hypergraphs with orthogonal hyperedges. In: Liotta, G. (ed.) GD 2003. LNCS, vol. 2912, pp. 381–386. Springer, Heidelberg (2004)
21. Schaefer, M., Štefankovič, D.: Decidability of string graphs. Journal of Computer and System Sciences 68(2), 319–334 (2004)
22. Simonetto, P., Auber, D.: Visualise undrawable Euler diagrams. In: Proceedings of the 12th International Conference on Information Visualization (InfoVis 2008), pp. 594–599. IEEE Computer Society Press, Los Alamitos (2008)
23. Simonetto, P., Auber, D.: An heuristic for the construction of intersection graphs. In: Proceedings of the 13th International Conference on Information Visualization (InfoVis 2009), pp. 673–678. IEEE Computer Society Press, Los Alamitos (2009)
24. Tarjan, R.E., Yannakakis, M.: Simple linear-time algorithms to test chordality of graphs, test acyclicity of hypergraphs, and selectively reduce acyclic hypergraphs. SIAM Journal on Computing 13(3), 566–579 (1984)
25. van Cleemput, W.M.: On the planarity of hypergraphs. Proceedings of the IEEE 66(4), 514–515 (1978)
26. Verroust, A., Viaud, M.-L.: Ensuring the drawability of extended Euler diagrams for up to 8 sets. In: Blackwell, et al. (eds.) [3], pp. 128–141
27. Walsh, T.R.S.: Hypermaps versus bipartite maps. Journal of Combinatorial Theory, Series B 18, 155–163 (1975)
28. Zykov, A.A.: Hypergraphs. Uspekhi Matematicheskikh Nauk 6, 89–154 (1974)

Testing the Simultaneous Embeddability of Two Graphs Whose Intersection Is a Biconnected Graph or a Tree*

Patrizio Angelini[1], Giuseppe Di Battista[1], Fabrizio Frati[1], Maurizio Patrignani[1], and Ignaz Rutter[2]

[1] Dipartimento di Informatica e Automazione, Università Roma Tre, Italy
{angelini,gdb,frati,patrigna}@dia.uniroma3.it
[2] Institute of Theoretical Informatics, Karlsruhe Institute of Technology (KIT), Germany
rutter@kit.edu

Abstract. In this paper we study the time complexity of the problem *Simultaneous Embedding with Fixed Edges* (SEFE), that takes two planar graphs $G_1 = (V, E_1)$ and $G_2 = (V, E_2)$ as input and asks whether a planar drawing Γ_1 of G_1 and a planar drawing Γ_2 of G_2 exist such that: (i) each vertex $v \in V$ is mapped to the same point in Γ_1 and in Γ_2; (ii) every edge $e \in E_1 \cap E_2$ is mapped to the same Jordan curve in Γ_1 and Γ_2. First, we show a polynomial-time algorithm for SEFE when the *intersection graph* of G_1 and G_2, that is the planar graph $G_{1\cap 2} = (V, E_1 \cap E_2)$, is *biconnected*. Second, we show that SEFE, when $G_{1\cap 2}$ is a *tree*, is equivalent to a suitably-defined *book embedding* problem. Based on such an equivalence and on recent results by Hong and Nagamochi, we show a linear-time algorithm for the SEFE problem when $G_{1\cap 2}$ is a *star*.

1 Introduction

Let $G_1 = (V, E_1), \dots, G_k = (V, E_k)$ be k graphs on the same set of vertices. A *simultaneous embedding* of $G_1, \dots, G_k$ consists of k planar drawings $\Gamma_1, \dots, \Gamma_k$ of $G_1, \dots, G_k$, respectively, such that any vertex $v \in V$ is mapped to the same point in every drawing Γ_i. Because of the applications to several visualization methods and of the interesting related theoretical problems, constructing simultaneous graph embeddings has recently grown up as a distinguished research topic in Graph Drawing.

The two main variants of the simultaneous embedding problem are the *geometric simultaneous embedding* and the *simultaneous embedding with fixed edges*. The former requires the edges to be straight-line segments, while the latter relaxes such a constraint by just requiring the edges that are common to distinct graphs to be represented by the same Jordan curve in all the drawings. Geometric simultaneous embedding turns out to have limited usability, as geometric simultaneous embeddings do not always exist if the input graphs are three paths [4], if they are two outerplanar graphs [4], if they are two trees [14], and even if they are a tree and a path [3]. Further, testing whether two planar graphs admit a geometric simultaneous embedding is $\mathcal{NP}$-hard [9].

On the other hand, a simultaneous embedding with fixed edges (SEFE) always exists for much larger graph classes. Namely, a tree and a path always have a SEFE with few

* Supported in part by MIUR (Italy), Projects AlgoDEEP no. 2008TFBWL4 and FIRB "Advanced tracking system in intermodal freight transportation", no. RBIP06BZW8.

C.S. Iliopoulos and W.F. Smyth (Eds.): IWOCA 2010, LNCS 6460, pp. 212–225, 2011.

bends per edge [8]; an outerplanar graph and a path or a cycle always have a SEFE with few bends per edge [7]; a planar graph and a tree always have a SEFE [12].

The main open question about SEFE is whether testing the existence of a SEFE of two planar graphs is doable in polynomial time or not. A number of known results are related to this problem. Namely, Gassner *et al.* proved that testing whether three planar graphs admit a SEFE is $\mathcal{NP}$-hard and that SEFE is in $\mathcal{NP}$ for any number of input graphs [13]; Fowler *et al.* characterized the planar graphs that always have a SEFE with any other planar graph and proved that testing whether two outerplanar graphs admit a SEFE is in $\mathcal{P}$ [11]; Fowler *et al.* showed how to test in polynomial time whether two planar graphs admit a SEFE if one of them contains at most one cycle [10]; Jünger and Schulz characterized the graphs $G_{1\cap 2}$ that allow for a SEFE of any two planar graphs G_1 and G_2 whose intersection graph is $G_{1\cap 2}$ [17]; Angelini *et al.* showed how to test whether two planar graphs admit a SEFE if one of them has a fixed embedding [1].

In this paper, we show the following results. In Sect. 3 we show a cubic-time algorithm for the SEFE problem when the intersection graph $G_{1\cap 2}$ of G_1 and G_2 is biconnected. Our algorithm exploits the SPQR-tree decomposition of $G_{1\cap 2}$ in order to test whether a planar embedding of $G_{1\cap 2}$ exists that allows the edges of G_1 and G_2 not in $G_{1\cap 2}$ to be drawn in such a way that no two edges of the same graph intersect. In Sect. 4 we show that the SEFE problem, when $G_{1\cap 2}$ is a tree, is equivalent to a suitably-defined book embedding problem. Namely, we show that, for every instance G_1, G_2 of SEFE such that $G_{1\cap 2}$ is a tree, there exist a graph G', whose edges are partitioned into two sets E'_1 and E'_2, and a set of hierarchical constraints on the set of vertices of G', such that G_1 and G_2 have a SEFE if and only if G' admits a 2-page book embedding in which the edges of E'_1 are in one page, the edges of E'_2 are in another page, and the order of the vertices of G' along the spine respects the hierarchical constraints. Based on this characterization and on recent results by Hong and Nagamochi [16] concerning 2-page book embeddings with the edges assigned to the pages in the input, we prove that linear time suffices to solve the SEFE problem when $G_{1\cap 2}$ is a star.

Several proofs are omitted because of space limitations. Further details are in [2].

2 Preliminaries

A *drawing* of a graph is a mapping of each vertex to a distinct point of the plane and of each edge to a simple Jordan curve connecting its endpoints. A drawing is *planar* if the curves representing its edges do not cross but, possibly, at common endpoints. A graph is *planar* if it admits a planar drawing. Two drawings of the same graph are *equivalent* if they determine the same circular ordering around each vertex. A *planar embedding* is an equivalence class of planar drawings. A planar drawing partitions the plane into topologically connected regions, called *faces*. The unbounded face is the *outer face*.

A *Simultaneous Embedding with Fixed Edges* (SEFE) of k planar graphs $G_1 = (V, E_1), G_2 = (V, E_2), \ldots, G_k = (V, E_k)$ consists of k drawings $\Gamma_1, \Gamma_2, \ldots, \Gamma_k$ such that: (i) Γ_i is a planar drawing of G_i, for $1 \leq i \leq k$; (ii) any vertex $v \in V$ is mapped to the same point in every drawing Γ_i, for $1 \leq i \leq k$; (iii) any edge $e \in E_i \cap E_j$ is mapped to the same Jordan curve in Γ_i and in Γ_j, for $1 \leq i, j \leq k$. The problem of testing whether k graphs admit a SEFE is called the SEFE *problem*. Given two planar graphs $G_1 = (V, E_1)$ and $G_2 = (V, E_2)$, the *intersection graph* of G_1 and G_2 is the planar graph $G_{1\cap 2} = (V, E_1 \cap E_2)$; further, the *exclusive subgraph* of G_1 (resp. of G_2) is the graph $G_{1\setminus 2} = (V, E_1 \setminus E_2)$ (resp. $G_{2\setminus 1} = (V, E_2 \setminus E_1)$). The *exclusive edges* of G_1 (of G_2) are the edges in $G_{1\setminus 2}$ (resp. in $G_{2\setminus 1}$).

A *book embedding* of a graph $G = (V, E)$ consists of a total ordering $\prec$ of the vertices in V and of an assignment of the edges in E to *pages* of a book, in such a way that no two edges (a, b) and (c, d) are assigned to the same page if $a \prec c \prec b \prec d$. A *k-page book embedding* is a book embedding using k pages. A *constrained k-page book embedding* is a k-page book embedding in which the assignment of edges to the pages is part of the input.

A graph is *connected* if every pair of vertices is connected by a path. A graph G is *biconnected* (resp. *triconnected*) if removing any vertex (resp. any two vertices) leaves G connected. In order to handle the decomposition of a biconnected graph into its triconnected components, we use the *SPQR-trees*, a data structure introduced by Di Battista and Tamassia (see, e.g., [5,6]). Definitions about SPQR-trees can be found in [5,6,15]. Here we give some notation. Given a biconnected graph G and its SPQR-tree $\mathcal{T}$, we say that a vertex v of G *belongs* to a node μ of $\mathcal{T}$ if v is a vertex of the pertinent graph $G(\mu)$ of μ. In this case we also say that μ *contains* v. We denote by $skel(\mu)$ the *skeleton* of a node μ of $\mathcal{T}$, that is, the graph representing the arrangement of the triconnected components composing $G(\mu)$. The edges of $skel(\mu)$ are called *virtual edges*. The skeleton of a node μ contains a virtual edge representing the *rest of the graph*, that is, the graph obtained from G by removing all the vertices of $G(\mu)$, except for its poles, together with their incident edges. In the following, we will only refer to the SPQR-tree of the intersection graph $G_{1\cap 2}$ of two graphs G_1 and G_2. However, with a slight abuse of notation, we will denote by $G_1(\mu)$ (by $G_2(\mu)$) the subgraph of G_1 (of G_2) induced by the vertices in $G_{1\cap 2}(\mu)$.

3 The Intersection Graph Is Biconnected

In this section we describe an algorithm for computing a SEFE of two planar graphs $G_1 = (V, E_1)$ and $G_2 = (V, E_2)$ when the intersection graph $G_{1\cap 2}$ is biconnected. Denote by $\mathcal{T}$ the SPQR-tree of $G_{1\cap 2}$.

To ease the description of the algorithm, we assume that $\mathcal{T}$ is rooted at any edge e of $G_{1\cap 2}$. Such an assumption implies that e is adjacent to the outer face of any computed embedding of $G_{1\cap 2}$. Observe that this does not preclude the possibility of finding a SEFE of G_1 and G_2. Namely, consider any SEFE in the plane; "wrap" the SEFE around a sphere; project the SEFE back to the plane from a point in a face incident to e, thus obtaining a SEFE of G_1 and G_2 in which e is incident to the outer face of the embedding of $G_{1\cap 2}$. Furthermore, if e is the parent in $\mathcal{T}$ of an S-node, subdivide the edge of $\mathcal{T}$ connecting e to its only child by inserting a P-node. Observe that the described insertion of an "artificial" P-node ensures that the parent of any S-node is either an R-node or a P-node.

For every P-node and R-node μ of $\mathcal{T}$, the *visible nodes* of μ are the children of μ that are not S-nodes plus the children of each child of μ that is an S-node.

An exclusive edge e of G_1 or of G_2 is an *internal edge* of a node $\mu \in \mathcal{T}$ if both the end-vertices of e belong to μ, at least one of them is not a pole of μ, and there exists no

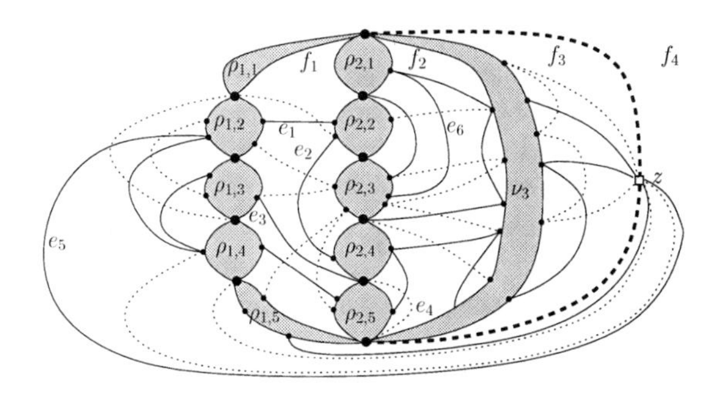

Fig. 1. A SEFE of graphs $G_1(\mu)$ and $G_2(\mu)$ when μ is a P-node with three children ν_1, ν_2, and ν_3. Also, ν_1 and ν_2 have children $\rho_{1,1}, \dots, \rho_{1,5}$ and $\rho_{2,1}, \dots, \rho_{2,5}$, respectively. For each visible node τ of μ, the interior of the cycle delimiting the outer face of $G_{1\cap 2}(\tau)$ is gray. Solid (dotted) edges are exclusive edges of G_1 (G_2). The dashed edge represents the rest of the graph.

descendant of μ containing both the end-vertices of e. An exclusive edge e of G_1 or of G_2 is an *outer edge* of a node $\mu \in \mathcal{T}$ if exactly one end-vertex of e belongs to μ and this end-vertex is not a pole of μ. An exclusive edge e of G_1 or of G_2 is an *intra-pole edge* of a node $\mu \in \mathcal{T}$ if its end-vertices are the poles of μ. Observe that an exclusive edge e of G_1 or of G_2 can be an outer edge of a linear number of nodes of $\mathcal{T}$; further, e is an internal edge of at most one node of $\mathcal{T}$; moreover, e can be an intra-pole edge of a linear number of nodes of $\mathcal{T}$; however, e can be an intra-pole edge of at most one P-node of $\mathcal{T}$. In Fig. 1, edge e_1 is an internal edge of μ and an outer edge of $\rho_{1,2}$, of $\rho_{2,2}$, of ν_1, and of ν_2; edge e_2 is an internal edge of ν_2 and an outer edge of $\rho_{2,2}$ and $\rho_{2,4}$; edge e_3 is an internal edge of μ and an outer edge of $\rho_{1,3}$, ν_1, and ν_2; edge e_4 is an intra-pole edge of $\rho_{2,5}$; edge e_5 is an outer edge of $\rho_{1,2}$, of ν_1, and of μ.

We have the following lemmata.

Lemma 1. *Let $\mathcal{E}_{1\cap 2}(\mu)$ be an embedding of $G_{1\cap 2}(\mu)$, with $\mu \in \mathcal{T}$, and let e be an internal edge of μ. Then, G_1 and G_2 have a* SEFE *in which the embedding of $G_{1\cap 2}(\mu)$ is $\mathcal{E}_{1\cap 2}(\mu)$ only if both end-vertices of e are incident to the same face of $\mathcal{E}_{1\cap 2}(\mu)$.*

Proof: Suppose, for a contradiction, that G_1 and G_2 have a SEFE in which the embedding of $G_{1\cap 2}(\mu)$ is $\mathcal{E}_{1\cap 2}(\mu)$ and the end-vertices of e are not both incident to the same face of $\mathcal{E}_{1\cap 2}(\mu)$. Then e crosses $G_{1\cap 2}(\mu)$, hence either two edges of G_1 or two edges of G_2 cross (depending on whether $e \in G_1$ or $e \in G_2$), a contradiction. □

Lemma 2. *Let $\mathcal{E}_{1\cap 2}(\mu)$ be an embedding of $G_{1\cap 2}(\mu)$, with $\mu \in \mathcal{T}$, and let e be an outer edge incident to μ in a vertex $u(e)$. Then, G_1 and G_2 have a* SEFE *in which the embedding of $G_{1\cap 2}(\mu)$ is $\mathcal{E}_{1\cap 2}(\mu)$ only if $u(e)$ is on the outer face of $\mathcal{E}_{1\cap 2}(\mu)$.*

Proof: Suppose, for a contradiction, that G_1 and G_2 have a SEFE in which the embedding of $G_{1\cap 2}(\mu)$ is $\mathcal{E}_{1\cap 2}(\mu)$ and $u(e)$ is not incident to the outer face of $\mathcal{E}_{1\cap 2}(\mu)$. Then e crosses $G_{1\cap 2}(\mu)$, hence either two edges of G_1 or two edges of G_2 cross (depending on whether $e \in G_1$ or $e \in G_2$), a contradiction. □

The algorithm performs a bottom-up traversal of $\mathcal{T}$. When the algorithm visits a node μ of $\mathcal{T}$, either it concludes that a SEFE of G_1 and G_2 does not exist, or it determines a SEFE $\Gamma(\mu)$ of $G_1(\mu)$ and $G_2(\mu)$ such that, if a SEFE of G_1 and G_2 exists, there exists one in which the SEFE of $G_1(\mu)$ and $G_2(\mu)$ is $\Gamma(\mu)$. The embedding $\Gamma(\mu)$ is computed by composing and possibly flipping the already computed embeddings of the descendants of μ. The rest of the graph, that is, the union of the graphs obtained from G_1 and G_2 by respectively removing the vertices of $G_1(\mu)$ and $G_2(\mu)$, except for $u(\mu)$ and $v(\mu)$, and their incident edges, will be placed in the same connected region of $\Gamma(\mu)$. Such a region is called the *outer face* of $\Gamma(\mu)$. The computed SEFE $\Gamma(\mu)$ of $G_1(\mu)$ and $G_2(\mu)$ is such that all the outer edges of μ can be *drawn towards the outer face*, that is, a vertex z can be inserted into the outer face of $\Gamma(\mu)$ and all the outer edges of μ can be drawn with z replacing their end-vertex not in μ, still maintaining the planarity of the drawings of $G_1(\mu)$ and $G_2(\mu)$. An example of insertion of z in a SEFE of $G_1(\mu)$ and $G_2(\mu)$ is shown in Fig. 1.

We are now ready to state the algorithm and to prove its correctness. We will later show that it has a polynomial-time implementation.

If μ is a Q-node, then $G_1(\mu)$, $G_2(\mu)$, and $G_{1\cap 2}(\mu)$ have exactly one embedding, hence no embedding choices have to be done.

If μ is an S-node, some information about the parent of μ in $\mathcal{T}$ are needed in order to decide an embedding for $G_1(\mu)$, $G_2(\mu)$, and $G_{1\cap 2}(\mu)$. Hence, such a decision is deferred to the step in which the parent of μ is analyzed.

If μ is a P-node, then, since for each visible node τ of μ the embeddings of $G_1(\tau)$ and $G_2(\tau)$ are already decided (up to a flip), an embedding of $G_1(\mu)$ and $G_2(\mu)$ is specified by an embedding of $skel(\mu)$, that is, an ordering of the nodes ν_i around the poles of μ, by a flip for each visible node τ of μ, and by an embedding of all the exclusive edges of $G_1(\mu)$ and $G_2(\mu)$ that have not yet been embedded (that is, the outer edges of μ, the internal edges of μ, the internal edges of the S-nodes children of μ, and the intra-pole edge of μ).

First, we determine an embedding $\mathcal{E}(skel(\mu))$ of $skel(\mu)$. Consider any child ν_i of μ that has an outer edge e. If e is an internal edge of μ, then e is an outer edge of a child ν_k of μ, with $k \neq i$; hence, by Lemma 1, ν_i and ν_k have to be consecutive around the poles of μ. If e is not an internal edge of μ, then e is an outer edge of μ; hence, by Lemma 2, ν_i and the virtual edge representing the rest of the graph in $skel(\mu)$ have to be consecutive around the poles of μ. Consider the graph O which has a vertex for each virtual edge of $skel(\mu)$, and which has an edge between two vertices if the corresponding virtual edges have to be consecutive around the poles of μ. If O is not a simple cycle and is not a collection of paths and isolated vertices, then we conclude that G_1 and G_2 have no SEFE. Otherwise, consider as the embedding $\mathcal{E}(skel(\mu))$ of $skel(\mu)$ any ordering of the virtual edges of $skel(\mu)$ around the poles of μ such that any two adjacent vertices in O are consecutive.

Second, we determine a flip for each visible node τ of μ and an embedding of all the exclusive edges of $G_1(\mu)$ and $G_2(\mu)$ that have not been embedded when processing the visible nodes of μ. In order to do this, we will use some auxiliary graphs. For each face f_k of $\mathcal{E}(skel(\mu))$, denote by $\nu_1(f_k)$ and by $\nu_2(f_k)$ the nodes of $\mathcal{T}$ corresponding to the two virtual edges adjacent to f_k in $skel(\mu)$ (recall that one of such virtual edges might be the one representing the rest of the graph), and construct two graphs F_k^1 and F_k^2 as follows. The nodes of F_k^1 (resp. of F_k^2) are the edges e of $G_{1\setminus 2}$ (resp. of $G_{2\setminus 1}$)

such that: (i) e is an internal edge of μ, an outer edge of $\nu_1(f_k)$, and an outer edge of $\nu_2(f_k)$, or (ii) e is an internal edge of an S-node (either $\nu_1(f_k)$ or $\nu_2(f_k)$) child of μ, or (iii) e is an outer edge of μ and f_k is incident in $\mathcal{E}(skel(\mu))$ to the virtual edge of $skel(\mu)$ representing the rest of the graph and to the virtual edge containing the end-vertex of e in μ, or (iv) e is an intra-pole edge of μ. Informally speaking, the nodes of F_k^1 (resp. of F_k^2) are the edges of $G_{1\setminus 2}$ (resp. of $G_{2\setminus 1}$) that have not yet been embedded by the algorithm after processing all the visible nodes of μ and that could be embedded into f_k. The edges of F_k^1 (resp. of F_k^2) connect two vertices of F_k^1 (resp. of F_k^2) whose corresponding edges cross if they are both embedded inside f_k. In the example of Fig. 1, vertices e_2 and e_6 and edge (e_2, e_6) belong to graphs F_1^1 and F_2^1.

Denote by τ any visible node of μ. Observe that:

- Deciding the flip for $G_{1\cap 2}(\tau)$ determines the face of $\mathcal{E}(skel(\mu))$ into which the outer edges of τ have to be embedded. Namely, once a flip for $G_{1\cap 2}(\tau)$ has been fixed, there is exactly one face of $\mathcal{E}(skel(\mu))$ into which each of its outer edges can be embedded without crossing $G_{1\cap 2}(\mu)$. In the example of Fig. 1, fixing the flip for $\rho_{2,3}$ determines that e_6 is embedded into f_1 or into f_2.
- Embedding an outer edge e of τ into a face f_k determines a flip for τ. Namely, there is exactly one flip of τ that brings the end-vertex of e in τ to be incident to f_k. In the example of Fig. 1, fixing the embedding of e_2 into f_1 or into f_2 determines the flip of $\rho_{2,4}$.
- Embedding an edge e of $G_{1\setminus 2}$ (resp. of $G_{2\setminus 1}$), that is represented by a node in F_k^1 (resp. in F_k^2), into a face f_k determines that an edge e' such that (e, e') belongs to F_k^1 (resp. to F_k^2) can not be embedded into f_k. Observe that, if e' can not be embedded into f_k and e' is not an intra-pole edge of μ, then there is at most one face $f_{k'}$ with $k' \neq k$ into which e' can be embedded without crossing $G_{1\cap 2}(\mu)$. In the example of Fig. 1, fixing the embedding of e_2 into f_1 determines that e_6 is embedded into f_2.

Our algorithm uses two sets E_f and T_f. Set E_f contains the edges of $G_{1\setminus 2}$ and of $G_{2\setminus 1}$ that belong to some graph F_k^1 or F_k^2, that have been already embedded into a face of $\mathcal{E}(skel(\mu))$, and that have not yet been processed by the algorithm. Set T_f contains the visible nodes of μ whose flip has already been decided and that have not yet been processed by the algorithm. When the algorithm processes the elements of E_f and T_f, it propagates to other edges and visible nodes the embedding choices already performed on such elements.

We initialize E_f as follows. For every exclusive edge e of G_1 or of G_2 in μ that is an outer edge of two components $\nu_1(f_k)$ and $\nu_2(f_k)$, embed e into f_k and add e to E_f. Note that the embedded edges are all the internal edges and the outer edges of μ. Further, if there exists an intra-pole edge $e \in G_1$ (resp. $e \in G_2$) of μ, then embed e into any face f_k of $\mathcal{E}(skel(\mu))$ such that no edge e' has already been embedded into f_k, with $(e, e') \in F_k^1$ (resp. with $(e, e') \in F_k^2$). If no such a face f_k exists, then conclude that G_1 and G_2 have no SEFE, otherwise add e to E_f.

Next, we repeatedly apply the following procedure, called *Embedding-Flipping Step*, till the flip of every visible node of μ and the embedding of every edge represented by a node in some graph F_k^1 or F_k^2 have been decided, or till the algorithm returns that there is no SEFE of G_1 and G_2.

Embedding-Flipping Step

- Case 1: $E_f \neq \emptyset$. Consider any edge $e \in E_f$. Remove e from E_f.
For each edge e' such that e and e' are adjacent in some graph F_k^1 or F_k^2 the following operations are performed: (i) if e has been embedded into f_k, if e' has not yet been embedded into any face of $\mathcal{E}(skel(\mu))$, and if e' belongs to a graph $F_{k'}^1$ or $F_{k'}^2$, with $k' \neq k$, then embed e' into $f_{k'}$ and add e' to E_f; (ii) if e has been embedded into f_k, if e' has not yet been embedded into any face of $\mathcal{E}(skel(\mu))$, and if e' does not belong to a graph $F_{k'}^1$ or $F_{k'}^2$, with $k' \neq k$, then conclude that G_1 and G_2 have no SEFE; (iii) if e has been embedded into a face $f_{k'} \neq f_k$ and if e' has not yet been embedded into any face of $\mathcal{E}(skel(\mu))$, then embed e' into f_k and add e' to E_f.
If e is the outer edge of a visible node τ of μ, then: (i) if no flip has yet been decided for τ, then flip τ so that the end-vertex of e in τ is incident to the face of $\mathcal{E}(skel(\mu))$ into which e has been embedded and add τ to T_f; (ii) if a flip has already been decided for τ such that the end-vertex of e in τ is not incident to the face of $\mathcal{E}(skel(\mu))$ into which e has been embedded, then conclude that G_1 and G_2 have no SEFE.
- Case 2: $E_f = \emptyset$ and $T_f \neq \emptyset$. Consider any node $\tau \in T_f$. Remove τ from T_f.
For each outer edge e of τ: (i) if e has not yet been embedded into any face of $\mathcal{E}(skel(\mu))$, then embed e into the face of $\mathcal{E}(skel(\mu))$ the end-vertex of e in τ is incident to and add e to E_f; (ii) if e has already been embedded into a face of $\mathcal{E}(skel(\mu))$ and the end-vertex of e in τ is not incident to such a face, then conclude that G_1 and G_2 have no SEFE.
- Case 3: $E_f = \emptyset$ and $T_f = \emptyset$. If all the edges of $G_{1\setminus 2}$ and $G_{2\setminus 1}$ that are represented by nodes in some graph F_k^1 or F_k^2 have been embedded into a face of $\mathcal{E}(skel(\mu))$ and if all the visible nodes of μ have been flipped, then the procedure stops. Otherwise, if there is a visible node of μ that still has to be flipped, then flip it either way and insert such a node into T_f. If there is no visible node of μ that still has to be flipped, then there is an edge e of $G_{1\setminus 2}$ and $G_{2\setminus 1}$ that is represented by a node in some graph F_k^1 or F_k^2 that still has to be embedded. Embed e into any face incident to both the visible nodes of μ the end-vertices of e belong to and add e to E_f.

If μ is an R-node, then the algorithm behaves exactly as in the P-node case, except that the phase in which the embedding $\mathcal{E}(skel(\mu))$ of $skel(\mu)$ is chosen is missing, as $skel(\mu)$ has exactly one planar embedding (up to a flip of the entire embedding).

We now prove the correctness of the algorithm.

Lemma 3. *A* SEFE *of G_1 and G_2 exists if and only if the described algorithm returns a* SEFE *of G_1 and G_2.*

Proof: One implication is trivial: If the algorithm returns a SEFE of G_1 and G_2, then a SEFE of G_1 and G_2 exists. We now prove the other implication.

Consider any SEFE Γ of G_1 and G_2 and, for any node μ of the SPQR-tree $\mathcal{T}$ of $G_{1\cap 2}$, consider the outer face of $G_{1\cap 2}(\mu)$ in Γ. Such a face is delimited by two paths $P^a(\Gamma,\mu)$ and $P^b(\Gamma,\mu)$ connecting $u(\mu)$ and $v(\mu)$. Consider such paths as starting at

$u(\mu)$ and ending at $v(\mu)$. Denote by $L_1^a(\Gamma, \mu)$ the ordered list of vertices incident to the outer edges of μ that belong to G_1 and whose end-vertex in μ belongs to $P^a(\Gamma, \mu)$. Such vertices are ordered in $L_1^a(\Gamma, \mu)$ as they are ordered in $P^a(\Gamma, \mu)$. List $L_1^b(\Gamma, \mu)$ is defined analogously, with $P^b(\Gamma, \mu)$ replacing $P^a(\Gamma, \mu)$. Lists $L_2^a(\Gamma, \mu)$ and $L_2^b(\Gamma, \mu)$ are defined analogously to $L_1^a(\Gamma, \mu)$ and $L_1^b(\Gamma, \mu)$, with G_2 replacing G_1.

Suppose that μ is not an S-node, the following claim asserts that each of $L_1^a(\Gamma, \mu)$, $L_1^b(\Gamma, \mu)$, $L_2^a(\Gamma, \mu)$, and $L_2^b(\Gamma, \mu)$ is the same in any SEFE Γ of G_1 and G_2, that is, the structure of the outer face of $G_{1\cap 2}(\mu)$ does not depend on the choices made by the algorithm.

Claim 1. *In any* SEFE Γ *of* G_1 *and* G_2*, for any node* $\mu \in \mathcal{T}$ *that is not an* S*-node, lists* $L_1^a(\Gamma, \mu)$, $L_1^b(\Gamma, \mu)$, $L_2^a(\Gamma, \mu)$*, and* $L_2^b(\Gamma, \mu)$ *are the same, up to simultaneous swaps of* $L_1^a(\Gamma, \mu)$ *with* $L_1^b(\Gamma, \mu)$ *and of* $L_2^a(\Gamma, \mu)$ *with* $L_2^b(\Gamma, \mu)$.

Proof: Suppose, for a contradiction, that there exist two SEFEs and a node of $\mathcal{T}$ that is not an S-node for which the statement does not hold. We will show that this implies that one of the two SEFEs is not correct.

Consider a node $\mu \in \mathcal{T}$ that is not an S-node, for which the statement does not hold, and such that for all the descendants of μ in $\mathcal{T}$ the statement holds.

If μ is a Q-node, then $L_1^a(\Gamma, \mu)$, $L_1^b(\Gamma, \mu)$, $L_2^a(\Gamma, \mu)$, and $L_2^b(\Gamma, \mu)$ are empty lists and the statement holds, thus obtaining a contradiction.

If μ is an R-node, consider any two SEFEs Γ and Γ' of G_1 and G_2 such that not all the following four equalities hold $L_1^a(\Gamma, \mu) = L_1^a(\Gamma', \mu)$, $L_1^b(\Gamma, \mu) = L_1^b(\Gamma', \mu)$, $L_2^a(\Gamma, \mu) = L_2^a(\Gamma', \mu)$, and $L_2^b(\Gamma, \mu) = L_2^b(\Gamma', \mu)$, and such that not all the following four equalities hold $L_1^a(\Gamma, \mu) = L_1^b(\Gamma', \mu)$, $L_1^b(\Gamma, \mu) = L_1^a(\Gamma', \mu)$, $L_2^a(\Gamma, \mu) = L_2^b(\Gamma', \mu)$, and $L_2^b(\Gamma, \mu) = L_2^a(\Gamma', \mu)$. Since the statement holds for every visible node of μ and since $skel(\mu)$ has one planar embedding, up to a reversal of the adjacency lists of all the vertices, there exists a visible node of μ that is flipped differently in Γ and in Γ' and that has an outer edge e that is also an outer edge of μ. Denote by $u(e)$ the end-vertex of e in μ. Suppose that $u(e)$ is incident to the outer face of $G_{1\cap 2}(\mu)$ in Γ. Then, $u(e)$ is not incident to the outer face of $G_{1\cap 2}(\mu)$ in Γ'. It follows that edge e crosses $G_{1\cap 2}(\mu)$ in Γ', a contradiction.

If μ is a P-node, then the (at most) two children ν_x and ν_y of μ that contain end-vertices of outer edges of μ are incident to the outer face of $G_{1\cap 2}(\mu)$ in any SEFE of G_1 and G_2. The flips of ν_x and ν_y (if they are not S-nodes) or the flips of the children of ν_x and ν_y (if they are S-nodes) determine lists $L_1^a(\Gamma, \mu)$, $L_1^b(\Gamma, \mu)$, $L_2^a(\Gamma, \mu)$, and $L_2^b(\Gamma, \mu)$ in any SEFE Γ of G_1 and G_2. Then, consider any two SEFE Γ and Γ' of G_1 and G_2 such that not all the following four equalities hold $L_1^a(\Gamma, \mu) = L_1^a(\Gamma', \mu)$, $L_1^b(\Gamma, \mu) = L_1^b(\Gamma', \mu)$, $L_2^a(\Gamma, \mu) = L_2^a(\Gamma', \mu)$, and $L_2^b(\Gamma, \mu) = L_2^b(\Gamma', \mu)$, and such that not all the following four equalities hold $L_1^a(\Gamma, \mu) = L_1^b(\Gamma', \mu)$, $L_1^b(\Gamma, \mu) = L_1^a(\Gamma', \mu)$, $L_2^a(\Gamma, \mu) = L_2^b(\Gamma', \mu)$, and $L_2^b(\Gamma, \mu) = L_2^a(\Gamma', \mu)$. Similarly to the R-node case, if a visible node of μ has an outer edge e that is also an outer edge of μ and such a node is flipped differently in Γ and in Γ', then the end-vertex $u(e)$ of e in μ is not incident to the outer face of $G_{1\cap 2}(\mu)$ either in Γ or in Γ'. It follows that edge e crosses $G_{1\cap 2}(\mu)$ in Γ or in Γ', a contradiction. □

The following claim asserts that the structure of the outer face of $G_{1\cap 2}(\mu)$ (and of the exclusive edges of G_1 and G_2 embedded into it) is the only property coming from an embedding of $G_1(\mu)$ and $G_2(\mu)$ that affects the possibility of constructing a SEFE of the rest of the graph.

Claim 2. *Suppose that a* SEFE *of G_1 and G_2 exists. Then, for any node $\mu \in \mathcal{T}$ that is not an S-node, any* SEFE *of $G_1(\mu)$ and $G_2(\mu)$ in which the outer edges of μ can be drawn towards the outer face can be extended into a* SEFE *of G_1 and G_2.*

Proof: Consider any SEFE Γ of G_1 and G_2 and consider any SEFE $\Gamma(\mu)$ of $G_1(\mu)$ and $G_2(\mu)$ in which the outer edges of μ can be drawn towards the outer face.

Similarly to the proof of Claim 1, it can be proved that if neither $L_1^a(\Gamma,\mu) = L_1^a(\Gamma(\mu),\mu)$, $L_1^b(\Gamma,\mu) = L_1^b(\Gamma(\mu),\mu)$, $L_2^a(\Gamma,\mu) = L_2^a(\Gamma(\mu),\mu)$, and $L_2^b(\Gamma,\mu) = L_2^b(\Gamma(\mu),\mu)$ nor $L_1^a(\Gamma,\mu) = L_1^b(\Gamma(\mu),\mu)$, $L_1^b(\Gamma,\mu) = L_1^a(\Gamma(\mu),\mu)$, $L_2^a(\Gamma,\mu) = L_2^b(\Gamma(\mu),\mu)$, and $L_2^b(\Gamma,\mu) = L_2^a(\Gamma(\mu),\mu)$ holds, then there is an end-vertex of an outer edge of μ that either is not incident to the outer face of $G_{1\cap 2}(\mu)$ in $\Gamma(\mu)$, thus contradicting the fact that the outer edges of μ can be drawn towards the outer face in $\Gamma(\mu)$, or is not incident to the outer face of $G_{1\cap 2}(\mu)$ in Γ, thus contradicting the fact that Γ is a SEFE. By suitably flipping Γ, we can hence assume that $L_1^a(\Gamma,\mu) = L_1^a(\Gamma(\mu),\mu)$, $L_1^b(\Gamma,\mu) = L_1^b(\Gamma(\mu),\mu)$, $L_2^a(\Gamma,\mu) = L_2^a(\Gamma(\mu),\mu)$, and $L_2^b(\Gamma,\mu) = L_2^b(\Gamma(\mu),\mu)$.

Remove from Γ the drawing of $G_1(\mu)$ and $G_2(\mu)$, except for $u(\mu)$ and $v(\mu)$. Insert $\Gamma(\mu)$ inside the face of the modified Γ into which the previous drawing of $G_1(\mu)$ and $G_2(\mu)$ used to lie; the modified Γ is scaled up till the insertion of $\Gamma(\mu)$ does not cause crossings among the edges of the modified Γ and those of $\Gamma(\mu)$. Continuously deform the edges incident to $u(\mu)$ and $v(\mu)$ in Γ so that they end at the points where $u(\mu)$ and $v(\mu)$ are drawn in $\Gamma(\mu)$. This can always be done since $u(\mu)$ and $v(\mu)$ are both incident to the face where $\Gamma(\mu)$ has been inserted. Finally, insert the outer edges of μ. This can always be done so that no outer edge of μ in G_1 (resp. in G_2) crosses an edge of $G_1(\mu)$ (resp. of $G_2(\mu)$), by the assumption that the outer edges of μ can be drawn towards the outer face, and so that no outer edge of μ in G_1 (resp. in G_2) crosses an edge of the graph obtained from G_1 be removing $G_1(\mu)$, except for its poles (resp. of the graph obtained from G_2 be removing $G_2(\mu)$, except for its poles), since the drawing of such outer edges in Γ used to exist and $L_1^a(\Gamma,\mu) = L_1^a(\Gamma(\mu),\mu)$, $L_1^b(\Gamma,\mu) = L_1^b(\Gamma(\mu),\mu)$, $L_2^a(\Gamma,\mu) = L_2^a(\Gamma(\mu),\mu)$, and $L_2^b(\Gamma,\mu) = L_2^b(\Gamma(\mu),\mu)$. □

Finally, the following claim asserts that the algorithm computes a SEFE of $G_1(\mu)$ and $G_2(\mu)$, if it exists, in which the structure of the outer face of $G_{1\cap 2}(\mu)$ is the one that allows for an extension into a SEFE of G_1 and G_2.

Claim 3. *If, for any node $\mu \in \mathcal{T}$ that is not an S-node, a* SEFE *of $G_1(\mu)$ and $G_2(\mu)$ exists in which the outer edges of μ can be drawn towards the outer face, then the algorithm computes one.*

Claims 1–3 prove the second implication of the lemma. Namely, when μ is the child of the root of $\mathcal{T}$, the algorithm computes a SEFE $\Gamma(\mu)$ of $G_1(\mu)$ and $G_2(\mu)$, by Claim 3. Observe that μ is not an S-node, by construction of $\mathcal{T}$. Also, observe that μ has no outer edge. By Claim 2, $\Gamma(\mu)$ can be extended into a SEFE Γ of G_1 and G_2, if such a SEFE exists. To this end, however, it is sufficient to draw the edge that is the root of $\mathcal{T}$. □

We get the following.

Theorem 1. *Let $G_1 = (V, E_1)$ and $G_2 = (V, E_2)$ be two planar graphs on the same set of n vertices such that the intersection graph $G_{1\cap 2}$ is biconnected. Then, it is possible to test whether G_1 and G_2 admit a* SEFE *in $O(n^3)$ time.*

Proof sketch: We sketch how to implement the algorithm described in this section in $O(n^3)$ time. The correctness of the algorithm has already been proved in Lemma 3.

First, we compute labels for the edges of G_1 and G_2 indicating whether the edge belongs to $G_{1\cap 2}$, to $G_{1\setminus 2}$, or to $G_{2\setminus 1}$. Second, for each node μ of $\mathcal{T}$ and for each exclusive edge e, we compute a label indicating whether e is an outer edge of μ, or an internal edge of μ, or an intra-pole edge of μ, or none of the previous ones. This can be done in total $O(n^2)$ time by traversing $\mathcal{T}$ once for each exclusive edge. Third, we choose an embedding for the skeletons of the nodes of $\mathcal{T}$. This is trivial for Q-, R-, and S-nodes and can be done in $O(n)$ time for each P-node μ by computing the graph O representing the constrained adjacencies among virtual edges of $skel(\mu)$ and by checking whether O is a cycle or a collection of paths. Fourth, we perform the Embedding-Flipping step in $O(n^2)$ time for each node $\mu \in \mathcal{T}$ that is not an S-node. For this sake, we construct graphs F_k^1 and F_k^2 for each face f_k of the computed embedding of $skel(\mu)$. Such graphs might have a total size that is quadratic in the size of $skel(\mu)$; however, once they have been constructed, the Embedding-Flipping step can be performed in a time that is linear in the size of such graphs. □

4 The Intersection Graph Is a Tree

In this section we show that the SEFE problem, when the intersection graph is a tree, is equivalent to a 2-page book embedding problem defined in the following.

Let G be a graph, let (E_1, E_2) be a partition of its edge set, and let T be a rooted tree whose leaves are the vertices of G. Problem PARTITIONED T-COHERENT 2-PAGE BOOK EMBEDDING with input (G, E_1, E_2, T) asks: Does a 2-page book embedding of G exist in which the edges of E_1 lie in one page, the edges of E_2 lie in the other page, and, for every internal vertex $t \in T$, the vertices of G in the subtree of T rooted at t appear consecutively in the vertex ordering of G defined in the book embedding?

We now show how to transform an instance $G_1 = (V, E_1), G_2 = (V, E_2)$ of SEFE in which $G_{1\cap 2}$ is a tree into an instance of PARTITIONED T-COHERENT 2-PAGE BOOK EMBEDDING. Such a transformation consists of two steps.

In the first step, we transform instance G_1, G_2 of SEFE into an equivalent instance G'_1, G'_2 of SEFE such that $G'_{1\cap 2}$ is a tree and all the exclusive edges of G'_1 and of G'_2 are incident only to leaves of $G'_{1\cap 2}$. To this end, we modify every edge $(u, v) \in G_{1\setminus 2}$ such that u is not a leaf of $G_{1\cap 2}$ as follows. We subdivide edge (u, v) with a new vertex u'; we add edge (u, u') to E_2, so that u' is a leaf in the intersection graph of the two modified graphs. Symmetrically, we subdivide every edge $(u, v) \in G_{2\setminus 1}$ such that u is not a leaf of $G_{1\cap 2}$ with a new vertex u' and we add edge (u, u') to E_1, so that u' is a leaf in the intersection graph of the two modified graphs. Note that the exclusive edges of G_1 and G_2 that are incident to two non-leaf vertices are subdivided twice. Denote by G'_1 and by G'_2 the resulting graphs. We have the following:

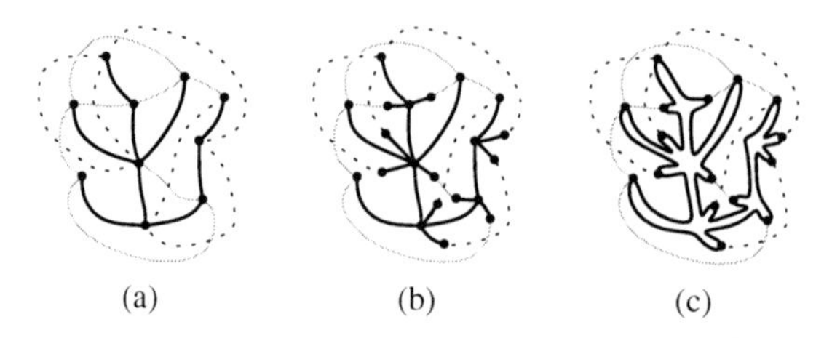

Fig. 2. (a) A SEFE Γ of G_1, G_2. (b) A SEFE Γ' of G'_1, G'_2. The edges of $G_{1\cap 2}$ and of $G'_{1\cap 2}$ are thick lines, the edges of $G_{1\setminus 2}$ and of $G'_{1\setminus 2}$ are drawn gray, and the edges of $G_{2\setminus 1}$ and of $G'_{2\setminus 1}$ are black dotted lines. (c) Euler Tour $\mathcal{E}$ of $G'_{1\cap 2}$ and exclusive edges.

Lemma 4. *G_1, G_2 is a positive instance of* SEFE *if and only if G'_1, G'_2 is a positive instance of* SEFE. *Further, $G'_{1\cap 2}$ is a tree and all the exclusive edges of G'_1 and of G'_2 are incident only to leaves of $G'_{1\cap 2}$. Moreover, $G'_{1\cap 2}$ has $O(n)$ vertices.*

Proof: $G_{1\cap 2}$ is a tree, by assumption. When an exclusive edge (u, v) in G_1 (resp. in G_2) such that u is not a leaf of $G_{1\cap 2}$ is subdivided with a vertex u' and edge (u, u') is added to E_2 (resp. to E_1), an edge is inserted into $G_{1\cap 2}$ connecting an internal vertex of $G_{1\cap 2}$ with a new leaf of $G_{1\cap 2}$, namely u'. Hence, $G_{1\cap 2}$ remains a tree after such a modification and thus $G'_{1\cap 2}$ is a tree. When an exclusive edge (u, v) in G_1 (resp. in G_2) such that u is not a leaf of $G_{1\cap 2}$ is subdivided with a vertex u' and edge (u, u') is added to E_2 (resp. to E_1), the number of incidences between exclusive edges and internal vertices of $G_{1\cap 2}$ decreases by one. Hence, after all such modifications have been performed, all the exclusive edges are incident only to leaves of $G'_{1\cap 2}$. Each exclusive edge is subdivided at most twice. Since the number of edges of $G_{1\setminus 2}$ and $G_{2\setminus 1}$ is $O(n)$, then $G'_{1\cap 2}$ has $O(n)$ vertices. We now prove that G_1, G_2 is a positive instance of SEFE if and only if G'_1, G'_2 is a positive instance of SEFE.

First, suppose that a SEFE Γ of G_1, G_2 exists. Modify Γ to obtain a SEFE Γ' of G'_1, G'_2 as follows (see Figs. 2(a) and 2(b)). When an exclusive edge (u, v) in G_1 (resp. in G_2) such that u is not a leaf of $G_{1\cap 2}$ is subdivided with a vertex u' and edge (u, u') is added to E_2 (resp. to E_1), insert u' in Γ along edge (u, v) arbitrarily close to u. Since the drawing of G_1 in Γ is not modified and since the drawing of G_2 in Γ is modified by inserting an arbitrarily small edge incident to a vertex, the resulting drawing is a SEFE of the current graphs and hence Γ' is a SEFE of G'_1, G'_2. Second, suppose that a SEFE Γ' of G'_1, G'_2 exists. A SEFE Γ of G_1, G_2 can be obtained by drawing each edge (u, v) of G_1 (resp. of G_2) exactly as in Γ'. Observe that (u, v) is subdivided never, once, or twice in G'_1 (resp. in G'_2); then, its drawing in Γ is composed of the concatenation of the one, two, or three curves representing the parts of (u, v) in Γ'. That no two edges of G_1 (resp. of G_2) intersect in the resulting drawing Γ directly descends from the fact that no two edges of G'_1 (resp. of G'_2) intersect in Γ'. □

In the second step, we transform an instance G_1, G_2 of SEFE such that $G_{1\cap 2}$ is a tree and all the exclusive edges of G_1 and of G_2 are incident only to leaves of $G_{1\cap 2}$ into an equivalent instance of PARTITIONED T-COHERENT 2-PAGE BOOK EMBEDDING.

The input of PARTITIONED T-COHERENT 2-PAGE BOOK EMBEDDING consists of the graph G composed of all the vertices which are leaves of $G_{1\cap 2}$, of all the exclusive

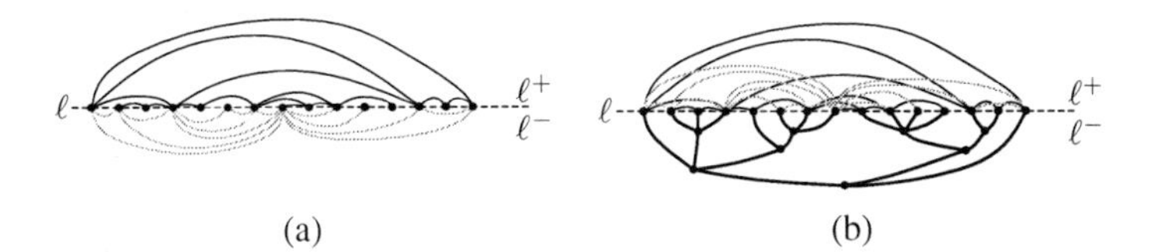

Fig. 3. (a) A Partitioned T-coherent 2-page book embedding of $(G, E_{1\setminus 2}, E_{2\setminus 1}, T)$. The edges in $E_{1\setminus 2}$ are thin black lines, the edges in $E_{2\setminus 1}$ are thin gray lines. (b) The SEFE of G_1, G_2 obtained from the book embedding of $(G, E_{1\setminus 2}, E_{2\setminus 1}, T)$.

edges $E_{1\setminus 2}$ of $G_{1\setminus 2}$, and of all the exclusive edges $E_{2\setminus 1}$ of $G_{2\setminus 1}$. The partition of the edges of G is $(E_{1\setminus 2}, E_{2\setminus 1})$. Finally, tree T is $G_{1\cap 2}$. We have the following:

Lemma 5. *G_1, G_2 is a positive instance of* SEFE *if and only if $(G, E_{1\setminus 2}, E_{2\setminus 1}, T)$ is a positive instance of* PARTITIONED T-COHERENT 2-PAGE BOOK EMBEDDING.

Proof: Suppose that $(G, E_{1\setminus 2}, E_{2\setminus 1}, T)$ is a positive instance of PARTITIONED T-COHERENT 2-PAGE BOOK EMBEDDING. See Fig. 3. An ordering of the vertices of G along a line ℓ exists such that the edges in $E_{1\setminus 2}$ are drawn on one side ℓ^+ of ℓ, the edges in $E_{2\setminus 1}$ are drawn on the other side ℓ^- of ℓ, no two edges in $E_{1\setminus 2}$ cross, and no two edges in $E_{2\setminus 1}$ cross. Move all the edges in $E_{2\setminus 1}$ to ℓ^+. Since such edges do not cross in ℓ^- and since the ordering of the vertices of G is not modified, the edges in $E_{2\setminus 1}$ still do not cross. Finally, construct a planar drawing of $G_{1\cap 2}$ in ℓ^-. This can always be done since, for each internal vertex t of $G_{1\cap 2}$, the vertices in the subtree of $G_{1\cap 2}$ rooted at t appear consecutively on ℓ. The resulting drawing is hence a SEFE of G_1, G_2.

Suppose that G_1, G_2 is a positive instance of SEFE. Consider any SEFE Γ of G_1, G_2 and consider an Euler Tour $\mathcal{E}$ of $G_{1\cap 2}$. Construct a planar drawing of $\mathcal{E}$ in Γ as follows. Each edge of $\mathcal{E}$ is drawn arbitrarily close to the corresponding edge in $G_{1\cap 2}$. Each end-vertex t of an edge of $\mathcal{E}$ that is a leaf in $G_{1\cap 2}$ is drawn at the same point where it is drawn in Γ. Each end-vertex t of an edge of $\mathcal{E}$ that is not a leaf in $G_{1\cap 2}$ and that has two adjacent edges (t, t_1) and (t, t_2) in $\mathcal{E}$ (observe that $t_1 \neq t_2$ as t is an internal vertex of $G_{1\cap 2}$) is drawn arbitrarily close to the point where t is drawn in Γ, in the region "between" edges (t, t_1) and (t, t_2). Clearly, the resulting drawing of $\mathcal{E}$ is planar (see Figs. 2(b) and 2(c)). Further, all the leaf vertices of $G_{1\cap 2}$ are drawn at the same point in Γ and in the drawing of $\mathcal{E}$. Moreover, all the exclusive edges of $G_{1\setminus 2}$ and all the exclusive edges of $G_{2\setminus 1}$ lie entirely outside $\mathcal{E}$, except for their end-vertices. Remove all the internal vertices and all the edges of $G_{1\cap 2}$ from the drawing. Move all the edges of $G_{2\setminus 1}$ inside $\mathcal{E}$. The resulting drawing is a Partitioned T-coherent 2-page book embedding of $(G, E_{1\setminus 2}, E_{2\setminus 1}, T)$. Namely, all the edges in $E_{1\setminus 2}$ are on one side of $\mathcal{E}$ and all the edges in $E_{2\setminus 1}$ are on the other side of $\mathcal{E}$. No two edges in $E_{1\setminus 2}$ cross as they do not cross in Γ. No two edges in $E_{2\setminus 1}$ cross as they do not cross in Γ. Finally, all the leaf vertices in a subtree of $G_{1\cap 2}$ rooted at an internal vertex t of $G_{1\cap 2}$ appear consecutively in $\mathcal{E}$, as the drawing of $G_{1\cap 2}$ in Γ is planar. □

Given an instance (G, E_1, E_2, T) of PARTITIONED T-COHERENT 2-PAGE BOOK EMBEDDING, it is possible to construct an equivalent instance of SEFE as follows. Let G_1 be the graph whose vertex set is composed of the vertices of G and of the internal vertices of T, and whose edge set is composed of the edges of E_1 and of the edges of T. Analogously, let G_2 be the graph whose vertex set is composed of the vertices of G and of the internal vertices of T, and whose edge set is composed of the edges of E_2 and of the edges of T. Analogous to Lemma 5, we can prove the following lemma.

Lemma 6. *(G, E_1, E_2, T) is a positive instance of* PARTITIONED T-COHERENT 2-PAGE BOOK EMBEDDING *if and only if G_1, G_2 is a positive instance of* SEFE.

Since both reductions can easily be performed in linear time we obtain the following.

Theorem 2. PARTITIONED T-COHERENT 2-PAGE BOOK EMBEDDING *and* SEFE *have the same time complexity.*

The problem PARTITIONED T-COHERENT 2-PAGE BOOK EMBEDDING has been recently studied by Hong and Nagamochi [16] when T is a star. That is, the graph has the edges partitioned into two pages as part of the input, but there is no constraint on the order of the vertices in the required book embedding. In such a case, Hong and Nagamochi proved that the problem is $O(n)$-time solvable [16]. While their motivation was a connection to the c-planarity problem, Lemmata 4 and 5 together with Hong and Nagamochi's result imply that deciding whether a SEFE exists for two graphs whose intersection graph is a star is a linear-time solvable problem.

Theorem 3. *The* SEFE *problem is solvable in linear time when the intersection graph is a star.*

5 Conclusions

In this paper we have shown new results on the time complexity of the problem of deciding whether two planar graphs admit a SEFE.

First, we have shown that the SEFE problem can be solved in cubic time if the intersection graph $G_{1\cap 2}$ of the input graphs G_1 and G_2 is biconnected. We believe that a refined implementation of our approach could reduce such a time bound to quadratic. More in general, with similar techniques we can solve in polynomial time the SEFE problem if $G_{1\cap 2}$ consists of one biconnected component plus a set of isolated vertices. Also, the following generalization of the SEFE problem with $G_{1\cap 2}$ biconnected seems worth to be tackled: What is the time complexity of computing a SEFE when $G_{1\cap 2}$ is *edge-biconnected*?

Second, we have shown that when $G_{1\cap 2}$ is a tree the SEFE problem can be equivalently stated as a 2-page book embedding problem with edges assigned to the pages and with hierarchical constraints. Hence, pursuing an $\mathcal{NP}$-hardness proof for such a book embedding problem is a possible direction for trying to prove the $\mathcal{NP}$-hardness for the SEFE problem.

Acknowledgments

Thanks to Seok-Hee Hong for presenting the results in [16] to the authors.

References

1. Angelini, P., Di Battista, G., Frati, F., Jelínek, V., Kratochvíl, J., Patrignani, M., Rutter, I.: Testing planarity of partially embedded graphs. In: Symposium on Discrete Algorithms (SODA 2010), pp. 202–221 (2010)
2. Angelini, P., Di Battista, G., Frati, F., Patrignani, M., Rutter, I.: Testing the simultaneous embeddability of two graphs whose intersection is a biconnected graph or a tree. Tech. Report 175, Dipartimento di Informatica e Automazione, Università Roma Tre (2010)
3. Angelini, P., Geyer, M., Kaufmann, M., Neuwirth, D.: On a tree and a path with no geometric simultaneous embedding. CoRR (2010)
4. Braß, P., Cenek, E., Duncan, C.A., Efrat, A., Erten, C., Ismailescu, D., Kobourov, S.G., Lubiw, A., Mitchell, J.S.B.: On simultaneous planar graph embeddings. Comput. Geom. 36(2), 117–130 (2007)
5. Di Battista, G., Tamassia, R.: On-line maintenance of triconnected components with SPQR-trees. Algorithmica 15(4), 302–318 (1996)
6. Di Battista, G., Tamassia, R.: On-line planarity testing. SIAM J. Comput. 25(5), 956–997 (1996)
7. Di Giacomo, E., Liotta, G.: Simultaneous embedding of outerplanar graphs, paths, and cycles. Int. J. Comput. Geometry Appl. 17(2), 139–160 (2007)
8. Erten, C., Kobourov, S.G.: Simultaneous embedding of planar graphs with few bends. J. Graph Algorithms Appl. 9(3), 347–364 (2005)
9. Estrella-Balderrama, A., Gassner, E., Jünger, M., Percan, M., Schaefer, M., Schulz, M.: Simultaneous geometric graph embeddings. In: Hong, S.H., Nishizeki, T., Quan, W. (eds.) GD 2007. LNCS, vol. 4875, pp. 280–290. Springer, Heidelberg (2008)
10. Fowler, J.J., Gutwenger, C., Jünger, M., Mutzel, P., Schulz, M.: An SPQR-tree approach to decide special cases of simultaneous embedding with fixed edges. In: Tollis, I.G., Patrignani, M. (eds.) GD 2008. LNCS, vol. 5417, pp. 157–168. Springer, Heidelberg (2009)
11. Fowler, J.J., Jünger, M., Kobourov, S.G., Schulz, M.: Characterizations of restricted pairs of planar graphs allowing simultaneous embedding with fixed edges. In: Broersma, H., Erlebach, T., Friedetzky, T., Paulusma, D. (eds.) WG 2008. LNCS, vol. 5344, pp. 146–158. Springer, Heidelberg (2008)
12. Frati, F.: Embedding graphs simultaneously with fixed edges. In: Kaufmann, M., Wagner, D. (eds.) GD 2006. LNCS, vol. 4372, pp. 108–113. Springer, Heidelberg (2007)
13. Gassner, E., Jünger, M., Percan, M., Schaefer, M., Schulz, M.: Simultaneous graph embeddings with fixed edges. In: Fomin, F.V. (ed.) WG 2006. LNCS, vol. 4271, pp. 325–335. Springer, Heidelberg (2006)
14. Geyer, M., Kaufmann, M., Vrt'o, I.: Two trees which are self-intersecting when drawn simultaneously. Discrete Mathematics 307, 1909–1916 (2009)
15. Gutwenger, C., Mutzel, P.: A linear time implementation of SPQR-trees. In: Marks, J. (ed.) GD 2000. LNCS, vol. 1984, pp. 77–90. Springer, Heidelberg (2001)
16. Hong, S.H., Nagamochi, H.: Two-page book embedding and clustered graph planarity. Tech. Report 2009-004, Department of Applied Mathematics & Physics, Kyoto University (2009)
17. Jünger, M., Schulz, M.: Intersection graphs in simultaneous embedding with fixed edges. J. Graph Alg. & Appl. 13(2), 205–218 (2009)

Skip Lift: A Probabilistic Alternative to Red-Black Trees

Prosenjit Bose, Karim Douïeb, and Pat Morin*

School of Computer Science, Carleton University, Herzberg Building
1125 Colonel By Drive, Ottawa, Ontario, K1S 5B6 Canada
{jit,karim,morin}@cg.scs.carleton.ca
http://cg.scs.carleton.ca

Abstract. We present the *Skip lift*, a randomized dictionary data structure inspired by the skip list [Pugh '90, Comm. of the ACM]. Similar to the skip list, the skip lift has the finger search property: Given a pointer to an arbitrary element f, searching for an element x takes expected $O(\log \delta)$ time where δ is the rank distance between the elements x and f. The skip lift uses nodes of $O(1)$ worst-case size and it is one of the few efficient dictionary data structures that performs an $O(1)$ worst-case number of structural changes during an update operation. Given a pointer to the element to be removed from the skip lift the deletion operation takes $O(1)$ worst-case time.

1 Introduction

The dictionary problem is fundamental in computer science. It asks for a data structure in the pointer machine model that stores a totally ordered set S of n elements and supports the operations search, insert and delete. A large number of data structures optimally solve this problem in worst-case $O(\log n)$ time per operation. Some of them guarantee an $O(1)$ worst-case number of structural changes (pointers/fields modifications) after an insertion or a deletion operation [12,19,11,13,10,6].

Typically the update operations, i.e., insert and delete, are performed in two phases: First, search for the position where the update has to take place. Second, perform the actual update and restore the balance of the structure. When the position where the new element has to be inserted or deleted is already known then the first phase of an update could be avoided. In general the first phase is considered to be part of the search operation. A dictionary that guarantees an $O(1)$ worst-case number of structural changes per update does not necessary quickly perform the second phase of the update. For example after inserting a new item in a red-black tree [12], $\Omega(\log n)$ steps may be required to find where the $O(1)$ number of rotations have to be performed in order to restore the balance. Much research effort has been aimed at improving the worst-case time taken by the second phase of the update: Levcopoulos and Overmars [13]

* Research partially supported by NSERC and MRI.

C.S. Iliopoulos and W.F. Smyth (Eds.): IWOCA 2010, LNCS 6460, pp. 226–237, 2011.

presented the first search tree that takes $O(1)$ worst-case time for this second phase of the update. Later Fleischer [10] simplified this result. Brodal *et al.* [6] additionally guaranteed that such structures can also have the finger search property in worst-case time. These structures however are quite complicated and not really practical.

On the other hand, most randomized dictionaries are simple, practical and achieve the same performance as the result of Brodal *et al.* [6] in the expected sense. In the worst case though their performance is far from optimal. Here we develop a simple randomized dictionary, called a skip lift, inspired by the skip list [18], that improves the worst-case performance of the second phase of the update operations. Namely we obtain a structure that has the finger search property in expectation and performs an $O(1)$ worst-case number of structural changes per update. Given a pointer to the element to be removed from the skip lift, the deletion operation takes $O(1)$ worst-case time.

In Section 1.1 we describe the original skip list dictionary. In Section 1.2 we mention some work related to the skip list dictionary. In Section 2 we introduce our new skip lift data structure. In Section 3 we show how to enhance the skip lift structure to allow a simple finger search. Finally in Section 4 we give an overview of some classical randomized dictionary data structures. For each of them we briefly describe its construction and how the dictionary operations are performed. We show that in some situations $\Omega(n)$ structural changes are necessary to perform the update operations.

1.1 Skip List

The *skip list* of Pugh [18] was introduced as a probabilistic alternative to balanced trees. It is a dictionary data structure storing a totally ordered set S of n elements that supports insertion, deletion and search operations in $O(\log n)$ expected time. Additionally the expected number of structural changes (pointer modifications) performed on the skip list during an update is $O(1)$. A skip list is built in levels, the bottom level (level 1) is a sorted linked list of all elements in S. The higher levels of the skip list are build iteratively. Each level is a sublist of the previous one where each element of a level is copied to the level above with (independent) probability p. The copies of an element are linked between adjacent levels (see Fig. 1.a).

The *height* $h(s)$ of an element s is defined as the highest level where s appears. The height $H(\mathcal{L})$ of a skip list $\mathcal{L}$ is defined as $\max_{s \in \mathcal{L}} h(s)$ and the *depth* $d(s)$ of s is $H(\mathcal{L}) - h(s)$. The expected height of a skip list is by definition $O(\log_{1/p} n)$. Adjacent elements on the same level are connected by their left and right pointers. The copies of the same element from two adjacent levels are connected by their up and down pointers.

Search: To search for a given element x in a skip list we start from the highest level of the sentinel element which has a key value $-\infty$. We follow the right pointers on a same level until we are about to overshoot the element x, i.e., until the element on the right has a key value strictly greater than x. Then we go down one level and we iterate the process until x is found or until we have reached the lowest level (in this case we know that x is not in S and we have found its predecessor).

Updates: To insert an element x in a skip list we first determine its height in the structure. Then we start a search for x in the list to find the position where x has to be inserted. During the search we update the pointers of the copies of the elements that are adjacent to a newly created copy of x.

The deletion of an element x from a skip list is straightforward given the insertion process. We first search for x and we delete one by one all its copies while updating the pointers of the copies of elements that are adjacent to a copy of x.

1.2 Related Work

Precise analysis of the expected search cost in a skip list has been extensively studied, we refer to the thesis of Papadakis for more information [17]. Several variants of the skip list have been considered: Munro *et al.* [16] developed a deterministic version of the skip list, based on B-trees [3], that performs each dictionary operation in worst-case $O(\lg n)$ time. Under the assumption that the distribution of access probabilities is given, Martínez and Roura [14] developed an algorithm that minimizes the expected access time by either building an optimal static skip list in $O(n^2 \lg n)$ time or a nearly optimal one in $O(n)$ time. Bagchi *et al.* [2] developed the biased skip list; it manages a biased dictionary, i.e., an ordered set S of elements x associated with a weight $w(x)$ and performs search, insert, delete, join, split, finger search and reweight operations in worst-case running times similar to those of biased search trees [4,9]. In the general case where access probabilities are unknown, Bose et al. [5] prove that for a class of skip lists that satisfies a weak balancing property, the working-set bound is

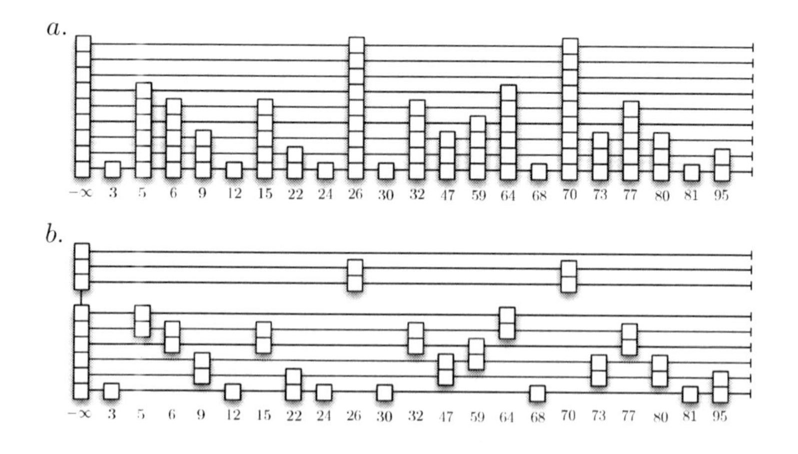

Fig. 1. (a) Skip list, (b) Skip lift

a lower bound on the time to access any sequence. Furthermore, they develop a deterministic self-adjusting skip list whose running time matches the working-set bound, thereby achieving dynamic optimality in this class (both in internal and external memory).

2 Skip Lift

The average amount of extra information per element (number of copies) in a standard skip list [18] is constant. In the worst-case this number can reach $\Omega(\log n)$. Hence the number of structural changes in a skip list during an update is $\Omega(\log n)$ in the worst-case. Here we present a slight modification of the skip list data structure (as the title of the paper suggests) which guarantees, in the worst-case, a constant amount of extra information per element and a constant number of structural changes per update.

A skip lift is a light version of the skip list where copies of elements have been removed from specific levels. A skip lift only keeps the copies of an element in the two highest levels where it would appear in the skip list. Every other copy of an element is removed. The copies of the elements at the same level are connected with their left and right pointers. Additionally the two copies of an element are connected with their up and down pointers (see Figure 1.b). Each copy stores its height in an extra height field.

A level of the skip lift is empty if no element of the set S appears in it. The copies of the sentinel element appearing in an empty level are deleted. The remaining copies of the sentinel element are connected with their up and down pointers. A copy of the sentinel element at height $+\infty$ is explicitly maintained, this copy is called the *header* of the skip lift.

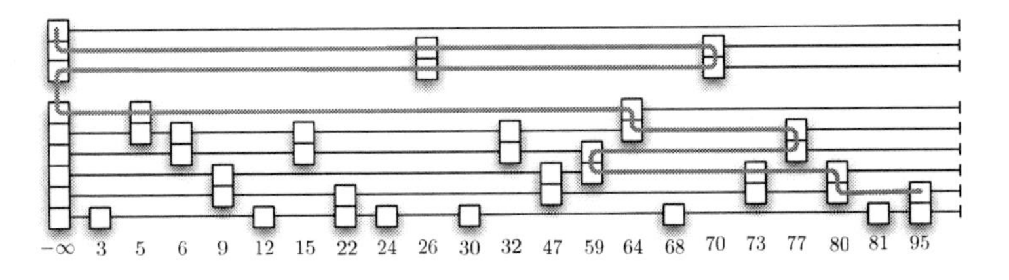

Fig. 2. Search Path for the element 95

Search: To search for a given element x in a skip lift we start at the header of the list. We follow the right pointers on the same level until we see that we are about to overshoot the element x, i.e., until the element on the right has a key value strictly greater than x. If it is possible we go down to the next non-empty level. Otherwise we follow the left pointers until we find an element which allows us to go down to the next non-empty level. Then we iterate the process until x is found or when we have reached the lowest level (in this case x is not in S and we know its predecessor). This procedure is described in detail in Algorithm 1 and illustrated in Figure 2.

Algorithm 1. Search(x)

```
c ⇐ header
pred ⇐ −∞
while c ≠ x and height[c]>1 do
  while down[c] = NIL do
    c ⇐ left[c]
  end while
  c ⇐ down[c]
  while right[c] ≠ NIL and right[c] ≤ x do
    c ⇐ right[c]
  end while
  if pred < c then
    pred ⇐ c
  end if
end while
return pred
```

Lemma 1. *A skip lift supports a search operation in* $O\left(\frac{1}{p}\log_{1/p} n\right)$ *expected time, where* n *is the number of elements in the skip lift and* p *is the probability for an element in level* i *to appear in level* $i+1$.

Proof. The expected length of the search path in a skip lift $\mathcal{L}$ corresponds to the expected number of vertical steps plus the expected number of horizontal steps. The number of vertical steps performed during a search is upper bounded by the height $H(\mathcal{L})$ of the skip lift which has an expected value of $\log_{1/p} n + \frac{1}{1-p}$. The expected height of a skip lift corresponds exactly to the expected height of a skip list [18].

Now we are going to bound the number of horizontal steps. At any level i of $\mathcal{L}$ only elements of height i and $i+1$ can appear with probability $1/(1+p)$ and $p/(1+p)$, respectively. This means that from any position in level i the expected number of horizontal steps required to reach an element of height i is at most

$$\sum_{j=1}^{\infty} j\left(\frac{p}{1+p}\right)^{j-1}\frac{1}{(1+p)} = 1+p. \tag{1}$$

Similarly the expected number of horizontal steps required to reach an element of height $i+1$ in level i is at most

$$\sum_{j=1}^{\infty} j\left(\frac{1}{1+p}\right)^{j-1}\frac{p}{(1+p)} = \frac{1+p}{p}. \tag{2}$$

Consider $e(i,x)$ the element of height i that has the greatest key value smaller than x. The search path to an element x in $\mathcal{L}$ traverses all elements $e(i,x)$ with $h(x) \leq i \leq H(\mathcal{L})$. These are the only elements where the search path performs a down step. Between each of these $e(i,x)$ elements, the search path traverses horizontally a certain number of other elements. On expectation this number differs depending on whether the path goes from left to right or right to left. If the path goes from right to left this expected number corresponds to eq.(1) otherwise it corresponds to eq.(2). The probability that the search path goes from left to right on level i is $1/(p+1)$. This corresponds to the probability of seeing $e(i,x)$ before $e(i+1,x)$ from the position of x on level i which also corresponds to the probability that an element of height i appears on level i. Respectively the probability that the search path goes from right to left on level i is $p/(p+1)$. Hence the expected number of horizontal steps performed between each element $e(i,x)$ is

$$p+1\frac{p}{p+1}+\frac{1+p}{p}\frac{1}{1+p}=p+\frac{1}{p}.$$

The expected cost to access the first element $e(H(\mathcal{L}),x)$ is smaller than the expected number of elements of height greater or equal to $\log_{1/p} n$ which is $1/p$. Thus total expected number of horizontal steps is upper bounded by

$$H(\mathcal{L})\left(p+\frac{1}{p}\right)+\frac{1}{p}.$$

Therfore the expected length of a search path in a skip lift is

$$H(\mathcal{L})+H(\mathcal{L})\left(p+\frac{1}{p}\right)+\frac{1}{p}=\frac{H(\mathcal{L})+1}{p}+(p+1)H(\mathcal{L})=O\left(\frac{\log_{1/p} n}{p}\right). \quad \square$$

Updates: To insert an element x in a skip lift we first determine its height $h(x)$ in the structure. Then we start to search for x in the list to find the position where x has to be inserted, i.e., its position in levels $h(x)$ and $h(x)-1$. Once we find these positions, the copies of the element x are inserted in the corresponding level. This is performed similarly to the insertion of an element in a standard doubly-linked list. If the level where the copy of x has to be insereted is empty then we create a new copy of the sentinel element and insert it in the skip lift (seeing all copies of the sentinel element as a doubly-linked list). This process is described in detail in Algorithm 2. We assume that x is not in the set S (otherwise we could simply search for x before performing the actual insert operation).

To delete an element x from a skip lift we first search the two copies of x using the search operation decribed above. Once we found the copies of x we delete them from their corresponding level. This is performed similar to the deletion of an element in a standard doubly-linked list. If the deletion of the copies of x creates an empty level, we remove the corresponding copy of the sentinel element. This process is described in detail in Algorithm 3.

Theorem 1. *The skip lift supports search, insert and delete operations in* $O\left(\frac{1}{p}\log_{1/p} n\right)$ *expected time and requires* $O(n)$ *worst-case space. The total number of structural changes performed during an update is* $O(1)$ *in worst-case.*

Algorithm 2. Insert(x)

```
c ⇐ header
h ⇐ randomLevel()
while height[c] ≥ h do
   while down[c] = NIL do
      c ⇐ left[c]
   end while
   if c = −∞ and height[down[c]] < h and h < height[c] then
      e ⇐ new element(−∞, h)
      down[e] ⇐ down[c]
      down[c] ⇐ e
   end if
   c ⇐ down[c]
   while right[c] ≠ NIL and right[c] ≤ x do
      c ⇐ right[c]
   end while
   if height[c] = h then
      right[x] ⇐ right[c]
      left[x] ⇐ c
      left[right[c]] ⇐ x
      right[c] ⇐ x
      x ⇐ down[x]
      if x ≠ NIL then
         h ⇐ h − 1
      end if
   end if
end while
```

3 Finger Search

A data structure satisfies the finger search property if searching for an element x given a pointer, called *finger*, to an arbitrary element f requires logarithmic time in the rank distance between x and f in the set of ordered elements. It is possible to describe a finger search operation on the skip lift (as described in the previous section) but it is a bit complicated. Instead we show how to enhance the skip lift structure in order to simplify the description of the finger search. The *enhanced skip lift* maintains an extra copy of each element at the bottom level. This copy is linked to the lowest copy of the corresponding element above the bottom level with the up and down pointers.

We can search for an element x in an enhanced skip lift starting at the bottom copy of any element f to which we are given an initial pointer. Assume without loss of generality that the key value of the element x is greater than that of f (the opposite case is symmetric). The finger search can be decomposed into an *up phase* and a *down phase*. The up phase behaves as the inverse of the search operation described in Alg. 1 and the down phase is similar to Alg. 1.

We start the search from the bottom copy of f then from any current position we follow the left pointers on the same level until the element on the left has a key value strictly smaller than f. If it is possible we go one level up (if the

Algorithm 3. Delete(x)

```
c ⇐ header
while height[c] > 1 do
   while down[c] = NIL do
      c ⇐ left[c]
   end while
   c ⇐ down[c]
   while right[c] ≠ NIL and right[c] ≤ x do
      c ⇐ right[c]
   end while
   while c = x do
      right[left[x]] ⇐ right[x]
      if right[x] ≠ NIL then
         left[right[x]] ⇐ left[x]
      else if left[x] = −∞ then
         down[up[left[x]]] ⇐ down[left[x]]
         if down[left[x]] ≠ NIL then
            up[down[left[x]]] ⇐ up[left[x]]
         end if
         delete left[x]
      end if
      c ⇐ down[c]
      delete(up[c])
   end while
end while
```

up pointer jumps over more than one level then we do not take it). Otherwise we follow the right pointers until we find an element which allows us to go one level up or when the element on the right has a key value greater than x (this last case corresponds to the end of the up phase). From the current position, the down phase consists of following the right pointers on the same level until the element on the right has a key value strictly greater than x. If it is possible we go down by one level (if the down pointer jumps over more than one level then we do not take it). Otherwise we follow the left pointers until we find an element which allows us to go down by one level. Then we iterate the process until x is found or until we have reached the lowest level (in this case x is not in S and we know its predecessor).

Theorem 2. *Finger searching for an element x given a finger pointing to an arbitrary element f in an enhanced skip lift takes $O\left(\frac{1}{p}\log_{1/p}\delta\right)$ time where δ is the rank distance between the finger and the search element x.*

Proof. The search path traverses only elements that are between f and x in the skip lift. The sublist between f and x contains δ elements by definition. Thus the expected height of this sublist is $O(\log_{1/p}\delta)$. In each level we perform $O(1/p)$ expected steps since this corresponds to the expected number of steps needed to find an element of height i or $i+1$ from any position on level i. Therefore the total length of the search path is $O\left(\frac{1}{p}\log_{1/p}\delta\right)$. □

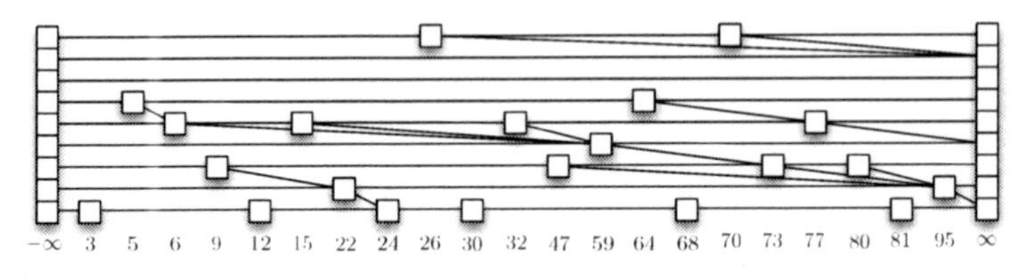

Fig. 3. Modified skip list

4 Overview of Randomized Dictionaries

We present an overview of some classical randomized dictionary data structures. For each of them we briefly describe its construction and how search, insertion and deletion operations are performed. It is easy to realize that the structural changes performed during an update operation can in some situations involve $\Omega(n)$ elements of the structure. Of course those situations are very unlikely to happen but are not impossible. This means that the skip lift is the first efficient randomized dictionary that guarantees a $O(1)$ number of structural changes per update.

4.1 Modified Skip List

A *modified skip list* is a variant of the skip list introduced by Cho and Sahni [8] that uses nodes of constant worst-case size (containing $O(1)$ pointers). The modified skip list structure is a skip list where all copies of an element are deleted except for its highest copy. Thus an element x only appears on the level $h(x)$. Each element x has three pointers: $\mathsf{right}[x]$, $\mathsf{left}[x]$ and $\mathsf{down}[x]$. The pointers $\mathsf{right}[x]$ and $\mathsf{left}[x]$ point to the elements on level $h(x)$ to the right and the left of x, respectively. The pointer $\mathsf{down}[x]$ points to the element on level $h(x) - 1$ that has the smallest key value greater than x. Two sentinel elements with key value $-\infty$ and ∞ are maintained, a copy of these elements appear in every level. The down pointer of a copy of a sentinel element points to the copy of itself on the level below (see Fig. 3).

Search: To search for a given element x in a modified skip list we start from the highest level of the sentinel element with key value $-\infty$. We follow the right pointers on a same level until the element on the right has a key value strictly greater than x. From this point we follow the left pointer then we immediately go down one level by following the down pointer from this left element. The process is iterated until x is found or when we have reached in the lowest level (in this case we know that x is not in S and we have found its predecessor).

Updates: The insert and delete operations require to search the position of x in the list. When inserting an element x in a modified skip list only one copy is created in the level $h(x)$ and the down pointer of x is set to the element in level $h(x) - 1$ that has the smallest key value greater then x. When deleting an element x from a modified skip list we have to update the down pointer of all the elements from level $h(x) + 1$ that are pointing to x by setting them to the element on the right of x.

A degenerate situation would be when all elements in the structure have height 2 except for the very last one (with the greatest key value). Deleting the last element would force the modification of the down pointer all elements, implying an $\Omega(n)$ number of structural changes in the structure. A similar situation can occurs if we insert an element just before the last one.

4.2 Treap

A *treap* is a randomized data structure introduced by Aragon and Seidel [1]. It is structured as a binary search tree structure, so the left and the right subtrees of any node only contain elements of smaller or greater key value, respectively. Each element of S is given a random priority. The treap is built such that the root is the minimum-priority node and the priority of any non-root node must be greater than or equal to the priority of its parent (heap-ordering property).

Search: To search for a given element x, we use the standard binary search algorithm in a binary search tree independently of the priorities.

Updates: To insert a new element x into the treap, we first generate a random priority for x. We perform a search for x in the treap. If $x \in S$ we do nothing otherwise we make x a child of the last element visited during the search. Then x is rotated up as long as its priority is smaller than the priority of its parent or when x becomes the new root.

To delete a node x from the treap three cases are considered. If x is a leaf, we simply remove it. If x has a single child, we remove x from the treap and make the child of x the new child of its the parent (or make the child of x the root if x had no parent). Finally, if x has two children, swap its position in the treap with its predecessor, resulting in one of the previously discuss cases. In this final case, the swap may violate the heap-ordering property, so additional rotations may need to be performed to restore it.

A degenerate situation would be when the tree is a path of n elements. Inserting an element at the end of the path with a given priority that is smaller than any priority in the tree would bring the new inserted element to the root. This is performed by a sequence of $\Omega(n)$ rotations, i.e., an $\Omega(n)$ number of structural changes in the tree. A similar situation could occur when deleting an element.

4.3 Randomized Binary Search Tree

A *Randomized binary search tree* is another dictionary data structure developed by Martínez and Roura [15]. Each subtree of a random search tree is itself a random search tree. The root of such a tree is chosen uniformly at random among the elements of S, i.e., with probability $1/n$. The remaining of the tree is defined iteratively.

Search: To search for a given element x, we use the standard binary search algorithm in a binary search tree.

Updates: To insert a new element x into a random search tree T we proceed as follows: With probability $1/(|T|+1)$ the element x has to be theroot of the

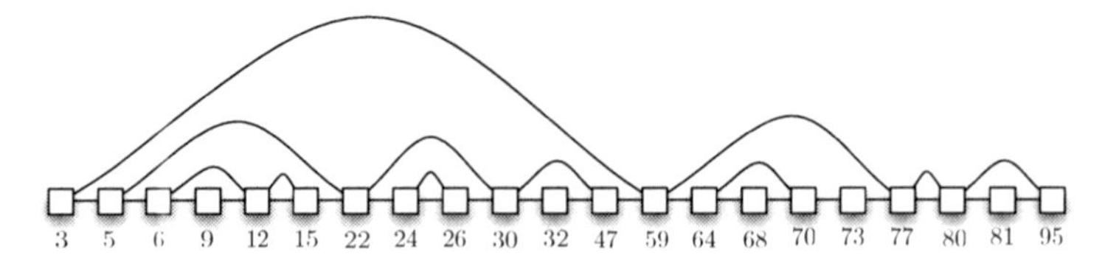

Fig. 4. Jumplist

new tree. In this case the tree T is split at x and the two obtained subtrees are attached as the children of x. Otherwise we iterate the process on the left (right) subtree if x is smaller (greater) than the key value of the root.

To delete an element x from a random search tree T, we search for it in T. Once it is found we remove it and we replace the subtree rooted at x by a newly created subtree obtained by joining the left and right subtree of x (this joining procedure is fully described in [15]).

A degenerate situation would be when the tree is a path of n elements so that the key of the elements from the root to the leaf are alternatively greater and smaller x. Assume we insert a new element with key value x, it could be that x has to be inserted has the root of the tree. In this case we split the tree at x which requires an $\Omega(n)$ number of structural changes in the tree. The inverse situation could occur when deleting an element.

4.4 Jumplist

A *jumplist* of Brönnimann *et al.* [7] is a randomized data structure inspired by the randomized tree. It is an linked list data structure ordered by key value whose nodes are endowed with an additional pointer, the *jump pointer* (see Fig. 4). An element x of a jumplist has a $\mathsf{next}[x]$ pointer which points to the immediate successor of x in S. Additionally an element has a $\mathsf{jump}[x]$ pointer which points to an element further on the list to the right of x. The jumplist is constructed as follows: The element j pointed to by the jump pointer of the head of the list is chosen uniformly at random among the elements in the list. This assignment divides the list into two independent sublists that are built recursively using the same random process. This construction ensures that the jump pointers do not cross.

Search: The jumplist is based on the *jump-and-walk* strategy: whenever possible use the jump pointer to speed up the search, and walk along the list otherwise. So to search for an element x we use the jump pointer until we are about to overshoot x in which case we follow the next pointer. We iterate this process until we find the element x or until the next pointer leads us to an element with greater key value than x (in this case we know that x is not in S and we have found its predecessor).

Updates: To insert an element x in a jumplist J we proceed as follows: With probability $1/|J|$ the element x has to be the element pointed by the jump pointer of the head of the list. In this case the whole list is rebuilt from scratch. Otherwise x is inserted in one of its sublists. In the case where x has to be inserted as the new head of a sublist, a process that does not rebuild the sublist from scratch is called to maintain the randomness of the structure.

Since an insertion could cause the reconstruction of the entire jumplist, this operation requires an $\Omega(n)$ number of structural changes in the list.

References

1. Aragon, C.R., Seidel, R.: Randomized search trees. Algorithmica 16, 464–497 (1996)
2. Bagchi, A., Buchsbaum, A.L., Goodrich, M.T.: Biased skip lists. Algorithmica 42(1), 31–48 (2005)
3. Bayer, R., McCreight, E.: Organization and maintenance of large ordered indexes. Acta Informatica 1, 173–189 (1972)
4. Bent, S.W., Sleator, D., Tarjan, R.: Biased search trees. SIAM Journal on Computing 14(3), 545–568 (1985)
5. Bose, P., Douïeb, K., Langerman, S.: Dynamic optimality for skip lists and B-trees. In: Proceedings of the Nineteenth Annual ACM-SIAM Symposium on Discrete Algorithms (SODA 2008), pp. 1106–1114 (2008)
6. Brodal, G.S., Lagogiannis, G., Makris, C., Tsakalidis, A.K., Tsichlas, K.: Optimal finger search trees in the pointer machine. J. Comput. Syst. Sci. 67(2), 381–418 (2003)
7. Brönnimann, H., Cazals, F., Durand, M.: Randomized jumplists: A jump-and-walk dictionary data structure. In: Alt, H., Habib, M. (eds.) STACS 2003. LNCS, vol. 2607, pp. 283–294. Springer, Heidelberg (2003)
8. Cho, S., Sahni, S.: Weight-biased leftist trees and modified skip lists. J. Exp. Algorithmics 3, 2 (1998)
9. Feigenbaum, J., Tarjan, R.: Two new kinds of biased search trees. Bell System Technical Journal 62(10), 3139–3158 (1983)
10. Fleischer, R.: A simple balanced search tree with $O(1)$ worst-case update time. International Journal of Foundations of Computer Science 7, 137–149 (1996)
11. Haeupler, B., Sen, S., Tarjan, R.E.: Rank-balanced trees. In: Dehne, F., Gavrilova, M., Sack, J.-R., Tóth, C.D. (eds.) WADS 2009. LNCS, vol. 5664, pp. 351–362. Springer, Heidelberg (2009)
12. Leonidas, R.S., Guibas, J.: A dichromatic framework for balanced trees. In: Proc. 19th IEEE Symp. on Foundations of Computer Science, pp. 8–21 (1978)
13. Levcopoulos, C., Overmars, M.: A balanced search tree with $O(1)$ worst-case update time. Acta Informatica 26(3), 269–277 (1988)
14. Martinez, C., Roura, S.: Optimal and nearly optimal static weighted skip lists. Technical report, LSI-95-34-R, Dept. Llenguatges i Sistemes Informàtics (Universitat Politèchnica de Catalunya) (1995)
15. Martínez, C., Roura, S.: Randomized binary search trees. J. ACM 45(2), 288–323 (1998)
16. Munro, I., Papadakis, T., Sedgewick, R.: Deterministic skip lists. In: Proceedings of the Third Annual ACM-SIAM Symposium on Discrete Algorithms, pp. 367–375 (1992)
17. Papadakis, T.: Skip lists and probabilistic analysis of algorithms. PhD thesis, University of Waterloo, Department of Computer Science and Faculty of Mathematics (Available as Tech. Report CS-93-28) (1993)
18. Pugh, W.: Skip lists: a probabilistic alternative to balanced trees. Communications of the ACM 33(6), 668–676 (1990)
19. Tarjan, R.E.: Updating a Balanced Search Tree in $O(1)$ Rotations. Inf. Process. Lett. 16(5), 253–257 (1983)

On a Relationship between Completely Separating Systems and Antimagic Labeling of Regular Graphs

Oudone Phanalasy[1,2], Mirka Miller[1,3,4,5], Leanne Rylands[6], and Paulette Lieby[7]

[1] School of Electrical Engineering and Computer Science, The University of Newcastle, NSW 2308, Australia
[2] Department of Mathematics, National University of Laos, Vientiane, Laos
[3] Department of Mathematics, University of West Bohemia, Pilsen, Czech Republic
[4] Department of Computer Science, King's College London, UK
[5] Department of Mathematics, ITB Bandung, Indonesia
[6] School of Computing and Mathematics, University of Western Sydney, NSW, Australia
[7] NICTA, Canberra ACT 2001, Australia

oudone.phanalasy@uon.edu.au, mirka.miller@newcastle.edu.au, l.rylands@uws.edu.au, paulette.lieby@rsise.anu.edu.au

Abstract. A completely separating system (CSS) on a finite set $[n]$ is a collection $\mathcal{C}$ of subsets of $[n]$ in which for each pair $a \neq b \in [n]$, there exist $A, B \in \mathcal{C}$ such that $a \in A$, $b \notin A$ and $b \in B$, $a \notin B$.

An antimagic labeling of a graph with p vertices and q edges is a bijection from the set of edges to the set of integers $\{1, 2, \ldots, q\}$ such that all vertex weights are pairwise distinct, where a vertex weight is the sum of labels of all edges incident with the vertex. A graph is antimagic if it has an antimagic labeling.

In this paper we show that there is a relationship between CSSs on a finite set and antimagic labeling of graphs. Using this relationship we prove the antimagicness of various families of regular graphs.

Keywords: completely separating system, vertex antimagic edge labeling, antimagic labeling, regular graph.

1 Introduction

The concept of completely separating system was first introduced in 1969 by Dickson [7]. Let $[n] = \{1, 2, \ldots, n\}$. A *completely separating system* (CSS) on $[n]$ is a collection $\mathcal{C}$ of subsets of $[n]$ in which for each pair $a \neq b \in [n]$, there exist $A, B \in \mathcal{C}$ such that $a \in A$, $b \notin A$ and $b \in B$, $a \notin B$. For example, the collection $\{\{1,2\},\{1,3\}\}$ is not a CSS. However, the collection $\{\{1,2\},\{1,3\},\{2,3\}\}$ is a CSS on $[3]$.

The sets in the (n)CSS are usually called *blocks* and the elements of these sets are usually called *points*. Let k be a positive integer and let $\mathcal{C}$ be an (n)CSS.

C.S. Iliopoulos and W.F. Smyth (Eds.): IWOCA 2010, LNCS 6460, pp. 238–241, 2011.

If $|A| = k$ for all $A \in \mathcal{C}$, then $\mathcal{C}$ is said to be an (n,k)CSS. A *d-element* in a collection of sets is an element which occurs in exactly d sets in the collection. For any n, k fixed positive integers, $R(n,k)$ is defined as follows: $R(n,k) = \min\{|\mathcal{C}| : \mathcal{C}$ is an $(n,k)\text{CSS}\}$. An (n,k)CSS for which $|\mathcal{C}| = R(n,k)$ is a *minimal* (n,k)CSS.

Subsequently, several variants have been explored in [10], [11], [12], [13] and [14], among others. Ramsay and Roberts [11], and Roberts [13] have explored minimal (n)CSSs, (n,h,k)CSSs and (n,k)CSSs. Roberts [13] gave a method for the construction of a class of minimal (n,k)CSSs. The special case of this construction given below is our main tool for the study of antimagic labelings of graphs, so it is restated here.

Roberts' construction [13]
Assume that $k \geq 2$, $n \geq \binom{k+1}{2}$ and $k|2n$, and let $R = R(n,k) = 2n/k$. An $(R \times k)$-array L is constructed, where each row of L forms a subset of $[n]$ and the R rows of L form an $(n,k)CSS$. Let e_{ij} denote the element of L in row i and column j. Initialize all elements of L to zero. For e from 1 to n, in order, include e in the two positions of L defined by

$$\min_j \min_i \{e_{ij} : e_{ij} = 0\},$$
$$\min_i \min_j \{e_{ij} : e_{ij} = 0\}.$$

That is, e is placed in the first row of L containing a 0, in the first 0-valued place in that row, e is then also placed in the first column of L containing a 0, in the first 0-valued place in that column. Each of the integers 1 to n appears in L in two positions, and the array L is the array of an (n,k)CSS. This concludes Roberts' construction.

In this paper we consider only graphs that are finite, simple and undirected.

The concept of labeling of graphs is becoming increasingly popular, partly because it contains many interesting mathematical challenges, and partly also because of the wide range of applications to other branches of science, for example, see [3] and [4].

The notion of a *vertex antimagic edge labeling*, known as an *antimagic labeling* of graphs was introduced in 1990 by Hartsfield and Ringel [9]. An antimagic labeling of a graph with p vertices and q edges is a bijection from the set of edges to the set of integers $\{1, 2, \ldots, q\}$ such that all vertex weights are pairwise distinct, where a vertex weight is the sum of labels of all edges incident with the vertex. A graph is antimagic if it has an antimagic labeling.

Hartsfield and Ringel [9], proposed

Conjecture 1. [9] Every connected graph, except K_2, is antimagic.

During two decades many papers on antimagicness of particular classes of graphs have been published, for example, see [1], [2], [5], [6], [15], [16] and [17]. For a detailed survey, see [8].

In this paper we introduce a new approach for obtaining antimagic labeling for some classes of regular graphs by using results on completely separating systems (CSSs). Using CSSs, we can determine that some classes of regular graphs are antimagic and we give an antimagic labeling for such classes of graphs. To the best of our best knowledge, this is the first time that CSSs have been applied to produce antimagic graph labelings.

2 Results

In this section we establish a powerful relationship between the combinatorics of finite sets and graph labeling. This relationship is used together with a known construction for (q, k)CSSs to produce a family of regular antimagic graphs.

Theorem 1. *Let* $V = \{v_1, v_2, \ldots, v_p\}$ *be a collection of* k*-subsets of* $[q]$. *Then* V *is a* (q, k)*CSS consisting of* 2*-elements if and only if a* k*-regular graph* $G(V, E)$ *with* q *edges and* p *vertices has an edge labeling.*

In view of the converse implication in Theorem 1, the existence of an edge labeling of a graph could be exploited to provide new results in the study of CSSs. We do not explore this possibility in this paper.

An edge labeling of a graph will often be represented by an array as follows.

- Each vertex is represented by a row (block) of the array;
- Each row (block) consists of the labels of all edges incident with the vertex represented by that row.

Hereafter we denote by $G(V, E, L)$, a graph $G(V, E)$ having an edge labeling L.

Theorem 2. *Let* L *be the array of a* (q, k)*CSS obtained by Roberts' construction. Then the* k*-regular graph* $G(V, E, L)$*, where* $|V| = p = 2q/k$, $|E| = q$*, is antimagic.*

Example 1. Let L be the array of the $(12, 4)$CSS obtained using Roberts' construction. Then we have the array L and the corresponding antimagic 4-regular graph $G(V, E, L)$ are shown in Fig. 1.

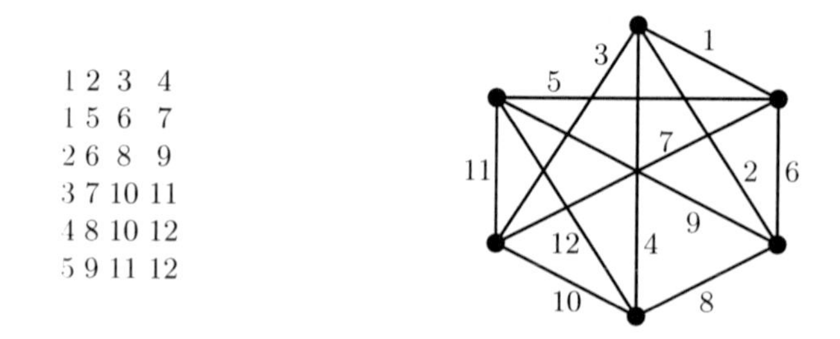

Fig. 1. The graph $G(V, E, L)$ and its antimagic labeling

3 Conclusion

To summarize, we have introduced a new method to study the antimagicness of graphs. Using this method we have proved that some families of k-regular graphs are antimagic. However, in general, Hartsfield and Ringel's conjecture remains open.

References

1. Alon, N., Kaplan, G., Lev, A., Roditty, Y., Yuster, R.: Dense graphs are antimagic. J. Graph Theory 47(4), 297–309 (2004), http://dx.doi.org/10.1002/jgt.20027
2. Bača, M., Miller, M.: Super Edge-Antimagic Graphs: a Wealth of Problems and Some Solutions. BrownWalker Press, Boca Raton (2008)
3. Bloom, G.S., Golomb, S.W.: Applications of numbered undirected graphs. Proc. IEEE 65, 562–570 (1977)
4. Bloom, G.S., Golomb, S.W.: Numbered complete graphs, unusual rulers, and assorted applications. In: Theory and Applications of Graphs (Proc. Internat. Conf., Western Mich. Univ., Kalamazoo, Mich., 1976). Lecture Notes in Math., vol. 642, pp. 53–65. Springer, Berlin (1978)
5. Cheng, Y.: A new class of antimagic Cartesian product graphs. Discrete Math. 308(24), 6441–6448 (2008), http://dx.doi.org/10.1016/j.disc.2007.12.032
6. Cranston, D.W.: Regular bipartite graphs are antimagic. J. Graph Theory 60(3), 173–182 (2009), http://dx.doi.org/10.1002/jgt.20347
7. Dickson, T.J.: On a problem concerning separating systems of a finite set. J. Combinatorial Theory 7, 191–196 (1969)
8. Gallian, J.A.: A dynamic survey of graph labeling. Electron. J. Combin. 16(♯DS6) (2009)
9. Hartsfield, N., Ringel, G.: Pearls in graph theory: a comprehensive introduction. Academic Press Inc., Boston (1990)
10. Phanalasy, O., Roberts, I., Rylands, L.: Covering separating systems and an application to search theory. Australas. J. Combin. 45, 3–14 (2009)
11. Ramsay, C., Roberts, I.T.: Minimal completely separating systems of sets. Australas. J. Combin. 13, 129–150 (1996)
12. Ramsay, C., Roberts, I.T., Ruskey, F.: Completely separating systems of k-sets. Discrete Math. 183(1-3), 265–275 (1998)
13. Roberts, I.T.: Extremal Problems and Designs on Finite Sets. Ph.D. thesis, Curtin University of Technology (1999)
14. Roberts, I., D'Arcy, S., Gilbert, K., Rylands, L., Phanalasy, O.: Separating systems, Sperner systems, search theory. In: Ryan, J., Manyem, P., Sugeng, K., Miller, M. (eds.) Proceedings of the Sixteenth Australasian Workshop on Combinatorial Algorithms, pp. 279–288 (September 2005)
15. Wang, T.M.: Toroidal grids are anti-magic. In: Wang, L. (ed.) COCOON 2005. LNCS, vol. 3595, pp. 671–679. Springer, Heidelberg (2005)
16. Wang, T.M., Hsiao, C.C.: On anti-magic labeling for graph products. Discrete Math. 308(16), 3624–3633 (2008)
17. Zhang, Y., Sun, X.: The antimagicness of the cartesian product of graphs. Theor. Comput. Sci. 410(8-10), 727–735 (2009)

Parameterized Complexity of k-Anonymity: Hardness and Tractability

Paola Bonizzoni[1], Gianluca Della Vedova[2], Riccardo Dondi[3], and Yuri Pirola[1]

[1] DISCo, Univ. Milano-Bicocca
bonizzoni@disco.unimib.it, pirola@disco.unimib.it
[2] Dip. Statistica, Univ. Milano-Bicocca
gianluca.dellavedova@unimib.it
[3] Dipartimento di Scienze dei Linguaggi, della Comunicazione e degli Studi Culturali, Università degli Studi di Bergamo
riccardo.dondi@unibg.it

Abstract. The problem of publishing personal data without giving up privacy is becoming increasingly important. A precise formalization that has been recently proposed is the k-anonymity, where the rows of a table are partitioned in clusters of size at least k and all rows in a cluster become the same tuple after the suppression of some entries. The natural optimization problem, where the goal is to minimize the number of suppressed entries, is hard even when the stored values are over a binary alphabet or the table consists of a bounded number of columns. In this paper we study how the complexity of the problem is influenced by different parameters. First we show that the problem is W[1]-hard when parameterized by the value of the solution (and k). Then we exhibit a fixed-parameter algorithm when the problem is parameterized by the number of columns and the number of different values in any column.

1 Introduction

In epidemic studies the analysis of large amounts of personal data is essential. At the same time the dissemination of the results of those studies, even in a compact and summarized form, can provide some information that can be exploited to identify the row pertaining to a certain individual. For instance, ZIP code, gender and date of birth can uniquely identify 87% of individuals in the U.S. [17]. Therefore when managing personal data it is of the utmost importance to effectively protect individuals' privacy.

One approach to deal with such problem is the k-anonymity model [15,17,14,11]. Each row of a given table represents all data regarding a certain individual. Then different rows are clustered together, and some entries of the rows in each cluster are suppressed (i.e. they are replaced with a $*$) so that each cluster consists of at least k identical rows. Therefore each row r in the resulting table is clustered with at least other $k-1$ rows identical to r, hence the resulting data do not allow to identify any individual. While such formulation is not really sophisticated and has some practical limitations, it is definitely interesting from a theoretical point of view, as witnessed by the rich literature available.

C.S. Iliopoulos and W.F. Smyth (Eds.): IWOCA 2010, LNCS 6460, pp. 242–255, 2011.

We will focus on separating the cases that can be solved efficiently from those that are intractable, therefore hinting at which strategies are likely or not going to be successfully employed when studying more sophisticated formalizations. Notice that different formulations of the problem have also been proposed [1], for example allowing the generalization of entry values, that is an entry value can be replaced with a less specific value [3], or considering a notion of proximity among values [9].

A parsimonious principle leads to the optimization problem where we want to minimize the number of entries in the table to be suppressed. The k-anonymity problem is known to be APX-hard even when the matrix entries are over a *binary* alphabet and $k = 3$ [5], as well as when the matrix has 8 columns and $k = 4$ (this time on arbitrary alphabets) [5]. Recently, a polynomial time algorithm for 2-anonymity has been given in [4].

Furthermore, a polynomial-time $O(k)$-approximation algorithm on arbitrary input alphabet, as well as approximation algorithms for restricted cases are known [2]. Recently, two polynomial-time approximation algorithms with factor $O(\log k)$ have been independently proposed [13,10].

In this paper we investigate the parameterized complexity [7,12] of the problem, unveiling how different parameters are involved in the complexity of the problem. A first systematic study of the parameterized complexity of the k-anonymity problem has been proposed in [6]. Here, we follow the same direction, showing that the problem is W[1]-hard when parameterized by the size of the solution and k, and we provide a fixed-parameter algorithm, when the problem is parameterized by the number of columns and the maximum number of different values in any column. These problems were left open in [6].

In Table 1 we report the status of the parameterized complexity of the k-anonymity problem, where in bold we have emphasized the new results presented in this paper. We recall that a problem P parameterized by a set S of parameters is in the class FPT [7] if it admits an exact algorithm with complexity $f(S)n^{O(1)}$, where f is an arbitrary function, and n is the size of the input problem, while it is W[i]-hard [7], for some $1 \leq i \leq p$ if it is unlikely to be fixed-parameter tractable. We recall that XP [7] is a superclass of all sets W[p]. Moreover, proving that a problem Π with parameter set S is NP-hard when all parameters in S are some constants, implies that $(\Pi, S) \notin$ XP unless P = NP.

Table 1. Summary of the parameterized complexity status of the k-anonymity problem; $|\Sigma|$ represents the maximum number of different values in a column, m represents the number of columns, n represents the number of rows, k represents the minimum size of a cluster, e represents the size of the solution

	–	k	e	k, e
–	NP-hard [11]	$\notin XP$ [5,2]	**W[1]-hard** *new*	**W[1]-hard** *new*
$\vert\Sigma\vert$	$\notin XP$ [5]	$\notin XP$ [5]	???	???
m	$\notin XP$ for $m \geq 8$ [5]	$\notin XP$ for $m \geq 8$, $k \geq 4$ [5]	FPT [6]	FPT [6]
n	FPT [6]	FPT [6]	FPT [6]	FPT [6]
$\vert\Sigma\vert, m$	**FPT** *new*	FPT [6]	FPT [6]	FPT [6]
$\vert\Sigma\vert, n$	FPT [6]	FPT [6]	FPT [6]	FPT [6]

The rest of the paper is organized as follows. In Section 2 we introduce some preliminary definition and we give the formal definition of the k-anonymity problem. In Section 3 we show that the k-anonymity is W[1]-hard when parameterized by the size of the solution and k. In Section 4 we give a fixed parameter algorithm, when the problem is parameterized by the number of columns and the maximum number of different values in any column. Due to space constraint, some proofs in the paper are omitted.

2 Preliminary Definitions

Let us introduce some preliminary definitions that will be used in the rest of the paper. Given a graph $G = (V, E)$, and $V' \subseteq V$, the *subgraph induced* by V' is denoted by $G[V'] = (V', E')$, where $E' = E \cap (V' \times V')$.

Given an alphabet Σ, a row r is a vector of elements taken from the set Σ, and the j-th element of r is denoted by $r[j]$. Notice that it is equivalent to consider a row as a vector over alphabet Σ. Let r_1, r_2 be two equal-length rows. Then $H(r_1, r_2)$ is the Hamming distance of r_1 and r_2, i.e. $|\{i : r_1[i] \neq r_2[i]\}|$. Let R be a set of l rows, then a *clustering* of R is a partition $\Pi = (P_1, \ldots, P_t)$ of R. Given a clustering $\Pi = (P_1, \ldots, P_t)$ of R, we define the *cost* of a row r belonging to the set P_i of Π as $c_\Pi(r) = |\{j : \exists r_1 \in P_i,\ r_1[j] \neq r[j]\}|$, that is the number of entries of r that have to be suppressed so that all rows in P_i are identical. The cost of a set P_i, denoted by $c_\Pi(P_i)$, is defined as $|P_i| c_\Pi(r)$, for some row $r \in P_i$. The cost of Π, denoted by $c(\Pi)$, is defined as $\sum_{P_i \in \Pi} c_\Pi(P_i)$.

We are now able to formally define the k-Anonymity Problem (k-AP).

Problem 1 *k-AP.*
Input: *a set R of equal length rows over an alphabet Σ_R.*
Output: *a clustering $\Pi = (P_1, \ldots, P_t)$ of R such that $|P_i| \geq k$ for each set P_i, and $c(\Pi)$ is minimum.*

Given a set S of parameters, we denote by $\langle S \rangle$-AP the k-AP problem parameterized by S, thus omitting k. The following parameters are considered:

- m is the number of columns of the rows in R;
- n is the number of rows in R;
- $|\Sigma|$ is the maximum number of different values in any column of the table;
- k is the minimum size of a cluster;
- e is the maximum number of entries that can be suppressed in a solution.

Let $\Pi = (P_1, \ldots, P_z)$ be a solution of the k-AP problem. Notice that a suppression at position j of a row r is represented by replacing the symbol $r[j]$ with a $*$. Given a set P_j of Π, some entries of the rows clustered in P_j are eventually suppressed, so that the resulting rows are all identical to a vector r over alphabet $\Sigma_R \cup \{*\}$; such a vector is the *resolution vector* associated with P_j. Given a resolution vector r, we define $del(r)$ as the number of entries suppressed in r, that is $del(r) = |\{j : r[j] = *\}|$. Given a resolution vector r and a row $r_i \in R$, we say that r is *compatible* with row r_i iff $r[j] \neq r_i[j]$ implies $r[j] = *$. Given a row r_i of R and a set of resolution vectors S', we define the set $comp(r_i, S') = \{r \in S' : r \text{ is compatible with } r_i\}$.

Given a set R of rows, a *group* of rows of R is a maximal set of identical rows. Given a group g, the *representative* of g, denoted by $r(g)$, is any row of g, while $s(g)$ is the number of rows in g and $exc(g) = \max\{0, s(g) - k\}$. We say that $r(g)$ is *compatible* with a resolved vector r iff $r(g) \neq r$ implies $r[j] = *$. Furthermore, $comp(r(g), S')$ is the set $\{r \in S' : r \text{ is compatible with } r(g)\}$. A set R of rows can be partitioned in groups of identical rows in polynomial time [6], therefore we can compute in polynomial time whether R can be partitioned into groups of size at least k. If this is not possible, then at least k entries of R must be suppressed to get a solution of the k-AP problem, that is $e \geq k$.

Hence the following property holds.

Proposition 1. *[6]*
The $\langle e\rangle$-AP is in FPT if and only if $\langle e, k\rangle$-AP is in FPT.

Consequently our parameterized reduction [7, 12] will show the fixed-parameter intractability of $\langle e\rangle$-AP and $\langle e, k\rangle$-AP.

3 $\langle e\rangle$-AP and $\langle e, k\rangle$-AP Are W[1]-Hard

We show that $\langle e\rangle$-AP and $\langle e, k\rangle$-AP are W[1]-hard. Given an set R of equal length rows, $\langle e\rangle$-AP and $\langle e, k\rangle$-AP ask if there exists a clustering $\Pi = (P_1, \ldots, P_t)$ of R such that $|P_i| \geq k$ for each set P_i, and $c(\Pi) \leq e$. We present a parameter preserving reduction from the h-Clique problem, which is known to be W[1]-hard [8], to the $\langle e\rangle$-AP problem. Given a graph $G = (V, E)$, an h-clique is a set $V' \subseteq V$ where each pair of vertices in V' are connected by an edge of G, and $|V'| = h$. The h-Clique problem asks for a subset V' of the vertices of a given graph G inducing an h-clique in G.

Clearly the vertices of a h-clique are connected by $\binom{h}{2}$ edges. Given a graph $G = (V, E)$, we use m_G and n_G to denote respectively the number of edges and of vertices of G. We construct the instance R of $\langle e\rangle$-AP associated with G. First, let us define $k = 2h^2$. The set R consists of $(k+1)m_G + (k - \binom{h}{2})$ rows and $2h + n_G$ columns over alphabet $\Sigma_R = \{0, 1\} \cup \{\sigma_{i,j} : (v_i, v_j) \in E\}$. More precisely, for each edge $e(i, j) = (v_i, v_j)$ in E, there is a group $R(i, j)$ of $k+1$ identical rows $r_x(i, j)$, $1 \leq x \leq k+1$, where

- $r_x(i, j)[l] = \sigma_{i,j}$, for $1 \leq l \leq 2h$;
- $r_x(i, j)[2h + i] = 1$, $r_x(i, j)[2h + j] = 1$;
- $r_x(i, j)[2h + l] = 0$, for $l \neq i, j$ and $1 \leq l \leq n$.

Moreover, R also contains a group R_0 made of $k - \binom{h}{2}$ identical rows equal to 0^{2h+n_G}.

Lemma 2. *Let R be the instance of $\langle e\rangle$-AP associated with G and consider two rows $r, r_x(i, j)$ of R, such that $r \in R_0$ and $r_x(i, j) \in R(i, j)$. Then, $r[t] \neq r_x(i, j)[t]$, for each $1 \leq t \leq 2h$.*

Proof. By construction, $r[t] = 0$ for all t with $1 \leq t \leq 2h$, while $r_x(i,j)[t] = \sigma_{i,j}$. □

Lemma 3. *Let $G = (V, E)$ be a graph, let V' be a h-clique of G and let R be the instance of $\langle e \rangle$-AP associated with G. Then we can compute in polynomial time a solution Π of $\langle e \rangle$-AP over instance R with cost at most $6h^3$.*

Lemma 4. *Let $G = (V, E)$ be an instance of h-Clique, let R be the instance of $\langle e \rangle$-AP associated with G and let Π be a solution of $\langle e \rangle$-AP over instance R with cost at most $6h^3$. Then we can compute in polynomial time a h-clique V' of G.*

Proof. First we will prove that Π must have a set $R'_0 \supset R_0$. Assume to the contrary that in Π there are two sets A, B containing at least a row of R_0. Notice that $|R_0| < k$ while $|A|, |B| \geq k$. Moreover, by Lemma 2, all rows in A or B must have suppressed the first $2h$ entries, which results in at least $4hk > 6h^3$ suppressions, contradicting the assumption on the cost of the solution. Hence, R_0 is properly contained in a set R'_0 of Π, as $|R_0| < k$. Moreover, let r' be a row of $R'_0 \setminus R_0$ and let r be a row of $\in R_0$. By Lemma 2 $r'[t] \neq r[t]$ for each column t, $1 \leq t \leq 2h$, therefore all entries in the first $2h$ columns of each row in R'_0 must be suppressed.

Now, let us prove that, for each set $R(i,j)$ of R, there exists a set $R'(i,j)$ of Π such that $R'(i,j) \subseteq R(i,j)$. Assume to the contrary that no such set $R'(i,j)$ exists, for a given $R(i,j)$. Then either $R(i,j) \subseteq R'_0$ or there exists a row of $R(i,j)$ clustered together with a row of $R(x,y)$ in Π, with $(x,y) \neq (i,j)$. In the first case, that is $R(i,j) \subseteq R'_0$, $|R'_0| \geq 2k + 1 - \binom{h}{2}$, by construction all entries of the first $2h$ columns of the rows in R'_0 must be suppressed, resulting in at least $2h(4h^2 - \binom{h}{2}) > 6h^3$ suppressions and thus contradicting the assumption on the cost of the solution. Consider now the second case, that is there is a set A in Π containing at least a row of two different sets $R(i,j)$ and $R(x,y)$ of R. Observe that given $r' \in R'_0 \setminus R_0$ and $r \in R_0$, r and r' differ in the first $2h$ columns. Thus the entries of the first $2h$ columns of the rows of R'_0 must be suppressed, resulting in at least $4hk > 6h^3$ suppressed entries and thus contradicting the assumption on the cost of the solution. Hence, for each set $R(i,j)$ of R, there exists a set $R'(i,j)$ of Π such that $R'(i,j) \subseteq R(i,j)$.

By our previous arguments we can assume that Π consists of the clusters R'_0 and $R'(i,j)$, for each $R(i,j) \in R$, and that $|R(i,j)| - 1 \leq |R'(i,j)| \leq |R(i,j)|$. Notice that only R'_0 can contain some suppressed entries. Also $|R'_0| = k$, for otherwise we can improve the cost of Π by moving a row in $R(i,j) \cap R'_0$ from R'_0 to $R'(i,j)$. Now let E' be the set of edges (v_i, v_j) of G such that a row of $R(i,j)$ is in R'_0 and let V' be the set of vertices incident on at least an edge in E'. Then we can show that $G[V']$ is a h-clique. Notice that the entries in the first $2h$ columns of R'_0 must be suppressed, as well as all columns with index $2h + l$ such that $v_l \in V'$, since in those columns all rows in R_0 have value 0 while some row in $R'_0 \setminus R_0$ have value 1. An immediate consequence is that the overall number of suppressed entries is at least $2hk + k|V'|$. Since, by hypothesis, the number of suppressed entries is at most $6h^3 = 3kh$, then $|V'| \leq h$. Notice that, since $|R_0| = k - \binom{h}{2}$ and $|R'_0| = k$, then $R'_0 \setminus R_0$ contains exactly $\binom{h}{2}$ distinct rows corresponding to edges in E' incident on V' vertices. Hence V' induces a h-clique in G. □

From Lemma 3 and 4, our reduction is parameter preserving, therefore $\langle e \rangle$-AP is W[1]-hard.

Theorem 5. *$\langle e \rangle$-AP and $\langle e, k \rangle$-AP are W[1]-hard.*

Corollary 6 is a consequence of Theorem 5 and Proposition 1.

Corollary 6. *$\langle e, k \rangle$-AP is W[1]-hard.*

4 An FPT Algorithm for $\langle |\Sigma|, m \rangle$-AP

In this section we present a fixed-parameter algorithm for the $\langle |\Sigma|, m \rangle$-AP problem, that is the instance of the AP problem, where the number m of columns and the maximum number $|\Sigma|$ of different values in any column are two parameters. Notice that k-AP parameterized by exactly one of $|\Sigma|$ or m is not in FPT, as k-AP is APX-hard (hence NP-hard) even when one of $|\Sigma|$ or m is a constant [5].

Before giving the details of the algorithm, let us first introduce some preliminary definitions. Let R be an instance of $\langle |\Sigma|, m \rangle$-AP, and for each column of R with index j, $1 \leq j \leq m$, let Σ_j be the set of different values that the rows of R have in column j. Notice that $|\Sigma_j| \leq |\Sigma|$, for each $1 \leq j \leq m$. Let $\Sigma^*_j = \Sigma_j \cup \{*\}$ and $\Sigma^* = \Sigma \cup \{*\}$. Assume $\Pi = \{P_1, \cdots, P_z\}$ is a feasible solution of $\langle |\Sigma|, m \rangle$-AP over instance R. The set S' consisting of a resolution vector for each set $P_i \in \Pi$ is called *candidate set* for solution $\langle |\Sigma|, m \rangle$-AP. Let S be the set of possible rows of length m and having value over alphabet Σ^*_j for the position j, $1 \leq j \leq m$, then $|S|$ is bounded by $|\Sigma^*|^m$. Given a candidate set S', notice that $S' \subseteq S$ and that each row $r \in R$ must be compatible with at least one resolution vector in S'.

Given a row r and the set S' of resolution vectors, recall that we denote by $Comp(r, S')$ the set of resolution vectors of S' compatible with r. Moreover, given a resolution vector $r' \in S'$, we denote by $del(r')$ the number of suppressions in r'. For each row $r \in R$ we define its weight as $w(r) = \max_{r_x \in Comp(r,S')}\{m - del(r_x)\}$. Notice that $w(r) = m$ whenever r is compatible with a row without suppressions. Informally, the weight of a row is equal to the maximum number of its entries that might be preserved in a solution where S' is the set of resolution vectors. Finally, we define $W = \sum_{r \in R} w(r)$ and $w'(r_x) = W + m - del(r_x) + 1$ for each row $r_x \in S'$. Notice that $w'(r_x) \geq \sum_{r \in R} w(r)$, for each $r_x \in R$. The weights defined above will be used later in Section 4.1 to define the weight function w_h.

Let us first describe the general idea of the algorithm. Given a candidate set S', the algorithm computes an optimal solution $\Pi_{S'}$ associated with a candidate set $S' \subseteq S$ (see Algorithm 1). The algorithm consists of two main phases. In the first phase (Section 4.1), given the set R of input rows and the candidate set S', the algorithm builds a weighted bipartite graph $G_{S',R}$ associated with R and S'.

Algorithm 1. Solving $\langle|\Sigma|, m\rangle$-AP

Input: An instance R of $\langle|\Sigma|, m\rangle$-AP made of a set of n rows, each one consisting of m symbols, and an integer e

Output: a solution of $\langle|\Sigma|, m\rangle$-AP over instance R, if $\langle|\Sigma|, m\rangle$-AP admits a solution that suppresses at most e entries;

1 $S \leftarrow$ the set of resolved vectors of length m, where each j-th symbol, $1 \leq j \leq m$, is taken from the alphabet Σ_j^*;
2 $W = \sum_{r \in R} w(r)$;
3 **foreach** *subset* S' *of* S **do**
4 $\quad G_{R,S'} \leftarrow$ the graph associated with R, S';
5 $\quad M \leftarrow$ a maximum matching of $G_{R,S'}$; $w \leftarrow$ the weight of M;
6 $\quad$ **if** M *is feasible and* $w \geq (W+1)k|S'| + m|R^l_{dist} \cup R^l_{safe}| - e$ **then**
7 $\quad\quad$ **return** *the solution* $\Pi_{S'}(M)$ *of* R *associated with* M;
8 **return** *No such solution exists*

In the second phase (Section 4.2) a solution of $\langle|\Sigma|, m\rangle$-AP is computed starting from a maximum weighted matching of the graph $G_{S',R}$. Section 4.3 is devoted to prove that the solution computed by the algorithm is optimal.

4.1 Building the Graph $G_{R,S'}$

Let us consider a candidate set S' of vectors for an optimal solution of $\langle|\Sigma|, m\rangle$-AP. Since $S' \subseteq S$, there exist at most $2^{|\Sigma^*|^m}$ possible candidate sets of rows S', therefore our FPT algorithm computes each candidate set S' and verifies if there exists a solution $\Pi_{S'}$ with cost at most e. In order to verify if such a solution exists, the algorithm builds a bipartite graph $G_{R,S'}$, as described in this section. The intuitive idea behind the graph is that edges of the graph correspond to possible ways of assigning each row in R to a resolution vector $x \in S'$. Rows assigned to the same resolution vector $x \in S'$ are clustered in the solution $\Pi_{S'}$.

The construction of the vertex set of the graph is based on a partition of R into two disjoint sets called R_{safe} and R_{dist} (that is $R_{dist} = R \setminus R_{safe}$). The set R_{safe} consists of those rows $r \in R$ belonging to the group g such that:

- $s(g) \geq k$, that is r belongs to a group of at least k identical rows, and
- there exists a row $r_j \in S'$, such that r_j and $r(g)$ are the same vector.

Notice that only rows in R_{safe} might have no suppressed entry in a solution $\Pi_{S'}$.

The vertex set of $G_{R,S'} = (V, E)$ has 6 sets. Two sets (R^l_{dist}, R^r_{dist}) consist of vertices associated with the rows in R_{dist}, three sets (R'^l_{safe}, R^l_{safe}, R^r_{safe}) consist of vertices associated with the rows in R_{safe}, and a final set called T consists of vertices associated with the rows in S'. In the latter case notice that for each row x in S' there exist k vertices in T to ensure that the cluster associated with x has size at least k. The vertex set is defined as follows:

- for each row $x \in R_{dist}$, there is a corresponding vertex $R^l_{dist}(x)$ in R^l_{dist} and a corresponding vertex $R^r_{dist}(x)$ in R^r_{dist};

- for each group g consisting of the set of rows $\{x_1, x_2, \ldots, x_{s(g)}\}$, where each $x_i \in R_{safe}$, $1 \leq i \leq s(g)$, there are k corresponding vertices in R'^{l}_{safe}, (such vertices are denoted by $R'^{l}_{safe}(g,1), \ldots, R'^{l}_{safe}(g,k)$), $exc(g)$ corresponding vertices in R^{l}_{safe} (such vertices are denoted by $R^{l}_{safe}(g,1)$, ..., $R^{l}_{safe}(g, exc(g))$, and $exc(g)$ corresponding vertices in R^{r}_{safe} (such vertices are denoted by $R^{r}_{safe}(g,1)$, ..., $R^{r}_{safe}(g, exc(g))$);
- for each row $x \in S'$, there are k corresponding vertices in T (such vertices are denoted by $T(x,1), \ldots, T(x,k)$).

Notice that our graph $G_{R,S'}$ is edge-weighted. Let w_h be the weight function assigning a positive weight to each edge of $G_{R,S'}$. Given the set of edges $E' \subseteq E$, we denote by $w_h(E') = \sum_{e \in E'} w_h(e)$.

First, notice that the set S' consists of two disjoint sets: the set S'_{safe}consists of those rows in S' that have no suppressions, while $S'_{cost} = S' \setminus S'_{safe}$. Each edge connects a vertex of $R'^{l}_{safe} \cup R^{l}_{safe} \cup R^{l}_{dist}$ with a vertex of $R^{r}_{safe} \cup R^{r}_{dist} \cup T$, hence the graph $G_{R,S'}$ is bipartite. The set S' consists of two disjoint sets: the set S'_{safe}consists of those rows in S' that have no suppressions, while $S'_{cost} = S' \setminus S'_{safe}$. Intuitively, each edge represents a possible assignment of a row in R to a resolution vector in S'.

Algorithm 2. From a matching to a feasible solution of $\langle|\Sigma|, m\rangle$-AP.

Input: A graph $G_{R,S'}$ associated with an instance R and a maximum weight matching M of $G_{R,S'}$

Output: A solution $\Pi_{S'}(M)$ of $\langle|\Sigma|, m\rangle$-AP over instance R

1 **foreach** *edge* y *of* M **do**
2   **if** $y = (R^{l}_{dist}(r), T(x,j))$ **then** `/* edges defined at point 1 */`
3   row r is assigned to a set whose resolution row is x, $x \in S'$
4   **if** $y = (R^{l}_{dist}(r), R^{r}_{dist}(r))$ **then** `/* edges defined at point 2 */`
5   row r is assigned to a set whose resolution row is $r_y = \arg\ \max w(r)$, $r_y \in S'$;
6   **if** $y = (R'^{l}_{safe}(g,i), T(x,j))$ **then** `/* edges defined at point 3 */`
7   assign the i-th row of g to a set whose resolution row is x, $x \in S'$;
8   **if** $y = (R^{l}_{safe}(g,i), T(x,j))$ **then** `/* edges defined at point 4 */`
9   assign the i-th exceeding row of g to a set whose resolution row is x, $x \in S'$;
10   **if** $y = (R^{l}_{safe}(g,i), R^{r}_{safe}(g,i))$ **then** `/* edges defined at point 5 */`
11   assign the i-th exceeding row of group g to the set whose resolution row is $r(g)$, with $r(g) \in S'$ and $r \in R_{safe}$;

Now we are ready to define formally the set of edges E of $G_{R,S'}$ and the weight function w_h. There are five possible kinds of edges.

1. Let r be a row of R_{dist}, and let x be a row in $Comp(r, S') \cap S'_{cost}$. Then there is an edge $y = (R^{l}_{dist}(r), T(x,j))$, for each $1 \leq j \leq k$, with weight $w_h(y) = w'(x)$.
2. Let r be a row in R_{dist}. Then there is an edge $y = (R^{l}_{dist}(r), R^{r}_{dist}(r))$ with weight $w_h(y) = w(r)$.

3. Let g be a group consisting of rows $\{r_1, \ldots, r_{s(g)}\}$, where r_i, for each i with $1 \leq i \leq s(g)$, is a row of R_{safe}; let r' be the resolution vector of S'_{safe} identical to $r(g)$. Then there is an edge $y_i = (R'^{l}_{safe}(g, i), T(r', i))$, for each i with $1 \leq i \leq k$. All edges y_i have weight $w_h(y_i) = w'(r')$.
4. Let g be a group consisting of rows $\{r_1, \ldots, r_{s(g)}\}$, where r_i, for each i with $1 \leq i \leq s(g)$, is a row of R_{safe}; let x be a row in $Comp(r(g), S') \cap S'_{cost}$. Then there is an edge $y_{i,j} = (R^l_{safe}(g, i), T(x, j))$, for each i with $1 \leq i \leq exc(g)$ and for each j with $1 \leq j \leq k$. All edges $y_{i,j}$ have weight $w_h(y_{i,j}) = w'(x)$.
5. Let g be a group consisting of rows $\{r_1, \ldots, r_{s(g)}\}$, where r_i, $1 \leq i \leq s(g)$, is a row of R_{safe}. Then there is an edge $y_i = (R^l_{safe}(g, i), R^r_{safe}(g, i))$ for each i with $1 \leq i \leq exc(g)$. All edges y_i have weight $w_h(y_i) = w(r(g))$.

4.2 Computing a Solution of $\langle |\Sigma|, m \rangle$-AP

In this section we prove in Lemma 9 that $\Pi_{S'}(M)$ is a clustering of the rows in R that is a feasible solution for the $\langle |\Sigma|, m \rangle$-AP problem. See Fig. 1 for an example.

Since $G_{R,S'}$ is bipartite, we can efficiently compute a maximum weight matching M of $G_{R,S'}$ [16]. Given a matching M of the graph $G_{R,S'}$, Algorithm 2 computes in polynomial time a clustering $\Pi_{S'}(M)$ of the rows in R. Informally, the clustering is computed by assigning the rows in R to the resolution vector in S', using the edges in the matching M.

Notice that, each vertex $R^l_{safe}(r, i)$ has only the edge $(R^l_{safe}(r, i), T(r, i))$ on it, hence we can always add those edges to any matching[1]. Let M be a matching of $G_{R,S'}$ and let v be a vertex of $G_{R,S'}$, then we say that v is *covered* by a matching M if there exists an edge of M for which v is one of its endpoints. Moreover, we will say that M is *feasible* if all vertices in T are covered by M. When a matching M covers all vertices in $R^l_{dist} \cup R^l_{safe}$ and is feasible, it is defined as a *complete matching*. Let Π be a clustering of an instance R of the $\langle |\Sigma|, m \rangle$-AP problem. Then Π is *feasible* if and only if each set of the partition Π contains at least k rows. The next part of this section is devoted to show that every maximum weight matching M is complete and that clustering $\Pi_{S'}(M)$ is *feasible*. First, we will show in the next two lemmata that, given $W' = k \sum_{r_x \in T} w'(r_x)$, W' is a threshold that distinguishes between matchings that are feasible and those that are not.

Lemma 7. *Let M be a matching of $G_{R,S'}$, let X be the subset of T consisting of the vertices of T that are covered by M, and let M_1 be the subset of the edges of M that have one endpoint in X. Then the total weight of the edges in M_1 is exactly $\sum_{T(t,i) \in X} w'(t)$.*

Proof. It is an immediate consequence of the observation that all edges where an endpoint is $T(t, j)$ have the same weight $w'(t)$, with $t \in S'$. □

[1] Notice that these connected components are introduced only to simplify the relationship between a matching M and the corresponding solution $\Pi_{S'}(M)$ of $\langle |\Sigma|, m \rangle$-AP.

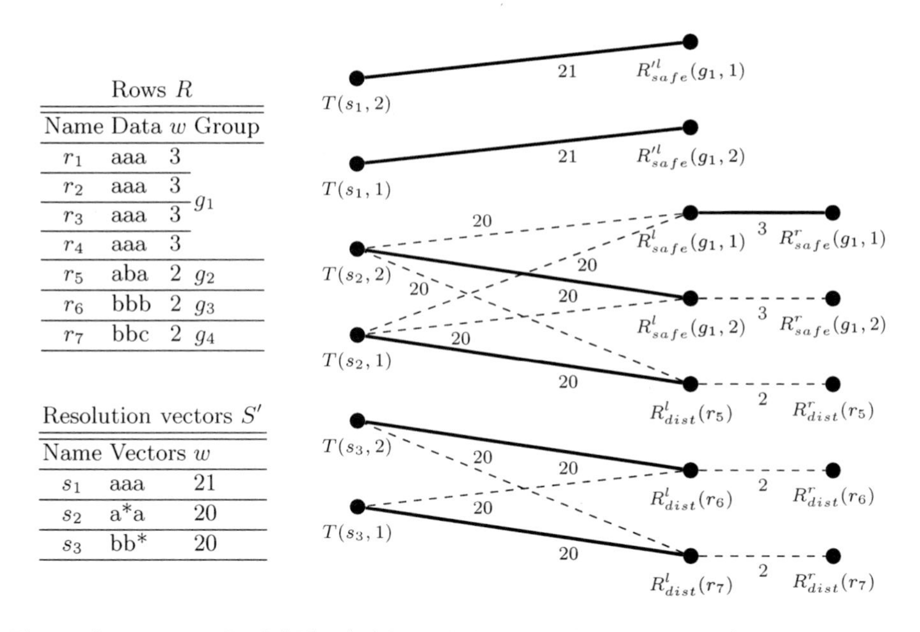

Rows R

Name	Data	w	Group
r_1	aaa	3	g_1
r_2	aaa	3	g_1
r_3	aaa	3	g_1
r_4	aaa	3	g_1
r_5	aba	2	g_2
r_6	bbb	2	g_3
r_7	bbc	2	g_4

Resolution vectors S'

Name	Vectors	w
s_1	aaa	21
s_2	a*a	20
s_3	bb*	20

Fig. 1. An instance R of $\langle|\Sigma|, m\rangle$-AP, with $k = 2$ and $m = 3$, a resolution vector set S' and the associated graph $G_{R,S'}$. The thick edges are a maximum weight matching of $G_{R,S'}$. The corresponding solution is made of the sets $\{r_1, r_2, r_3\}$ (cost 0), $\{r_4, r_5\}$ (cost 2), $\{r_6, r_7\}$ (cost 2).

Lemma 8. *Let M be a matching of $G_{R,S'}$ and let M_1 be the subset of the edges of M that have one endpoint in T. Then the total weight of the edges in M_1 is at least $W' = k \sum_{r \in S'} w'(r)$ if and only if M is feasible.*

Proof. Let M_1 be the subset of the edges of M that have one endpoint in T, and let W_1 be the total weight of edges in M_1. An immediate consequence of Lemma 7 is that $W_1 = W'$ if and only if M_1 is feasible. Assume now that M is not feasible, then there exists at least one vertex $S'(x, j) \in T$ that is not covered by M. Again, a consequence of Lemma 7 is that $W_1 \leq W' - w'(x)$. Let M_2 be the set $M \setminus M_1$. By construction, $w'(x) > W$ and W is an upper bound on the total weight of M_2, therefore $W_1 + w_h(M_2) < W'$, completing the proof. □

Using Lemmata 7 and 8, we can prove Lemma 9.

Lemma 9. *Let M be a maximum weight matching of $G_{R,S'}$, then M is complete and the solution $\Pi_{S'}(M)$ computed by Algorithm 2 is feasible.*

4.3 Proving the Optimality of $\Pi_{S'}(M)$

This section is devoted to prove that, starting from a maximum weight matching M, Algorithm 2 computes an optimal solution $\Pi_{S'}(M)$ of $\langle|\Sigma|, m\rangle$-AP. In order to prove that any maximum weight matching M of the graph $G_{R,S'}$ leads to an optimal solution of $\langle|\Sigma|, m\rangle$-AP over instance R, we are going to prove that $\sum_{(u,v)\in M} w_h((u,v)) \geq (W+1)k|S'| + m|R^l_{dist} \cup R^l_{safe} \cup R'^l_{safe}| - e$ if and only if $\langle|\Sigma|, m\rangle$-AP over instance R admits a solution with cost not greater than e, and such solution is computed by applying Algorithm 2. Such result will be obtained through a sequence of technical lemmata.

Since M is a maximum weighted matching, we can assume by Lemma 9 that M is complete. Given a complete matching M, we denote by $M(T)$ the set of edges of M with one endpoint in $R^l_{dist} \cup R^l_{safe} \cup R'^l_{safe}$ and one endpoint in T, while we denote by $M(L)$ the set of those edges of M that have one endpoint in $R^l_{dist} \cup R^l_{safe}$ and one endpoint in $R^r_{dist} \cup R^r_{safe}$. Furthermore, let us denote by $V(T)$ the set of vertices of $R^l_{dist} \cup R^l_{safe} \cup R'^l_{safe}$ that are endpoints of an edge in $M(T)$ and by $V(L)$ the set of vertices of $R^l_{dist} \cup R^l_{safe}$ that are endpoints of an edge in $M(L)$. Notice that by definition of $V(L)$ and, by definition of complete matching, $V(T) \cup V(L) = R^l_{dist} \cup R^l_{safe} \cup R'^l_{safe}$. Finally, let us denote by $R(L)$ the set of rows in R associated with the vertices in $V(L)$. Lemma 10 shows how the weight of a complete matching M is related to the edge weights of $G_{R,S'}$.

Lemma 10. *Let M be a complete matching of $G_{R,S'}$, and let $w_h(M)$ be the total weight of M. Then*

$$w_h(M) = k\sum_{r\in S'}(W + m - del(r) + 1) + \sum_{r\in R(L)}(m - del(r)) =$$

$$= (W+1)k|S'| + m|R^l_{dist} \cup R^l_{safe} \cup R'^l_{safe}| - (k\sum_{r\in S'} del(r) + \sum_{r\in R(L)} del(r)).$$

Proof. The total weight $w_h(M)$ of the matching M is defined as

$$w_h(M) = \sum_{(u,v)\in M(T)} w_h((u,v)) + \sum_{(u,v)\in M(L)} w_h(u,v).$$

By Lemma 8 and by definition of the weight function w_h,

$$w_h(M) = k\sum_{r\in S'} w'(r) + \sum_{r\in R(L)} (m - del(r))$$

and by definition of $w'(r)$ it holds

$$w_h(M) = k\sum_{r\in S'}(W + m - del(r) + 1) + \sum_{r\in R(L)}(m - del(r)).$$

Hence

$$w_h(M) = (W + m + 1)k|S'| - k\sum_{r\in S'} del(r) + \sum_{r\in R(L)} m - \sum_{r\in R(L)} del(r).$$

By definition of feasible matching and by construction of graph $G_{R,S'}$, $|V(T)| = |T|$. Furthermore, since $|T| = k|S'|$, then $mk|S'| = m|T| = m|V(T)|$. By construction $\sum_{r\in R(L)} m = m|V(L)|$ and $V(T) \cup V(L) = R^l_{dist} \cup R^l_{safe}$. Hence

$$w_h(M) = (W+1)k|S'| + m|R^l_{dist} \cup R^l_{safe}| - (k\sum_{r\in S'} del(r) + \sum_{r\in R(L)} del(r)). \quad \square$$

In the next two lemmata, we will show that: (i) given an instance R of $\langle|\Sigma|, m\rangle$-AP, if there exists a solution of $\langle|\Sigma|, m\rangle$-AP over R that suppresses at most e entries then the graph $G_{R,S'}$ associated with R admits a complete matching of $G_{R,S'}$ with total weight $w_G(M) \geq (W+1)k|S'| + m|R^l_{dist} \cup R^l_{safe} \cup R'^l_{safe}| - e$; (ii) given a complete matching of the graph $G_{R,S'}$ of total weight $w_G(M) \geq (W+1)k|S'| + m|R^l_{dist} \cup R^l_{safe} \cup R'^l_{safe}| - e$, Algorithm 2 returns a solution $\Pi_{S'}(M)$ of $\langle|\Sigma|, m\rangle$-AP that suppresses at most e entries. These lemmata, coupled with Lemma 9, prove the correctness of Algorithm 2 in Theorem 13.

Lemma 11. *Let R be an instance of $\langle|\Sigma|, m\rangle$-AP, let $\Pi_{S'}$ be a feasible solution of $\langle|\Sigma|, m\rangle$-AP over instance R that suppresses at most e entries, let $G_{R,S'}$ be the graph associated with R and S'. Then there exists a complete matching of $G_{R,S'}$ with total weight $w_G(M) \geq (W+1)k|S'| + m|R^l_{dist} \cup R^l_{safe} \cup R'^l_{safe}| - e$.*

Lemma 12. *Let R be an instance of $\langle|\Sigma|, m\rangle$-AP, let $G_{R,S'}$ be the graph associated with R, and let M be a complete matching of $G_{R,S'}$ of weight $w_h(M) \geq (W+1)k|S'| + m|R^l_{dist} \cup R^l_{safe} \cup R'^l_{safe}| - e$. Then, starting from the matching M of $G_{R,S'}$, Algorithm 2 computes a feasible solution $\Pi_{S'}(M)$ of $\langle|\Sigma|, m\rangle$-AP over instance R, where there are at most e suppressions.*

Proof. Since M is complete, for each vertex $T(x, j)$ of T, with $1 \leq j \leq k$, there exists an edge $(v, T(x, j)) \in M$ for some $v \in (R^l_{dist} \cup R^l_{safe} \cup R'^l_{safe})$. Then Algorithm 2 defines a solution $\Pi_{S'}(M)$ for $\langle|\Sigma|, m\rangle$-AP assigning, for each edge $(v, T(x, j))$, the row r corresponding to vertex v to the set that has resolution vector x. More precisely, row r is defined by Algorithm 2 as the j-th element of the set that has resolution vector x. Therefore each set associated with a resolution row $x \in S'$ will consist of at least k rows compatible with x. Hence $\Pi_{S'}(M)$ is a feasible solution.

Recall that M has a total weight of at least $(W+1)k|S'| + m|R^l_{dist} \cup R^l_{safe} \cup R'^l_{safe}| - e$. We will prove that $\Pi_{S'}(M)$ induces at most e suppressions. By Lemma 10, $w_h(M) = k\sum_{r \in S'}(W + m - del(r) + 1) + \sum_{r \in R(L)} m - del(r) = (W+1)k|S'| + m|R^l_{dist} \cup R^l_{safe} \cup R^l_{safe}| - (k\sum_{r \in S'} del(r) + \sum_{r \in R(L)} del(r)) \geq (W+1)k|S'| + m|R^l_{dist} \cup R^l_{safe} \cup R^l_{safe}| - e$ where $k\sum_{r \in S'} del(r) + \sum_{r \in R(L)} del(r) \leq e$. Notice that, by definition of $\Pi_{S'}(M)$, each vertex of $V(T)$ corresponds to a row in R assigned to a set with a resolution vector in S'. Such rows associated with $V(T)$ induce a cost in $\Pi_{S'}(M)$ of $k\sum_{r \in S'} del(r)$. Furthermore, the vertices of $V(L)$ corresponds to rows of R inducing a cost of at most $\sum_{r \in R(L)} del(r)$. Therefore $\Pi_{S'}(M)$ induces $k\sum_{r \in S'} del(r) + \sum_{r \in R(L)} del(r) \leq e$ suppressions. □

We can conclude that Algorithm 1 is indeed correct.

Theorem 13. *Let R be an instance of $\langle|\Sigma|, m\rangle$-AP. Then Algorithm 1 returns a solution $\Pi_{S'}(M)$ of cost at most e if and only if such a solution exists.*

If $\langle|\Sigma|, m\rangle$-AP admits a solution that suppresses at most e entries, then there exists a set S^* of resolution vectors such that Π_{S^*} is a solution for $\langle|\Sigma|, m\rangle$-AP with resolution vectors S^* with the property that Π_{S^*} suppresses at most

e entries. Now, there exist $O(2^{(|\Sigma|+1)^m})$ possible sets of resolution vectors and the construction of graph $G_{R,S'}$ requires $O(k|S^*||R|) \leq O(ke|R|) \leq O(kmn^2)$. A maximum matching M of a bipartite graph can be computed in polynomial time [16] and starting from M, we can compute a solution of the $\langle|\Sigma|, m\rangle$-AP in time $O(|M|) \leq O(m)$. Hence the overall time complexity of the algorithm is $O(2^{(|\Sigma|+1)^m} kmn^2)$.

5 Conclusions

We have studied the tractability of the k-anonymity problem depending on different parameters. We have shown that the problem is W[1]-hard when parameterized by the size of the solution e and k, while it admits a fixed parameter algorithm when parameterized by the number of columns and the maximum number of different values in any column.

Some problems remain open: the computational complexity of k-anonymity when the input matrix consists of two columns, and the parameterized complexity of $\langle e, |\Sigma|\rangle$-anonymity.

Acknowledgments

PB, GDV and YP have been partially supported by FAR 2008 grant "Computational models for phylogenetic analysis of gene variations". PB has been partially supported by the MIUR PRIN grant "Mathematical aspects and emerging applications of automata and formal languages". RD has been partially supported by FAR 2009 grant "Algoritmi per il trattamento di sequenze".

References

1. Aggarwal, G., Feder, T., Kenthapadi, K., Khuller, S., Panigrahy, R., Thomas, D., Zhu, A.: Achieving anonymity via clustering. In: Vansummeren, S. (ed.) PODS, pp. 153–162. ACM, New York (2006)
2. Aggarwal, G., Feder, T., Kenthapadi, K., Motwani, R., Panigrahy, R., Thomas, D., Zhu, A.: Anonymizing tables. In: Eiter, T., Libkin, L. (eds.) ICDT 2005. LNCS, vol. 3363, pp. 246–258. Springer, Heidelberg (2005)
3. Aggarwal, G., Kenthapadi, K., Motwani, R., Panigrahy, R., Thomas, D., Zhu, A.: Approximation algorithms for k-anonymity. J. Privacy Technology (2005)
4. Blocki, J., Williams, R.: Resolving the complexity of some data privacy problems. In: Abramsky, S., Gavoille, C., Kirchner, C., auf der Heide, F.M., Spirakis, P.G. (eds.) ICALP 2010. LNCS, vol. 6199, pp. 393–404. Springer, Heidelberg (2010)
5. Bonizzoni, P., Della Vedova, G., Dondi, R.: The k-anonymity problem is hard. In: Kutylowski, M., Gebala, M., Charatonik, W. (eds.) FCT 2009. LNCS, vol. 5699, pp. 26–37. Springer, Heidelberg (2009)
6. Chaytor, R., Evans, P.A., Wareham, T.: Fixed-parameter tractability of anonymizing data by suppressing entries. J. Comb. Optim. 18(4), 362–375 (2009)
7. Downey, R., Fellows, M.: Parameterized Complexity. Springer, Heidelberg (1999)
8. Downey, R.G., Fellows, M.R.: Fixed-parameter tractability and completeness ii: On completeness for $W[1]$. Theoretical Computer Science 141, 109–131 (1995)

9. Du, W., Eppstein, D., Goodrich, M.T., Lueker, G.S.: On the approximability of geometric and geographic generalization and the min-max bin covering problem. In: Dehne, F.K.H.A., Gavrilova, M.L., Sack, J.R., Tóth, C.D. (eds.) WADS 2009. LNCS, vol. 5664, pp. 242–253. Springer, Heidelberg (2009)
10. Gionis, A., Tassa, T.: k-anonymization with minimal loss of information. TKDD 21(2), 206–219 (2009)
11. Meyerson, A., Williams, R.: On the complexity of optimal k-anonymity. In: Deutsch, A. (ed.) PODS, pp. 223–228. ACM, New York (2004)
12. Niedermeier, R.: Invitation to Fixed-Parameter Algorithms. Oxford University Press, Oxford (2006)
13. Park, H., Shim, K.: Approximate algorithms for k-anonymity. In: Chan, C.Y., Ooi, B.C., Zhou, A. (eds.) SIGMOD Conference, pp. 67–78. ACM, New York (2007)
14. Samarati, P.: Protecting respondents' identities in microdata release. IEEE Trans. Knowl. Data Eng. 13, 1010–1027 (2001)
15. Samarati, P., Sweeney, L.: Generalizing data to provide anonymity when disclosing information (abstract). In: PODS, p. 188. ACM, New York (1998)
16. Schwartz, J., Steger, A., Weißl, A.: Fast algorithms for weighted bipartite matching. In: Nikoletseas, S.E. (ed.) WEA 2005. LNCS, vol. 3503, pp. 476–487. Springer, Heidelberg (2005)
17. Sweeney, L.: k-anonymity: a model for protecting privacy. International Journal on Uncertainty, Fuzziness and Knowledge-based Systems 10, 557–570 (2002)

On Fast Enumeration of Pseudo Bicliques

Zareen Alamgir, Saira Karim, and Syed Husnine

National University of Computer and Emerging Sciences
Lahore, Pakistan

Abstract. Pseudo bicliques relax the rigid connectivity requirement of bicliques to effectively deal with missing data. In this paper, we propose an algorithm based on reverse search to generate all pseudo bicliques in a given graph. We introduce various enhancements to our algorithm based on the structure of pseudo bicliques and underlying bipartite graph. We perform composite analysis using theoretical bounds and computational experiments, to show that these improvements significantly reduce the running time of our algorithm. Our algorithm is optimal in the sense that it takes average linear time to generate each pseudo biclique.

1 Introduction

Bicliques are used to model various real-world problems: document and words co-clustering, discovery of web communities and protein interactions [5,3]. Due to the rigid connectivity requirement of biclique, it is not suitable for dealing with missing data. Therefore, researchers are now considering pseudo bicliques to model more natural interactions in real world problems [5].

There are many ways to define pseudo biclique, we consider density based model of pseudo bicliques. The benefit of this model is that the restraint on the number of edges changes with the size of the subgraph. Thus, small subgraphs are classified as pseudo bicliques only if they are bicliques. The generation of density based pseudo biclique is a non-trivial task because straightforward back-tracking and branch-and-bound schemes involve a NP-complete problem [7]. Secondly, the monotone property does not hold in the family of density based pseudo bicliques.

Some schemes have been devised to enumerate pseudo bicliques [6,5] but no significant work is conducted for density based pseudo bicliques. David Gibson [2] proposed an algorithm for finding as many disjoint dense subgraphs in a given graph as possible, but his algorithm skips some useful dense graphs. In [4], the scheme to find quasi cliques in a given graph is extended to deal with quasi bicliques, but it can only list balanced quasi bicliques.

In this paper, we design an efficient algorithm for listing all pseudo bicliques in a given graph G. The framework of our algorithm is based on reverse search [1,7]. We evaluated the performance of our algorithm on randomly generated bipartite graphs. The results are very promising and shows that average linear time is incurred for generating a pseudo biclique.

C.S. Iliopoulos and W.F. Smyth (Eds.): IWOCA 2010, LNCS 6460, pp. 256–259, 2011.

2 Pseudo Biclique Generation Algorithm

In this section, we develop an algorithm to list all pseudo bicliques in a given bipartite graph $G = (V_1 \cup V_2, E)$ using reverse search. Let us first formally define density and pseudo biclique subgraphs.

Definition 1. *For a bipartite graph $G = (V_1 \cup V_2, E)$, the density $\rho(G)$ is given by $\rho(G) = \frac{|E|}{|V_1||V_2|}$*

Definition 2. *A pseudo-biclique B_{U_1,U_2} is a bipartite subgraph of graph G, if $\rho(B) \geq \theta$, where $0 < \theta \leq 1$.*

Here, θ is a density threshold. We denote the degree of vertex v in B_{U_1,U_2} by $deg_{B_{U_1,U_2}}(v)$, the maximum degree by $\triangle(B_{U_1,U_2})$ and minimum by $\delta(B_{U_1,U_2})$.

In reverse search technique, we construct a tree-shaped traversal route on the family of the combinatorial object under consideration. In order to form the tree, we define a parent for each element and ensure that definition of the parent is unique and acyclic. Reverse search algorithm traverses the tree in a depth first manner to list each structure. In a bipartite graph, the removal of a minimum vertex does not decrease the density of the resulting subgraph. If there are more than one such vertices, consider the minimum index one. We use this observation to define a parent-child relationship on the set of pseudo bicliques.

Lemma 2.1. *Let $G = (V_1 \cup V_2, E)$ be a bipartite subgraph, and vertex $v \in (V_1 \cup V_2)$. If $deg_G(v) = \delta(G)$ then $\rho(G \setminus v) \geq \rho(G)$.*

Proof. We have to show that $\rho(G \setminus v) - \rho(G) \geq 0$. Here, $\rho(G \setminus v)$ is the density of the graph $G \setminus v$, thus we have $\rho(G \setminus v) - \rho(G) = \frac{|E| - deg_G(v)}{(|V_1|-1)|V_2|} - \frac{|E|}{|V_1||V_2|}$. Note that in a bipartite graph, $|E| = \Sigma_{i \in V_1} deg_G(i) = \Sigma_{j \in V_2} deg_G(j)$, so we have

$$
\begin{aligned}
&= \frac{\Sigma_{i \in V_1} deg_G(i) - deg_G(v)}{(|V_1| - 1)|V_2|} - \frac{\Sigma_{i \in V_1} deg_G(i)}{|V_1||V_2|} \\
&= \frac{|V_1|\Sigma_{i \in V_1} deg_G(i) - |V_1| deg_G(v) - (|V_1| - 1)\Sigma_{i \in V_1} deg_G(i)}{(|V_1| - 1)|V_1||V_2|} \\
&= \frac{\Sigma_{i \in V_1} deg_G(i) - |V_1| deg_G(v)}{(|V_1| - 1)|V_1||V_2|} \\
&\geq \frac{|V_1| deg_G(v) - |V_1| deg_G(v)}{(|V_1| - 1)|V_1||V_2|} \\
&= 0
\end{aligned}
$$

Using the lemma above, we can establish that each pseudo biclique has density no more than its parent, thus a parent is a pseudo biclique if B_{U_1,U_2} is a pseudo biclique. In other words, we can say that for any pseudo biclique B_{U_1,U_2}, $B_{U_1,U_2} \setminus v$ will also be a pseudo biclique if v is a minimum vertex in B_{U_1,U_2}.

Now, we outline our pseudo biclique generation algorithm. Given G and θ, the routine GenPseudoBiclique is called for each edge in E, and it enumerates all the pseudo biclique in that branch of the enumeration tree.

Algorithm 1. GenPseudoBiclique(B_{U_1,U_2})

Require: Graph $G(V_1 \cup V_2, E)$, density threshold θ
1: **for** each $v \in \{V_1 \cup V_2\} \setminus \{U_1 \cup U_2\}$ **do**
2:   **if** $\rho(B_{U_1,U_2} \cup v) \geq \theta$ **then**
3:     **if** $B_{U_1,U_2} \cup v$ is a child of B_{U_1,U_2} **then**
4:       Output B_{U_1,U_2}
5:       GenPseudoBiclique($B_{U_1,U_2} \cup v$)
6:     **end if**
7:   **end if**
8: **end for**

We traverse the search space in a way that allows straightforward pruning of non-dense pseudo bicliques. According to the defined adjacency relationship, a non-dense pseudo biclique will always has non-dense descendants. Thus, during the traversal we prune the path whenever the density check fails at a node. In our algorithm, we compute minimum vertex and density before adding each vertex. In simple implementation, each of these operations will take $O(V)$ time. These operations are performed at most V times in an iteration. Thus, time to compute a pseudo biclique is $O(V^2)$.

Improvements for Efficient Computation: To improve the time requirements of the algorithm, we keep information about minimum vertex and degrees of all vertices of G in current B_{U_1,U_2}. This allows us to calculate density in constant time. Furthermore, we observe that in most cases, a comparison between $deg_{B_{U_1,U_2}}(v)$ and $\delta(B_{U_1,U_2})$ is sufficient to verify the parent-child relationship. We divide the task of determining child of B_{U_1,U_2} in one of the following three cases,where m denotes the minimum vertex in B_{U_1,U_2}.

1. If $deg_{B_{U_1,U_2}}(v)$ is less than $\delta(B_{U_1,U_2})$, then $B_{U_1,U_2} \cup \{v\}$ is a child of B_{U_1,U_2}
2. If $deg_{B_{U_1,U_2}}(v)$ is greater than $\delta(B_{U_1,U_2}) + 1$, then it is not a child of B_{U_1,U_2}
3. Otherwise one of the two possibilities can occur
   (a) If v is connected to m, then verify the parent-child relationship
   (b) If v is not connected m, then a comparison between label of v and m completes the task

In all the above cases except $3(a)$, verification of child can be done in constant time. Only the case $3(a)$ takes $O(V)$ time.

The cost incurred on constant-checkings can be distributed to pseudo bicliques as overhead. When a pseudo biclique is generated, it takes $O(V)$. This is because when a vertex is added to B_{U_1,U_2}, the degrees of all of its adjacent vertices in the array are updated. This operation takes $O(\Delta(G))$ time. For each B_{U_1,U_2}, the number of constant-checkings that does not yield any child are at most $O(V)$. We include this overhead in the generation cost of B_{U_1,U_2}. In section 3, we have estimated the non-constant-checkings using experiments and found that they are $O(\beta(G))$, where $\beta(G)$ is the total number of pseudo bicliques in G. Apart from this, we propose to start the algorithm with an edge. This will avoid trivial pseudo bicliques that have all vertices in one partition.

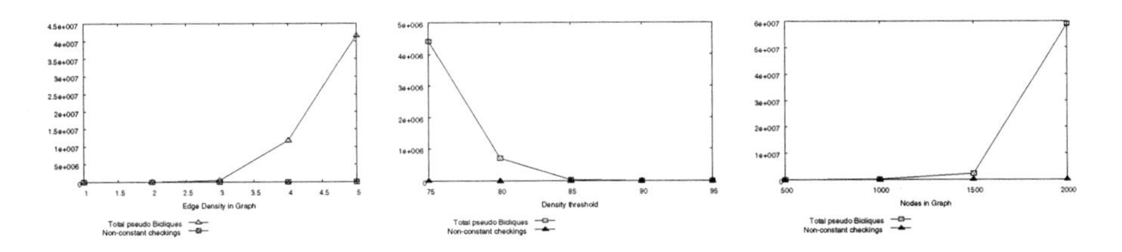

Fig. 1. Results obtained from computational experiments

3 Computational Results

We carry out experiments to show that average linear cost is incurred for generating each pseudo biclique. The edges in randomly generated graphs are uniformly distributed according to the defined edge density. In the experiments, we have estimated the ratio of the total pseudo bicliques in G to the number of non-constant-checkings. The number of pseudo bicliques depends on the given θ or on the graph size. For this purpose, we have evaluated the performance of our algorithm on three parameters: edge density, density threshold, and number of vertices. We conducted three different experiments to estimate the desired ratios. The results of these experiments are shown in Figure 1.

In all experiments, we observe that the number of non-constant-checkings is less than the pseudo bicliques generated. Thus, we can distribute the cost of these checkings to pseudo bicliques and hence, amount of work done per structure is $O(V)$. Secondly, we observe that the growth rate of the number of non-constant-checking is far less than that of pseudo bicliques when graph size is increased or θ is decreased. From this, we deduce that the average cost of computing a pseudo biclique decreases as the search space of algorithm increases.

References

1. Avis, D., Fukuda, K.: Reverse search for enumeration. Discrete Applied Mathematics 65, 21–46 (1996)
2. Gibson, D., Kumar, R., Tomkins, A.: Discovering large dense subgraphs in massive graphs. In: Proceedings of the 31st Conference on VLDB (2006)
3. Dhillon, I.S., Mallela, S., Modha, D.S.: Information-theoretic co-clustering. In: Proceedings of the 9th ACM SIGKDD (2003)
4. Abello, J., Resende, M.G.C., Sudarsky, S.: Massive quasi-clique detection. In: Rajsbaum, S. (ed.) LATIN 2002. LNCS, vol. 2286, p. 598. Springer, Heidelberg (2002)
5. Li, J., Sim, K., Liu, G., Wong, L.: Maximal quasi-bicliques with balanced noise tolerance: Concepts and co-clustering applications. In: Proceedings of SDM 2008 (2008)
6. Mishra, N., Ron, D., Swaminathan, R.: A new conceptual clustering framework. Machine Learning 56, 115–151 (2004)
7. Uno, T.: An efficient algorithm for solving pseudo clique enumeration problem. Algorithmica 56, 3–16 (2010)

Efficient Chaining of Seeds in Ordered Trees

Julien Allali[1,2,3], Cédric Chauve[3],
Pascal Ferraro[1,2,4], and Anne-Laure Gaillard[1]

[1] LaBRI, Université Bordeaux 1, IPB, CNRS
{julien.allali,anne-laure.gaillard,pascal.ferraro}@labri.fr
[2] Pacific Institute for Mathematical Sciences and CNRS UMI3069
[3] Department of Mathematics, Simon Fraser University
cedric.chauve@sfu.ca
[4] Department of Computer Science, University of Calgary

Abstract. We consider here the problem of chaining seeds in ordered trees. Seeds are mappings between two trees Q and T and a chain is a subset of non overlapping seeds that is consistent with respect to postfix order and ancestrality. This problem is a natural extension of a similar problem for sequences, and has applications in computational biology, such as mining a database of RNA secondary structures. For the chaining problem with a set of m constant size seeds, we describe an algorithm with complexity $O(m^2 \log(m))$ in time and $O(m^2)$ in space.

1 Introduction

Comparing sequences is a basic task in computational biology, either for mining genomics database, or for filtering large sequence datasets. A fundamental application of sequence comparison is to search efficiently in a database a set of sequences close to a query sequence. The exponential increase of available sequence data motivates the need for very efficient sequence comparison algorithms. In particular, pairwise comparison relying on computing the exact edit distance between the query and every every sequence of the database can not practically be applied due to the quadratic time complexity of edit distance computation. A typical approach to tackle this issue is to rely on short sequences, called *seeds*, present in the query. Seeds can be detected very quickly in the database using indexing techniques; then an optimal set of seeds, called a *chain*, that tiles both the query and a sequence of the database, must be identified while conserving the same order in both sequences. Widely used programs such as BLAST [2] and FASTA [11,14] rely on such an approach. We refer the reader to [3,7] for surveys of sequence comparison in computational biology. From an algorithmic point of view, an optimal chain between two sequences, given m seeds, can be computed in $O(m \log(m))$ time and $O(m)$ space [10] (see [13] for a recent survey).

With the recent development of high-throughput genome annotation methods, similar problems appear to be relevant for the analysis of more complex biological structures [15]. For instance, an RNA secondary structures can be

C.S. Iliopoulos and W.F. Smyth (Eds.): IWOCA 2010, LNCS 6460, pp. 260–273, 2011.

represented by a tree or a graph whose nodes are the nucleotides and whose edges are the chemical bonds between them [16]. Mining large RNA secondary structure databases, such as Rfam [6], is now an important computational biology problem. An initial approach, adapting the notion of edit distance to ordered trees, was pioneered by Zhang and Shasha [17]. The tree edit approach has been extended in several ways since then, leading either to hard problems, when a comprehensive set of edit operations is considered [9], or to algorithms with a worst-case time complexity at best cubic, even with a minimal set of edit operations [5,17].

Recently, Heyne *et al.* [8] introduced a chaining problem on an alternative representation of ordered trees called arc annotated sequences, motivated by pairwise RNA secondary structure comparison: once an optimal chain of seeds between two given RNA secondary structures is detected, the regions between successive seeds are processed independently using an edit distance algorithm, which speeds up significantly the comparison process. They considered seeds defined as *exact common patterns* and designed a dynamic programming algorithm to solve the seeds chaining problem. To the best of our knowledge, [8] is the first paper addressing a chaining problem in trees.

After some preliminaries (Sections 2 and 3), we describe in Section 4 an algorithm for finding the score of an optimal chain between two ordered trees (Maximal Chaining Problem) in $O(m^2 \log(m))$ time and $O(m^2)$ space when there are m seeds of constant size, thus improving on the result of Heyne *et al.* [8]. We conclude with further research avenues.

2 Background and Problem Statement

Let T be an ordered rooted tree of size n. Nodes of T are identified with their postfix-order index from 0 to $n-1$. Thus, $n-1$ represents the root of T. T_i is the subtree of T rooted at i. We denote by $T[i,j]$ the forest induced by the nodes that belong to the interval $[i,j]$; if $i > j$, then $T[i,j]$ is empty. The partial relationship "i is an ancestor of j" is denoted by $i \prec j$. For a tree T and a node i of T, the first leaf visited during a postfix traversal of T_i is denoted by $l(i)$ and called the *leftmost leaf* of the node i. The ordered forest induced by the proper descendants of i is denoted by $\widehat{T_i} = T[l(i), i-1]$.

Definition 1. Let T be an ordered rooted tree:

1. Let $G = \{g_0, \dots, g_{k-1}\}$ be an ordered set of k nodes of T, with $0 \leq g_j < n$. If the subgraph of T induced by G is connected, then G is called an *internal tree* rooted at g_{k-1} also referred to as r_G.
2. The set of leaves of the internal tree G is denoted by $L(G)$.
3. A node g_j of G is said to be *completely inside* G if g_j is not a leaf of T and all its children belong to G. The set of nodes of G that are not completely inside G is called the *border of* G and is denoted by $B(G)$.
4. Two internal trees G^1 and G^2 *overlap* if they share at least one node, *i.e.* $G^1 \cap G^2 \neq \emptyset$.

We now recall the central notion of *valid mapping* between two trees introduced in [16] for the tree edit distance. Given two trees Q and T, a valid mapping P between Q and T is a set of pairs of $Q \times T$ such that, if (q_i, t_i) and (q_j, t_j) belong to P, then

1. $q_i = q_j$ if and only if $t_i = t_j$,
2. $q_i < q_j$ if and only if $t_i < t_j$,
3. $q_i \prec q_j$ if and only if $t_i \prec t_j$.

From now we use the term *mapping* to refer to a valid mapping. Given a mapping P between Q and T, the smallest internal tree of Q (resp. T) that contains all nodes of Q (resp. T) belonging to a pair of P is denoted by Q_P (resp. T_P). Q_P and T_P are respectively called the internal trees of Q and T induced by P.

Definition 2. Let Q and T be two ordered trees.

1. A *seed* P between Q and T is a mapping between Q and T such that $(r_{Q_P}, r_{T_P}) \in P$ and all the nodes of the border of Q_P (resp. T_P) belong to a pair of P.
2. The border (resp. leaves) $B(P)$ (resp. $L(P)$) of the seed P is the set of pairs $(x, y) \in P$ such that $x \in B(Q_P)$ and $y \in B(T_P)$ (resp. $x \in L(Q_P)$ and $y \in L(T_P)$).
3. The *size* $|P|$ of the seed P is the number of pairs its mapping contains.
4. For a set S of seeds, $\|S\|$ is the sum of the sizes of the $|S|$ seeds in S.

Note that, theoretically, the number of seeds between Q and T can be exponential in the size of Q and T, although in applications such as RNA secondary structure comparison, this exponential upper bound is unlikely to be reached (see [8] for example).

Definition 3. Let Q and T be two ordered trees.

1. A pair (P^1, P^2) of seeds between Q and T is *chainable* if Q_{P^1} does not overlap Q_{P^2}, T_{P^1} does not overlap T_{P^2}, and $P^1 \cup P^2$ is a mapping.
2. A *chain* is a set $C = \{P^0, P^1, \ldots, P^{\ell-1}\}$ of seeds between Q and T such that any pair (P^i, P^j) of distinct seeds in C is chainable.
3. Given a scoring function v for the seeds P^i, the score of a chain C is the sum of the scores of its seeds: $v(C) = \sum_i v(P^i)$.
4. Given a set S of possibly overlapping seeds between Q and T, $\mathcal{C}_S(Q, T)$ denotes the set of all possible chains between Q and T included in S.

We can now define the main problem we consider in the present paper (illustrated in Fig. 1).

Problem. Maximum Chaining Problem (MCP):
Input: A pair (Q, T) of ordered rooted trees, a set $S = \{P^0, \ldots, P^{m-1}\}$ of m possibly overlapping seeds between Q and T, a scoring function v on the seeds P^i.
Output: The maximum score chain C included in S:

$$MCP(Q, T, S) = \max\{v(C); C \in \mathcal{C}_S(Q, T)\}.$$

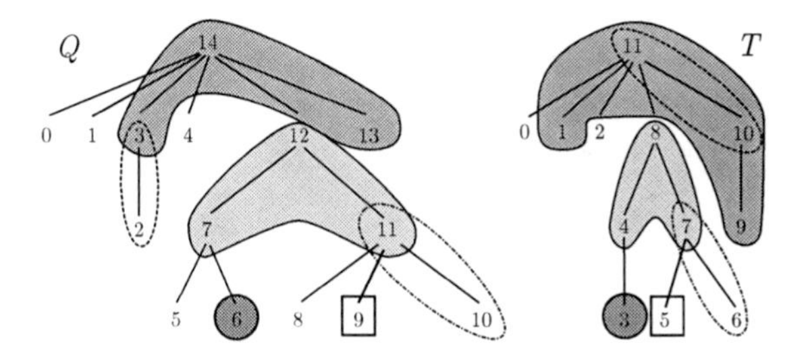

Fig. 1. An instance of the MCP with 6 seeds: $P^0 = \{(2,10),(3,11)\}$, $P^1 = \{(6,3)\}$, $P^2 = \{(9,5)\}$, $P^3 = \{(10,6),(11,7)\}$, $P^4 = \{(7,4),(11,7),(12,8)\}$, $P^5 = \{(3,1),(13,9),(14,11)\}$. If for every seed $v(P^i) = |P^i|$, an optimal chain is composed of $\{P^1, P^2, P^4, P^5\}$ and has a score of 8.

Remark 1. The notion of mapping extends naturally to ordered forests. Hence, if S is a set of seeds such that each seed is a seed between a tree of F_1 and a tree of F_2, then the MCP can naturally be extended to ordered forests.

Remark 2. To compare with chaining algorithms for sequences, we represent a sequence $u = (u_0, \ldots, u_{n-1})$ by a unary tree, rooted at a node labeled by u_{n-1}, where every internal node has a single child and u_0 is the unique leaf: the sequence of nodes visited by the postfix-order traversal of this tree is exactly u.

Motivation and background. As far as we know, [8] is the only work that attacks the MCP in tree structures, although the authors describe the problem in terms of arc-annotated sequences. They proposed a dynamic programming algorithm to solve the maximum chaining problem with some restrictions on the seeds (precisely, seeds are maximal exact pattern common to the considered sequences). This dynamic programming technique is different from the approach used for the currently best known algorithms for Maximum Chaining Problem in sequences [10,13]. Moreover, when applied to arc-annotated sequences with no arc (*i.e.* sequences) and m seeds, it can be shown this algorithm has a worst-case time complexity in $O(m^2)$.

Result statement. Our main result is the following:

Theorem 1. *Let S be a set of m seeds between two ordered trees Q and T. After an $O(\|S\|)$ time preprocessing of the m seeds of S one can solve the Maximum Chaining Problem in $O(\|S\| \log(\|S\|) + m\|S\| \log(m))$ time and $O(m\|S\|)$ space.*

Note that we described the complexity of our algorithm using uniquely the set of seeds S, unlike Heyne *et al.* [8], who, for the same problem, also consider the sizes of Q and T (see [1] for a detailed analysis of the complexity of the algorithm of [8]). We prove in Section 4, that our algorithm solves the maximal chaining problem on sequences (*i.e.* unary trees as described in Remark 2 above) in $O(m \log(m))$ time and $O(m)$ space complexity.

Remark 3. Without loss of generality, from now we assume that the seeds P^i are sorted increasingly according to the postfix number of their roots in Q, that is: $r_{Q_{P^0}} \leq \cdots \leq r_{Q_{P^i}} \leq \cdots \leq r_{Q_{P^{m-1}}}$. For a given chain C, the *last* seed of C is then the seed with the highest postfix index in Q.

3 Combinatorial Properties of Seeds and Chains

We first describe combinatorial properties of seeds and chains, that naturally lead to a recursive scheme to compute a maximum chain. Indeed, we show that given a chain C and its last seed P, the root and border of P define a partition of both $Q - Q_P$ and $T - T_P$ into pairs of forests that contain the seeds $C - \{P\}$ and form sub-chains of C. More precisely, for every border nodes (x, y) of a seed P, we define the couples of forests included in $(\widehat{Q_x}, \widehat{T_y})$, that is composed of descendants of (x, y), such that any seed included into such couple of forest is chainable with P.

Definition 4. Let P be a seed on two trees Q and T and $(a, b; c, d)$ be a quadruple such that $l(r_{Q_P}) \leq a < b < r_{Q_P}$, $l(r_{T_P}) \leq c < d < r_{T_P}$ and the pair of forests $(Q[a, b], T[c, d])$ does not contain any node involved in P ($Q_P \cap Q[a, b] = \emptyset$ and $T_P \cap T[c, d] = \emptyset$). $(a, b; c, d)$ is a *chainable area* if for all $i \in [a, b]$ and all $j \in [c, d]$, $P \cup (i, j)$ is a valid mapping. $(a, b; c, d)$ is a *maximal chainable area* for P if neither $(a-1, b; c, d)$ or $(a, b+1; c, d)$ or $(a, b; c-1, d)$ or $(a, b; c, d+1)$ are chainable areas for P.

For example, in Fig. 1, let us consider the seed $P = P^5$; then, $(4, 12; 2, 8)$ is a maximal chainable area. See also Figure 2.

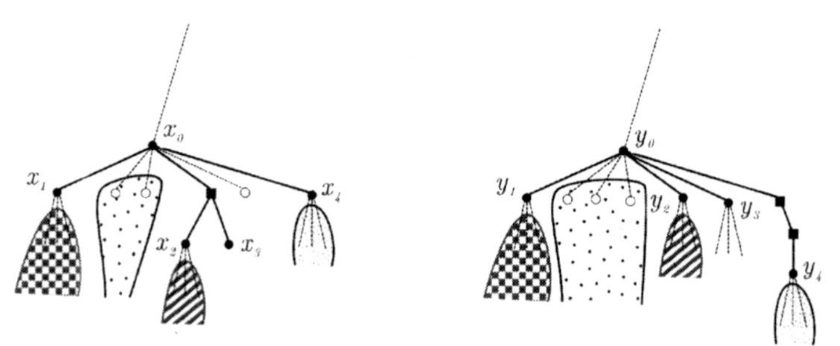

Fig. 2. Illustration of the notion of chainable areas of a seed of size 5: $P = \{(x_0, y_0), \ldots, (x_4, y_4)\}$ and there are 4 chainable areas for P each indicated by a different filling pattern

Definition 5. Let $(x, y) \in B(P)$ for a seed P between Q and T. We define by $F(x, y) = \{(a_i, b_i; c_i, d_i)\}$ the set of all maximal chainable areas for P included in $(Q_x; T_y)$ such that there is no border node of P in Q (resp. T) on the path from b to x (resp. d to y). We call this set the *chainable areas* of (x, y).

For example, let us consider a pair (x, y) in $L(P)$ such that x and y are not a leaf of respectively Q and T, then $F(x, y)$ represents the couple of forests $\widehat{Q_x}$ and $\widehat{T_y}$, $F(x, y) = \{(l(x), x-1; l(y), y-1)\}$. In Fig. 1, with $P = P^4$ and $(x, y) = (11, 7)$, $F(x, y) = \{(8, 10; 5, 6)\}$); if $(x, y) = (14, 11) \in B(P^5) - L(P^5)$, $F(x, y) = \{(0, 1; 0, 0), (4, 12; 2, 8)\}$. See also Figure 3.

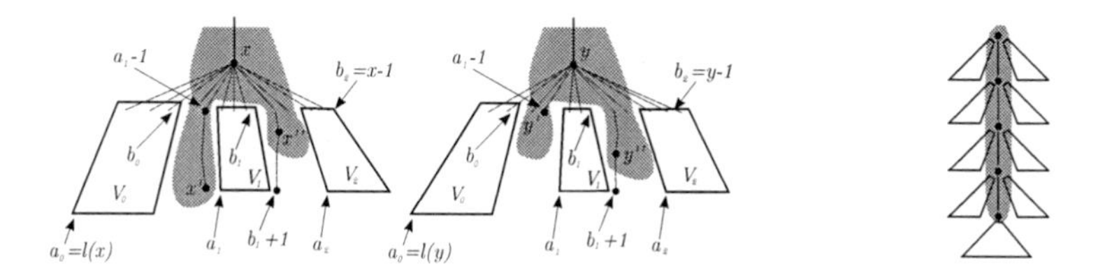

Fig. 3. (Left) Illustration of Definition 5 for a seed P (the shaded zone) and $(x, y) \in B(P) - L(P)$: $F(x, y) = \{(a_0, b_0, c_0, c_1), (a_1, b_1, c_1, c_1), (a_2, b_2, c_2, c_2)\}$. (Right) Illustration of the maximum number of chainable areas of a seed.

Definition 6. The *chainable areas* of a seed P, denoted by $CA(P)$, is the union of the sets of quadruples $F(x, y)$ for all pairs $(x, y) \in B(P)$.

Notation. For a seed P (resp. chain C) and a chainable area $(a, b; c, d)$, we say that $P \subset (a, b; c, d)$ (resp. $C \subset (a, b; c, d)$) if $a \leq r_{Q_P} \leq b$ and $c \leq r_{T_P} \leq d$.

The following property is a relatively straightforward consequence of the definitions of seeds and chainable areas (Fig. 3).

Property 1. Given a seed P between trees Q and T, $|CA(P)| \leq 2 \times |B(P)| - 1$.

From now, for every (x, y) of a seed P^j, we denote by x^j the unique node y of T associated with x in P^j. We also denote by $F_j(x)$ the set of quadruples $F(x, x^j)$ for the pair of nodes $(x, x^j) \in P^j$.

The next property describes the structure of any chain between two forests $Q[a, b]$ and $T[c, d]$ included in a set of m seeds $S = \{P^0, \ldots, P^{m-1}\}$. It is a direct consequence of the constraints that define a valid mapping and the fact that seeds are non-overlapping in a chain.

Property 2. Let P^j be the last seed of a chain C included into two forests $Q[a, b]$ and $T[c, d]$.

1. C can be decomposed into $|CA(P^j)|+2$ (possibly empty) distinct sub-chains: P^j itself, $|CA(P^j)|$ chains: for each $(e, f; g, h) \in CA(P^j)$ a (possible empty) chain included into $Q[e, f]$ and $T[g, h]$ and a chain included into the forests $Q[a, l(r_j) - 1]$ and $T[c, l(r_j^j) - 1]$.
2. Moreover, C is a chain of maximum score among all chains in $Q[a, b]$ and $T[c, d]$ that contain P^j if and only if all of its sub-chains described above are chains of maximum score with respect to the corresponding forests defined by $CA(P^j)$.

Property 2.2 naturally leads to a recursive scheme to compute an optimal chain between two forests $Q[a,b]$ and $T[c,d]$ that ends by the last seed of a set. If $MCP'(Q[a,b],T[c,d],\{P^0 \dots P^j\})$ is the score of a maximum chain between $Q[a,b]$ and $T[c,d]$ and that contains P^j:

$$MCP'(Q[a,b],T[c,d],\{P^0 \dots P^j\}) = \tag{1}$$

$$\begin{cases} 0 & \text{if } P^j \not\subset (a,b;c,d), \\ v(P^j) + \sum_{(e,f;g,h)\in CA(P^j)} MCP(Q[e,f],T[g,h],\{P^0 \dots P^{j-1}\}) & \text{otherwise.} \\ \quad +MCP(Q[a,l(r_j)-1],T[c,l(r_j^j)-1],\{P^0 \dots P^{j-1}\}) & \end{cases}$$

and thus $MCP(Q,T,S)$ can be computed using MCP' as follow[1]:

$$MCP(Q[a,b],T[c,d],\{P^0 \dots P^j\}) = \max_{i=0\dots j} MCP'(Q[a,b],T[c,d],\{P^0 \dots P^i\}) \tag{2}$$

$$MCP(Q,T,S) = MCP(Q[0,r_Q],T[0,r_T],S) \tag{3}$$

The main challenge in designing an algorithm for the MCP is then to implement efficiently this recursive formula, that was already central in the dynamic programming algorithm of [8]. In Section 4, we will rely on the fact that for every seed P^j, $CA(P^j)$ and, for every border node x of P^j, $F_j(x)$, have been computed during a preprocessing phase. A journal version will discuss the issues related to this preprocessing and will show that it can be done in $O(\|S\|)$ time and space (see also [1]).

4 Algorithms for the Maximum Chaining Problem

From now, we consider that we are given two ordered trees Q and T, a set $S = \{P^0,\dots,P^{m-1}\}$ of seeds and a scoring score v on S. Furthermore, we assume that the score $v(j)$ of a seed P^j can be accessed in constant time and the seeds of S are given as a list I of triples (i,f,j) such that: (1) i is the postfix number of either the root of P^j_Q or a border node of P^j_Q (*ie.* $i \in B(P^j_Q) \cup \{r_j\}$) and (2) f is a flag indicating if i is either border ($f=0$) or root ($f=1$) for P^j_Q. Thus if i is both in $B(P^j_Q)$ and the root of P^j_Q then i appears in two distinct triples[2]. Moreover, for a node i in Q belonging to a seed P^j, we assume that the corresponding node in T, i^j (or more precisely its postfix number in T) can be accessed in constant time. Finally, for every node i in Q and T, its leftmost leaf $l(i)$ is also supposed to be accessed in constant time.

As a preprocessing, I is sorted in lexicographic order. Thus, if a node is both in the border and root of P^j, it first appears in I as a border, then as a root. This sorting can be done in $O(||S|| \log(||S||))$ time. In our algorithms, we visit

---

[1] We remind that the seeds are supposed to be sorted incrementally (see Remark 3).

[2] Hence, we do not require as input the whole seeds mappings but just the borders and roots of the seeds, as it is usual when chaining seeds in sequences.

successively the elements of I in increasing order, and a seed P^j is said to be *processed* after its root has been processed (*i.e.* the current element of I is greater than $(r_j, 1, j)$ for the order defined above).

In the following, we first introduce a simple but non optimal algorithm to compute the MCP between Q and T which does not require any special data structure. In a second step, we will present a more efficient method based on a simple modification of this algorithm.

4.1 A Simple Non Optimal Algorithm

In order to compute in constant time the partial MCP for any pair of forests in $CA(P^j)$ as described in equation (1), we introduce a data structure M indexed by quadruples of integers $(a, b; c, d)$ defining the forests $Q[a, b]$ and $T[c, d]$. These quadruples $(a, b; c, d)$ belong to a set $Y = Y_1 \cup Y_2 \cup Y_3$ defined as follows:

$$Y_1 = \bigcup_{j=0}^{m-1} CA(P^j), \; Y_2 = \{(0, r_Q, 0, r_T)\},$$

$$Y_3 = \{(a, l(r_j) - 1; c, l(r^j_j) - 1) \mid \exists (b, d); (a, b; c, d) \in Y_1 \cup Y_2 \text{ and } P^j \subset (a, b; c, d)\}$$

In algorithm 1, the function `Update` replaces the value of $M[a, b, c, d]$ by a real number w if w is greater than $M[a, b, c, d]$. We also use an array V of m integers to store the intermediate quantities of MCP'. The correctness of the algorithm relies on the following invariants for the two data structures V and M, that we prove later:

M1. After P^j has been processed, then $M[a, b, c, d] = MCP(Q[a, b], T[c, d], \{P^0, \ldots, P^j\})$ for every $(a, b; c, d) \in Y$.
V1. After P^j has been processed, then $V[j] = MCP'(Q, T, \{P^0, \ldots, P^j\})$.

Correctness of the algorithm. Obviously, V1 implies that $\max_j V[j]$ contains the score of the maximum chain (equations (2) and (3)). Let us assume now that M1 is satisfied. If the seed P^j has been processed, then $V[j]$ contains the sum of $v(j)$ (line 1), the MCP scores of the chainable areas of all its border nodes (line 5) and the MCP score between forests $Q(0, l(r_j) - 1)$ and $T(0, l(r^j_j) - 1)$ (line 11). From Property 2 and (1), $V[j] = MCP'(Q, T, \{P^0, \ldots, P^j\})$ and V1 is satisfied.

We prove M1 by induction. Initially, since no seed has been processed, line 2 ensures that M1 is satisfied. Now let us assume that M1 is satisfied for all processed seeds $\{P^0, \ldots, P^{j-1}\}$ and the input $(i, 1, j)$ is being processed. If $P^j \not\subset (a, b; c, d)$, then by induction, M1 is satisfied for $M[a, b, c, d]$. Otherwise, the loop in lines 7 and 8 ensures that M1 is satisfied for all entries $M[a, b, c, d]$ such that $(a, b; c, d) \in Y_1 \cup Y_2$, as $(a, l(r_j) - 1; c, l(r^j_j) - 1)$ does not contain P^j ; thus by induction M1 is satisfied for this index. Finally, the loop in line 9 update all $(a, b; c, d) \in Y_3$ including P^j,and M1 is satisfied for all entries of M.

Algorithm 1. MCP_1: compute the score of a maximum chain.

```
1 for j from 0 to m − 1 do V[j] = v(j)
2 foreach (a, b; c, d) ∈ Y do M[a, b, c, d] = 0
3 foreach (i, f, j) in I do
4   if f = 0 then # i.e. (i, i^j) ∈ B(P^j)
5      foreach (a, b; c, d) ∈ F_j(i) do V[j] = V[j] + M[a, b, c, d]
6   else # i.e. f = 1 and i is the root of Q_{P^j}, i = r_j
7     foreach (a, b; c, d) ∈ Y_1 ∪ Y_2 s.t. P^j ⊂ (a, b; c, d) do
8       Update M[a, b, c, d] with w = V[j] + M[a, l(r_j) − 1, c, l(r^j_j) − 1]
9       foreach P^g ⊂ (r_j + 1, b; r^j_j + 1, d) do
10        Update M[a, l(r_g) − 1, c, l(r^g_g) − 1] with w
11    V[j] = V[j] + M[0, l(r_j) − 1, 0, l(r^j_j) − 1]
12 return max_j V[j]
```

Complexity analysis. From Property 1, the space required to encode the entries of M indexed by Y_1 is in $O(\|S\|)$. The space required to encode the entries of M indexed by Y_3 is in $O(m^2)$, as for every pair of seeds P^i and P^j, there is at most one chainable area of $CA(P^i)$ that contains P^j.

We now address the worst-case time complexity. We do not factor the preprocessing required to compute the F_j and CA and we assume I has been sorted in time $O(\|S\| \log(\|S\|))$. The amortized cost of lines 4–5 is $O(\|S\|)$, as each chainable area is considered once, there are $O(\|S\|)$ such areas, and we assumed we can access them in amortized constant time. A naive implementation of lines 6–11 would require $O(m^2\|S\|)$ operations: indeed, there are m iterations of the loop in line 6, the loop in line 7 considers only entries indexed by $Y_1 \cup Y_2$ (there are $O(\|S\|)$ such entries) and the loop on line 9 iterates $O(m)$ times. However, we can notice that there are $O(m)$ entries $(a, b; c, d) \in Y_1 \cup Y_2$ such that $P^j \subset (a, b; c, d)$, and it is possible to preprocess I in time and space $O(m\|S\|)$ in such a way that the loop in line 7 can be implemented to perform $O(m)$ iterations, leading to a total time complexity of $O(\|S\| \log(\|S\|) + m\|S\| + m^3)$ (respectively for sorting the input, preprocessing and then the main algorithm).

4.2 A More Efficient Algorithm

We describe and analyze now a more efficient algorithm, which proves our main result, Theorem 1.

The key ideas are to access less entries from M (while maintaining property M1 on the remaining entries though) and to complement M with a data structure R that can be queried in $O(\log(m))$ instead of $O(1)$, but whose maintenance does not require a loop with $O(m^2)$ iterations. Formally, let $X = \{(a, c) \text{ s.t. } \exists (a, b; c, d) \in Y_1 \cup Y_2\}$ and R be a data structure indexed by X such that, for a given index $(a, c) \in X$, $R[a, c]$ is a set of pairs (j, s) where j is the index of the seed P^j and s is the maximum score of chains in $Q[a, r_j], T[c, r^j_j]$

that ends with P^j. Roughly, M is used to access, still in $O(1)$ time, the values $MCP(a, l(r_j) - 1, c, l(r_j^j) - 1, \{P^0 \dots P^{j-1}\})$ required to compute MCP' in equation (1) and $R[a, c]$ is used to access, in time $O(\log(m))$, the scores of the best chains included in $(Q[a, r_Q], T[c, r_T])$ (the values $MCP(Q[e, f], T[g, h], \{P^0 \dots P^{j-1}\})$ in equation (1)) and replace the entries $M[a, b, c, d]$ with $(a, b; c, d) \in Y_1 \cup Y_2$, which were used in the previous algorithm.

Finally, the algorithm iterates on a list of triples $J = I \bigcup \left(\cup_{j=0}^{m-1} (l(r_j), -1, j)\right)$, sorted using the lexicographic order used in the previous section, with the following modification: if we have two seeds P^j and P^g with $g > j$ such that $(l(r_j), l(r_j^j)) = (l(r_g), l(r_g^g))$ then only $(l(r_j), -1, j)$ occurs in J. This preprocessing requires $O(||S|| \log(||S||))$ time.

Algorithm 2. $MCP_2(Q, T, S, v)$: compute a maximum chaining from S.

1 **for** j from 0 to $m - 1$ **do** $V[j] = v(j)$
2 **foreach** $(a, b; c, d) \in Y_3$ **do** $M[a, b, c, d] = 0$
3 **foreach** $(a, c) \in X$ **do** $R[a, c] = \emptyset$
4 **foreach** (i, f, j) **in** J **do**
5   **if** $f = -1$ **then** # $i = l(r_j)$
6   **foreach** $(a, c) \in X$ s.t. $a, c < l(r_j), l(r_j^j)$ **do**
7       $M[a, l(r_j) - 1, c, l(r_j^j) - 1] =$ value s of the last (y, s) of $R[a, c]$ s.t. $r_y^y < l(r_j^j)$
8   **else if** $f = 0$ **then** # $(i, i^j) \in B(P^j)$
9   **foreach** $(a, b; c, d) \in F_j(i)$ **do**
10   Add to $V[j]$ the value s of the last entry (y, s) of $R[a, c]$ s.t. $r_y^y \leq d$
11   **else** # $f = 1$ *and* i *is the root of* Q_{P^j}, $i = r_j$
12   **foreach** $(a, c) \in X$ s.t. $a, c < l(r_j), l(r_j^j)$ **do**
13       $w = V[j] + M[a, l(r_j) - 1, c, l(r_j^j) - 1]$
14   Insert entry (j, w) into $R[a, c]$ and update $R[a, c]$ as follow:
15   Find the last entry (y, s) s.t. $r_y^y < r_j^j$
16   **if** $s < w$ **then**
17   Insert (j, w) just after (y, s) **in** $R[a, c]$
18   Remove from $R[a, c]$ all entries (z, t) s.t. $r_j^j \leq r_z^z$ and $t < w$
19     $V[j] = V[j] + M[0, l(r_j) - 1, 0, l(r_j^j) - 1]$
20 **return** $\max_j V[j]$

Correctness of the algorithm. We consider the following invariants.

M2. After P^j has been processed, then $M[a, b, c, d] = MCP(Q[a, b], T[c, d], \{P^0, \dots, P^j\})$ for every $(a, b; c, d) \in Y_3$.

V1. After P^j has been processed, then $V[j] = MCP'(Q, T, \{P^0, \dots, P^j\})$.

R1. After P^j has been processed, then for all $(a, c) \in X$, $R[a, c]$ contains all (y, s) that satisfies
 a. $y \leq j$ and $s = MCP'(Q[a, r_y], T[c, r_y^y], \{P^0, \dots, P^y\})$.
 b. $\forall (z, t) \in R[a, c]$, $r_z^z < r_y^y \Rightarrow t < s$.

R2. $\forall (a, c) \in X$, $R[a, c]$ is totally ordered as follows: $(y, s) < (z, t)$ iff $r_y^y < r_z^z$.

We first assume that R1 and R2 are satisfied. As previously, if V1 is satisfied, then the algorithm computes $MCP(Q,T,S)$. The initialization line 1 ensures that $V[j]$ contains $v(j)$. Next to prove V1 we only need to show that, when we process a border i of a seed P^j, in line 10 we add to $V[j]$ the best chain of each chainable area $(a,b;c,d)$ of the border; it follows from (1) the fact that every seed P^{j+e} with $e>0$ does not belong to the forest $Q[a,b]$ (because $b<i\leq r_{j+e}$) and thus can not belong to a chain in the $(a,b;c,d)$ area, (2) the fact that the score of this chain is present in $R[a,c]$ (from R1) and (3) the fact that it is the last entry (y,s) such that $r_y^y\leq d$ (from R2).

M2 is similar to M1 but restricted to entries $M[a,b,c,d]$ such that $(a,b;c,d)\in Y_3$. To check it is satisfied, we only need to focus on line 7, as it is the only line that updates M. For entries $M[a,b,c,d]$ such that $a\geq l(r_j)$ or $c\geq l(r_j^j)$, then $M[a,b,c,d]=0$ due to the initialization in line 1. For all other entries, M2 follows immediately from R1 and R2, using argument similar to the previous ones.

Finally, we need to check that R1 and R2 are satisfied. First, as previously, in the case where $a\geq l(r_j)$ or $c\geq l(r_j^j)$, $R[a,c]=\emptyset$ which is ensured by the initialization in line 3. So we need only to consider the case where $a,c<l(r_j),l(r_j^j)$, that is handled in lines 11 to 18. Every seed P^y such that $y<j$ has already been processed and $s=MCP'(Q[a,r_y],T[c,r_y^y],\{P^0,\ldots,P^y\})$ can not be modified after P^y has been processed, so lines 12 and 13, together with M2, ensure that (y,s) has been inserted into $R[a,c]$ previously, and the same argument applies if $y=j$. Entries (z,t) removed at line 18 do not belong to any of these (y,s), which implies that R1.a and R1.b, and so R1, are satisfied. R2 is obviously satisfied from the position where (j,w) is inserted into $R[a,c]$ in line 17.

Complexity analysis. The space complexity is given by the space required for structures M and R. M requires a space in $O(m^2)$ as it is indexed by Y_3. R requires a space in $O(m\|S\|)$, as $|Y_1\cup Y_2|\in O(\|S\|)$ and for each seed P^j, an entry (j,s) is inserted at most once in each $R[a,c]$. All together, the space complexity is then $O(m^2+m\|S\|)=O(m\|S\|)$.

We now describe the time complexity. First, note that following the technique used for computing maximum chains in sequence [7,10,13], the structures $R[l_Q,l_T]$ can be implemented using classical data structures such as AVL or concatenable queues supporting query requests, insertions and deletions, successor and predecessor, in a set of n totally ordered elements in $O(\log(n))$ worst-case time.

Now, we analyze the complexity of lines 5 to 7. The loop of line 6 is performed at most $O(m\|S\|)$ times and each iteration requires $O(\log(m))$ in time (line 7), which gives an amortized time complexity of $O(m\|S\|\log(m))$.

Line 10 is applied at most once for each of the $O(\|S\|)$ chainable area $F_j(i)$ (Property 1), and each iteration requires $O(\log(m))$, which gives an $O(\|S\|\log(m))$ amortized time complexity.

Finally, we analyze the complexity of lines 11 to 19. First, we do not consider the operation in line 18. The loop starting in line 12 is performed in $O(m)$, and the complexity of each loop is in $O(\|S\|)$. The cost of the operations performed during each iteration is $O(\log(\|S\|))$ (lines 13 and 16 are both performed in

$O(1)$ and lines 14 and 15 in time $O(\log(\|S\|))$. The total time complexity of this part, without considering line 18, is then $O(m\|S\|\log(\|S\|))$. To complete the time complexity analysis, we show that the amortized complexity of line 18 is in $O(m\|S\|)$. Indeed, it follows from R2 that all entries removed in one step are consecutive in the total order on $R[a, c]$ defined in R2. Hence, if one call to line 18 removes k elements from $R[a, c]$, it can be done in $O((k + 2)\log(m))$ time, as the successor of a given element can be retrieved in $O(\log(m))$ time. As every element of R is removed at most once during the whole algorithm, this leads to an amortized complexity of $O(m\|S\|\log(m))$ for line 18. Altogether, our algorithm solves computes $MCP(Q, T, S)$ in time $O(m\|S\|\log(m))$, using standard data structures and after a preprocessing in time $O(\|S\|\log(\|S\|))$ to compute the chainable areas and to sort J.

Additional remarks. If we consider that Q and T are sequences, or, as described in Section 2, unary trees, then each of the two trees has a single leaf and each seed is unambiguously defined by its root and border, which implies that$\|S\| = m$. There is only one $R[a, c]$, as $a = c = 0$, that contains $O(m)$ entries. Hence, all loops that were iterating on R have now a single iteration, which reduces the time complexity by a factor m to $O(\|S\|\log(m)) = O(m\log(m))$.

In the complexity analysis above, we followed the approach used for expressing the complexity of chaining in sequences, as we expressed the complexity only in terms of the size of the seeds. To express the complexity of our algorithm in terms of the size of Q and T, a finer analysis of the data structure R and of the number of different chainable areas leads to the following result: the worst-case space complexity of our algorithm is $O(|Q|^2|T|^2)$ (similar to the algorithm of Heyne et al.), and its worst-case time complexity is in $O(\|S\|\log(\|S\|) + |Q||T|\log(|T|)(|Q||T| + m))$, to compare with the complexity of the Heyne et al algorithm, which is in $O(\|S\|\log(\|S\|)+|Q|^2|T|^2(|Q||T|+m))$ [1]. This alternative complexity analysis is mostly of theoretical interest as in practice, for RNA analysis, one can expect that $m \ll |Q||T|$.

5 Conclusion

The current paper describes algorithms to solve chaining problems in ordered trees. With respect to similar problems in sequences, these methods exhibit a linear factor increase both in time and space. Chains so obtained can be used to speed-up RNA structure comparisons, as illustrated in [8,12].

A natural question related to chaining problems, that, as far as we know, has not been considered in the case of sequences, is to decide whether a given seed P of a set of seeds S belongs to *any* optimal chains or not. However a trade-off between quality and speed needs to be found. Indeed, identifying these *always optimal* seeds would probably ensure a chain of good quality, whereas the high complexity of these identifications might slow down the detection of similar structures in a large database.

Acknowledgements. Pacific Institute for Mathematical Sciences (PIMS, UMI CNRS 3069), Agence Nationale pour la Recherche project BRASERO (ANR-06-BLAN-0045), Natural Sciences and Engineering Research Council of Canada (NSERC), Multiscale Modeling of Plants associated team (INRIA).

References

1. Allali, J., Chauve, C., Ferraro, P., Gaillard, A.-L.: Efficient chaining of seeds in ordered trees. arXiv:1007.0942v1 [q-bio.QM] (2010)
2. Altschul, S.F., Gish, W., Miller, W., Myers, E.W., Lipman, D.J.: Basic local alignment search tool. J. Mol. Biol. 215(3), 403–410 (1990)
3. Aluru, S. (ed.): Handbook of Computational Molecular Biology. CRC Press, Boca Raton (2005)
4. Backofen, R., Will, S.: Local sequence-structure motifs in RNA. J. Bioinform. Comput. Biol. 2(4), 681–698 (2004)
5. Demaine, E.D., Mozes, S., Rossman, B., Weimann, O.: An optimal decomposition algorithm for tree edit distance. ACM Trans. Algorithms 6(1), Article 2 (2009)
6. Gardner, P.P., Daub, J., Tate, J.G., et al.: Rfam: updates to the RNA families database. Nucleic Acids Res. 37(Database issue), D136–D140 (2009)
7. Gusfield, D.: Algorithms on Strings, Trees and Sequences. Cambridge University Press, Cambridge (1997)
8. Heyne, S., Will, S., Beckstette, M., Backofen, R.: Lightweight comparison of RNAs based on exact sequence-structure matches. Bioinformatics 25(16), 2095–2102 (2009)
9. Jiang, T., Lin, G., Ma, B., Zhang, K.: A general edit distance between RNA structures. J. Comput. Biol. 9(2), 371–388 (2002)
10. Joseph, D., Meidanis, J., Tiwari, P.: Determining DNA sequence similarity using maximum independent set algorithms for interval graphs. In: Nurmi, O., Ukkonen, E. (eds.) SWAT 1992. LNCS, vol. 621, pp. 326–337. Springer, Heidelberg (1992)
11. Lipman, D.J., Pearson, W.R.: Rapid and sensitive protein similarity searches. Science 227(4693), 1435–1441 (1985)
12. Lozano, A., Pinter, R.Y., Rokhlenko, O., Valiente, G., Ziv-Ukelson, M.: Seeded tree alignment. IEEE/ACM TCBB 5(4), 503–513 (2008)
13. Ohlebusch, E., Abouelhoda, M.I.: Chaining Algorithms and Applications in Comparative Genomics. In: Handbook of Computational Molecular Biology. CRC Press, Boca Raton (2005)
14. Pearson, W.R., Lipman, D.J.: Improved tools for biological sequence comparison. PNAS 85(8), 2444–2448 (1988)
15. Pedersen, J.S., et al.: Identification and classification of conserved RNA secondary structures in the human genome. PLoS Comput. Biol. 2(4), e33 (2006)
16. Shapiro, B.A., Zhang, K.: Comparing multiple RNA secondary structures using tree comparisons. CABIOS 6, 309–318 (1990)
17. Zhang, K., Shasha, D.: Simple fast algorithms for the editing distance between trees and related problems. SIAM J. Comput. 18(6), 1245–1262 (1989)

Appendix

Time Complexity of Algorithm 2

To establish the worst-case complexity of Algorithm 2, we have to study the cost of the algorithm for each f values. To ease the reading, we denote by n_1 the size of Q and n_2 the size of T. Without loss of generality, we furthermore assume that $n_2 \leq n_1$.

Following invariants $R1$ and $R2$, each list of R contains at most $min(m, n_2)$ elements, as there are no $(y, s), (y', s') \in R[a, c]$ s.t. $r^y_y = r^{y'}_{y'}$, and $|X| \leq min(\|S\|, n_1 n_2)$. Thus, in the worst-case, we have at most $O(n_1^2 n_2^2)$ different chainable areas, $|R| = O(n_1 n_2)$, for all (a, c): $|R[a, c]| = O(n_2)$ and $|X| = O(n_1 n_2)$.

- $f = -1$ line 5: Over the whole execution of the algorithm each $M[a, l(r_j) - 1, c, l(r^j_j) - 1]$ is computed only once for all possible quadruplets as there is no (i, f, j), $(i', f', j') \in J$ such that $(l(r_j), l(r^j_j)) = (l(r_{j'}), l(r^{j'}_{j'}))$. Each computation require a search in $R[a, c]$ that can be done in $O(log$ $(n_2))$. Thus, the total time complexity for this case is $O(n_1^2 n_2^2 log(n_2))$.
- $f = 0$ line 8: The computation line 10 can be store in a dedicated array M' such that the best chain of the area (a, b, c, d) is computed only once. Thus, over all the execution of the algorithm, each different chainable area requires a search into a $R[a, c]$ and the total time complexity for this case is $O(\|S\| + n_1^2 n_2^2 \log(n_2))$.
- $f = 1$ line 11: This case is run once peer seeds, so $O(m)$ times. Each run costs $O(n_1 n_2 \log(n_2))$ and the total time complexity is $O(m n_1 n_2 \log(n_2))$.

From above, we conclude that the worst-case time complexity of our algorithm is

$$
\begin{aligned}
&O(\|S\| \log(\|S\|) + n_1^2 n_2^2 log(n_2) + \|S\| + n_1^2 n_2^2 \log(n_2) + m n_1 n_2 \log(n_2)) \\
&= O(\|S\| \log(\|S\|) + n_1 n_2 \log(n_2)(n_1 n_2 + m) + \|S\|) \\
&= O(\|S\| \log(\|S\|) + n_1 n_2 \log(n_2)(n_1 n_2 + m))
\end{aligned}
$$

which represents an improvement of the worst-case complexity of Heyne et al. algorithm [8].

To conclude, we can merge the worst-case complexity analysis with the time complexity analysis of section 4.2 leading to the following time complexity for Algorithm 2:

$$
\begin{aligned}
O(\ &\|S\| && \text{computing the chainable areas} \\
&+\|S\| \log(\|S\|) && \text{sorting the areas} \\
&+ \min(m, n_1 n_2) \times \min(\|S\|, n_1 n_2) \times log(\min(m, n_2)) && f = -1 \text{ case} \\
&+\|S\| + \min(\|S\|, n_1^2 n_2^2) \times \log(\min(m, n_2)) && f = 0 \text{ case} \\
&+m \times \min(\|S\|, n_1 n_2) \log(\min(m, n_2)) && f = 1 \text{ case}
\end{aligned}
$$

as $|X| \leq \min(\|S\|, n_1 n_2)$ and $|R[a, c]| \leq \min(m, n_2)$ for all a, c.

On the Computational Complexity of Degenerate Unit Distance Representations of Graphs

Boris Horvat[1,⋆], Jan Kratochvíl[2,⋆⋆], and Tomaž Pisanski[3]

[1] IMFM, University of Ljubljana, Slovenia
boris.horvat@fmf.uni-lj.si
[2] Charles University, Prague, Czech Republic
honza@kam.mff.cuni.cz
[3] IMFM, University of Ljubljana and University of Primorska, Slovenia
tomaz.pisanski@fmf.uni-lj.si

Abstract. Some graphs admit drawings in the Euclidean k-space in such a (natural) way, that edges are represented as line segments of unit length. Such embeddings are called k-dimensional unit distance representations. The embedding is strict if the distances of points representing nonadjacent pairs of vertices are different than 1. When two non-adjacent vertices are drawn in the same point, we say that the representation is degenerate. Computational complexity of nondegenerate embeddings has been studied before. We initiate the study of the computational complexity of (possibly) degenerate embeddings. In particular we prove that for every $k \geq 2$, deciding if an input graph has a (possibly) degenerate k-dimensional unit distance representation is NP-hard.

Keywords: unit distance graph; the dimension of a graph; the Euclidean dimension of a graph; degenerate representation; complexity;

Mathematics Subject Classification: 05C62, 05C12, 68Q17.

1 Introduction and Background

1.1 Notation

In this paper we consider finite undirected graphs without loops or multiple edges. The vertex set of a graph G is denoted by $V(G)$ and its edge set is denoted by $E(G)$. Edges are two-element subsets of the vertex set and we use the simplified notation uv for an edge $\{u, v\}$. When the graph G is clear from the concept, we also write $u \sim v$ when vertices u and v are adjacent, i.e., when $uv \in E(G)$.

⋆ This work was supported in part by the research program P1-0294 of the Slovenian Agency for Research and in part by grant L1-0696 from Ministry of high education, science and technology of the Republic of Slovenia.

⋆⋆ Research supported by Czech research grants 1M0545 and MSM0021620838.

C.S. Iliopoulos and W.F. Smyth (Eds.): IWOCA 2010, LNCS 6460, pp. 274–285, 2011.

The union $G_1 \cup G_2$ of graphs G_1 and G_2 is the graph with vertex set $V(G_1) \cup V(G_2)$ and edge set $E(G_1) \cup E(G_2)$. The disjoint union of two graphs G_1 and G_2, denoted by $G_1 \tilde{\cup} G_2$, is the graph obtained by taking the union of G_1 and G_2 on disjoint vertex sets, $V(G_1)$ and $V(G_2)$. The join of two simple graphs G_1 and G_2, written $G_1 * G_2$ is the graph obtained by taking the disjoint union of G_1 and G_2 and adding all edges joining $V(G_1)$ and $V(G_2)$.

1.2 Unit Distance Representations

Given a set S of points in the plane, the *unit distance graph of* S is the graph whose vertices are these points and edges connect vertices whose Euclidean distance is 1. A graph is called a *unit distance graph* if it is isomorphic to the unit distance graph of some set of points in the plane. Unit distance graphs play an important role in the famous question of the chromatic number of the plane, the so called Hadwiger-Nelson problem, also popularized by P. Erdős (cf. a survey [16]).

From the computational complexity point of view, David Eppstein recently remarked [6]: "I'm pretty sure that the Eades-Whitesides logic engine technique can be used to show that it's NP-hard to test whether a graph is a unit distance graph, but I haven't worked through the details carefully and I haven't succeeded in finding a paper that states the hardness of this problem explicitly". This indeed is true. In fact, the proof of Eades and Whitesides [4] of NP-hardness of the nearest neighbor graph in the plane can be adapted to reach this goal, and even more, cf. Theorem 1. Moreover, Eades and Wormald [5] proved that it is NP-hard to decide if a given graph has a planar embedding such that edges are non-crossing, are represented by straight line segments and have unit length (they call such graphs *matchstick graphs*). It should be noted that membership in NP is not known. The recognition problem is in PSPACE, as we will argue in a broader sense below. Recently Cabello, Demaine and Rote proved that it is NP-hard to find straight-line embeddings of planar 3-connected infinitesimally rigid graphs with unit edge lengths; see [3].

Many more authors have considered questions related to unit distance representations of graphs, their generalizations to higher dimensions or further restrictions; see for example [2,9,10,12,13,14]. The terminology is not fully uniform. For instance, the definition we used above requires that non-edges have lengths different than 1, while some authors do not require this. More importantly, an embedding of a graph in the plane (or higher dimensional space) implicitly assumes that different vertices are represented by distinct points. Viewing an embedding as a mapping from the vertex set into the Euclidean space and lifting the requirement of injectivity of this mapping opens a new view on unit distance representations. The aim of our paper is to study these problems from the computational complexity view. For better orientation in the different variants considered, we offer the following unifying definitions:

Definition 1. *A k-dimensional* unit distance representation *of a graph G is a mapping $\rho : V(G) \to \mathbb{R}^k$ such that for every edge $uv \in E(G)$, the Euclidean distance of $\rho(u)$ and $\rho(v)$ equals 1. The representation is* strict *if for any pair of* *nonadjacent vertices* u *and* v*, the distance of* $\rho(u)$ *and* $\rho(v)$ *is different from 1. The representation is* nondegenerate *if* ρ *is injective.*

Definition 2. *For a graph* G*, we set*

$$\dim_{NS}(G) = \min\ k, \text{ such that } G \text{ has a nondegenerate strict } k\text{-dimensional unit distance representation,}$$

$$\dim_{NW}(G) = \min\ k, \text{ such that } G \text{ has a nondegenerate } k\text{-dimensional unit distance representation,}$$

$$\dim_{DS}(G) = \min\ k, \text{ such that } G \text{ has a strict } k\text{-dimensional unit distance representation}$$

and

$$\dim_{DW}(G) = \min\ k, \text{ such that } G \text{ has a } k\text{-dimensional unit distance representation.}$$

In accordance with the above notation we will sometimes call a representation *weak* in the sense of *not necessarily strict*, and *degenerate* in the sense of *not necessarily nondegenerate*.

A graph G is a unit distance graph (in the plane) if and only if $\dim_{NS}(G) \leq 2$. Erdös *et al.* [7] defined the *dimension* of a graph G, they denoted as $\dim(G)$, as the minimum integer k such that G has a nondegenerate k-dimensional unit distance representation; hence their dimension corresponds to our $\dim_{NW}(G)$. Other authors (e.g. [2,12,13,14]) defined the *Euclidean dimension* of G as what we call $\dim_{NS}(G)$. A graph with $\dim_{NS}(G) = k$ is also called a k-dimensional *strict unit distance graph* (or, following Boben *et al.* [1], k-dimensional *unit distance coordinatization*).

We also note the connection to graph homomorphisms (i.e., edge-preserving vertex-mappings between graphs):

Observation 1. *A graph* G *has* $\dim_{DW}(G) \leq k$ *if and only if there exits a homomorphism from* G *to a graph* H *with* $\dim_{NS}(H) \leq k$*, and this happens if and only if there is a homomorphism from* G *to a graph* H' *with* $\dim_{NW}(H') \leq k$*.*

Proof. Let $\rho : V(G) \to \mathbb{R}^k$ be a unit distance representation of G. Let H be the unit distance graph defined by $\rho(V(G))$. Then $\dim_{NS}(H) \leq k$ and ρ is a homomorphism from G to H. Since $\dim_{NW}(H) \leq \dim_{NS}(H) \leq k$, H serves as H' as well. If, on the other hand, $\dim_{NW}(H') \leq k$ and $\sigma : G \to H'$ is a homomorphism, a weak k-dimensional unit distance representation of G is obtained as the composition of σ and $\tau : V(H') \to \mathbb{R}^k$ (the embedding witnessing $\dim_{NW}(H') \leq k$). □

Another easy but useful observation is the following.

Observation 2. *Let* $\rho : V(G) \to \mathbb{R}^k$ *be a strict unit representation of a graph* G*. If* $\rho(u) = \rho(v)$ *for some pair of (necessarily nonadjacent) vertices* u *and* v*, then these vertices have the same open neighborhoods.*

Proof. If x is adjacent to u but not to v, then $\rho(x)$ should be at distance 1 from $\rho(u)$, but at different distance from $\rho(v) = \rho(u)$, what is impossible. □

The following observation is clear from the definition:

Observation 3. *For every connected graph G, it holds*

$$\dim_{DW}(G) \leq \dim_{DS}(G) \leq \dim_{NS}(G),$$

and

$$\dim_{DW}(G) \leq \dim_{NW}(G) \leq \dim_{NS}(G).$$

The wheel graph W_n on $n \geq 4$ vertices is defined as $W_n = W_{1,n-1} = K_1 * C_{n-1}$, where K_1 is the one vertex graph and C_{n-1} is the cycle on $n-1$ vertices. It is well known [2] that $\dim_{NW}(W_n) = 3$ for $n \neq 7$ and $\dim_{NW}(W_7) = 2$. Even though $\dim_{NS}(W_5) = \dim_{NW}(W_5) = 3$, there exist degenerate planar unit distance representations of W_5; see Fig. 1.

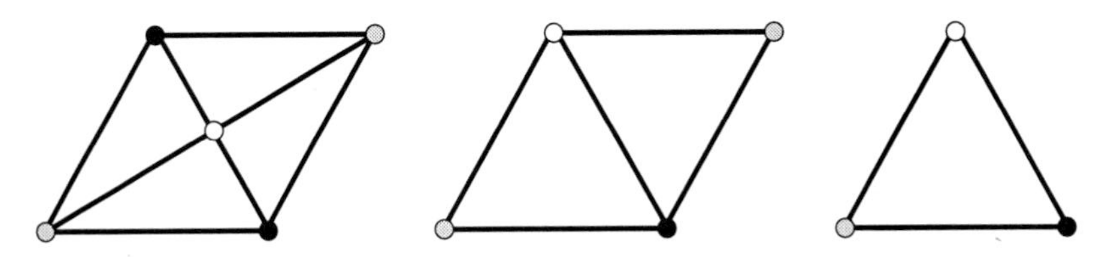

Fig. 1. The wheel graph W_5 (on the left), which admits a proper 3 coloring, is not a planar unit distance graph. Degenerate planar unit distance representations of the wheel graph W_5 are obtained using black vertex identification (in the middle) and possibly additional gray vertex identification (on the right).

The only connected graph with zero dimension (in all four variants) is K_1. For dimension one we have an easy observation:

Observation 4. *For a connected graph G, $\dim_{NS}(G) = 1$ if and only if G is a path with at least two vertices. The same holds true for $\dim_{NW}(G) = 1$.*

For any graph G, $\dim_{DW}(G) = 1$ if and only if G is a bipartite graph with at least one edge.

For a connected graph G, $\dim_{DS}(G) = 1$ if and only if G can be obtained from a path P with at least two vertices by replacing its vertices by independent sets, and replacing the edges of the path P by complete bipartite graphs between the corresponding independent sets.

All these graphs can be recognized in polynomial time. The problems become much harder for dimension two and higher. The logic engine construction of Eades and Whitesides [4] implicitly shows NP-hardness of several variants.

Theorem 1. *Deciding $\dim_{NS}(G) \leq 2$, $\dim_{NW}(G) \leq 2$, and $\dim_{DS}(G) \leq 2$ are NP-hard problems.*

Proof. The crucial building block of their construction of the logic engine is the wheel W_7 which has a unique nondegenerate unit distance representation in the plane. It follows from the construction that if the formula Φ one reduces from is not NAE-satisfiable, then the constructed graph G_Φ satisfies $\dim_{NW}(G_\Phi) > 2$, and hence also $\dim_{NS}(G_\Phi) > 2$. But since in G_Φ no two vertices have the same neighborhood, every strict unit distance representation is nondegenerate by Observation 1.2, and hence $\dim_{DS}(G_\Phi) > 2$ as well. If, on the other hand, Φ is NAE-satisfiable, G_Φ has a non-crossing strict nondegenerate unit distance representation in the plane. □

The aim of our paper is to discuss the complexity of degenerate representations. In Sections 3 and 4 we will prove the following theorem.

Theorem 2. *For every $k \geq 2$, the problem of deciding whether $\dim_{DW}(G) \leq k$ is NP-hard.*

2 PSPACE Membership

The goal of this section is to prove an upper bound on the complexity of the dimension problems. We prove it in a stronger form, when k is part of the input.

Theorem 3. *All four problems of deciding whether $\dim_{NS}(G) \leq k$, $\dim_{NW}(G) \leq k$, $\dim_{DS}(G) \leq k$, and $\dim_{DW}(G) \leq k$ belong to PSPACE (even when both G and k are part of the input and the size of the input is measured as $n = |V(G)|$).*

Proof. If $k > 2n$ we answer "yes" without any computation, since every graph on n vertices has $\dim_{NS}(G) \leq 2n$ [14]. Hence we may assume $k = O(n)$.

We reduce to solvability of polynomial inequalities with integral coefficients in the reals, a problem that is known to be in PSPACE [15]. Given a graph G, we introduce k variables $u_1, u_2, \ldots, u_k$ for every vertex $u \in V(G)$. The total number of variables is $O(n^2)$.

For every edge $uv \in E(G)$, we add inequalities

$$(u_1 - v_1)^2 + (u_2 - v_2)^2 + \ldots + (u_k - v_k)^2 \geq 1$$

and

$$(u_1 - v_1)^2 + (u_2 - v_2)^2 + \ldots + (u_k - v_k)^2 \leq 1.$$

Obviously, since the values of a solution correspond to coordinates of a k-dimensional unit distance representation, this system of inequalities has a real solution if and only if $\dim_{DW}(G) \leq k$.

If we do not want to allow degenerate representations, we add an inequality

$$(u_1 - v_1)^2 + (u_2 - v_2)^2 + \ldots + (u_k - v_k)^2 > 0$$

for every two distinct vertices u and v. Solvability of such enlarged system becomes equivalent to $\dim_{NW}(G) \leq k$.

If instead, we aim at strict representations, we add inequalities

$$((u_1 - v_1)^2 + (u_2 - v_2)^2 + \ldots + (u_k - v_k)^2 - 1)^2 > 0$$

for all nonedges $uv \notin E(G)$. In such a way the enlarged system of inequalities describes $\dim_{DS}(G) \leq k$.

If we consider the conjunction of all above introduced inequalities, we obtain a system whose solvability is equivalent to $\dim_{NS}(G) \leq k$.

Note that in each of the cases we have $O(n^2)$ inequalities. □

3 2-Dimensional Unit Distance Representations

In this section we prove the case $k = 2$ of Theorem 2.

Theorem 4. *Deciding if* $\dim_{DW}(G) \leq 2$ *is NP-hard.*

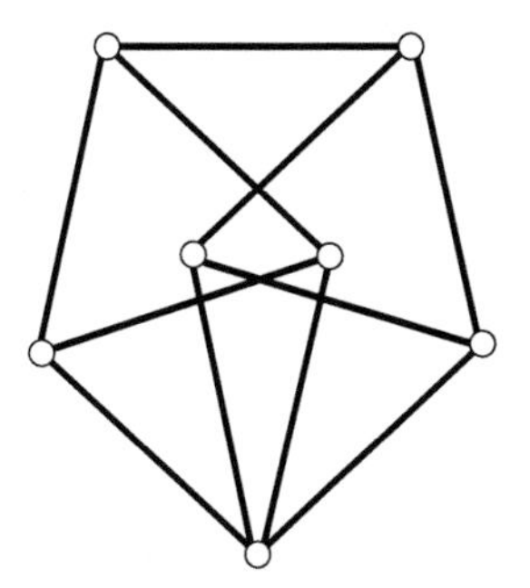

Fig. 2. The well known Moser graph (Moser spindle) is a 2-dimensional strict unit distance graph

Proof. The proof is based on the Moser graph M (see Fig. 2) which is a well known 2-dimensional strict unit distance graph. It has the property that every homomorphism to a unit distance graph H is injective and the image H is isomorphic to the Moser graph itself. In particular, adding any edge to the Moser graph results in a graph with dimension at least 3. Hence $\dim_{NW}(M) = \dim_{NS}(M) = \dim_{DW}(M) = \dim_{DS}(M) = 2$. Using Laman's Theorem, see e.g. [8], it is not hard to show that the unit distance coordinatization of Moser graph is rigid.

The NP-hardness reduction goes from 3SAT. Given a formula Φ with a set of variables X and a set of clauses C (each clause containing exactly 3 literals), we construct a graph G_Φ, such that $\dim_{DS}(G_\Phi) \leq 2$ if and only if Φ is satisfiable.

The cornerstone of the construction is a copy of the Moser graph with vertices $\{u, T, F, u', u'', v, w\}$ and edges as drawn in Fig. 3. For every variable $x \in X$, we add an edge $x'x''$, where x' represents the positive literal and x'' the negation of x. Also each x' and x'' are made adjacent to both u and u'. The part of the construction of G_Φ that corresponds to a variable $x \in X$ is illustrated by bold edges in Fig. 3.

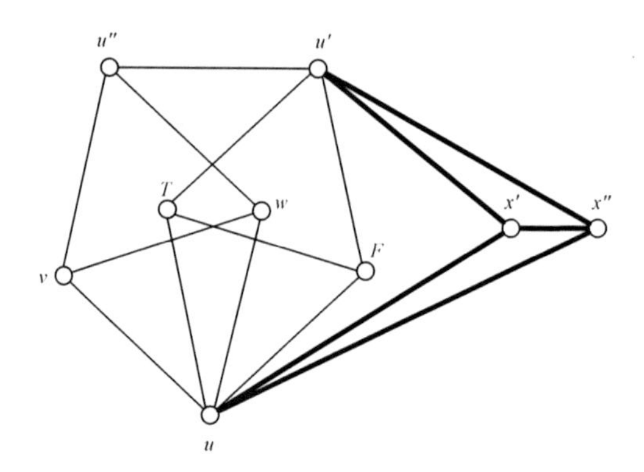

Fig. 3. The variable part of the construction of G_Φ

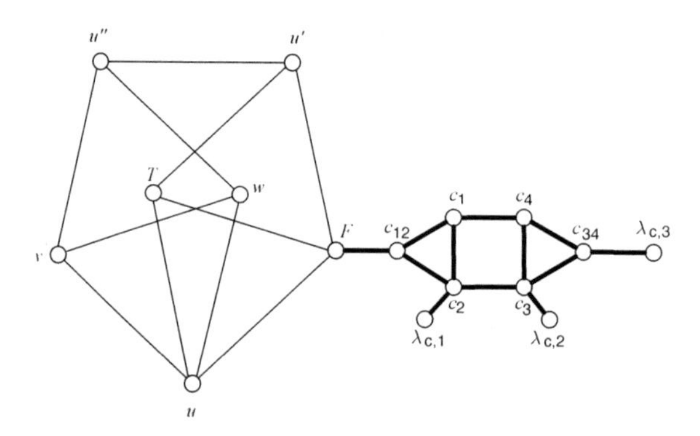

Fig. 4. The clause part of G_Φ

For every clause $c \in C$, say $c = (\lambda_{c,1} \vee \lambda_{c,2} \vee \lambda_{c,3})$, the clause gadget is depicted in Fig. 4. It contains six vertices, three of them being connected to the literals appearing in c (one to each) and one adjacent to the special vertex F of the Moser graph.

Formally,

$$V(G_\Phi) = \{u, v, w, u', u'', T, F\} \cup \bigcup_{x \in X} \{x', x''\} \cup \bigcup_{c \in C} \{c_1, c_2, c_3, c_4, c_{12}, c_{34}\}$$

and

$$\begin{aligned} E(G_\Phi) = \{&uT, uF, TF, Tu', Fu', uv, uw, vw, vu'', wu'', u'u''\} \cup \\ &\bigcup_{x \in X} \{x'x'', x'u, x''u, x'u', x''u'\} \cup \\ &\bigcup_{c \in C} \{c_1c_2, c_2c_3, c_3c_4, c_4c_1, c_1c_{12}, c_2c_{12}, c_3c_{34}, c_4c_{34}, \\ &\quad c_{12}F, c_2\lambda_{c,1}, c_3\lambda_{c,2}, c_{34}\lambda_{c,3}\}. \end{aligned}$$

Suppose first that $\dim_{DW}(G_\Phi) \leq 2$ (i.e., G_Φ is homomorphic to a graph H with $\dim_{NS}(H) = 2$). Consider a strict 2-dimensional unit distance representation of H. There are at most two paths of length two between two vertices of a planar unit distance graph [12]. Since there are four paths of length two between u and u' in G_Φ, the homomorphism either places x' in T and x'' in F, or vice versa. This placement defines a truth assignment on the variables of Φ - we say that x is true if x' is placed in T, and that it is false otherwise. We claim that Φ is satisfied by this assignment. Suppose there is an unsatisfied clause, say

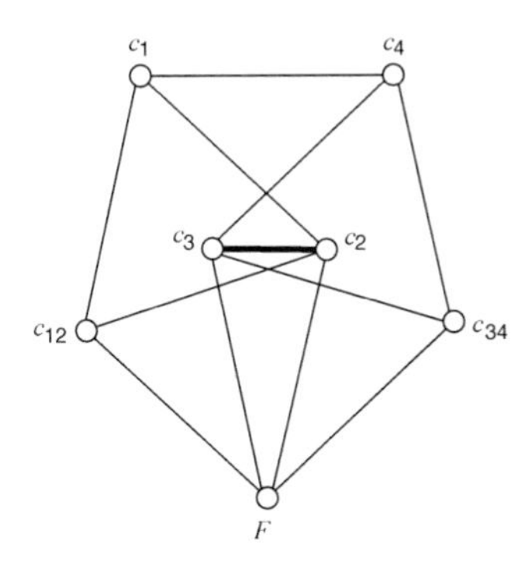

Fig. 5. The unsatisfied clause gadget, where $c = (\lambda_{c,1} \vee \lambda_{c,2} \vee \lambda_{c,3})$ and $\lambda_{c,1} = \lambda_{c,2} = \lambda_{c,3} =$ false. In the plane the edge c_2c_3 can not be represented by line segment of unit length.

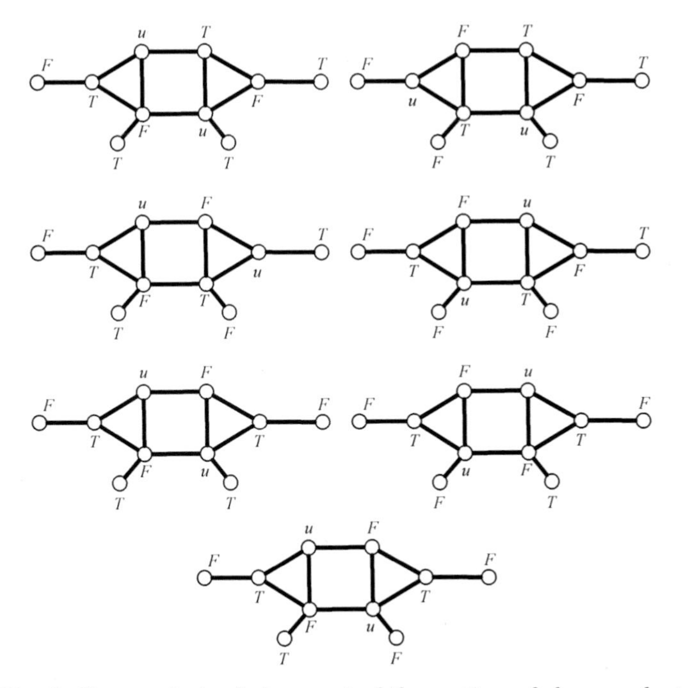

Fig. 6. Case analysis of placement of the vertices of clause gadgets

$c = (\lambda_{c,1} \vee \lambda_{c,2} \vee \lambda_{c,3})$. Then all three vertices $\lambda_{c,1}, \lambda_{c,2}, \lambda_{c,3}$ must be placed in F, and the clause gadget must map onto another copy of the Moser graph (with F being its degree four vertex). But then, in the plane, the edge c_2c_3 cannot have a unit length; see Fig. 5. We have a contradiction.

Suppose now that Φ is satisfied by a truth assignment $\phi : X \longrightarrow \{\mathsf{true}, \mathsf{false}\}$. We fix a strict planar unit distance realization of the Moser graph, place x' in T and x'' in F whenever $\phi(x) = \mathsf{true}$ (and vice versa if $\phi(x) = \mathsf{false}$). Following the case analysis in Fig. 6, the vertices of the clause gadgets can be placed in vertices u, T and F. Hence, we have constructed a well defined surjective homomorphism of G_Φ onto the Moser graph, which has a strict unit distance representation in the Euclidean plane. Therefore, $\dim_{DW}(G_\Phi) \leq 2$. □

4 *k*-Dimensional Unit Distance Representations

In this section, namely in Theorem 6, we prove the $k \geq 3$ part of Theorem 2.

Let $\mathsf{S}_k(\boldsymbol{s}, r)$ denote the k-dimensional (hyper)sphere in $\mathbb{R}^k$ with center in $\boldsymbol{s}$ and radius r. When the center and the radius of a sphere are not important, the abbreviation S_k will be used. The following two lemmas are obvious:

Lemma 1. *Let $k > 1$ be an integer. A non-empty non-degenerated intersection of two k-dimensional spheres with distinct centers is a $(k-1)$-dimensional sphere.*

Lemma 2. *Let S_2 denote the circle in the Euclidean plane that is circumscribed to a unit distance representation of the complete graph K_3 on three vertices. Let G be a connected graph with (possibly degenerate) planar unit distance representation which places all vertices of G into points that lie on S_2. Then $\chi(G) \leq 3$.*

Lemma 2 can be generalized. Let $G_{k,r,\alpha}$ denote the graph with vertices being points of a k-dimensional sphere $S_k(\mathbf{0}, r)$ with radius r in $\mathbb{R}^k$, where two vertices are connected if and only if they are at distance α. Lovász [11] proved the following inequalities.

Theorem 5. *Let $k \geq 3$ be a natural number. Then, for $0 \leq \alpha \leq 2$ holds*

$$k \leq \chi(G_{k,1,\alpha})$$

and

$$\chi\left(G_{k,1,\sqrt{\frac{2(k+1)}{k}}}\right) \leq k+1.$$

Corollary 1. *Let $k \geq 3$ be a natural number and let S_k denote the k-dimensional sphere in $\mathbb{R}^k$ that is circumscribed to the unit distance representation of the complete graph K_{k+1} on $k+1$ vertices. Let G be a connected graph with (possibly degenerate) k-dimensional unit distance representation which places all vertices of G into points that lie on S_k. Then $\chi(G) \leq k+1$.*

Proof. It is known that the circumradius of the (hyper)sphere that is circumscribed to the regular simplex with $k+1$ vertices and all sides of length ℓ, equals to $\ell\sqrt{\frac{k}{2(k+1)}}$. Thus, the circumradius of the sphere S_k is equal to $\sqrt{\frac{k}{2(k+1)}}$. The representation of the graph $G_{k,\sqrt{\frac{k}{2(k+1)}},1}$ on the sphere S_k can be (down)scaled to obtain a representation on a k-dimensional unit sphere (with radius one) that is circumscribed to the regular simplex with all sides of length $\sqrt{\frac{2(k+1)}{k}}$. Following Theorem 5, $\chi\left(G_{k,1,\sqrt{\frac{2(k+1)}{k}}}\right) \leq k+1$. A proper $(k+1)$-coloring of $G_{k,1,\sqrt{\frac{2(k+1)}{k}}}$ gives rise to a proper $(k+1)$-coloring of $G_{k,\sqrt{\frac{k}{2(k+1)}},1}$. Since G is a subgraph of $G_{k,\sqrt{\frac{k}{2(k+1)}},1}$, $\chi(G) \leq k+1$. □

Theorem 6. *Let $k \geq 3$ be an integer. Deciding if* $\dim_{DW}(G) \leq k$ *is NP-hard.*

Proof. Let K_k' and K_k'' be two copies of the complete graph with $k \geq 3$ vertices. Let v, w', w'' be additional vertices, such that $v, w', w'' \notin V(K_k') \cup V(K_k'')$. Denote $M_k' = K_k' * \{v, w'\}$, $M_k'' = K_k'' * \{v, w''\}$ and $M_k = M_k' \cup M_k'' \cup \{w'w''\}$, cf. Fig. 7. The graph M_k is well known Moser-Raiskii spindle.

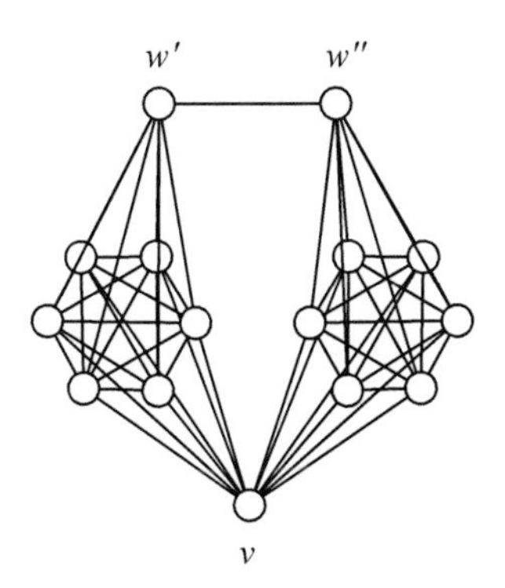

Fig. 7. The rigid graph M_6 contains two similar subgraphs, which are obtained as a graph join of the complete graph on six vertices and two disconnected vertices

Note that $\dim_{NS}(K_k') = \dim_{NW}(K_k') = k-1$. Since $K_k' * v$ is the complete graph on $k+1$ vertices, $\dim_{NS}(K_k' * v) = \dim_{NW}(K_k' * v) = k$. Applying another graph join does not change any of the dimensions. Thus, $\dim_{NS}(M_k') = \dim_{NS}(M_k'') = \dim_{NW}(M_k') = \dim_{NW}(M_k'') = k$. Both M_k' and M_k'' are rigid in $\mathbb{R}^k$. Consider a unit distance representation of $M_k' \cup M_k''$ in $\mathbb{R}^k$. We can rotate M_k'' around the representation of the vertex v, such that w' and w'' end up distance one apart. Hence $\dim_{NW}(M_k) = k$. It can be easily checked that all fixed points of the rotation of M_k'' around the axes of the rotation vw'' lie on the axes vw'', and at the same time none of the vertices of K_k'' lies on the axes vw''. Since there exist infinitely many angles of rotation and only finitely many possible non-strict situations and degeneracies, we can easily avoid problems. Hence, we can rotate the subgraph K_k'' in such a way that $\dim_{NS}(M_k) = k$. Note that M_k

has the property that every k-dimensional unit distance representation in $\mathbb{R}^k$ places the vertices v and w' in distinct points (this is guaranteed by the edge $w'w''$).

Suppose a connected graph G is given and its k-colorability is questioned. Let us construct the graph H_k such that $V(H_k) = V(G) \cup V(M_k)$ and

$$E(H_k) = E(G) \cup E(M_k) \cup \bigcup_{u \in V(G)} \{uv, uw'\},$$

where vertices v, w' are vertices from M_k.

We reduce from graph k-colorability. We will prove that $\dim_{DW}(H_k) \leq k$ if and only if $\chi(G) \leq k$.

Suppose H_k has a (possibly degenerate and possibly weak) unit distance representation ρ in $\mathbb{R}^k$. All vertices of G are one apart from $\rho(v)$ and from $\rho(w')$, and they all belong to the intersection of two unit k dimensional spheres with centers in $\rho(v)$ and $\rho(w')$. Using Lemma 1, a non-empty non-degenerated intersection of two k dimensional spheres with distinct centers is a $(k-1)$-dimensional one. All vertices of the complete graph K'_k belong to the same intersection and hence to the same $(k-1)$-dimensional sphere. Since G is connected, we can use Corollary 1 and $\chi(G) \leq k$.

Assume now that $\chi(G) \leq k$. Using a proper k-coloring of G we can map vertices of G into vertices of K'_k. Hence M_k is a homomorphic image of H_k and $\dim_{DS}(H_k) \leq k$. □

5 Conclusion

We have proven that deciding the existence of a weak degenerate k-dimensional unit representation is NP-hard for every $k > 1$, and noted that NP-hardness of 2-dimensional strict u.d. representations, 2-dimensional nondegenerate u.d. representations, and 2-dimensional strict nondegenerate u.d. representations are all NP-hard problems. The latter three results should be extendable to every $k \geq 2$.

Acknowledgment

We thank Ondra Suchý and Daniel Paulusma for useful discussions leading to improvement of Theorem 3.

References

1. Boben, M., Pisanski, T.: Polycyclic configurations. European J. Combin. 24(4), 431–457 (2003)
2. Buckley, F., Harary, F.: On the Euclidean dimension of a Wheel. Graphs Combin. 4, 23–30 (1988)
3. Cabello, S., Demaine, E., Rote, G.: Planar embeddings of graphs with specified edge lengths. In: Liotta, G. (ed.) GD 2003. LNCS, vol. 2912, pp. 283–294. Springer, Heidelberg (2004)

4. Eades, P., Whitesides, S.: The logic engine and the realization problem for nearest neighbor graphs. Theor. Comput. Sc. 169(1), 23–37 (1996)
5. Eades, P., Wormald, N.C.: Fixed edge-length graph drawing is NP-hard. Discrete Appl. Math. 28(2), 111–134 (1990)
6. Eppstein, D.: Blog entry (January 2010), `http://11011110.livejournal.com/188807.html`
7. Erdös, P., Harary, F., Tutte, W.T.: On the dimension of a graph. Mathematika 12, 118–122 (1965)
8. Hendrickson, B.: Conditions For Unique Graph Realizations. SIAM J. Comput. 21(1), 65–84 (1992)
9. Horvat, B., Pisanski, T.: Unit distance representations of the Petersen graph in the plane. Ars Combin. (to appear)
10. Žitnik, A., Horvat, B., Pisanski, T.: All generalized Petersen graphs are unit-distance graphs (submitted)
11. Lovász, L.: Self-dual polytopes and the chromatic number of distance graphs on the sphere. Acta Sci. Math. (Szeged) 45(1-4), 317–323 (1983)
12. Maehara, H.: On the Euclidean dimension of a complete multipartite graph. Discrete Math. 72, 285–289 (1988)
13. Maehara, H.: Note on Induced Subgraphs of the Unit Distance Graph. Discrete Comput. Geom. 4, 15–18 (1989)
14. Maehara, H., Rödl, V.: On the Dimension to Represent a Graph by a Unit Distance Graph. Graphs Combin. 6, 365–367 (1990)
15. Renegar, J.: On the computational complexity and geometry of the first-order theory of the reals, part I: Introduction. Preliminaries. The geometry of semi-algebraic sets. The decision problem for the exitential theory of the reals. J. Symb. Comput. 13, 255–300 (1992)
16. Soifer, A.: The Mathematical Coloring Book: Mathematics of Coloring and the Colorful Life of its Creators. Springer, Heidelberg (2008)

Recognition of Probe Ptolemaic Graphs*

(Extended Abstract)

Maw-Shang Chang and Ling-Ju Hung

Department of Computer Science and Information Engineering
National Chung Cheng University, Chiayi 62102, Taiwan
{mschang,hunglc}@cs.ccu.edu.tw

Abstract. Let $\mathcal{G}$ denote a graph class. An undirected graph G is called a *probe* $\mathcal{G}$ *graph* if one can make G a graph in $\mathcal{G}$ by adding edges between vertices in some independent set of G. By definition graph class $\mathcal{G}$ is a subclass of probe $\mathcal{G}$ graphs. *Ptolemaic graphs* are chordal and induced gem free. They form a subclass of both chordal graphs and distance-hereditary graphs. Many problems NP-hard on chordal graphs can be solved in polynomial time on ptolemaic graphs. We proposed an $O(nm)$-time algorithm to recognize probe ptolemaic graphs where n and m are the numbers of vertices and edges of the input graph respectively.

1 Introduction

A *probe graph* P is a two-tuple $(P_G = (P_V, P_E), P_L)$ where P_G is an undirected graph with vertex set P_V and edge set P_E and P_L is a function from P_V to the set $\{\mathbb{P}, \mathbb{N}, \mathbb{U}\}$ of labels, called *probes*, *nonprobes*, and *primes*, respectively and satisfying the condition that the set of nonprobes is an independent set of P_G. There are three classes of probe graphs: (i) *fully partitioned*: a fully partitioned probe graph has no primes; (ii) *unpartitioned*: all vertices are primes in an unpartitioned probe graphs; (iii) *partially partitioned*: all probe graphs that are neither fully partitioned nor unpartitioned are partially partitioned. A probe graph P^* is an *embedding* of probe graph P if it is obtained from P by two steps: (i) relabeling all primes in P as probes or nonprobes such that all nonprobes in P^* form an independent set in P_G and (ii) adding some edges between nonprobes after relabeling. Let $\mathcal{G}$ be a class of graphs. Probe graph P is called a *probe* $\mathcal{G}$ *graph* if there exists an embedding P^* of P such that $P^*_G \in \mathcal{G}$. Determine whether a fully partitioned probe graph is a probe $\mathcal{G}$ graph is a special case of the $\mathcal{G}$ graph sandwich problem [11]. The recognition of fully partitioned interval graphs arose from the physical mapping problem in the human genome project [15].

Many graph problems NP-hard on general graphs become solvable in polynomial time on some graph classes. For example, the Hamiltonian cycle problem is solvable in polynomial time on ptolemaic graphs. We consider the class of probe $\mathcal{G}$ graphs as a promising extension of graph class $\mathcal{G}$ to identify more graphs on

* This research is supported by National Science Council of Taiwan under grant no. NSC 95-2221-E-194-038-MY3.

C.S. Iliopoulos and W.F. Smyth (Eds.): IWOCA 2010, LNCS 6460, pp. 286–290, 2011.

Table 1. Some results and open problems on probe graphs. Here n and m are the number of vertices and the number of edges in a given probe graph and $|\mathbb{P}|$ denotes the number of vertices labeled $\mathbb{P}$ in a fully partitioned probe graph.

Graph class	Fully partitioned	Unpartitioned
probe chordal	$O(\|\mathbb{P}\|m)$ [3]	$O(m^2)$ [3]
probe strongly chordal	Poly. [5]	**Open**
probe chordal bipartite	Poly. [5]	**Open**
probe interval	$O(n+m)$ [14]	Poly. [8]
Probe DHG	$O(n^2)$ [4]	$O(nm)$ [10]
probe cographs	$O(n+m)$ [13]	$O(n+m)$ [13]
Probe bipartite DHG	$O(n^2)$ [4]	$O(nm)$ [9]
probe ptolemaic	$O(nm)$ [**this paper**]	$O(nm)$ [**this paper**]
probe comparability	$O(nm)$ [6]	**Open**
probe co-comparability	$O(n^3)$ [6]	**Open**
probe permutation	$O(n^2)$ [7]	**Open**
probe trivially-perfect	$O(n+m)$ [2]	$O(n+m)$ [2]
probe threshold	$O(n+m)$ [2]	$O(n+m)$ [2]

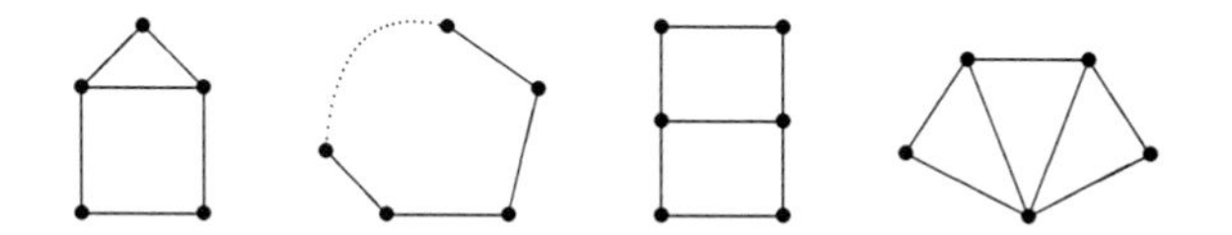

Fig. 1. A house, a hole, a domino, and a gem

which some NP-hard problems can be solved in polynomial time. As the first step of the study of probe $\mathcal{G}$ graphs, we developed polynomial time recognition algorithms. In Table 1, we list some recent results and open problems on the recognition of probe graphs of different graph classes. In this paper we give an $O(nm)$-time algorithm to recognize *partially partitioned probe ptolemaic graphs.*

Distance-hereditary graphs are those graphs having no house, hole, domino, or gem as an induced subgraph [1]. Ptolemaic graphs are those graphs that are chordal and have no gem as an induced subgraph [12]. Ptolemaic graphs are a subclass of chordal graphs and distance-hereditary graphs.

Let $G = (V, E)$ be a graph. For $v \in V$, $N_G(v)$ consists of all vertices adjacent to v in G, $N_G[v] = N_G(v) \cup \{v\}$, and for $X \subseteq V$, $N_G(X) = \bigcup_{v\in X} N_G(v) - X$. A *module* X is a vertex set of G such that $N_G(v) - X$ is the same for every $v \in X$. A *twin* is a pair of vertices u and v such that $\{u, v\}$ is a module, the pair is a *true twin* if u and v are adjacent and a *false twin* otherwise. A *clique* X in G is a vertex subset such that every two distinct vertices in X are adjacent. A *clique module* X of G is a vertex set of G that is both a module and a clique of G. A clique module is *trivial* if it consists of exactly a vertex. A *universal vertex* is a vertex adjacent to all other vertices. Suppose v is a vertex of ptolemaic graph $G = (V, E)$ where v is a non-universal vertex. Let C be a minimal component of $G[V - N_G[v]]$ where $G[V - N_G[v]]$ is a subgraph of G induced by $V - N_G[v]$ and $N_G(C)$ does not properly contain $N_G(C')$ for any component C' of $G[V - N_G[v]]$. Then either $N_G(C)$ is a non-trivial clique module of G or $N_G(C)$ consists of exactly a vertex that is a cut vertex of G. Notice that $N_G(C) \subseteq N_G(v)$. If $N_G(C)$ is a non-trivial clique module, then there is at least a pair of true twins in $N_G(C)$.

2 A Recognition Algorithm

To determine whether a probe graph P is a probe ptolemaic graph, we have to determine a label $\mathbb{P}$ or $\mathbb{N}$ for every prime of P and then add some edges between nonprobes to make P_G a member of ptolemaic graphs. The technique used in our algorithm is reduction, *i.e.*, reducing the problem into the same problem of smaller input size. Our algorithm is based upon the following characterization of probe ptolemaic graphs.

Theorem 1. *Suppose P is a partially or fully partitioned probe graph, every neighbor of a nonprobe is a probe and a non-universal vertex p is a probe. If P is a probe ptolemaic graph, then one of the following conditions holds:*

1. *A vertex in $N_{P_G}(p)$ is a cut vertex of P_G.*
2. *Two vertices in $N_{P_G}(p)$ are true twins in P_G and at least one of them is not a nonprobe.*
3. *Two vertices in $N_{P_G}(p)$ are false twins in P_G and none of them is a probe. If both of them are primes, then both of them are adjacent to a non-neighbor of p that is not a nonprobe.*
4. *Two vertices u and v in $N_{P_G}(p)$ satisfy all following conditions:*
   (a) $N_{P_G}[v] \subset N_{P_G}[u]$.
   (b) u is not a nonprobe.
   (c) v is not a probe.
   (d) If one of them is a nonprobe, then v is a nonprobe.

Reducing a cut vertex. A cut vertex of a probe ptolemaic graph is also a cut vertex of its minimal ptolemaic embedding. Suppose v is a cut vertex of P_G and C is a connected component of $P_G - v$ of minimum size. Then P is a probe ptolemaic graph if and only if $P - C$ has a ptolemaic embedding P' and $P[C + v]$ has a ptolemaic embedding P'' such that either $P'_L(v) = P''_L(v) = \mathbb{P}$ or $P'_L(v) = P''_L(v) = \mathbb{N}$.

Reducing true twins. Suppose u and v are true twins in P_G and at least one of them is not a nonprobe. Let u be a probe if at least one of u and v is a probe and let u be a prime and change its label from $\mathbb{U}$ to $\mathbb{P}$ if neither of them are probes. Then P is a probe ptolemaic graph iff $P - v$ is a probe ptolemaic graph.

Reducing probe true twins. A probe u and a nonprobe v in P are called *probe true twins* if $N_{P_G}[v] \subset N_{P_G}[u]$ and all vertices in $N_{P_G}[u] - N_{P_G}[v]$ are nonprobes. Suppose probe u and nonprobe v in P are probe true twins of P. Then P is a probe ptolemaic graph iff $P - v$ is a probe ptolemaic graph.

Reducing false twins. Suppose u and v are false twins in P_G and one of them is a nonprobe. Let u be the vertex that is a prime and change the label of

u from $\mathbb{U}$ to $\mathbb{N}$ if exactly one of them is a prime. Then P is a probe ptolemaic graph if and only if $P - v$ is a probe ptolemaic graph. Suppose two primes u and v are false twins in P_G and both of them are adjacent to two distinct vertices that are not adjacent to each other in P_G and one of them is a probe. Change the label of u from $\mathbb{U}$ to $\mathbb{N}$. Then P is a probe ptolemaic graph iff $P - v$ is a probe ptolemaic graph.

We sketch the algorithm in the following:

1. If the size of P is small, we solve the problem by brute force.
2. If P is unpartitioned, arbitrarily select an edge (u, v). Generate two partially partitioned probe graphs. They are obtained from P by relabeling u and v as probes, respectively. Then P is a probe ptolemaic graph iff one of them is. They can be checked by the algorithm for partially partitioned case.
3. If P is partially or fully partitioned, there is a probe p. If p is a universal vertex, then P is a probe trivially perfect graph and can be determined in linear time [2]. Otherwise we reduce the problem instance to smaller partially or fully partitioned probe graphs according to the cases given in Theorem 1.

Theorem 2. *There exists an $O(nm)$-time algorithm to check if a probe graph P is a probe ptolemaic graph.*

References

1. Bandelt, H.J., Mulder, H.M.: Distance-hereditary graphs. Journal of Combinatorial Theory, Series B 41, 182–208 (1986)
2. Bayer, D., Le, V.B., de Ridder, H.N.: Probe threshold and probe trivially perfect graphs. Theoretical Computer Science 410, 4812–4822 (2009)
3. Berry, A., Golumbic, M.C., Lipshteyn, M.: Recognizing Chordal Probe Graphs and Cycle-Bicolorable Graphs. SIAM J. Discrete Math. 21, 573–591 (2007)
4. Chandler, D.B., Chang, M.-S., Kloks, T., Liu, J., Peng, S.-L.: Recognition of probe cographs and partitioned probe distance hereditary graphs. In: Cheng, S.-W., Poon, C.K. (eds.) AAIM 2006. LNCS, vol. 4041, pp. 267–278. Springer, Heidelberg (2006)
5. Chandler, D.B., Guo, J., Kloks, T., Niedermeier, R.: Probe matrix problems: totally balanced matrices. In: Kao, M.-Y., Li, X.-Y. (eds.) AAIM 2007. LNCS, vol. 4508, pp. 368–377. Springer, Heidelberg (2007)
6. Chandler, D.B., Chang, M.-S., Kloks, T., Liu, J., Peng, S.-L.: Partitioned probe comparability graphs. Theoretical Computer Science 396, 212–222 (2008)
7. Chandler, D.B., Chang, M.-S., Kloks, A.J.J., Liu, J., Peng, S.-L.: On probe permutation graphs. Discrete Applied Mathematics 157, 2611–2619 (2009)
8. Chang, G.J., Kloks, A.J.J., Liu, J., Peng, S.-L.: The PIGs full monty - a floor show of minimal separators. In: Diekert, V., Durand, B. (eds.) STACS 2005. LNCS, vol. 3404, pp. 521–532. Springer, Heidelberg (2005)
9. Chang, M.-S., Hung, L.-J., Rossmanith, P.: Probe bipartite distance-hereditary graphs. In: Proceedings of NCS 2009: Workshop on Algorithms and Bioinformatics, pp. 16–27 (2009)

10. Chang, M.-S., Hung, L.-J., Rossmanith, P.: Probe distance-hereditary graphs. In: Proceedings of CATS 2010. CRPIT, vol. 109, pp. 55–64 (2010)
11. Golumbic, M.C., Kaplan, H., Shamir, R.: Graph sandwich problems. Journal of Algorithms 19, 449–473 (1995)
12. Howorka, E.: A characterization of ptolemaic graphs. Journal of Graph Theory 5, 323–331 (1981)
13. Le, V.B., de Ridder, H.N.: Characterisations and linear-time recognition of probe cographs. In: Brandstädt, A., Kratsch, D., Müller, H. (eds.) WG 2007. LNCS, vol. 4769, pp. 226–237. Springer, Heidelberg (2007)
14. McConnell, R.M., Nussbaum, Y.: Linear-time recognition of probe interval graphs. In: Fiat, A., Sanders, P. (eds.) ESA 2009. LNCS, vol. 5757, pp. 349–360. Springer, Heidelberg (2009)
15. Zhang, P.E., Schon, A., Fischer, S.G., Cayanis, E., Weiss, J., Kistler, S., Bourne, E.: An algorithm based on graph theory for the assembly of contigs in physical mapping of DNA. CABIOS 10, 309–317 (1994)

Graphs of Separability at Most Two: Structural Characterizations and Their Consequences

Ferdinando Cicalese[1] and Martin Milanič[2,⋆]

[1] Dipartimento di Informatica ed Applicazioni, University of Salerno, Fisciano, Italy
cicalese@dia.unisa.it
[2] FAMNIT and PINT, University of Primorska, Koper, Slovenia
martin.milanic@upr.si

Abstract. *Graphs of separability at most k* are defined as graphs in which every two non-adjacent vertices are separated by a set of at most k other vertices. For $k \in \{0,1\}$, the only connected graphs of separability at most k are complete graphs and block graphs, respectively. For $k \geq 3$, graphs of separability at most k form a rich class of graphs containing all graphs of maximum degree k. Graphs of separability at most 2 generalize complete graphs, cycles and trees. We prove several characterizations of graphs of separability at most 2 and examine some of their consequences.

1 Introduction

Let $G = (V,E)$ be a graph. The *separability* $\text{sep}_G(x,y)$ of two distinct non-adjacent vertices x,y in G is defined as the minimum cardinality of a set $S \subseteq V$ such that x and y are in different components of $G - S$. We define the *separability of a graph G*, denoted by $\text{sep}(G)$, as the maximum over all separabilities of non-adjacent vertex pairs (unless G is complete, in which case we define its separability to be 0). Notice that by definition, graphs of separability at most k are precisely the graphs in which every two non-adjacent vertices can be separated by removing a set of at most k other vertices. Hence, by Menger's Theorem, the separability of G is equal to the maximum number of internally vertex-disjoint paths connecting two non-adjacent vertices.

Graphs of separability at most k arise naturally in connection with the parsimony haplotyping problem from computational biology. We are interested in characterizations and structural properties of graphs of separability at most k, for small values of k. It can be easily seen that for every k, the set $\mathcal{G}_k$ of graphs of separability at most k is closed under vertex deletions; hence, with every graph G, the class $\mathcal{G}_k$ contains all induced subgraphs of G. Such graph classes are called *hereditary*. This family of graph classes is of particular interest, since hereditary (and only hereditary) classes admit a uniform description in terms of forbidden induced subgraphs. For a set $\mathcal{F}$ of graphs, we say that a graph G is $\mathcal{F}$*-free* if it does not contain an induced subgraph isomorphic to a member of $\mathcal{F}$. Given a hereditary class $\mathcal{G}$, denote by $\mathcal{F}$ the set of all graphs G with the property that

⋆ Supported in part by "Agencija za raziskovalno dejavnost Republike Slovenije", research program P1-0285.

C.S. Iliopoulos and W.F. Smyth (Eds.): IWOCA 2010, LNCS 6460, pp. 291–302, 2011.

$G \notin \mathcal{G}$ but $H \in \mathcal{G}$ for every proper induced subgraph H of G. The set $\mathcal{F}$ is said to be the set of forbidden induced subgraphs for $\mathcal{G}$, and $\mathcal{G}$ is precisely the class of $\mathcal{F}$*-free* graphs. The set $\mathcal{F}$ can be either finite or infinite, and many interesting classes of graphs can be characterized as being $\mathcal{F}$-free for some family $\mathcal{F}$. Such characterizations can be useful for establishing inclusion relations among hereditary graph classes, and were obtained, among others, for even-signable graphs [11], universally signable graphs [13], and for perfect graphs in the famous Strong Perfect Graph Theorem conjectured by Berge in 1961 [3] and proved by Chudnovsky, Robertson, Seymour and Thomas in 2006 [7].

There are also theorems that elucidate the structure of graphs in a certain hereditary class by showing that every graph in the class either belongs to one of a few basic classes (in which case it has a prescribed and relatively transparent structure) or it has one of a set of prescribed structural faults, along which it can be decomposed in a useful way. Several such decomposition results were obtained in recent years, including those for Meyniel graphs [5], perfect graphs [7], cap-free graphs [10], universally signable graphs [14], even-hole-free graphs [12], certain subclass of odd-hole-free graphs [9], and (diamond, even hole)-free graphs [23]. Few result of a stronger type are also known, in which the decomposition can also be reversed in the sense that a graph is in the class if and only if it can be constructed by gluing basic graphs along the decompositions prescribed. Such *composition* results are known for example for chordal graphs [19], claw-free graphs [8], graphs with no cycle with a unique chord [29] and bull-free graphs [6]. Decomposition results often have nice algorithmic consequences and provide means for obtaining bounds on certain graph parameters in terms of others.

We initiate the study of the structural properties of graphs of separability at most k, for small values of k. For $k \in \{0,1\}$, graphs of separability at most k are completely understood: Graphs of separability 0 are precisely the disjoint unions of complete graphs, and graphs of separability at most 1 are precisely the *block graphs*, that is, graphs every block of which is complete. From this description, a forbidden induced subgraph characterization is easy to obtain, and it is immediate how to build such graphs from the complete graphs. For $k \geq 3$, graphs of separability at most k form a rich class of graphs containing all graphs of maximum degree 3, as well as all pairwise k-separable graphs (defined by Miller [26]). The main focus of this paper is on the class of graphs of separability at most 2. These graphs form a common generalization of complete graphs, cycles and trees, and more generally, block graphs, cacti (graphs in which every edge belongs to at most one cycle), forests, and block-cactus graphs (graphs in which every block induces either a complete graph or a cycle).

Our results. We show in Section 2 that graphs of separability at most 2 are precisely the graphs that can be built from complete graphs and chordless cycles by an iterative application of the disjoint union operation and of pasting two disjoint graphs along a vertex or along an edge. In Section 3 we examine the unboundedness of the tree-width and the clique-width, when restricted to graphs of separability at most 2. We show that the structure theorem leads to polynomial-time solvability of several generally NP-hard problems, in this class. The structure theorem also implies the existence of an efficient recognition algorithm of graphs of separability at most 2. Interestingly, some well-known hard problems remain intractable when restricted to graphs of separability at most 2. In Section 4, we characterize the graphs of separability at most 2 in terms of minimal forbidden induced subgraphs and minimal forbidden induced minors; these characterizations imply that every graph of separability at most 2 is universally signable. Section 5 concludes the paper with some open problems. Due to the space limitations, some proofs are omitted.

Notation and definitions. All graphs considered are finite, simple and undirected. As usual, C_n and K_n denote the chordless cycle and the complete graph on n vertices, respectively, and $K_{s,t}$ the complete bipartite graph with parts of size s and t. For a vertex $x \in V(G)$, we denote by $N(x)$ the neighborhood of x, i.e., the set of vertices adjacent to x. The *degree* of x is the size of its neighborhood. For a set $A \subseteq V(G)$, we denote by $N(A)$ the set $\cup_{a\in A}\{u \in N(a) : u \notin A\}$, and for sets $A,B \subseteq V(G)$ we denote $N_B(A) := N(A) \cap B$. Unless stated otherwise, m and n will denote the number of edges and vertices of the graph under consideration. A graph G is *chordal* if every cycle in G on at least four vertices has a chord (an edge connecting two non-consecutive vertices of the cycle). We say that a graph G is obtained from two graphs G_1 and G_2 by *pasting along a k-clique*, and denote this by $G = G_1 \oplus_k G_2$, if for some $r \leq k$ there exist two r-cliques $K_1 = \{x_1,\ldots,x_r\} \subseteq V(G_1)$ and $K_2 = \{y_1,\ldots,y_r\} \subseteq V(G_2)$ such that G is isomorphic to the graph obtained from the disjoint union of G_1 and G_2 by identifying each x_i with y_i, for all $i = 1,\ldots,r$. In particular, if $k = 0$, then $G_1 \oplus_k G_2$ is the disjoint union of G_1 and G_2. For terms left undefined, we refer to [18].

2 A Structure Theorem for Graphs of Separability at Most Two

Complete graphs and (chordless) cycles are graphs of separability at most 2. The main result of this section is the following theorem, showing that complete graphs and cycles form the main building blocks for every graph of separability at most 2:

Theorem 1. *Let G be a graph. Then, G is of separability at most 2 if and only if G can be built from complete graphs and chordless cycles by an iterative application of pasting along 2-cliques.*

We start with a simple observation and a consequence of it.

Lemma 1. *The class $\mathcal{G}_2$ is closed under pasting along 2-cliques.*

Corollary 1. *Let $\mathcal{H}$ denote the minimal class of graphs that contains all complete graphs and chordless cycles, and is closed under pasting along 2-cliques. Then, $\mathcal{H} \subseteq \mathcal{G}_2$.*

A key result to the converse statement is the following lemma.

Lemma 2. *Let $G \in \mathcal{G}_2$. Then, either G is a complete graph, or G is a chordless cycle, or G has a separating clique of at most two vertices.*

Proof. Let $G \in \mathcal{G}_2$ which is neither complete nor a cycle. Suppose for contradiction that G has no separating cliques of at most two vertices. In particular, G is connected.

First, we show that G is chordal. Suppose for contradiction that G contains a chordless cycle C on at least 4 vertices, and let K denote a connected component of $G - V(C)$.

Then, the set $N_C(K)$ of neighbors of K in C is a clique, for otherwise any two non-adjacent vertices x and y in $N_C(K)$ would be connected by three internally vertex-disjoint paths in G (by two paths on the cycle C and another one with all its internal vertices in K). But now, the clique $N_C(K)$ is a separating clique in G of at most two vertices, contrary to the assumption.

It follows from a result of Dirac [19] that every connected chordal graph without separating cliques is a complete graph. Since G is chordal but not complete and using the above assumption, we conclude that G contains a (minimally) separating clique K of at least three vertices, say x_1, x_2, x_3. Let C_1 and C_2 be two connected components of $G-K$. By minimality of K, every vertex of K has a neighbor in C_j, for $j \in \{1,2\}$. For $i \in \{1,2,3\}$ and $j \in \{1,2\}$, let $v_j^i \in N(x_i) \cap C_j$, and let T_j be a minimal connected subgraph of C_j containing v_j^1, v_j^2 and v_j^3. Let $G' = (V', E')$ where $V' = V(T_1) \cup \{x_1, x_2, x_3\} \cup V(T_2)$ and $E' = E(T_1) \cup \{x_i v_j^i : i \in \{1,2,3\}, j \in \{1,2\}\} \cup E(T_2))$. Then G' is a subgraph of G that consists of three internally vertex disjoint paths, contrary to the fact that $G \in \mathcal{G}_2$. □

Proof of Theorem 1. We need to show that $\mathcal{G}_2 = \mathcal{H}$, where $\mathcal{H}$ is the class defined in Corollary 1. The inclusion $\mathcal{G}_2 \supseteq \mathcal{H}$ has been shown in Corollary 1.

Now we show the inclusion $\mathcal{G}_2 \subseteq \mathcal{H}$. Suppose that it fails, and let $G \in \mathcal{G}_2 \backslash \mathcal{H}$ be a minimal counterexample. By Lemma 2, either G is a complete graph, or G is a chordless cycle, or G has a separating clique of at most two vertices. Since complete graphs and chordless cycles belong to $\mathcal{H}$, we conclude that G has a separating clique K of at most two vertices. Thus, there exist graphs G_1 and G_2 such that $G = G_1 \oplus_2 G_2$. Both G_1 and G_2 are induced subgraphs of G, and therefore belong to $\mathcal{G}_2$. Since G is a minimal counterexample, both G_1 and G_2 belong to $\mathcal{H}$. But then, since $\mathcal{H}$ is closed under pasting along 2-cliques, it follows that $G = G_1 \oplus_2 G_2$ belongs to $\mathcal{H}$ too; a contradiction. □

As a relaxation of the concept of perfection, Gyárfás introduced in [22] the notion of χ-bounded classes. A hereditary class of graphs $\mathcal{G}$ is called *χ-bounded* if the chromatic number $\chi(G)$ of every graph in $\mathcal{G}$ can be bounded from above by a function of its maximum clique size $\omega(G)$: there exists a function f such that $\chi(G) \leq f(\omega(G))$ holds whenever $G \in \mathcal{G}$. The above structure theorem implies that graphs of separability at most 2 are χ-bounded:

Proposition 1. *For every graph $G \in \mathcal{G}_2$, it holds that $\chi(G) \leq \max\{3, \omega(G)\}$.*

Another immediate consequence of Theorem 1 is that every graph of separability at most 2 contains either a simplicial vertex (one whose neighborhood forms a clique), or a pair of adjacent vertices of degree 2. Further consequences of Theorem 1 will be discussed in the next section.

3 Algorithmic and Complexity Issues

In this section, we examine some algorithmic aspects of graphs of small separability. First, we observe that Theorem 1, together with a modified version of Tarjan's decomposition by clique separators [28] (in which we only consider separating cliques of size at most two), implies that graphs of separability at most 2 can be recognized in time $O(mn)$. An alternative approach, which may be faster for graphs with many edges, is to directly check, for all non-adjacent vertex pairs, whether there exist three internally vertex-disjoint paths connecting the two vertices. This can be done in time $O(n)$, using the algorithm by Nagamochi and Ibaraki [27], and the overall complexity of this approach is $O((\binom{n}{2} - m)n)$.

Combined with Tarjan's algorithm for decomposing a graph along its clique separators and its consequences [28], Theorem 1 implies the existence of polynomial-time solutions to several generally NP-hard problems, when restricted to graphs of separability at most 2:

Theorem 2. *The following problems are polynomially solvable:*

- *Given a vertex-weighted graph $G \in \mathcal{G}_2$, find a maximum-weight independent set in G.*
- *Given a vertex-weighted graph $G \in \mathcal{G}_2$, find a maximum-weight clique in G.*
- *Determine the chromatic number of a given graph $G \in \mathcal{G}_2$.*

In contrast to Theorem 2, it is NP-hard to find the maximum size of an independent set in a graph of separability at most 3. This follows from the fact that the maximum independent set problem is NP-hard for graphs of maximum degree 3. Moreover, using a reduction from the (NP-complete) 3-colorability problem in planar graphs of maximum degree 4, we show the following result.

Theorem 3. *The 3-colorability problem is NP-complete for planar graphs of separability 3 and of maximum degree 6.*

On the other hand, for every fixed k the maximum-weight clique problem is polynomial for graphs of separability at most k: Observe that every graph of separability at most k is $\{R_k\}$-free, where R_k is the graph obtained from the complete graph on $2\lfloor \frac{k}{2} \rfloor + 4$ vertices by deleting from it a perfect matching. It follows that every graph of separability at most k contains at most $O(n^{\lfloor k/2 \rfloor + 1})$ maximal cliques (see, e.g., [2]), which can be enumerated in polynomial time [30].

Some well-known problems remain NP-complete when restricted to graphs of separability at most 2. More specifically, using reductions from the vertex cover and the simple max cut problems in general graphs, we show the following result.

Theorem 4. *The dominating set problem and the (simple) max cut problem remain NP-complete when restricted to graphs of separability at most 2.*

Theorem 4 is best possible in the sense that both problems are solvable in polynomial time for graphs of separability at most 1. For the dominating set problem, this follows from the fact that graphs of separability at most 1 are of clique-width at most 4, and the results by Courcelle et al. about optimization problems expressible in Monadic Second Order Logic, on graphs of bounded clique-width [17]. Strictly speaking, to apply the theorem from [17], bounded clique-width does not suffice, there has to be a polynomial algorithm to construct a 4-expression for a given graph of separability 1. But this is not hard to obtain (see the paragraph following the proof of Poposition 3 below).

For the max cut problem, this follows from the fact that an optimal solution to the max cut problem for a given graph G can be easily obtained from optimal solutions to the max cut problem on the blocks of G.

Numerous problems that are NP-hard in general admit polynomial-time solutions when restricted to graphs of bounded tree-width or clique-width (see, e.g., [1, 16, 17, 20, 21, 24]). We now examine the (un)boundedness of these two graph parameters for graphs of separability at most 2. First, observe that since the tree-width of a complete graph K_n is equal to $n-1$, the tree-width of graphs in $\mathcal{G}_2$ is unbounded. However, it follows from Theorem 1 that the tree-width $tw(G)$ of a graph $G \in \mathcal{G}_2$ can only be large due to the presence of a large clique.

Proposition 2. *Let $G \in \mathcal{G}_2$. Then $tw(G) \leq \max\{2, \omega(G)-1\}$.*

Proposition 2 is best possible in the sense that the tree-width of graphs of separability at most 3 is not bounded from above by any function of their maximum clique size. (There exist triangle-free graphs of maximum degree 3 and of arbitrarily large tree-width.)

Proposition 3. *The clique-width of graphs of separability at most 2 is unbounded.*

Proof. For a graph G, let G^* denote the graph obtained from G by adding a new vertex v_e for each edge $e = xy \in E(G)$ together with the edges $\{v_e, x\}$ and $\{v_e, y\}$. Apply this transformation to a complete graph K_n to obtain the graph K_n^*. Then, $K_n^* \in \mathcal{G}_2$. It is known that the clique-width of graphs K_n^* is unbounded. In fact, it is at least $n/72$ for $n \geq 100$ [25]. Therefore, the clique-width is also unbounded in the class $\mathcal{G}_2$. □

By contrast, the clique-width of graphs of separability at most 1 is bounded. This follows from two facts: (i) the clique-width of complete graphs is at most 2, and (ii) the clique-width of every graph exceeds the maximum clique-width of its blocks by at most 2 [4]. Therefore, since every block of a graph $G \in \mathcal{G}_1$ is complete and hence of clique-width 2, the clique-width of G is at most 4 (and a 4-expression can be computed in polynomial time).

The complexity results for graphs of small separability discussed in this section are summarized in Table 1.

Table 1. Some complexity results for graphs of small separability
P stands for polynomial, NP-c for NP-complete, bdd for bounded, unbdd for unbounded.

	$\mathcal{G}_1$	$\mathcal{G}_2$	$\mathcal{G}_3$
recognition	P	P	P
CLIQUE	P	P	P
CHROMATIC NUMBER	P	P	NP-c
INDEPENDENT SET	P	P	NP-c
DOMINATING SET	P	NP-c	NP-c
MAX CUT	P	NP-c	NP-c
clique-width	bdd	unbdd	unbdd

4 Characterizations by Forbidden Substructures

In this section, we derive two characterizations of graphs of separability at most two by means of forbidden substructures: the one in terms of minimal forbidden induced subgraphs and the one in terms of minimal forbidden induced minors. We also show that for $k > 2$, graphs of separability at most k cannot be characterized by forbidden induced minors.

Minimal forbidden induced subgraphs. For every k, the set of graphs of separability at most k forms a hereditary graph class. For $k \leq 1$, the characterization of graphs in $\mathcal{G}_k$ in terms of minimal forbidden induced subgraphs is easy to obtain. Recall that for $k \in \{0,1\}$, the only connected graphs of separability at most k are complete graphs and block graphs, respectively. It follows that: (i) $\mathcal{G}_0$ coincides with the class of $\{P_3\}$-free graphs, where P_3 is a path on three vertices; (ii) $\mathcal{G}_1$ coincides with the class of $\{diamond, C_4, C_5, \ldots\}$-free graphs, where a *diamond* is a cycle of length 4 with exactly one chord.

We now provide the characterization of graphs in $\mathcal{G}_2$ in terms of minimal forbidden induced subgraphs. First, let us describe the forbidden induced subgraphs. The graph K_5^- is K_5 minus an edge. 3PC is an acronym for a *3-path configuration*, which is one of the graphs of type H_0, H_1 or H_2 depicted in Fig. 1. A graph of type H_0 is called a $3PC(x,y)$ where vertex x and vertex y are connected by three internally vertex-disjoint paths P_1, P_2 and P_3. A graph of type H_1 is called a $3PC(xyz,u)$, where xyz is a triangle in G and P_1, P_2 and P_3 are three internally vertex-disjoint paths with endpoints x, y and z respectively and a common endpoint u. A graph of type H_2 is called a $3PC(xyz,uvw)$ and consists of two vertex-disjoint triangles xyz and uvw and three disjoint paths P_1, P_2 and P_3 with endpoints x and u, y and v, and z and w, respectively. Furthermore in all three cases the vertices of $P_i \cup P_j$ $(i \neq j)$ must induce a *hole* (a chordless cycle of at least four vertices). This implies that all paths P_1, P_2, P_3 of H_0 have length greater than one, and at most one path of H_1 has length one. Wheels are graphs of type H_3 in Fig. 1; they consist of a hole called the *rim* together with a vertex called the center that has at least three neighbors on the rim.

Theorem 5. *Let G be a graph. Then, G is of separability at most 2 if and only if G contains no induced K_5^-, no induced 3PC and no induced wheel.*

Proof. Let $\mathcal{F}$ be the set of graphs consisting of the graph K_5^-, all 3PC's and all wheels. Let $\mathcal{F}'$ denote the set of minimal forbidden induced subgraphs for $\mathcal{G}_2$. We need to show that $\mathcal{F} = \mathcal{F}'$.

Let $F \in \mathcal{F}$ be one of the graphs shown in Fig. 1. It is straightforward to verify that F is of separability 3, while every proper induced subgraph of F is of separability at most 2 (this can be verified with the help of Theorem 1). Therefore, $F \in \mathcal{F}'$, and consequently $\mathcal{F} \subseteq \mathcal{F}'$.

It remains to show that $\mathcal{F}' \subseteq \mathcal{F}$, or equivalently, that if G is $\mathcal{F}$-free, then G is of separability at most 2. Suppose for contradiction that there exists an $\mathcal{F}$-free graph G such that the separability of G is at least 3. Among all non-adjacent vertex pairs $\{x,y\}$ in G with $\mathrm{sep}_G(x,y) \geq 3$, pick a pair $\{x,y\}$ such that the total length of three internally vertex-disjoint paths connecting x and y is as small as possible. Let P, Q, R be three

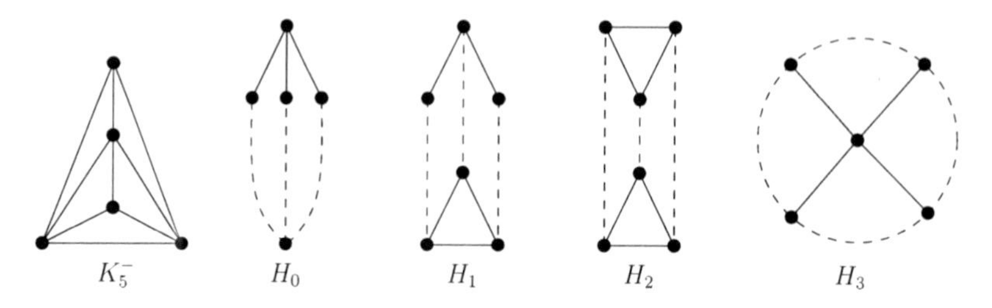

Fig. 1. Forbidden induced subgraphs for graphs of separability at most 2: K_5^-, 3-path configurations and wheels. A dotted line indicates a chordless path containing one or more edges. (Notice that the illustration of H_3 does not capture the definition of wheels in all its generality.)

internally vertex-disjoint paths connecting x and y with smallest total length, and let X be the subgraph of G induced by $V(P \cup Q \cup R)$. At least one of the paths P, Q, R contains at least three edges, for otherwise X would be either a $3PC(x,y)$ (if no edge connects internal vertices of the paths), a wheel (if there are at most two edges connecting internal vertices of the paths), or a K_5^- (otherwise) – contrary to the fact that G is $\mathcal{F}$-free.

A *PQ-chord* is an edge pq connecting an internal vertex p of P with an internal vertex q of Q. *PR*- and *QR*-chords are defined similarly. A *PQ*-, *PR*-, or a *QR*-chord will be simply called a chord. Clearly, there must exist a chord since otherwise G would contain an induced $3PC(x,y)$. The following claim restricts the set of possible chords.

Claim: Let pq be a *PQ*-chord. Then, either $p \in N(x)$ or $q \in N(y)$. Moreover, either $p \in N(y)$ or $q \in N(x)$.

Proof of claim: If pq is a *PQ*-chord such that $p \notin N(x)$ and $q \notin N(y)$, then $\{x,p\}$ would form a non-adjacent vertex pair in G with $\text{sep}_G(x,p) \geq 3$ and such that there exist three internally vertex-disjoint paths connecting x and p with total length shorter than that of $\{P,Q,R\}$, contrary to the choice of $\{x,y\}$. The other statement of the claim follows by symmetry.

Let p,q,r and p',q',r' denote the neighbors of x and y on the paths P,Q,R, respectively. The above claim implies that if both P and Q have at least three edges then every *PQ* chord is contained in the set $\{pq, p'q'\}$; similar statements hold for the other two pairs of paths. We split the rest of the proof into two exhaustive cases.

Case 1: Each of the three paths P,Q,R has at least three edges. Let C be the set of chords. By the above, $\emptyset \neq C \subseteq \{pq, pr, qr, p'q', p'r', q'r'\}$.

If $C = \{pq\}$, then G contains a $3PC(xpq,y)$, a contradiction.

If $C = \{pq, p'q'\}$, then G contains a $3PC(xpq, yp'q')$, a contradiction.

If $C = \{pq, p'r'\}$, then G contains a $3PC(xpq, r'p'y)$, a contradiction.

If $|C \cap \{pq, pr, qr\}| \geq 2$, say $\{pq, pr\} \subseteq C$, then $\{p,y\}$ forms a non-adjacent vertex pair in G with $\text{sep}_G(p,y) \geq 3$ and such that there exist three internally vertex-disjoint paths connecting p and y with total length shorter than that of $\{P,Q,R\}$, contrary to the choice of $\{x,y\}$.

Each remaining subcase is symmetric to one of the above subcases. This completes Case 1.

Case 2: P has only two edges. There must exist a QR-chord, since otherwise G would contain either an induced $3PC(x,y)$ or an induced wheel. We consider two further subcases.

Case 2.1. Both Q and R have at least three edges. Let C be the set of chords. Without loss of generality, $qr \in C$. Note that p (the internal vertex of P) must be contained in a chord, since otherwise G would contain either an induced $3PC(xqr,y)$ or an induced $3PC(xqr,yq'r')$. However, the only possible chords containing p are pq' and pr' since if pz is a PQ-chord with $z \neq q'$, then $\{z,y\}$ would form a non-adjacent vertex pair in G with $\text{sep}_G(z,y) \geq 3$ such that there exist three internally vertex-disjoint paths connecting z and y with total length shorter than that of $\{P,Q,R\}$, contrary to the choice of $\{x,y\}$. If $q'r' \in C$, then, by a similar argument as above, the only possible chords containing p would be pq and pr, a contradiction since $\{q,r\} \cap \{q',r'\} = \emptyset$. Therefore $q'r' \notin C$.

If $\{pq',pr'\} \subseteq C$, then $\{q',r'\}$ forms a non-adjacent vertex pair in G contradicting the choice of $\{x,y\}$. Therefore we may assume, without loss of generality, that $C \cap \{pq',pr'\} = \{pq'\}$. But now, G contains an induced $3PC(xqr,pq'y)$, a contradiction. This completes Case 2.1.

Case 2.2. Q has only two edges. Recall that there exists at least one QR-chord. By symmetry, there also exists a PR-chord. Suppose that there exists a PR-chord pw such that $w \notin \{r,r'\}$. Then, using an arbitrary QR-chord qz it is easy to see that either $\{w,x\}$ or $\{w,y\}$ forms a non-adjacent vertex pair in G contradicting the choice of $\{x,y\}$. Moreover, a similar argument shows that for every $w \in \{r,r'\}$, it is not possible that both pw and qw are chords. Therefore, we may assume that the only PR- and QR-chords are the chords pr and qr'.

If pq is also a chord then $\{p,r'\}$ forms a non-adjacent vertex pair in G contradicting the choice of $\{x,y\}$. On the other hand, if pq is not a chord, then G contains an induced $3PC(xpr,qyr')$. This contradiction completes the proof. □

Theorem 5, combined with the forbidden induced subgraph characterization of universally signable graphs from [13], implies that every graph of separability at most 2 is universally signable. (In conjunction with structural results about universally signable graphs [13], this also gives an alternative proof of Theorem 2.) On the other hand, it can be shown that for $k \geq 3$ not all graphs of separability at most k are universally signable.

Minimal forbidden induced minors. Given two graphs G and H, we say that G is an *induced minor* of H if G can be obtained from H by a sequence of vertex deletions and edge contractions. Clearly, every minor-closed graph class is closed under induced minors, and every class closed under induced minors is hereditary. In the following theorem, we characterize classes of graphs of bounded separability that are closed under induced minors.

Theorem 6. *The set of graphs of separability at most k is closed under induced minors if and only if $k \leq 2$.*

Proof. It is easy to check that for $k \in \{0,1\}$, the set $\mathcal{G}_k$ of graphs of separability at most k is closed under edge contraction. For $k = 2$, the fact that graphs of separability at most 2 are closed under induced minors can be shown by induction, using Theorem 1. First, observe that the set of cycles and complete graphs is closed under edge contraction. Suppose that a graph $G \in \mathcal{G}_2$ can be obtained from two smaller graphs $G_1, G_2 \in \mathcal{G}_2$ by pasting along a 2-clique, and let G' be the graph obtained from G by contracting an edge e. Then, either $e \in E(G_1)$ or $e \in E(G_2)$ (or both). Denoting by G_i' the graph obtained from G_i by contracting e, we see that G' can be obtained from G_1' and G_2' by pasting along a 2-clique. By the inductive hypothesis, $G_1', G_2' \in \mathcal{G}_2$, and since graphs of separability at most 2 are closed under pasting along 2-cliques, we also have $G' \in \mathcal{G}_2$.

Suppose now that $k \geq 3$. The graph G_6 depicted in Fig. 2 is of separability 3 but can be contracted to $K_{2,6}$—a graph of separability 6—contracting the "horizontal" edges.

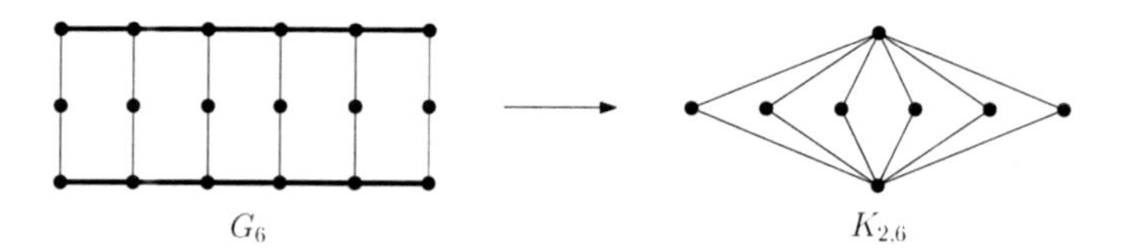

Fig. 2. A graph of separability 3 that can be contracted to a graph of separability 6

By generalizing the example from Fig. 2, it follows that for every $k \geq 3$, there exists a graph of separability 3 (and thus of separability at most k) that can be contracted to a graph of separability $k+1$. Therefore, for every $k \geq 3$, the set of graphs of separability at most k is not closed under induced minors. □

Just as every hereditary graph class can be uniquely characterized by the minimal set of forbidden induced subgraphs, every graph class closed under induced minors can be uniquely characterized by the minimal set of forbidden induced minors. Moreover, if $\mathcal{G}$ is a graph class closed under induced minors, and the set $\mathcal{F}$ of minimal forbidden induced subgraphs is known for graphs in $\mathcal{G}$, then it is not hard to obtain from $\mathcal{F}$ the set of $\mathcal{F}'$ of forbidden induced minors for $\mathcal{G}$. For a set of graphs $\mathcal{F}'$, we say that a graph is *$\mathcal{F}'$-induced-minor-free* if no induced minor of G is isomorphic to a graph in $\mathcal{F}'$. The following straightforward observation establishes the relation between the minimal sets of forbidden induced subgraphs and minimal forbidden induced minors.

Proposition 4. *Let $\mathcal{G}$ be a graph class closed under induced minors, and let $\mathcal{F}$ be the set of minimal forbidden induced subgraphs for $\mathcal{G}$. Then, $\mathcal{G}$ coincides with the set of all $\mathcal{F}'$-induced-minor-free graphs, where $\mathcal{F}'$ is the set of minimal elements in the poset on $\mathcal{F}$ partially ordered by contraction.*

Proposition 4 and the results on forbidden induced subgraphs yield:

Theorem 7. *(i) Graphs of separability 0 are precisely the $\{P_3\}$-induced-minor-free graphs. (ii) Graphs of separability at most 1 are precisely the $\{C_4, diamond\}$-induced-minor-free graphs. (iii) Graphs of separability at most 2 are precisely the $\{K_{2,3}, F_5, W_4, K_5^-\}$-induced-minor-free graphs, where $K_{2,3}, F_5, W_4, K_5^-$ are the four graphs depicted in Fig. 3.*

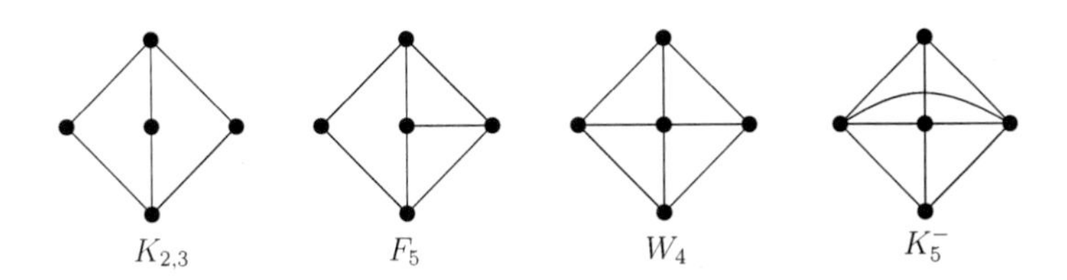

Fig. 3. Forbidden induced minors for graphs of separability at most 2

5 Open Problems

Problem 1. *For $k > 2$, characterize graphs of separability at most k that cannot be decomposed along separating cliques of size at most k. Are there other meaningful (de)composition operations for graphs of separability at most k?*

Problem 2. *For $k > 2$, determine whether graphs of separability at most k are χ-bounded.*

Problem 3. *For $k > 2$, characterize graphs of separability at most k in terms of minimal forbidden induced subgraphs.*

For a given hereditary graph class $\mathcal{G}$, the complexity of the independent dominating set problem is usually "sandwiched" between the complexities of the independent set and the dominating set problems. This motivates the following problem (which was suggested by Vadim Lozin):

Problem 4. *Determine the complexity of the independent dominating set problem for graphs of separability at most 2.*

Acknowledgements. We thank Marcin Kamiński for stimulating discussions and Vadim Lozin for suggesting Problem 4.

References

1. Arnborg, S., Proskurowski, A.: Linear time algorithms for NP-hard problems restricted to partial k-trees. Discrete Appl. Math. 23, 11–24 (1989)
2. Balas, E., Yu, C.S.: On graphs with polynomially solvable maximum-weight clique problem. Networks 19, 247–253 (1989)
3. Berge, C.: Färbung von Graphen, deren sämtliche bzw. deren ungerade Kreise starr sind. Wiss. Z. Martin-Luther-Univ. Halle-Wittenberg Math.-Natur. Reihe 10, 114 (1961)
4. Boliac, R., Lozin, V.: On the clique-width of graphs in hereditary classes. In: Bose, P., Morin, P. (eds.) ISAAC 2002. LNCS, vol. 2518, pp. 44–54. Springer, Heidelberg (2002)
5. Burlet, M., Fonlupt, J.: Polynomial algorithm to recognize a Meyniel graph. In: Berge, C., Chvátal, V. (eds.) Topics on Perfect Graphs. North-Holland Math. Stud., vol. 88, pp. 253–278. North-Holland, Amsterdam (1984)
6. Chudnovsky, M.: The structure of bull-free graphs I–III (submitted)
7. Chudnovsky, M., Robertson, N., Seymour, P., Thomas, R.: The strong perfect graph theorem. Ann. of Math. 164(2), 51–229 (2006)

8. Chudnovsky, M., Seymour, P.: Claw-free graphs. IV. Decomposition theorem. JCTB 98, 839–938 (2008)
9. Conforti, M., Cornuéjols, G.: Graphs without odd holes, parachutes or proper wheels: a generalization of Meyniel graphs and of line graphs of bipartite graphs. JCTB 87, 300–330 (2003)
10. Conforti, M., Cornuéjols, G., Kapoor, A., Vušković, K.: Even and odd holes in cap-free graphs. J. Graph Theory 30, 289–308 (1993)
11. Conforti, M., Cornuéjols, G., Kapoor, A., Vušković, K.: A Mickey-mouse decomposition theorem. In: Balas, E., Clausen, J. (eds.) IPCO 1995. LNCS, vol. 920, pp. 321–328. Springer, Heidelberg (1995)
12. Conforti, M., Cornuéjols, G., Kapoor, A., Vušković, K.: Even-hole-free graphs. I. Decomposition theorem. J. Graph Theory 39, 6–49 (2002)
13. Conforti, M., Cornuéjols, G., Kapoor, A., Vušković, K.: Universally signable graphs. Combinatorica 17, 67–77 (1997)
14. Conforti, M., Gerards, B., Kapoor, A.: A theorem of Truemper. Combinatorica 20, 15–26 (2000)
15. Corneil, D.G., Rotics, U.: On the relationship between clique-width and treewidth. SIAM J. Comput. 34, 825–847 (2005)
16. Courcelle, B.: The expression of graph properties and graph transformations in monadic second-order logic. In: Handbook of Graph Grammars and Computing by Graph Transformation, vol. 1, pp. 313–400. World Sci. Publishing, River Edge (1997)
17. Courcelle, B., Makowsky, J.A., Rotics, U.: Linear time solvable optimization problems on graphs of bounded clique-width. Theory Comput. Syst. 33, 125–150 (2000)
18. Diestel, R.: Graph Theory, 3rd edn. Springer, Heidelberg (2005)
19. Dirac, G.A.: On rigid circuit graphs. Abh. Math. Sem. Univ. Hamburg 25, 71–76 (1961)
20. Espelage, W., Gurski, F., Wanke, E.: How to solve NP-hard graph problems on clique-width bounded graphs in polynomial time. In: Brandstädt, A., Le, V.B. (eds.) WG 2001. LNCS, vol. 2204, pp. 117–128. Springer, Heidelberg (2001)
21. Gerber, M.U., Kobler, D.: Algorithms for vertex-partitioning problems on graphs with fixed clique-width. Theoret. Comput. Sci. 299, 719–734 (2003)
22. Gyárfás, A.: On Ramsey covering problems. Coll. Math. Soc. János Bolyai 10, 801–816 (1973)
23. Kloks, T., Müller, H., Vušković, K.: Even-hole-free graphs that do not contain diamonds: a structure theorem and its consequences. JCTB 99, 733–800 (2009)
24. Kobler, D., Rotics, U.: Edge dominating set and colorings on graphs with fixed clique-width. Discrete Applied Math. 126, 197–221 (2003)
25. Makowsky, J.A., Rotics, U.: On the clique-width of graphs with few P_4's. International J. Foundations of Computer Science 10, 329–348 (1999)
26. Miller, G.L.: Isomorphism of graphs which are pairwise k-separable. Information and Control 56, 21–33 (1983)
27. Nagamochi, H., Ibaraki, T.: A linear-time algorithm for finding a sparse k-connected spanning subgraph of a k-connected graph. Algorithmica 7, 583–596 (1992)
28. Tarjan, R.E.: Decomposition by clique separators. Discrete Math. 55, 221–232 (1985)
29. Trotignon, N., Vušković, K.: A structure theorem for graphs with no cycle with a unique chord and its consequences. J. Graph Theory 63, 31–67 (2009)
30. Tsukiyama, S., Ide, M., Ariyoshi, H., Shirakawa, I.: A new algorithm for generating all the maximal independent sets. SIAM J. Comput. 6, 505–517 (1977)

On Antimagic Labeling for Generalized Web and Flower Graphs

Joe Ryan[1], Oudone Phanalasy[1,2], Mirka Miller[1,3,4,5], and Leanne Rylands[6]

[1] School of Electrical Engineering and Computer Science,
The University of Newcastle, NSW, Australia
[2] Department of Mathematics, National University of Laos, Vientiane, Laos
[3] Department of Mathematics, University of West Bohemia, Pilsen, Czech Republic
[4] Department of Computer Science, King's College London, UK
[5] Department of Mathematics, ITB Bandung, Indonesia
[6] School of Computing and Mathematics, University of Western Sydney,
NSW, Australia
{joe.ryan,mirka.miller}@newcastle.edu.au,
oudone.phanalasy@uon.edu.au, l.rylands@uws.edu.au

Abstract. An antimagic labeling of a graph with p vertices and q edges is a bijection from the set of edges to the set of integers $\{1, 2, \ldots, q\}$ such that all vertex weights are pairwise distinct, where a vertex weight is the sum of labels of all edges incident with the vertex. A graph is antimagic if it has an antimagic labeling.

Completely separating systems arose from certain problems in information theory and coding theory. Recently these systems have been shown to be useful in constructing antimagic labelings of particular graphs.

Keywords: m-level generalized web graph, m-level generalized flower graph.

1 Introduction

All graphs in this paper are finite, simple, undirected and connected, unless stated otherwise. In 1990, Hartsfield and Ringel [6] introduced the concept of an antimagic labeling of graph, that is a vertex antimagic edge labeling. An antimagic labeling of a graph with q edges and p vertices is a bijection from the set of edges to the set of integers $\{1, 2, \ldots, q\}$ such that all vertex weights are pairwise distinct, where a vertex weight is the sum of labels of all edges incident with the vertex. A graph is antimagic if it has an antimagic labeling.

Hartsfield and Ringel [6] showed that paths, stars, cycles, complete graphs K_m, wheels W_m and bipartite graphs $K_{2,m}$, $m \geq 3$, are antimagic. They also conjectured that every connected graph, except K_2, is antimagic, a conjecture which remains open. Subsequently, several families of graphs have been proved to be antimagic, for example, see [1], [2] and [3]. Many other results concerning antimagic graphs are catalogued in the dynamic survey by Gallian [5].

In this paper, we give an overview of *completely separating systems* (CSSs), define two new families of graphs, the *generalized web* and *generalized flower*

C.S. Iliopoulos and W.F. Smyth (Eds.): IWOCA 2010, LNCS 6460, pp. 303–313, 2011.

and show how CSS may be used to construct antimagic labelings for these new families of graphs.

Separating systems were first introduced in 1961 by Renyi [9] in the context of solving certain problems in information theory. Let $[n]$ represent the set of integers from 1 to n and let $\mathcal{C}$ be a collection of subsets of $[n]$. An element $a \in [n]$ is said to be *separated* from $b \in [n]$ in $\mathcal{C}$ if there is a set in $\mathcal{C}$ which contains a but not b. A *separating system* (or SS) $\mathcal{C}$ on $[n]$ is a collection of subsets of $[n]$ in which for each pair of elements $a \neq b \in [n]$, either a is separated from b or b is separated from a.

A *completely separating system* (CSS) [4] on $[n]$, or (n)CSS, is a collection of subsets of $[n]$ in which for each pair of elements $a \neq b \in [n]$, a is separated from b and b is separated from a in $\mathcal{C}$. For example, in the collection $\{\{1,2\},\{1,3\}\}$, 1 is separated from 2 by $\{1,3\}$ but 2 is not separated from 1. The collection $\{\{1,2\},\{1,3\}\}$ is not a CSS. However, the collection $\{\{1,2\},\{1,3\},\{2,3\}\}$ is a CSS on $[3]$.

The sets in the (n)CSS are usually called *blocks* and the elements of these sets are usually called *points*. Let k be a positive integer and let $\mathcal{C}$ be an (n)CSS. If $|A| = k$ for all $A \in \mathcal{C}$, then $\mathcal{C}$ is said to be an (n,k)CSS.

The technique for assigning antimagic labelings using CSS is based on Roberts' construction for Completely Separating Systems [10], so it is restated here.

Roberts' construction

Assume that $k \geq 2$, $n \geq \binom{k+1}{2}$ and $k|2n$, and let $R = R(n,k) = 2n/k$. An $(R \times k)$-array L is constructed, where each row of L forms a subset of $[n]$ and the R rows of L form an $(n,k)CSS$. Let e_{ij} denote the element of L in row i and column j. Initialize all elements of L to zero. For e from 1 to n, in order, include e in the two positions of L defined by

$$\min_j \min_i \{e_{ij} : e_{ij} = 0\},$$
$$\min_i \min_j \{e_{ij} : e_{ij} = 0\}.$$

That is, e is placed in the first row of L containing a 0, in the first 0-valued place in that row, e is then also placed in the first column of L containing a 0, in the first 0-valued place in that column. Each of the integers 1 to n appears in L in two positions, and the array L is the array of an (n,k)CSS. This concludes Roberts' construction.

The following example illustrates this construction.

Example 1. Using Roberts' construction, we obtain the following array of a $(9,3)$CSS

$$\begin{matrix} 1 & 2 & 3 \\ 1 & 4 & 5 \\ 2 & 6 & 7 \\ 3 & 6 & 8 \\ 4 & 7 & 9 \\ 5 & 8 & 9 \end{matrix}$$

The following theorems will be useful when creating antimagic labelings of the generalized web graphs and generalized flower graphs families, so we recall them here.

Theorem 1. *[8] Let $V = \{v_1, v_2, \dots, v_p\}$ be a collection of k-subsets of $[q]$. Then V is a (q, k)CSS consisting of 2-elements if and only if a k-regular graph $G(V, E)$ with p vertices and q edges has an edge labeling.*

Hereafter we denote by $G(V, E, L)$, a graph $G(V, E)$ having an edge labeling L. Note that in this paper L can be seen as the array of edge labels.

Theorem 2. *[8] Let L be the array of a (q, k)CSS obtained using Roberts' construction. Then the k-regular graph $G(V, E, L)$, where $|V| = p = 2q/k$, $|E| = q$, is antimagic.*

Theorem 3. *[7] The Cartesian product graph $\prod_r K_2$, $r \geq 2$, is antimagic.*

Theorem 4. *[7] Let $G_h = G_h(V_h, E_h, L_h)$, $1 \leq h \leq t$, where L_h is the array of a (q_h, k_h)CSS obtained using Roberts' construction. Then the Cartesian product graph $\prod_h G_h$ is antimagic.*

Theorem 5. *[7] The Cartesian product $\prod_s K_2 \times \prod_h G_h$, $s, h \geq 1$, is antimagic.*

2 Results

We first define two new families of graphs. Let G be a k-regular graph with p vertices. The *generalized pyramid graph*, $P(G, 1)$, is the graph obtained from the graph G by joining each vertex of the graph G to a vertex called the apex, and the graph G is called the base. Note that the wheel is a special case of the generalized pyramid graph $P(G, 1)$ when $G = C_n$, $n \geq 3$. The *2-level generalized pyramid graph*, $P(G, 2)$, is the graph obtained from the graph $P(G, 1)$ by attaching a pendant edge at each vertex of the base G and then joining the pendant vertices to corresponding vertex of a copy of G. By iterating the process of adding the pendant vertices and joining them to form a new copy of G, it is called the *m-level generalized pyramid graph* (or simply, *generalized pyramid graph*) and denoted by $P(G, m)$, $m \geq 1$. Alternatively, to get $P(G, m)$ take the Cartesian product $G \times P_m$ adjoin a vertex to each vertex of one end copy of G.

The *m-level generalized web graph* (or simply, *generalized web graph*), $WB(G, m)$, is the graph obtained from the m-level generalized pyramid graph $P(G, m)$ by attaching a pendant edge at each vertex of the furthermost copy of G from the apex; and the graph G is called the base of the graph $WB(G, m)$. When G is a cycle, $WB(G, m)$ is simply called the m-level web graph (see Fig. 3).

The *m-level generalized flower graph with p petals* (or simply, *generalized flower graph with p petals*), $FL(G, m, p)$, is the graph obtained from the generalized web $WB(G, m)$ by connecting each pendant vertex to the apex with an edge. The *m-level generalized flower graph with mp petals* (or simply, *generalized flower graph with mp petals*), $FL(G,m,mp)$, is the graph obtained from the m-level generalized web $WB(G,m)$ by connecting each pendant vertex and each vertex of the $m-1$ copies of G to the apex with an edge, except the vertices of the the nearest copy of G to the apex. When G is a cycle, $FL(G,m,mp)$ is simply called the m-level flower graph (see Fig. 6).

We denote by T^t the transpose of the array T.

Theorem 6. *Let $G_h = G_h(V_h, E_h, L_h)$, $1 \leq h \leq t$, where L_h is the array of a $(q_h, k_h)CSS$ obtained using Roberts' construction. Let $G = G_h$ or $\prod_r K_2$, $r \geq 2$, or $\prod_h G_h$ or $\prod_s K_2 \times \prod_h G_h$, $s, h \geq 1$. Then the generalized web graph $WB(G,m)$, $m \geq 1$, is antimagic.*

Proof. Assume that G has p vertices and q edges. We divide the proof into two cases as follows.
Case 1: $G = K_p$, $p \geq 3$.

Let T_l, $1 \leq l \leq m+1$, be the $(p \times 1)$-array of edges e_i, $1 \leq i \leq p$. As in the construction given in the proof of Theorem 2, let M_j, $1 \leq j \leq m$, be the array of edge labels of the j-th graph G. We construct the array A of edge labels of the generalized web graph $WB(G,m)$, $m \geq 1$, as follows.

(1) Label the edge e_i, $1 \leq i \leq p$, in the block i of the array T_l, $1 \leq l \leq m+1$, with $i + (l-1)p$, $1 \leq l \leq 2$, and $i + (l-1)p + (l-2)q$, $2 < l \leq m+1$;
(2) Relabel the edge labels in the array M_j, $1 \leq j \leq m$, by adding $(j+1)p + (j-1)q$ to each of its edge labels;
(3) Form the array A into two subcases as shown below.

Subcase 1.1: $m = 1$.

$$\begin{array}{ccc} & & T_1 \\ & & T_2^t \\ T_1 & T_2 & M_1 \end{array}$$

By the construction of the array A, it is clear that the weight of each vertex (block) in the array is less than the weight of the vertex below.
Subcase 1.2: $m \geq 2$.

$$\begin{array}{ccc} & & T_1 \\ & & T_2^t \\ T_1 & M_1 & T_3 \\ T_2 & M_2 & . \\ \vdots & \vdots & \vdots \\ T_m & T_{m+1} & M_m \end{array}$$

The diagram below illustrates the construction used here.

By the construction of the array A, it is clear that the weight of each vertex (block) in the array is less than the weight of the vertex below.
Case 2: $G \neq K_p$, $p > 3$.

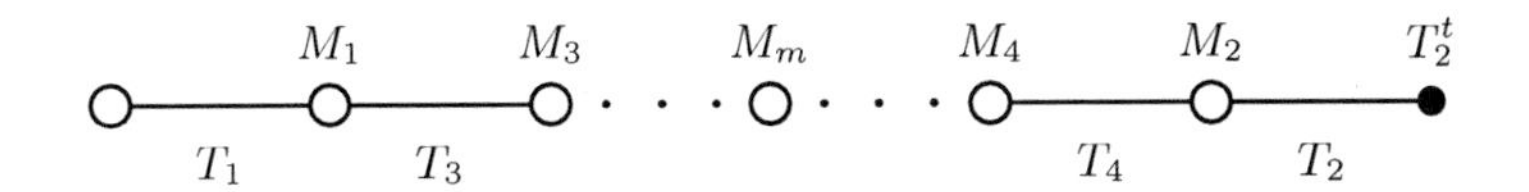

Fig. 1. Illustration of the construction of the generalized web graph $WB(K_p, m), p \geq 3$, $m > 3$

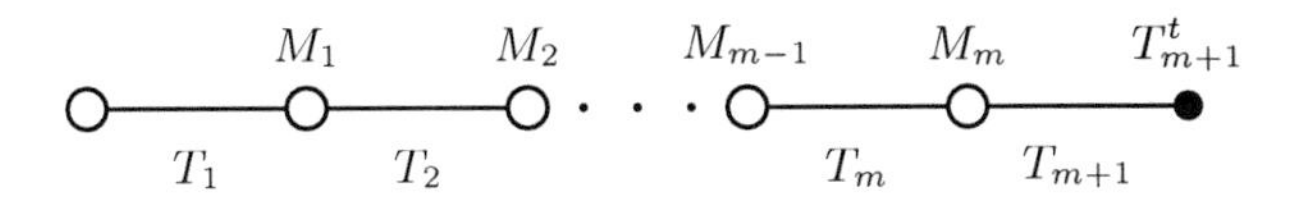

Fig. 2. Illustration of the construction of the generalized web graph $WB(G, m)$, $G \neq K_p$, $p > 3$, $m > 3$

Let T_l, $1 \leq l \leq m+1$, be the $(p \times 1)$-array of edges e_i, $1 \leq i \leq p$. As in the construction given in the proof of Theorems 2, 3, 4 and 5, let M_j, $1 \leq j \leq m$, be the array of edge labels of the j-th graph G. We construct the array A of edge labels of the generalized web graph $WB(G, m)$ as follows.

(1) Label the edge e_i, $1 \leq i \leq p$, in the block i of the array T_l, $1 \leq l \leq m+1$, with $i + (l-1)(p+q)$;
(2) Relabel the edge labels in the array M_j, $1 \leq j \leq m$, by adding $jp + (j-1)q$ to each of its edge labels;
(3) Form the array A into two subcases as shown below.

Subcase 2.1: $m = 1$.

$$\begin{array}{ccc} & & T_1 \\ T_1 & M_1 & T_2 \\ & & T_2^t \end{array}$$

By the construction of the array A, it is clear that the weight of each vertex (block) in the array is less than the weight of the vertex below.
Subcase 2.2: $m \geq 2$.

$$\begin{array}{ccc} & & T_1 \\ T_1 & M_1 & T_2 \\ T_2 & M_2 & . \\ \vdots & \vdots & \vdots \\ T_m & M_m & T_{m+1} \\ & & T_{m+1}^t \end{array}$$

The diagram below illustrates the construction used here.
By the construction of the array A, it is clear that the weight of each vertex (block) is less than the weight of the vertex below. □

The following example illustrates the construction of an antimagic labeling of the web graph $WB(K_3, 2)$.

Example 2. Using the construction in the proof of Subcase 1.2 of Theorem 6, we have the array A of edge labels of the web graph $WB(K_3, 2)$

$$\begin{array}{rrrr} & & & 1 \\ & & & 2 \\ & & & 3 \\ & 4 & 5 & 6 \\ 1 & 7 & 8 & 10 \\ 2 & 7 & 9 & 11 \\ 3 & 8 & 9 & 12 \\ 4 & 10 & 13 & 14 \\ 5 & 11 & 13 & 15 \\ 6 & 12 & 14 & 15 \end{array}$$

and the corresponding antimagic web graph $WB(K_3, 2)$ as shown in Fig. 3.

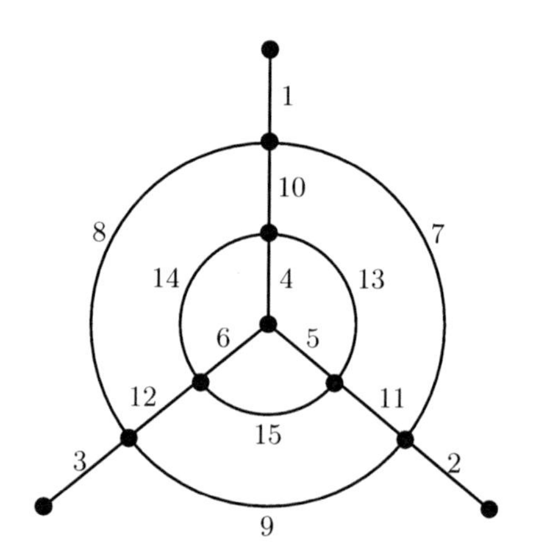

Fig. 3. The web graph $WB(K_3, 2)$ and its antimagic labeling

The only possible edge labeling of K_2 can be presented by the (2×1)-array, M_0, with both entries 1. Interestingly, the same construction as in the proof of Case 1 of Theorem 6 also works when the array M_j is replaced by M_{0j} (M_{0j} is the array of the edge labels of the j-th graph K_2), although K_2 itself is not antimagic, then we have

Theorem 7. *The generalized web graph $WB(K_2, m)$, $m \geq 1$, is antimagic.*

Moreover, the following theorems are the extensions of Theorems 6 and 7.

Theorem 8. *Let $G_h = G_h(V_h, E_h, L_h)$, $1 \leq h \leq t$, where L_h is the array of a (q_h, k_h)CSS obtained using Roberts' construction. Let $G = G_h$ or $\prod_r K_2$, $r \geq 2$, or $\prod_h G_h$ or $\prod_s K_2 \times \prod_h G_h$, $s, h \geq 1$. Then the generalized flower $FL(G, m, p)$, $m \geq 1$, is antimagic.*

Proof. Assume that G has p vertices and q edges. We divide the proof into two cases as follows.
Case 1: $G = K_p$, $p \geq 3$.

Let T_l, $1 \leq l \leq m+2$, be the $(p \times 1)$-array of edges e_i, $1 \leq i \leq p$. As in the construction given in the proof of Theorem 2, let M_j, $1 \leq j \leq m$, be the array of edge labels the j-th graph G. We construct the array A of edge labels of the generalized flower graph $FL(G, m, p)$, $m \geq 1$, as follows.

(1) Label the edge e_i, $1 \leq i \leq p$, in the block i of the array T_l, $1 \leq l \leq m+2$, with $i + (l-1)p$, $1 \leq l \leq 3$ and $i + (l-1)p + (l-3)q$, $3 < l \leq m+2$;
(2) Relabel the edge labels in the array M_j by adding $3p + (j-1)(p+q)$ to each of its edge labels;
(3) Form the array A as shown below.
Subcase 1.1: $m = 1$.

$$\begin{array}{ccc} & T_1 & T_3 \\ & T_1^t & T_2^t \\ T_2 & T_3 & M_1 \end{array}$$

By the construction of the array A, it is clear that the weight of each vertex (block) in the array is less than the weight of the vertex below, except in a few exceptional cases. These cases include the weights of the last block in the subarray T_1T_3, the block $T_1^tT_2^t$ and the first block in the subarray $T_1T_3M_1$ which need to be verified.

Let r_p, r_{p+1} and r_{p+2} be the last block in the subarray T_1T_3 and the block $T_1^tT_2^t$ and first block in the subarray $T_1T_3M_1$, respectively. Let $wt(r_p)$, $wt(r_{p+1})$ and $wt(r_{p+2})$ be the weights of r_p, r_{p+1} and r_{p+2}, respectively. We have the edge labels in the block r_p, r_{p+1} and r_{p+2} as shown below.

$$\begin{array}{lcccccc} r_p : & & & & & p & 3p \\ r_{p+1} : 1\ 2 \ldots & . & . & . & \ldots & 2p-1 & 2p \\ r_{p+2} : & 1+p & 1+2p & 1+3p & \ldots & (p-2)+3p & (p-1)+3p \end{array}$$

It is clear that $wt(r_p) = 4p < \frac{2p(2p+1)}{2} = wt(r_{p+1}) < \frac{7p^2-p+4}{2} = wt(r_{p+2})$.
Subcase 1.2: $m \geq 2$.

$$\begin{array}{ccc} & T_1 & T_3 \\ & T_1^t & T_2^t \\ T_2 & M_1 & T_4 \\ T_3 & M_2 & . \\ \vdots & \vdots & \vdots \\ T_{m+1} & T_{m+2} & M_m \end{array}$$

The diagram below illustrates the construction used here.

As in Subcase 1.1, we have $wt(r_p) = 4p < \frac{2p(2p+1)}{2} = wt(r_{p+1})$. We have the weight of the first block in the subarray $T_2M_1T_4$, this is $wt(r_{p+2}) = (1+p)+(1+3p)+\cdots+((p-1)+3p)+(1+3p+q) = \frac{7p^2+p+2q+4}{2}$. Hence $wt(r_{p+1}) < wt(r_{p+2})$.
Case 2: $G \neq K_p$, $p > 3$.

Let T_l, $1 \leq l \leq m+2$, be the $(p \times 1)$-array of edges e_i, $1 \leq i \leq p$. As in the construction given in the proof of Theorems 2, 3, 4 and 5, let M_j, $1 \leq j \leq m$,

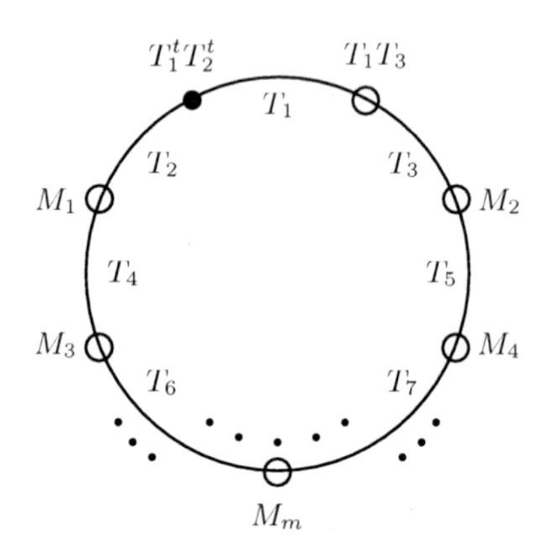

Fig. 4. Illustration of the construction of the generalized flower graph $FL(K_p, m, p)$, $p \geq 3$, $m \geq 3$

be the array of edge labels the j-th graph G. We construct the array A of edge labels of the generalized flower graph $FL(G, m, p)$ as follows.

(1) Label the edge e_i, $1 \leq i \leq p$, in the block i of the array T_l, $1 \leq l \leq m+2$, with $i + (l-1)(p+q)$, $1 \leq l \leq m+1$ and $i + (l-1)p + (l-2)q$, $l = m+2$;
(2) Relabel the edge labels in the array M_j, $1 \leq j \leq m$, by adding $jp + (j-1)q$ units to each of its edge labels;
(3) Form the array A as shown below;
Subcase 2.1: $m = 1$.

$$\begin{array}{ccc} & T_1 & T_2 \\ T_1 & M_1 & T_3 \\ & T_2^t & T_3^t \end{array}$$

By the construction of the array A, it is clear that the weight of each vertex (block) in the array is less than the weight of the vertex below.
Subcase 2.2: $m \geq 2$.

$$\begin{array}{ccc} & T_1 & T_2 \\ T_1 & M_1 & T_3 \\ M_2 & T_3 & T_4 \\ \vdots & \vdots & \vdots \\ M_m & T_{m+1} & T_{m+2} \\ & T_2^t & T_{m+2}^t \end{array}$$

The diagram below illustrates the construction used here.

By the construction of the array A, it is clear that the weight of each vertex (block) in the array is less than the weight of the vertex below. □

The same construction as in the proof of Case 1 of Theorem 8 also works when the array M_j is replaced by the array M_{0j}, although K_2 itself is not antimagic, then we have

Theorem 9. *The generalized flower graph $FL(K_2, m, 2)$, $m \geq 1$, is antimagic.*

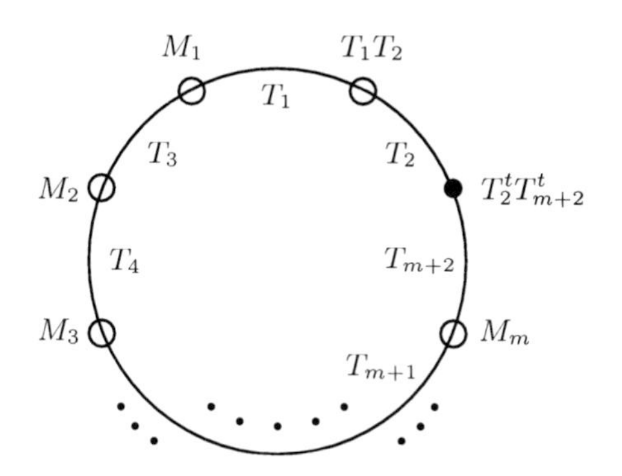

Fig. 5. Illustration of the construction of the generalized flower graph $FL(G, m, p)$, $G \neq K_p$, $p > 3$, $m > 3$

The following theorems are the extensions of the generalized flower graphs.

Theorem 10. *Let $G_h = G_h(V_h, E_h, L_h)$, $1 \leq h \leq t$, where L_h is the array of a (q_h, k_h)CSS obtained using Roberts' construction. Let $G = G_h$ or $\prod_r K_2$, $r \geq 2$, or $\prod_h G_h$ or $\prod_s K_2 \times \prod_h G_h$, $s, h \geq 1$. Then the generalized flower graph $FL(G, m, mp)$, $m \geq 2$, is antimagic.*

Proof. Assume that G has p vertices and q edges. Let T_l, $1 \leq l \leq 2m+1$, be the $(p \times 1)$-array of edges e_i, $1 \leq i \leq p$. As in the construction given in the proof of Theorems 2, 3, 4 and 5, let M_j, $1 \leq j \leq m$, be the array of edge labels the j-th graph G. We construct the array A of edge labels of the generalized flower graph $FL(G, m, mp)$, $m \geq 2$, as follows.

(1) Label the edge e_i, $1 \leq i \leq p$, in the block i of the array T_l, $1 \leq l \leq 2m+1$, with $i + (l-1)p$ for $1 \leq l \leq 3$, $i + (l-1)p + (l-3)q$ for $3 < l \leq m+3$ and $i + (l-1)p + mq$ for $m+4 \leq l \leq 2m+1$;
(2) Relabel the edge labels in the array M_j, $1 \leq j \leq m$, by adding $3p + (j-1)(p+q)$ to each of its edge labels;
(3) Form the array A in two cases.
Case 1: $m = 2$.

$$\begin{array}{llll} & T_1 & T_3 & \\ T_2 & M_1 & T_4 & \\ T_3 & T_4 & M_2 & T_5 \\ & T_1^t & T_2^t & T_5^t \end{array}$$

By the construction of the array A, it is clear that the weight of each vertex (block) in the array is less than the weight of the vertex below, except the weight of the last vertex (block) in the subarray $T_3T_4M_2T_5$ and the weight of the block $T_2^tT_2^tT_5^t$ that need to be verified.

Let $e_{f,g}$ be the edge label at the block f and the column g in the array A. Let r_{3p} and r_{3p+1} be the last vertex (block) of the subarray $T_3T_4M_2T_5$ and the block $T_2^tT_2^tT_5^t$, respectively. Let $wt(r_{3p})$ and $wt(r_{3p+1})$ be the weights of the r_{3p} and r_{3p+1}, respectively. We show in three subcases.

Subcase 1.1: $G = K_3$.
By exhaustion, we have $wt(r_9) = 80 < 81 = wt(r_{10})$.
Subcase 1.2: $G = K_p$, $p > 3$.
We have the edge labels in the blocks r_{3p} and r_{3p+1} as shown below.

$$\begin{array}{lccccc} r_{3p}: & & 3p & 4p+q & 5p+q-1 \ldots & 5p+2q \\ r_{3p+1}: & 1\ 2 \ldots\ 2p-1 & 2p & 1+4p+2q \ldots & & 5p+2p \end{array}$$

Since, for $p > 3$, $e_{3p,2p-1} + e_{3p,2p} = 7p + q < p(2p+1) = \sum_{h=1}^{2p} e_{3p+1,h}$ and for $2p + 1 \leq g \leq 3p$, $e_{3p,g} \leq e_{3p+1,g}$, hence $wt(r_{3p}) < wt(r_{3p+1})$.
Subcase 1.3: $G \neq K_p$, $p > 3$.
Similarly to the Subcase 1.2, hence $wt(r_{3p}) < wt(r_{3p+1})$.
Case 2: $m \geq 3$.

$$\begin{array}{ccccccccc} & & & & & T_1 & T_3 & \\ & & & & T_2 & M_1 & T_4 & \\ & & & & T_3 & M_2 & T_5 & T_{m+3} \\ & & & & T_4 & M_3 & . & T_{m+4} \\ & & & & \vdots & \vdots & \vdots & \vdots \\ & & & & T_{m+1} & T_{m+2} & M_m & T_{2m+1} \\ T_1^t & T_2^t & T_{m+3}^t & \cdots & T_{2m-2}^t & T_{2m-1}^t & T_{2m}^t & T_{2m+1}^t \end{array}$$

By construction of the array A, it is clear that the weight of each vertex (block) in the array is less than the weight of the vertex below.

Example 3. Using the construction in the proof of Subcase 1.1 of Theorem 10, we have the array A of edge labels of the generalized flower graph $FL(K_3, 2, 6)$

$$\begin{array}{cccccccccc} & & & & & 1 & 7 & & \\ & & & & & 2 & 8 & & \\ & & & & & 3 & 9 & & \\ & & & & 4 & 10 & 11 & 13 & \\ & & & & 5 & 10 & 12 & 14 & \\ & & & & 6 & 11 & 12 & 15 & \\ & & & & 7 & 13 & 16 & 17 & 19 \\ & & & & 8 & 14 & 16 & 18 & 20 \\ & & & & 9 & 15 & 17 & 18 & 21 \\ 1 & 2 & 3 & 4 & 5 & 6 & 19 & 20 & 21 \end{array}$$

and the corresponding antimagic generalized flower graph $FL(K_3, 2, 6)$ as shown in Fig. 6.

The same construction as in the proof of Theorem 10 also works when the array M_j is replaced by the array M_{0j}, although K_2 itself is not antimagic, then we have

Theorem 11. *The generalized flower graph $FL(K_2, m, 2m)$, $m \geq 2$, is antimagic.*

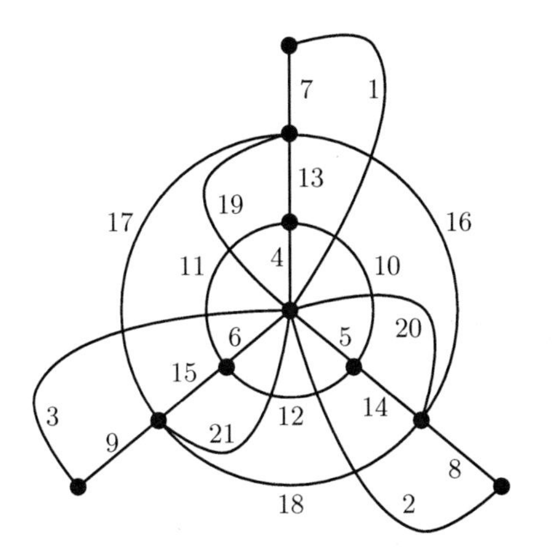

Fig. 6. The generalized flower graph $FL(K_3, 2, 6)$ and its antimagic labeling

References

1. Bača, M., Miller, M.: Super Edge-Antimagic Graphs: a Wealth of Problems and Some Solutions. BrownWalker Press, Boca Raton (2008)
2. Cheng, Y.: A new class of antimagic Cartesian product graphs. Discrete Math. 308(24), 6441–6448 (2008), `http://dx.doi.org/10.1016/j.disc.2007.12.032`
3. Cranston, D.W.: Regular bipartite graphs are antimagic. J. Graph Theory 60(3), 173–182 (2009), `http://dx.doi.org/10.1002/jgt.20347`
4. Dickson, T.J.: On a problem concerning separating systems of a finite set. J. Combinatorial Theory 7, 191–196 (1969)
5. Gallian, J.A.: A dynamic survey of graph labeling. Electron. J. Combin. 16(♯DS6) (2009)
6. Hartsfield, N., Ringel, G.: Pearls in graph theory: a comprehensive introduction. Academic Press Inc., Boston (1990)
7. Phanalasy, O., Miller, M., Iliopoulos, C.S., Pissis, S.P., Vaezpour, E.: Construction of antimagic labeling for the Cartesian product of regular graphs (preprint)
8. Phanalasy, O., Miller, M., Rylands, L., Lieby, P.: On a relationship between completely separating systems and antimagic labeling of regular graphs (submitted)
9. Rényi, A.: On random generating elements of a finite Boolean algebra. Acta Sci. Math. Szeged 22, 75–81 (1961)
10. Roberts, I.T.: Extremal Problems and Designs on Finite Sets. Ph.D. thesis, Curtin University of Technology (1999)

Chains-into-Bins Processes

Tuğkan Batu[1], Petra Berenbrink[2], and Colin Cooper[3]

[1] Department of Mathematics, London School of Economics, London WC2A 2AE, UK
t.batu@lse.ac.uk

[2] School of Computing Science, Simon Fraser University, Burnaby, BC V5A 1S6, Canada
petra@cs.sfu.ca

[3] Department of Computer Science, King's College London, London WC2R 2LS, UK
colin.cooper@kcl.ac.uk

Abstract. The study of *balls-into-bins processes* or *occupancy problems* has a long history. These processes can be used to translate realistic problems into mathematical ones in a natural way. In general, the goal of a balls-into-bins process is to allocate a set of independent objects (tasks, jobs, balls) to a set of resources (servers, bins, urns) and, thereby, to minimize the maximum load. In this paper, we analyze the maximum load for the *chains-into-bins* problem, which is defined as follows. There are n bins, and m objects to be allocated. Each object consists of balls connected into a chain of length ℓ, so that there are $m\ell$ balls in total. We assume the chains cannot be broken, and that the balls in one chain have to be allocated to ℓ consecutive bins. We allow each chain d independent and uniformly random bin choices for its starting position. The chain is allocated using the rule that the maximum load of any bin receiving a ball of that chain is minimized. We show that, for $d \geq 2$ and $m \cdot \ell = O(n)$, the maximum load is $((\ln \ln m)/\ln d) + O(1)$ with probability $1 - \tilde{O}(1/m^{d-1})$.

Keywords: Balls-into-bins processes, chains-into-bins processes, random processes, offline assignment.

1 Introduction

The study of *balls-into-bins processes* or *occupancy problems* has a long history. These models are commonly used to derive results in probability theory. Furthermore, balls-into-bins processes can be used as a means of translating realistic load-balancing problems into mathematical ones in a natural way. In general, the goal of a balls-into-bins process is to allocate a set of independent objects (tasks, jobs, balls) to a set of resources (servers, bins, urns). It is assumed that the balls are independent and do not know anything about the other balls. Each ball is allowed to choose a subset of the bins independently and uniformly at random (i.u.r.) in order to be allocated into one of these bins. The performance of these processes is usually analyzed in terms of the maximum load of any bin.

One extreme solution is to allow each ball to communicate with every bin. Thus, it is possible to query the load of every bin and to place the ball into

C.S. Iliopoulos and W.F. Smyth (Eds.): IWOCA 2010, LNCS 6460, pp. 314–325, 2011.

the bin that is least loaded. This allocation process always yields an optimum allocation of the balls. However, the time and the number of communications needed to allocate the balls are extremely large. The opposite approach is to allow every ball to communicate with only one bin. The usual model is for every ball to be thrown into one bin chosen independently and uniformly at random. For the case of m balls allocated to n bins i.u.r., it is well known that a bin that receives $m/n + \Theta\left(\sqrt{(m \log n)/n}\right)$ balls exists with high probability (w.h.p.).[1] An alternative approach which lies between these two extremes, is to allow every ball to select one of $d \geq 2$ i.u.r. chosen bins. The GREEDY[d] process, studied by Azar et al. [1], chooses d i.u.r. bins per ball, and the ball is allocated into the least loaded among these bins. For this process, and $m \geq n$, the maximum number of balls found in any bin, i.e., the *maximum load*, is $m/n + \ln \ln n / \ln d + O(1)$ w.h.p. (see, for example, [1], [2]). Thus, even a small amount of additional random choice can decrease the maximum load drastically compared to a single choice. This phenomenon is often referred to as the "power of two random choices" (see [7]).

In this paper we consider the chains-into-bins problem which can be regarded as a generalization of the balls-into-bins problem. We are given m chains consisting of ℓ balls each. The balls of any chain have to be allocated to ℓ consecutive bins. We allow each chain d i.u.r. bin choices, and allocate the chain using the rule that the maximum load of any bin receiving a ball of that chain is minimized. In this paper, we show that, for $d \geq 2$ and $m \cdot \ell = o(n \cdot (\ln \ln m)^{1/2})$, the maximum load achieved by this algorithm is at most $(\ln \ln m / \ln d) + O((m\ell/n)^2) + O(1)$, with probability $1 - O((\ln m)^d / m^{d-1})$. This result shows that for a fixed number of balls, the maximum load decreases with increasing chain length. The maximum load depends on the number of chains only in the following sense. Allocating $m = n/\ell$ chains of length ℓ (with a total number of n balls) into n bins will, w.h.p., result in a maximum load of at most $\ln \ln(n/\ell) / \ln d + O(1)$. It follows that if $\ell = O((\ln n)^a)$ for any $a > 0$, our result is asymptotically the same as that for allocating n/ℓ balls into n/ℓ bins using GREEDY[d] protocol of [1].

We also prove that the naive heuristic that 'allocates the chain headers using GREEDY[d] and hopes for the best' performs badly for some values of m, ℓ as one might expect. Indeed, if $m \geq \ln^2 n$ and $\ell \geq (\ln m)/\ln \ln m$, then the maximum occupancy of this heuristic is at least $(\ln m)/(2 \ln \ln m)$, w.h.p.

Clustering. It can be seen that, provided we can make some extra assumptions, the results of [1] can be applied to the chains-into-bins problem for m chains of length ℓ. Suppose we are *allowed to cluster the bins* into $N = n/\ell$ clusters of ℓ successive bins, and each chain *can be allocated directly to one cluster.* This is now equivalent to allocating m balls into N bins. Thus, we would get a maximum load of $\Theta((\ln \ln N)/\ln(d) + 1)$ with GREEDY[d] and $(\ln \ln N)/d + O(1)$ using the ALWAYS-GO-LEFT[d] protocol. However, this solution essentially ignores the model under consideration, and is equivalent in a hashing context to saying that we do not need to hash the data item at the given location, but rather somewhere in the next ℓ cells at our convenience. If we have this freedom to

[1] A sequence of events A_n occurs with high probability if $\lim_{n \to \infty} P(A_n) = 1$.

ignore the locations we are given and pack the balls into $N = n/\ell$ clusters of length ℓ, then why not $N = n/(2\ell)$ clusters of length 2ℓ, placing the chains one after the other? Indeed why not arbitrary N, or even $N = 1$ and pack the chains cyclically in a round-robin fashion? That would be even more efficient. Obviously if such clustering were available it would be easier to organize the behavior of the balls. We assume henceforth that we have to put the chains where we are instructed, rather than where we would like to. Finally, we remark that, provided $m = N = n/\ell$, we can get the same results without restructuring the problem, and thus, the extra provisions are unnecessary.

Applications. Our model can be viewed as a form of hashing in which the first data item of the chain is placed in the selected position of the hash table, and the remaining items overflow into the neighbouring positions of the table.

The chains-into-bins problem has several important applications. One example is data storage on disk arrays, such as RAID systems (see [9]). Here, each data item is stored on several neighbouring disks in order to increase the data transfer rate. In this case, the bins model the disks from the storage array, and the chains model data requests which are directed to several neighbouring "bins." A second application is the scheduling of reconfigurable embedded platforms (see [4,11]). Here, the tasks and the reconfigurable chip are modeled as rectangles with integral dimensions. All tasks have the same height but different length. The chip is modeled by a much larger rectangle that can hold several tasks in both dimensions. The goal is to allocate the tasks to a chip with a fixed length such that the required height is minimized. In this case, the tasks are modeled by the chains and the chip is modeled by the bins. The problem also models a scheduling problem where m allocated items persist in the system for ℓ time steps. For example, imagine a train traveling in a circle with n station stops. The bins represent stations and the length of a chain represent the number of stops traveled by a passenger.

1.1 Related Work

Azar et al. [1] introduced GREEDY[d] to allocate n balls into n bins. GREEDY[d] chooses d bins i.u.r. for each ball and allocates the ball into a bin with the minimum load. They show that after placing n balls, the maximum load is $\Theta((\ln\ln n)/\ln d)$, w.h.p. Compared to single-choice processes, this is an *exponential* decrease in the maximum load. For the case where $m < n$, their results can be extended to show a maximum load of at most $(\ln\ln n - \ln\ln(n/m))/\ln d + O(1)$, w.h.p. Vöcking [13] introduced the ALWAYS-GO-LEFT[d] protocol, which clusters the bins into d clusters of n/d consecutive bins each. Every ball now chooses i.u.r. one bin from every cluster and is allocated into a bin with the minimum load. If several of the chosen bins have the same minimum load, the ball is allocated into the "leftmost" bin. The protocol yields a maximum load of $(\ln\ln n)/d + O(1)$, w.h.p. In [5], Kenthapadi and Panigrahy suggest an alternative protocol yielding the same maximum load. They cluster the bins into $2n/d$ clusters of $d/2$ consecutive bins each. Every ball now randomly chooses 2 of these clusters and it is allocated into the cluster with the smallest total load. In the chosen cluster, the ball is then allocated into the bin with minimum load again. The authors also argue in that paper that clustering is essential to reduce the load to $(\ln \ln n)/d + O(1)$. In [2], the authors analyse GREEDY$[d]$ for $m \gg n$. It is shown that the maximum load is $m/n + \ln \ln(n)$, w.h.p. Mitzenmacher et al. [8] show that a similar performance gain occurs if the process is allowed to memorize a constant number of bins with small load.

In [10], Sanders and Vöcking consider the *random arc allocation problem*, which is closely related to the chains-into-bins problem. In their model, they allocate arcs of an arbitrary length to a cycle. Every arc is assigned a position i.u.r. on the cycle. The chains-into-bins problem with $d = 1$ can be regarded as a special discrete case of their problem, where the cycle represents the n bins and the arcs represent the chains (in [10], different arc lengths are allowed). Translated into the chains-into-bins setting, the authors show the following result. If $m = n/\ell$ chains of length ℓ are allocated to n bins ($m \to \infty$), then the maximum load is at most $(\ln(n/\ell))/(\ln \ln(n/\ell))$, w.h.p. Note that their result is asymptotically the same as that for allocating n/ℓ balls into n/ℓ bins, provided that $n/\ell \to \infty$. In [3], the author shows that the expected maximum load is smaller if we allocate $n/2$ chains of length 2 with one random choice per chain, compared to n balls into n bins with $d = 2$.

2 Model and Results

Assume m chains of length ℓ are allocated i.u.r. to bins wrapped cyclically round $1, ..., n$. A chain contains ℓ balls linked together sequentially. The first ball of a chain is called *header*; the remaining balls comprise the *tail* of the chain. If chain i (meaning the header of chain i) is allocated to bin j, then the balls of the chain occupy bins $j, j+1, \ldots, j+\ell-1$, where counting is modulo n. We define the *h-load* of a bin as the number of headers allocated to the bin. This is to be distinguished from the *load* of bin j. The load is the total number of balls allocated to bin j; that is, the number of chain headers allocated to bins $j - \ell + 1, \ldots, j - 1, j$.

We consider the case where each chain header randomly chooses d bins $j_1, \ldots, j_d$. For random choice j_k it computes the maximum load of bins $j_k, j_k + 1, \ldots j_k + \ell - 1$. The chain header is allocated to the bin $j_k \in j_1, \ldots, j_d$ such that the maximum load is minimized. This allocation process is called GREEDY_CHAINS$[d, \ell]$.

We show the following result, which is proved in Section 3.

Theorem 1. *Let $m \leq n$, $\ell \geq 1$, $d \geq 2$, and assume $m \cdot \ell \geq n/(2e)$ and $m \cdot \ell = o(n(\ln \ln m)^{1/2})$. Let m chains of length ℓ be allocated to n bins with d i.u.r. bin choices per chain header. The maximum load of any bin obtained by GREEDY_CHAINS[d, ℓ] is at most*

$$\frac{\ln \ln m}{\ln d} + O\left(\left(\frac{m \cdot \ell}{n}\right)^2\right),$$

with probability $1 - O((\ln m)^d / m^{d-1})$.

Note that when $m \cdot \ell = O(n)$, GREEDY_CHAINS$[d, \ell]$ achieves a maximum load of $(\ln \ln m)/(\ln d) + O(1)$, with high probability. In order to make a direct comparison, we extend of the results of [1] on the algorithm GREEDY$[d]$ to the case where $m < n$.

Theorem 2. *Assume that $d \geq 2$, $m \leq n$, and that c is an arbitrary constant. Then, the maximum load achieved by GREEDY[d] after the allocation of m balls is at most*

$$\frac{\ln \ln n - \ln \ln(n/m)}{\ln d} + c$$

with a probability of $1 - O(n^{-s})$, where s is a constant depending on c.

Theorem 2 gives the bin load arising from chain headers (ignoring the rest of the chain). Since other collisions can occur, for example, between chain headers and internal links of the chain, this will always be a lower bound on the maximum load. Then, provided $m\ell/n = O(1)$,

$$\frac{\ln \ln n - \ln \ln(n/m)}{\ln d} + O(1) \leq \text{ max load } \leq \frac{\ln \ln m}{\ln d} + O(1).$$

We see that, provided that $\ell = e^{(\ln n)^{o(1)}}$ (in particular, ℓ is poly-logarithmic in n), the ratio of the upper and lower bounds on the maximum load is $(1 + o(1))$.

Finally, suppose we allocate chain headers using GREEDY$[d]$ but ignore the effect of this allocation on the rest of the chain. The following theorem, proved in Section 4, shows that this approach leads to a large maximum occupancy.

Theorem 3. *Assume that $m \cdot \ell \geq n/(2e)$, that $m < n/(2e)$, that $m \geq \ln^2 n$, and that $\ell \geq (\ln m)/(\ln \ln m)$. Then, the maximum occupancy of any bin based on GREEDY[d] allocation of chain headers is at least $(\ln m)/(2 \ln \ln m)$, w.h.p.*

The proofs of Theorem 2and Theorem 3 can be found in Section 4.

3 Analysis of GREEDY_CHAINS$[d, \ell]$

In this section, we prove Theorem 1. The proof uses layered induction. In the case of GREEDY$[d]$, Azar et al. [1] use variables γ_i as a high-probability upper bound on the number of bins with i or more balls, where $\gamma_6 = n/2e$ and, for $i > 6$, $\gamma_i = e \cdot n \cdot (\gamma_{i-1}/n)^d$.

Since we allocate chains into bins, we cannot consider only the number of bins with i or more chain headers, we have to consider both the chain headers and tails. Hence, to calculate the load of a bin, we have to consider the chain headers allocated to neighbouring bins. To do so, we define the set S_i which can be thought of as the set of bins which will result in a maximum load of at least $i + 1$ if one of the bins in S_i is chosen for a chain header. The set S_i contains all bins j with load (at least) i and the bins at distance at most $\ell - 1$ in front of bin j. We emphasize that not all bins in S_i have load of i themselves. We use variables β_i as high-probability upper bounds and show that, for i large enough, $|S_i| \leq \beta_i = 2e \cdot m \cdot \ell \cdot (\beta_{i-1}/n)^d$, w.h.p., in our induction. In the following, we define some sets and random variables that are used in our analysis.

- Let $\lambda_j(t)$ be a random variable counting the *h-load* of bin j. That is, $\lambda_j(t)$ is the number of chain headers allocated to bin j at (the end of) step t for $t = 0, 1, ..., m$.
- For given $A \subseteq [n]$, define $\lambda_A(t) = \sum_{j \in A} \lambda_j(t)$. Thus, $\lambda_A(t)$ is a random variable counting the total *h-load* of the bins in A at the end of step t.
- Let $R_j = \{j - \ell + 1, ..., j - 1, j\}$, the set of bins that will increase the load of bin j if a chain header is allocated to them.
- Let $L_j(t)$ be a random variable counting the *load* of bin j at (the end of) step t. Thus, $L_j(t) = \lambda_{R(j)}(t)$, the load arising from the chain headers allocated to the ℓ bins of $R(j)$.
- Let $Q_i(t) = \{j : L_j(t) \geq i\}$ be the set of labels of bins whose load is at least i at (the end of) step t.
- Let $S_i(t) = \cup_{j \in Q_i(t)} R_j$. Thus, $S_i(t)$ contains the labels of bins such that an allocation of a chain header to one of these bins will increase the load of a bin with a load of at least i by 1.
- Let $\theta_{\geq i}(t) = |S_i(t)|$.
- Let h_t be a random variable counting the height of chain t. The algorithm GREEDY_CHAINS$[d, \ell]$ allocates the header to the bin which minimizes the maximum total load h_t, where

$$h_t = 1 + \min_{i=1,\ldots,d} \max \{L_{j_i+k}(t-1),\ k = 0, ..., \ell - 1\} \tag{1}$$

  and $j_1, ..., j_d$ are the bins chosen i.u.r. at step t.

Our method of proving Theorem 1 uses an approach developed in [1], but incorporates the added complexity of considering the maximum load over the chain length. For consistency, we have preserved notation as far as possible.

Let $\alpha = m\ell/n$ with $\alpha \geq 1/2e$ and $k = \lceil 8\alpha^2 e \rceil$. First, we show that i chains can contribute a block of bins of length at most $2\ell - 1$ to $S_i(m)$.

Lemma 1. *For $i \geq k$, (i) $\theta_{\geq i}(m) \leq 2m\ell/i$, (ii) $\theta_{\geq i}(m) \leq n/2$.*

Proof. To prove part (i) we first consider the following worst case scenario. Suppose at step t bin j contains i chain headers, bins $j - \ell + 1, ..., j - 1$ are empty, and bins $j + 1, ...j + \ell - 1$ do not contain any chain headers. Then, $\{j, j + 1, ...j + \ell - 1\} \subseteq Q_i(t)$, $\{j - \ell + 1, ..., j, j + 1, ...j + \ell - 1\} \subseteq S_i(t)$ and $|S_i(t)| = 2\ell - 1$.

Suppose that $S_i(t)$ contains r chain headers. By stacking the r headers on top of each other in blocks of i, at positions $p = k(2\ell - 1)$, $k = 0, 1, ..r/i$, we maximize the total size of $S_i(t)$.

Thus in general $r(2\ell - 1) \geq i|S_i(t)|$, which means that we need at least

$$r \geq \frac{i \cdot |S_i(t)|}{2\ell - 1}$$

chain headers. In particular, for $t \leq m$, we get

$$\lambda_{S_i(t)}(t) \geq \frac{i \cdot |S_i(t)|}{2\ell - 1} > \frac{i \cdot \theta_{\geq i}(t)}{2\ell}$$

since $\theta_{\geq i}(t) = |S_i(t)|$. For $t = m$, we have

$$\frac{i \cdot \theta_{\geq i}(m)}{2\ell} \leq \lambda_{S_i(m)}(m) \leq m. \tag{2}$$

Part (ii) now follows from part (i). Since $i \geq k$, $\alpha = m\ell/n$, and $\alpha \geq 1/(2e)$, we get

$$\theta_{\geq i}(m) \leq \theta_{\geq k}(m) \leq \frac{2m\ell}{k} \leq \frac{2m\ell}{8\alpha^2 e} = \frac{2m\ell}{8 \cdot (m\ell/n) \cdot \alpha e} = \frac{n}{4\alpha e} \leq \frac{n}{2}.$$

□

To prove Theorem 1, we define

$$\beta_i = \begin{cases} n & i = 1, ..., k-1; \\ \frac{n}{4e\alpha} & i = k; \\ 2em\ell \cdot \left(\frac{\beta_{i-1}}{n}\right)^d & i > k. \end{cases}$$

For $i \geq 0$ and $j = k + i$, it follows from the definition of β_i that

$$\beta_j = \beta_{k+i} = n \cdot \frac{(2e)^{\frac{d^i - 1}{d-1}}}{(4e\alpha)^{d^i}} = n \cdot 2^{-d^i} \cdot (2e\alpha)^{-(d^i(d-2)+1)/(d-1)}, \tag{3}$$

and, thus, provided $2e\alpha \geq 1$ (i.e., $m\ell/n \geq 1/2e$), we have $\beta_{k+i} \leq n \cdot 2^{-d^i}$.

Define $\mathcal{E}_i(t) = \{\theta_{\geq i}(t) \leq \beta_i\}$ and let

$$\mathcal{E}_i = \mathcal{E}_i(m) = \{\theta_{\geq i}(m) \leq \beta_i\} \tag{4}$$

be the event that $|S_i(t)|$ is bounded by β_i throughout the process. From the discussion following (2), we have that $\mathcal{E}_k$ holds with certainty. Our goal is to obtain a value for i such that $\mathbf{Pr}(\mathcal{E}_i)$ is close to 1 and, given $\mathcal{E}_i$, no bin receives more than i balls, with high probability.

We next state a standard lemma, proof of which is omitted.

Lemma 2. *Let $X_1, X_2, ..., X_m$ be a sequence of random variables with values in an arbitrary domain, and let $Y_1, Y_2, ..., Y_m$ be a sequence of binary random variables with the property that $Y_t = Y_t(X_1, ..., X_t)$. If*

$$\mathbf{Pr}(Y_t = 1 \mid X_1, ..., X_{t-1}) \leq p,$$

then

$$\mathbf{Pr}\left(\sum_{t=1}^{m} Y_t \geq k\right) \leq \mathbf{Pr}(B(m,p) \geq k),$$

where $B(m,p)$ denotes a binomially distributed random variable with parameters m and p.

As the d choices of bins for a chain header are independent, we have that

$$\mathbf{Pr}\left(h_t \geq i+1 \mid \theta_{\geq i}(t-1) = r\right) \leq \left(\frac{r}{n}\right)^d ,$$

where h_t is given by Eq. (1). For chain t and integer i, let $Y_t^{(i)}$ be an indicator variable given by

$$Y_t^{(i)} = 1 \iff \{h_t \geq i+1, \quad \theta_{\geq i}(t-1) \leq \beta_i\}.$$

Let $X_i = (x_i^1, \ldots, x_i^d)$ denote the set bin choices of ith chain header, and let $X_{1,t} = (X_1, ..., X_t)$ be the choices of the first t chains. We define $\boldsymbol{X}_{1,t}$ as the event $\{\boldsymbol{X}_{1,t} = (X_1, ..., X_t)\}$.

Assume $\boldsymbol{X}_{1,t-1} \in \mathcal{E}_i(t-1)$, meaning after the allocation of the first $t-1$ chains, we have at most β_i bins that would result in a load of at least $i+1$ when hit by chain t. Then,

$$\mathbf{Pr}(Y_t^{(i)} = 1 \mid \boldsymbol{X}_{1,t-1}) \leq \left(\frac{\beta_i}{n}\right)^d$$

and, if $\boldsymbol{X}_{1,t-1} \notin \mathcal{E}_i(t-1)$, then $\mathbf{Pr}(Y_t^{(i)} = 1 \mid \boldsymbol{X}_{1,t-1}) = 0$. Either way,

$$\mathbf{Pr}(Y_t^{(i)} = 1 \mid \boldsymbol{X}_{1,t-1}) \leq \left(\frac{\beta_i}{n}\right)^d \triangleq p_i.$$

We can apply Lemma 2 to conclude that

$$\mathbf{Pr}(\sum_{t=1}^{m} Y_t^{(i)} \geq r) \leq \mathbf{Pr}(B(m,p_i) \geq r). \tag{5}$$

Considering the extreme case discussed above in Lemma 1, we see that each event $\{Y_t^{(i)} = 1\}$ adds at most an extra $2\ell - 1$ bins to $S_{i+1}(t)$. Thus for $\boldsymbol{X}_{1,m} \in \mathcal{E}_i$,

$$\theta_{\geq i+1}(m) \leq 2\ell \cdot \sum_{t=1}^{m} Y_t^{(i)}. \tag{6}$$

Let $r_i = e \cdot m \cdot p_i$. Then, provided that $\sum_{t=1}^{m} Y_t^{(i)} \leq r_i$, we have

$$\theta_{\geq i+1}(m) \leq 2\ell r_i = 2\ell em \cdot p_i = 2em\ell \cdot \left(\frac{\beta_i}{n}\right)^d = \beta_{i+1}. \tag{7}$$

From (5) and (6), we have

$$\mathbf{Pr}\left(\theta_{\geq i+1}(m) > 2\ell \cdot r_i \;\middle|\; \mathcal{E}_i\right) \leq \mathbf{Pr}\left(\sum_{t=1}^{m} Y_t^{(i)} > r_i \;\middle|\; \mathcal{E}_i\right) \leq \frac{\mathbf{Pr}(B(m,p_i) \geq r_i)}{\mathbf{Pr}(\mathcal{E}_i)}. \tag{8}$$

Provided that $m \cdot p_i \geq 2\ln\omega$ (where the precise value of ω is established below in (13)), using the Chernoff bounds, we get

$$\mathbf{Pr}(B(m,p_i) \geq em \cdot p_i) \leq e^{-m \cdot p_i} \leq \frac{1}{\omega^2}. \tag{9}$$

Recall that $\mathbf{Pr}(\neg\mathcal{E}_k) = 0$. Assume inductively that $\mathbf{Pr}(\neg\mathcal{E}_i) \leq i/\omega^2$, for $i \geq k$. Since

$$\mathbf{Pr}(\neg\mathcal{E}_{i+1}) \leq \mathbf{Pr}(\neg\mathcal{E}_{i+1} \mid \mathcal{E}_i) \cdot \mathbf{Pr}(\mathcal{E}_i) + \mathbf{Pr}(\neg\mathcal{E}_i),$$

we have, from (4), (7), (8), and (9), that

$$\mathbf{Pr}(\neg\mathcal{E}_{i+1}) \leq \frac{i+1}{\omega^2}.$$

Choose i^* as the smallest i such that $p_i = \left(\frac{\beta_i}{n}\right)^d \leq \frac{2\ln\omega}{m}$. From (3),

$$i^* - k \leq \frac{\ln\ln(m/\ln\omega)}{\ln d} + O(1). \tag{10}$$

Also, as $\alpha = m\ell/n = o((\ln\ln m)^{1/2})$, we have that $k = o(i^*)$ so that the induction is not empty. Since $p_{i^*} \leq (2\ln\omega)/m$, we have that

$$\begin{aligned}\mathbf{Pr}(\theta_{\geq i^*+1}(m) \geq (2\ell)\cdot 6\ln\omega \mid \mathcal{E}_{i^*}) &\leq \frac{\mathbf{Pr}(B(m,p_{i^*}) \geq 6\ln\omega)}{\mathbf{Pr}(\mathcal{E}_{i^*})} \\ &\leq \frac{\mathbf{Pr}(B(m,(2\ln\omega)/m) \geq 6\ln\omega)}{\mathbf{Pr}(\mathcal{E}_{i^*})} \\ &\leq \frac{1}{\omega^2\cdot\mathbf{Pr}(\mathcal{E}_{i^*})},\end{aligned}$$

and, thus,

$$\mathbf{Pr}(\theta_{\geq i^*+1}(m) \geq (2\ell)\cdot 6\ln\omega) \leq \frac{1}{\omega^2\cdot\mathbf{Pr}(\mathcal{E}_{i^*})}\cdot\mathbf{Pr}(\mathcal{E}_{i^*}) + \mathbf{Pr}(\neg\mathcal{E}_{i^*}) \leq (i^*+1)/\omega^2. \tag{11}$$

Conditioned on $\theta_{\geq i^*+1}(m) \leq 12\ell\ln\omega$, the probability that a chain is placed at height at least i^*+2 is at most $(12\ell\ln\omega/n)^d$. Given that $Y \sim B(m,(12\ell\ln\omega/n)^d)$, $\mathbf{Pr}(Y \geq 1) \leq m(12\ell\ln\omega/n)^d$ by Markov's Inequality. Thus applying Lemma 2, we get

$$\mathbf{Pr}\left(\sum_{t=1}^{m} Y_t^{(i^*+1)} \geq 1 \;\middle|\; \theta_{\geq i^*+1}(m) \leq (2\ell)\cdot 6\ln\omega\right) \leq \frac{m\cdot\left(\frac{12\ell\cdot\ln\omega}{n}\right)^d}{\mathbf{Pr}(\theta_{\geq i^*+1} \leq 12\ell\cdot\ln\omega)}. \tag{12}$$

Let ω satisfy

$$\omega = \left(\frac{m^{d-1}\cdot\ln\ln m}{(\ln m)^d}\right)^{1/2}. \tag{13}$$

Using $m \cdot \ell = o(n(\ln\ln m)^{1/2})$, (13), (12) and (11), the probability that there is a bin with load at least $i^* + 2$ is bounded by a term of order

$$\frac{i^* + 1}{\omega^2} + O\left(\frac{(\ln \omega)^d}{m^{d-1}}\right) = O\left(\frac{(\ln m)^d}{m^{d-1}}\right).$$

By plugging in ω and k into (10), we get

$$i^* \leq \frac{\ln\ln m}{\ln d} + O(\alpha^2) + O(1).$$

Thus, w.h.p, no bin receives more than $i^* + 1$ balls. □

4 Allocating Chain Headers

In this section we present the proofs of Theorem 2 and Theorem 3.

4.1 Proof of Theorem 2

This theorem can be shown similarly to the proof of Theorem 4 in [1]. We define $\gamma_1 = \gamma_2 = \cdots = \gamma_5 = n$, $\gamma_6 = m/(2e)$, and

$$\gamma_i = em \cdot \left(\frac{\gamma_{i-1}}{n}\right)^d \text{ for } i > 6.$$

Thus $\gamma_i = C\, n(m/(n2e))^{d^i}$ for some $C > 1$ constant. Integer i^* is defined as the smallest i such that $em(\gamma_i/n)^d \leq 6 \ln n$, which holds for

$$i^* \leq \frac{\ln\ln n - \ln\ln(n/m)}{\ln d} + O(1).$$

It can be shown that the maximum load is bounded by $i^* + 2 = (\ln\ln n - \ln\ln(n/m))/\ln d + O(1)$, w.h.p.

4.2 Proof of Theorem 3

The idea of the proof is as follows. Let U_m be the number of bins with load at least one after the allocation of m chains by GREEDY[d] applied to the chain headers. We show that, with a good probability, there exists a strip of ℓ consecutive bins which i) is used by one chain, and ii) at least t of its bins are in U_m. This gives us a bin with load at least t.

First, we find a lower bound on U_m. Azar et al. [1] show that the protocol GREEDY[d] for $d \geq 1$ is majorized by GREEDY[1] in the following sense. Let x_i be the load of the bin with the ith largest load after allocation of m balls with GREEDY[d], and let x'_i be the load of the bin with the ith largest load after allocation of m balls with GREEDY[1]. It is shown in [1] that there exists a one-to-one mapping between the random choices of GREEDY[1] and GREEDY[d] such that for all $1 \leq j \leq n$,

$$\sum_{i=1}^{j} x_i \leq \sum_{i=1}^{j} x'_i.$$

From this it follows that the number of empty bins in an allocation with GREEDY[d] is smaller that the one in GREEDY[1]. Let $f(m)$ be the number of occupied bins in an allocation generated by GREEDY[1]. When $m = n/(2e)$ we get

$$E[f(m)] = n\left(1 - \left(1 - \frac{1}{n}\right)^m\right) \geq 0.9m.$$

As m decreases, $m - E[f(m)]$ decreases, too. Thus, provided $m \leq n/(2e)$, the expected number of occupied cells with $d \geq 1$ choices per ball is always at least $0.9m$. By concentration, we can assume $U_m \geq m/2 + 1$, provided $m \geq \ln^2 n$.

Given U_m, we can assume that the locations of the bins occupied by chain headers are sampled uniformly without replacement from $[1, \ldots, n]$. We fix one of the m chains and consider the strip of ℓ consecutive bins occupied by the chain. Assuming $\ell \geq 2t$, the probability of at least t bins occupied by additional chain headers in that strip is at least

$$\binom{U_m - 1}{t} \cdot (\ell)_t \cdot \left(\frac{1}{n}\right)^t \geq \binom{m/2}{t} \cdot (\ell)_t \cdot \left(\frac{1}{n}\right)^t \geq \left(\frac{m\ell}{2etn}\right)^t \geq \left(\frac{1}{4e^2 t}\right)^t,$$

as $m\ell \geq n/(2e)$ and $(\ell)_t = \ell(\ell - 1)\cdots(\ell - t + 1) \geq (\ell/e)^t$.

Let $c = 1/(4e^2)$. Then, the expected number of chains allocated into a strip with load at least $t = \ln m/(2 \ln\ln m)$ is at least

$$\begin{aligned} m \cdot \left(\frac{c}{t}\right)^t &= \exp\left(\ln m - t \ln t/c\right) \\ &\geq m^{1/3}. \end{aligned}$$

The probability that such an event does not occur tends to zero by the Chebychev's inequality. □

5 Conclusions and Open Problems

In this paper we analyse the maximum load for the *chains-into-bins* problem where m balls are connected in n/ℓ chains of length ℓ. We show that, provided $m\ell \geq n/2e$ and $m\ell = o(n(\ln\ln m)^{1/2})$, the maximum load is at most $\frac{\ln\ln m}{\ln d} + O(1)$, with probability $1 - O((\ln m)^d/m^{d-1})$. This shows that the maximum load is going down with increasing chain length.

Surprisingly, there are many open questions in the area of balls-into-bins processes. Only very few results are known for weighted balls-into-bins processes, where the balls come with weights and the load of a bin is the sum of the weights of the balls allocated to it. Here, it is even not known if two or more random choices improve the maximum load, compared to the simple process where every ball is allocated to a randomly chosen bin (see [12]). Also, it would be interesting to get tight results for the maximum load and results specifying "worst-case" weight distributions for the balls. Something in the flavor "given that the total weight of the balls is fixed, it is better to allocate lots of small balls, compared to fewer bigger ones." Another interesting problem is to show results relating the maximum load to the order in which the balls are allocated. For example, is it always better to allocate balls in the order of decreasing ball weight, compared to the order of increasing ball weight?

For chains-into-bins problem, an open question is to prove Knuth's [6] conjecture stating that breaking chains into two parts only increases the maximum load. This question still open for a single choice and also for several random choices per ball. See [3] for a first progress in this direction. Another question is if similar results to the one we showed in this paper for GREEDY[d] applied to chains also holds for the ALWAYS-GO-LEFT protocol from [13] applied on chains.

Finally, we note that it would be interesting to generalize the problem to two dimensional packing, and consider online allocation of m objects of length ℓ and width w to the cells of a toroidal grid of length n and width h.

References

1. Azar, Y., Broder, A.Z., Karlin, A.R., Upfal, E.: Balanced Allocations. SIAM J. Computing 29, 180–200 (1999)
2. Berenbrink, P., Czumaj, A., Steger, A., Vöcking, B.: Balanced Allocations: The Heavily Loaded Case. SIAM J. Computing 35, 1350–1385 (2006)
3. Englert, M.: Chains of Length Two into Bins. Manuscript, University of Aachen
4. Fekete, S., Köhler, E., Teich, J.: Optimal FPGA module placement with temporal precedence constraints. In: Proc. of the Conference on Design, Automation and Test in Europe (DATE 2001), pp. 658–667 (2001)
5. Kenthapadi, K., Panigrahy, R.: Balanced allocation on graphs. In: Proc. of 17th Annual Symposium on Discrete Algorithms (SODA 2006), pp. 434–443 (2006)
6. Knuth, D.E.: Sorting and Searching, 2nd edn. The Art of Computer Programming, vol. 3. Addison-Wesley, Reading (1998)
7. Mitzenmacher, M., Richa, A.W., Sitaraman, R.: The Power of Two Random Choices: A Survey of Techniques and Results. In: Handbook of Randomized Computing (2000)
8. Mitzenmacher, M., Prabhakar, B., Shah, D.: Load Balancing with Memory. In: Proc. of the 43rd Annual IEEE Symposium on Foundations of Computer Science (FOCS 2002), pp. 799–808 (2002)
9. Patterson, D.A., Gibson, G.A., Katz, R.H.: A Case for Redundant Arrays of Inexpensive Disks (RAID). In: Proc. of SIGMOD International Conference on Management of Data, pp. 109–116 (1988)
10. Sanders, P., Vöcking, B.: Tail Bounds And Expectations For Random Arc Allocation And Applications. Combinatorics, Probability & Computing 12(3) (2003)
11. Steiger, C., Walder, H., Platzner, M.: Operating Systems for Reconfigurable Embedded Platforms: Online Scheduling of Real-Time Tasks. IEEE Trans. Computers 53(11), 1393–1407 (2004)
12. Talwar, K., Wieder, U.: Balanced allocations: the weighted case. In: Proc. of the 39th Symposium on Theory of Computing (STOC), pp. 256–265 (2007)
13. Vöcking, B.: How Asymmetry Helps Load Balancing. J. ACM 50(4), 568–589 (2003)

Complexity of Locally Injective Homomorphism to the Theta Graphs*

Bernard Lidický and Marek Tesař

Department of Applied Mathematics, Charles University,
Malostranské nám. 25, 118 00 Prague, Czech Republic
{lidicky,tesar}@kam.mff.cuni.cz

Abstract. A Theta graph is a multigraph which is a union of at least three internally disjoint paths that have the same two distinct end vertices. In this extended abstract we show full computational complexity characterization of the problem of deciding the existence of a locally injective homomorphism from an input graph G to any fixed Theta graph.

Keywords: computational complexity; locally injective homomorphism; Theta graph.

1 Introduction

Let G be a graph. We denote its set of vertices by $V(G)$ and its set of edges by $E(G)$. Graphs in this extended abstract are generally simple. If they may have parallel edges or loops, we explicitly say so. We denote the degree of a vertex v by $\deg_G(v)$ and the set of all neighbors of v by $N_G(v)$. We omit G in the subscript if it is clear from the context. By $[n]$ we denote the set of integers $\{1, \ldots, n\}$.

Let G and H be graphs. A *homomorphism* is an edge preserving mapping $f : G \to H$. A homomorphism is *locally injective (resp. surjective, bijective)* if $N(v)$ is mapped to $N(f(v))$ injectively (resp. surjectively, bijectively). A locally bijective homomorphism is also known as a *covering projection* or simply a *cover*. Similarly, locally injective homomorphism is known as a *partial covering projection* and a *partial cover*.

We consider the following decision problem. Let H be a fixed graph and G be an input graph. Determine the existence of a locally injective (surjective, bijective) homomorphism $f : G \to H$. We denote the problem by H-LIHom (resp. H-LSHom, H-LBHom). If there is no local restriction on the homomorphism, the problem is called H-Hom.

In this extended abstract we consider the H-LIHom problem.

Problem: H-LIHom
Input: graph G
Question: Does there exist a locally injective homomorphism $f : G \to H$.

* Supported by Charles University as GAUK 95710.

C.S. Iliopoulos and W.F. Smyth (Eds.): IWOCA 2010, LNCS 6460, pp. 326–336, 2011.

Locally injective homomorphisms are closely related to $H(2,1)$-labelings, which have applications in frequency assignment. Let H be a graph. An $H(2,1)$-*labeling* of a graph G is a mapping $f : V(G) \to V(H)$ such that every pair of adjacent vertices are mapped to distinct and nonadjacent vertices. Moreover, image of every pair of vertices in distance two is two distinct vertices. The mapping f corresponds to a locally injective homomorphism to the complement of H.

The computational complexity of H-HOM was fully determined by Hell and Nešetřil [9]. They show that the problem is solvable in polynomial time if H is bipartite and it is NP-complete otherwise.

The study of H-LSHOM was initiated by Kristiansen and Telle [13] and completed by Fiala and Paulusma [8] who gave a full characterization by showing that H-LSHOM is NP-complete for every connected graph on at least three vertices.

The complexity of locally bijective homomorphisms was first studied by Bodlaender [2] and Abello et al. [1]. Despite the effort [10,11,12] the complete characterization is not known.

Similarly for the locally injective homomorphism the dichotomy for the complexity is not known. Some partial results can be found in [4,5,7]. Fiala and Kratochvíl [6] also considered a list version of the problem and showed dichotomy.

Fiala and Kratochvíl [5] showed, that H-LBHOM is reducible in polynomial time to H-LIHOM. Hence it makes sense to study the complexity of H-LIHOM where H-LBHOM is solvable in polynomial time. This is the case for Theta graphs, which we consider in this extended abstract. Note that no other direct consequences of complexity of H-HOM or H-SHOM to H-LIHOM are known.

Fiala and Kratochvíl [4] showed, that if Theta graph H contains only simple paths of length a, then H-LIHOM is always polynomial. They also showed that if H contains only simple paths of two different lengths a and b, then:

- if both a and b are odd, then H-LIHOM is polynomial,
- if a and b have different parity, then H-LIHOM is NP-complete,
- if both a and b are even, then H-LIHOM is as hard as H'-LIHOM, where H' is a Theta graph, that arise from H by replacing paths of length a, resp. b by paths of lengths $\frac{a}{2}$, resp. $\frac{b}{2}$.

The study of Theta graphs continues in the work of Fiala et al. [7], which proves NP-completeness for Theta graphs with exactly three odd different lengths of simple paths. We extend the last result to all Theta graphs, which finishes the complexity characterization of Theta graphs.

Theorem 1. *Let H be a Theta graph with simple paths of at least three distinct lengths. Then H-LIHOM problem is NP-complete.*

In the next section, we introduce several definitions and gadgets which we use in NP-hardness reductions. In Section 3 we state necessary Lemmas for the proof of Theorem 1. We omitted proofs of some Propositions, Lemmas and Theorems due to the page limit for this extended abstract.

2 Definitions and Gadgets

A graph G is a *Theta graph* (or *Θ-graph*) if it is the union of at least three internally disjoint paths that have the same two distinct end vertices. We denote the two vertices of degree at least three by A and B. Note that if two paths of the union are of length one, the resulting graph have parallel edges.

A Θ-graph $\mathcal{T}$ is denoted by $\Theta(a_1^{t_1}, a_2^{t_2}, \ldots, a_n^{t_n})$, where $1 \leq a_1 < a_2 < \cdots < a_n$ and $t_i \geq 1$ for all $1 \leq i \leq n$, if $\mathcal{T}$ is the union of paths of lengths $a_1, a_2, \ldots, a_n$ and t_i are the corresponding multiplicities. We write a_i instead of a_i^1. We assume that $n \geq 3$ as the case $n \leq 2$ is already solved [4].

Throughout this section we assume that $\mathcal{T} = \Theta(a_1^{t_1}, \ldots, a_n^{t_n})$ is some Θ-graph.

Let G be a graph and $v_1, v_2, \ldots, v_n$ be a path in G. The path is *simple* if v_1 and v_n are vertices of degree at least three and all inner vertices of the path have degree two. We denote a simple path of length n by SP_n.

Let G be a graph and f be a locally injective homomorphism from G to $\mathcal{T}$. Note that f must map all vertices of degree at least three to A or B in $\mathcal{T}$. Hence every end vertex of every simple path of G must be mapped to A or B. We call a vertex *special* if it has degree at least three or if we insist that it is mapped to A or B. Note that A and B are also special vertices and if v is a special vertex of degree less than three, then adding extra pendant leaves forces, that v must be mapped to A or B. We need to control what are the possible mappings of simple paths. Let $v_1, v_2, \ldots, v_{l-1}, v_l$ be a simple path P. For a locally injective homomorphism f, define a function $g_f^P(v_1, v_l) = a_i$ if the edge $v_1 v_2$ is mapped by f to an edge of SP_{a_i} in $\mathcal{T}$. We omit the superscript P if there is only one simple path containing v_1 and v_l.

We say that SP_n allows *decomposition* $a_i - a_j$ if there exists a graph H containing a simple path P of length n with end vertices u and v and a locally injective homomorphism $f : H \to \mathcal{T}$ such that $g_f^P(u, v) = a_i$ and $g_f^P(v, u) = a_j$. We denote the decomposition by $a_i -_k a_j$ (resp. $a_i -_c a_j$) if it forces that $f(u) = f(v)$ (resp. $f(u) \neq f(v)$).

In case of $x -_k y$ (resp. $x -_c y$) decomposition we say, that the decomposition keeps (resp. changes) the parity.

Proposition 1. *Every simple path SP_{a_i} always allows decomposition $a_i -_c a_i$ and does not allow decomposition $a_i -_k a_i$. Similarly, for $i \neq j$ holds that $SP_{a_i + a_j}$ always allows decomposition $a_i -_k a_j$ and never allows $a_i -_c a_j$.*

The proof of Proposition 1 as well as proofs of the other propositions are omitted due to the page limit.

Let M be a positive integer and $\mathcal{E} \subseteq \{a_1, a_2, \ldots a_n\}$. The following notation

$$\boldsymbol{M_{\mathcal{E}}^{\mathcal{T}}} : x_1 - y_1, x_2 - y_2, \ldots, x_s - y_s, (z_1 - w_1), (z_2 - w_2), \ldots, (z_t - w_t)$$

describes the list of all decompositions $x - y$ of SP_M where $x, y \in \mathcal{E}$. Decompositions $x_i - y_i$ must be possible and decompositions $z_j - w_j$ are optional for all $i \in [t]$ and $j \in [s]$. Moreover, $-_k$ and $-_c$ can be used instead of just $-$.

Now we introduce gadget $B_z^{\mathcal{T}}$, which can be used for blocking a simple path of length z at some vertex. It has a central vertex y which is for every $i \in [n]$ connected by paths of length a_i to vertices v_j^i where $j \in [t_i]$. Moreover, every vertex v_j^i except v_1^z has two extra pendant leaves (so v_j^i is special). If X is a copy of $B_z^{\mathcal{T}}$, we refer to the vertex v_1^z by $X(w)$ or w if X is clear from the context. Moreover, we demand that w is special. See Figure 1.

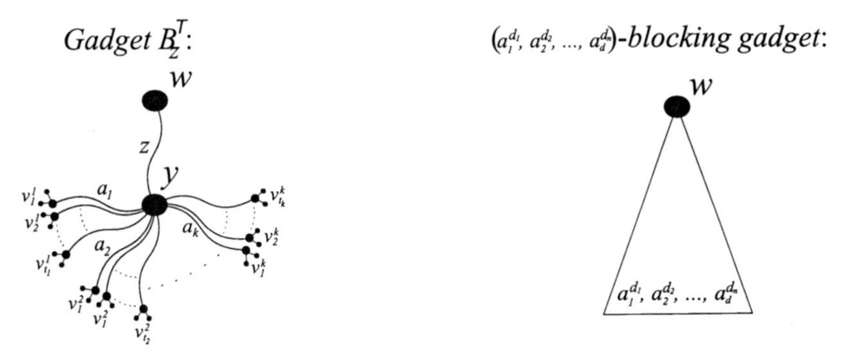

Fig. 1. $B_z^{\mathcal{T}}$ and a blocking gadget

Proposition 2. *Let G be a graph and let X be a copy of $B_z^{\mathcal{T}}$ in G. Moreover, w has degree at least three. Suppose, that there exists a locally injective homomorphism $f : G \to \mathcal{T}$. Then:*

$$g_f(w, y) = z = g_f(y, w)$$

The gadget $B_z^{\mathcal{T}}$ blocks usage of one z at w by forcing $\mathcal{T}$-LIHOM to map the path wy to SP_z in $\mathcal{T}$.

We usually need to use several copies of the gadget $B_z^{\mathcal{T}}$ at once. Let d_1, $d_2, \ldots, d_n$ be nonnegative integers such that $d_i \leq t_i$ for all $i \in [n]$. We define the $(a_1^{d_1}, \ldots, a_n^{d_n})$-*blocking gadget* to be the union of $t_i - d_i$ copies of $B_{a_i}^{\mathcal{T}}$ for every $i \in [n]$ where there is only one vertex w shared by all of them. If X is a copy of the blocking gadget, we refer to the vertex w by $X(w)$ or w if X is clear from the context. Note that we will consider only copies of the blocking gadget where vertex w is special.

In the notation we omit a^0 and the superscript d_i if $d_i = 1$. In further figures, we depict the $(a_1^{d_1}, \ldots, a_n^{d_n})$-blocking gadget by a triangle with one vertex corresponding to w and with inscribed text $a_1^{d_1}, \ldots, a_n^{d_n}$, see Figure 1.

Proposition 3. *Let G be a graph and X be a copy of $(a_1^{d_1}, a_2^{d_2}, \ldots, a_n^{d_n})$-blocking gadget in G where $\deg_G(w) \geq 3$. Let $P_1, P_2, \ldots, P_k$ be the all simple paths, starting at w with without any other intersection with the blocking gadget X and with end points $u_1, u_2, \ldots, u_k$. Suppose, that there exists a locally injective homomorphism $f : G \to \mathcal{T}$. Then $k \leq \sum_{i=1}^{n} d_i$ and*

$$\forall i \in [n] : \quad |\{u_j, g_f^{P_j}(w, u_j) = a_i\}| \leq d_i.$$

Note that the blocking gadget on its own is not sufficient for reducing $\Theta(a^k, b^l, c^m, a_4^{t_4}, \ldots, a_n^{t_n})$ to $\Theta(a, b, c)$. The obstacle is that a simple path may have different possible inner decompositions and the blocking gadget cannot be used inside paths in general.

Apart from blocking some paths we also need to force that several special vertices are mapped to the same vertex (to A or B). Hence we introduce the following gadget.

Definition 1. *Let $\mathcal{T} = \Theta(a^k, b^l, c^m, a_4^{t_4}, \ldots, a_n^{t_n})$ be a Θ-graph. Let $r \geq 2$ be an integer and N be the smallest power of two such that $N \geq 2r$. Define a graph $PC_a^{\mathcal{T}}(r)$ (see Figure 2) with special vertices $u_1, u_2, \ldots, u_{2N-1}, u_1', u_2', \ldots, u_{N-1}'$, $v_1, v_2, \ldots, v_{2N-1}, v_1', v_2', \ldots, v_{N-1}'$ to be a graph constructed in the following way:*

- *$\forall i \in \{1, 2, \ldots, N-1\}$, connect vertex u_i' with vertices u_i, u_{2i} and u_{2i+1} by paths of lengths c, a and b (in this order),*
- *$\forall i \in \{1, 2, \ldots, N-1\}$, connect vertex v_i' with vertices v_i, v_{2i} and v_{2i+1} by paths of lengths c, a and b (in this order),*
- *$\forall i \in \{2, 3, \ldots, N-1\}$, take copies U_i and V_i of (a, c)-blocking gadget if i is even and (b, c)-blocking gadget if i is odd and identify vertex u_i with $U_i(w)$ and vertex v_i with $V_i(w)$,*
- *$\forall i \in \{1, 2, \ldots, N-1\}$, take copies U_i' and V_i' of (a, b, c)-blocking gadget and identify vertex u_i' with $U_i'(w)$ and vertex v_i' with $V_i'(w)$,*
- *identify vertex u_1' with v_1 and vertex v_1' with u_1.*

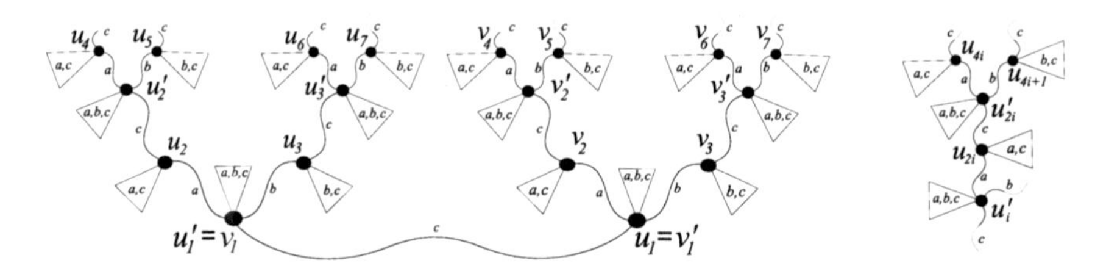

Fig. 2. Graph $PC_a^{\mathcal{T}}(N)$ and local neighborhood of vertices u_i and u_i'

Proposition 4. *Let $r \geq 2$ be an integer, $\mathcal{T} = \Theta(a^k, b^l, c^m, a_4^{t_4}, \ldots, a_n^{t_n})$ be a Θ-graph and let Z be a copy of graph $PC_a^{\mathcal{T}}(r)$ in a graph G. Let N be as in the definition of $PC_a^{\mathcal{T}}(r)$. Suppose, that there exists a locally injective homomorphism $f : G \to \mathcal{T}$, such that for all $i \in [2N-1] : f(Z(u_i)), f(Z(v_i)) \in \{A, B\}$. Then for all even $i, j \in \{N, N+1, \ldots, 2N-1\}$ the following hold:*

$$f(Z(u_i)) = f(Z(u_j)) \neq f(Z(v_j)) = f(Z(v_i)),$$

$$g_f(Z(u_i), Z(u_{i/2}')) = a = g_f(Z(v_j), Z(v_{j/2}')).$$

Let Z be a copy of $PC_a^{\mathcal{T}}(r)$. For $i \in [N]$ we define $Z(x_i)$ to be u_{N+2i-2} and $Z(y_i)$ to be v_{N+2i-2}. Similarly as the gadget $PC_a^{\mathcal{T}}(r)$, we define a gadget $PC_b^{\mathcal{T}}(r)$, with the only difference, that a and b are swapped in the construction. We call the graphs $PC_a^{\mathcal{T}}(r)$ and $PC_b^{\mathcal{T}}(r)$ *parity controllers.* With parity controllers we are able to create arbitrary many special vertices, which are mapped to the same vertex of $\mathcal{T}$ in every locally injective homomorphism to $\mathcal{T}$. Moreover, each of these special vertices is an end point of a path which must be mapped to a simple path of length a (resp. b) in $\mathcal{T}$.

For some $\mathcal{T}$, we reduce 3-SAT or NAE-3-SAT to $\mathcal{T}$-LIHOM. In the reduction we use copies the following gadget for representing clauses.

Let $\mathcal{T} = \Theta(a^k, b^l, c^m, a_4^{t_4}, \ldots, a_n^{t_n})$ be a Θ-graph. We define *$\mathcal{T}$-clause gadget* to be a graph with special vertices u_0, u_1, u_2, u_3 such that, for all $i \in \{1,2,3\}$, vertex u_i is connected to u_0 by a path of length $a+b+c$ and u_0 is identified with the vertex $X(w)$, where X is a copy of the (a,b,c)-blocking gadget. If Y is a copy of $\mathcal{T}$-clause gadget, we refer to the vertices u_j by $Y(u_j)$ or u_j if Y is clear from the context for all $j \in \{0,1,2,3\}$. Note that we will consider only such copies of $\mathcal{T}$-clause gadget, that vertices u_1, u_2 and u_3 are special. See Figure 3.

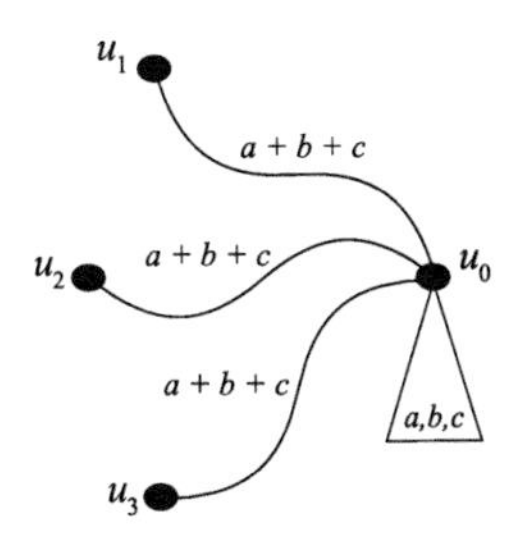

Fig. 3. $\mathcal{T}$-clause gadget

Let Y be a copy of the $\mathcal{T}$-clause gadget and $\gamma \in \{a,b\}$. We say, that $\mathcal{T}$ is *γ-positive* if and only if there exists a locally injective homomorphism $f : Y \to \mathcal{T}$ such that:

- $f(u_0) \neq f(u_1) = f(u_2) = f(u_3) \in \{A, B\}$,
- $g_f(u_1, u_0) = g_f(u_2, u_0) = g_f(u_3, u_0) = \gamma$.

Proposition 5. *Let $a < b < c$ be positive integers, such that $a + b \neq c$. Let $\mathcal{T} = \Theta(a^k, b^l, c^m, a_4^{t_4}, \ldots, a_n^{t_n})$ be a Θ-graph and Y be the $\mathcal{T}$-clause gadget. Let $\gamma \in \{a,b\}$ and $x, y, z \in \{\gamma, c\}$.*

Then there exists a locally injective homomorphism $f : Y \to \mathcal{T}$ satisfying:

- $f(u_0) \neq f(u_1) = f(u_2) = f(u_3) \in \{A, B\}$,
- $g_f(u_1, u_0) = x,\ g_f(u_2, u_0) = y,\ g_f(u_3, u_0) = z$

if and only if at least one of the following conditions hold:

- $\{x, y, z\} = \{\gamma, c\}$,
- $x = y = z = \gamma$ *and $\mathcal{T}$ is γ-positive.*

3 NP-Completeness Reductions

In this section we give several lemmas, which each show NP-completeness for some Θ-graphs. Together, they cover all Θ-graphs and hence they imply Theorem 1. We present the proof only of Lemma 1. Proofs of the other lemmas are omitted due to the page limit.

Note that the lemmas show only NP-hardness as H-LIHOM is clearly in NP for any H.

In this section we assume that $\mathcal{T} = \Theta(a^k, b^l, c^m, a_4^{t_4}, \ldots, a_n^{t_n})$.

Lemmas are grouped into three blocks, which reflect what type of reduction is used. Reductions in each group are similar. The first group shows NP-hardness from 3-SAT and NAE-3-SAT.

Lemma 1. *Let $\mathcal{T}$ be a Θ-graph such that $a + b \neq c$ and*

$$(\boldsymbol{a+b})^{\mathcal{T}}_{\boldsymbol{a,b,c}} : \; a-b, (a-a)$$
$$(\boldsymbol{a+c})^{\mathcal{T}}_{\boldsymbol{a,b,c}} : \; a-c, (a-a), (b-b)$$

then $\mathcal{T}$-LIHOM is NP-complete.

Proof. Let $\phi = \vee_{i=1}^{p}(c_i^1 \wedge c_i^2 \wedge c_i^3)$ be a boolean formula in conjunctive normal form with variables $s_1, s_2, \ldots, s_r$ (where every clause has exactly 3 literals). Let var, neg and ord be functions from the set of all literals of the formula ϕ, such that $var(c_i^j)$ is the variable corresponding to the literal c_i^j, $neg(c_i^j)$ is 0 if the literal c_i^j is a positive occurrence of the variable $var(c_i^j)$ and $neg(c_i^j) = 1$ otherwise, and $ord(c_i^j)$ is the order of occurrence of the literal of the variable $var(c_i^j)$ in ϕ.

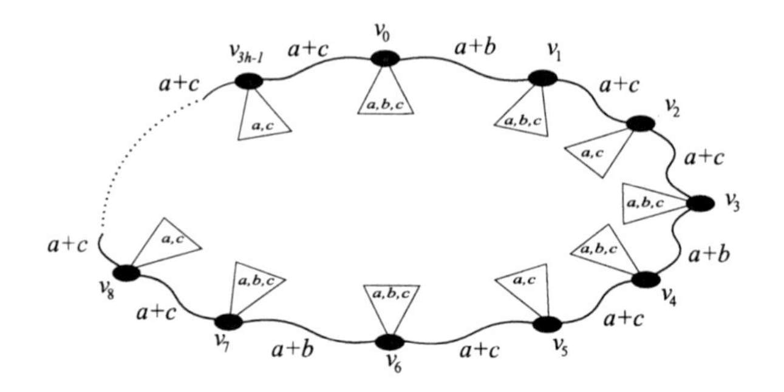

Fig. 4. Variable gadget α

Define *variable gadget α of order h* (see Figure 4) as a graph with special vertices $v_0, v_1, \ldots, v_{3h-1}$ such that for all $i \in \{0, \ldots, h-1\}$, vertices v_{3i} and v_{3i+1} are connected by a path of length $a+b$ and vertices v_{3i+1} and v_{3i+2} as well as vertices v_{3i+2} and v_{3i+3} are connected by a path of length $a+c$ (all indices are counted by modulo $3h$). For every $i \in \{0, \ldots, h-1\}$ we take two copies B_i^0 and B_i^1 of the (a, b, c)-blocking gadget and identify the vertex $B_i^0(w)$ with the vertex v_{3i} and the vertex $B_i^1(w)$ with the vertex v_{3i+1}, and for every $j = 0, \ldots, h-1$ we take a copy B_i^2 of the (a, c)-blocking gadget and identify the vertex $B_i^2(w)$ with the vertex v_{3j+2}.

For every $i \in [r]$, let n_i be the number of occurrences of the variable s_i in the formula ϕ, let X_i be a copy of the variable gadget α of order $n_i + 1$. For every $j \in [p]$ let Z_j be a copy of the the clause gadget and let Y be a copy of the parity controller $PC_b^{\mathcal{T}}(r)$. Now define a graph G_ϕ, which contains copies

$X_1, \ldots, X_r, Z_1, \ldots, Z_p, Y$ and for every literal c_j^d of the formula ϕ, if $var(c_j^d) = s_i$ then we identify the vertices $X_i(v_{3ord(c_j^d)+neg(c_j^d)-3})$ and $Z_j(u_d)$. For every $i \in [r]$ we replace the copy of the (a,c)-blocking gadget on vertex $X_i(v_{3n_i+2})$ by a copy of the (a,b,c)-blocking gadget and identify vertices $X_i(v_{3n_i+2})$ and $Y(x_i)$ (clearly the combination of the (a,b,c)-blocking gadget and Y creates for the vertex $X_i(v_{3n_i+2})$ the same constraints as the (a,c)-blocking gadget), and to every vertex $X_i(v_j)$ and $Y(y_i)$ of degree less then three we add new pendant leaves (so all vertices $X_i(v_j)$ and $Y(y_i)$ are special).

We claim, that if $\mathcal{T}$ is b-positive then ϕ is satisfiable if and only if there exists a locally injective homomorphism from G_ϕ to $\mathcal{T}$. And if $\mathcal{T}$ is not b-positive then ϕ is NAE-satisfiable if and only if there exists a locally injective homomorphism from G_ϕ to $\mathcal{T}$. The fact that 3-SAT and NAE-3-SAT are NP-complete problems and $\mathcal{T}$-LIHom is in NP imply that $\mathcal{T}$-LIHom is NP-complete.

At first suppose that $\mathcal{T}$ is b-positive and there exists a locally injective homomorphism $f : G_\phi \to \mathcal{T}$. Let X be one of the copies of the variable gadget α of order d in G_ϕ. Since $\mathbf{(a+b)}^{\mathcal{T}}_{\mathbf{a,b,c}}$: $a -_k b, (a-a)$, we know that $g_f(v_0, v_1) \in \{a, b\}$. If $g_f(v_0, v_1) = b$, then necessarily $g_f(v_1, v_0) = a$. But since there is a copy of the (a,b,c)-blocking gadget on vertex v_1 we know, that $g_f(v_1, v_2)$ is b or c. Since $\mathbf{(a+c)}^{\mathcal{T}}_{\mathbf{a,b,c}}$: $a -_k c, (a-a), (b-b)$ if $g_f(v_1, v_2) = b$, then $g_f(v_2, v_1) = b$, which is not possible because of the copy of the (a,c)-blocking gadget on v_2 and so $g_f(v_1, v_2) = c$ and necessary $g_f(v_2, v_1) = a, g_f(v_2, v_3) = c, g_f(v_3, v_2) = a$ and then necessarily $g_f(v_3, v_4) = b$. And since the gadget X is symmetric, we can continue in the same way until we reach the vertex v_0 again. Then $\forall i \in \{0, \ldots, d-2\}$ if there exists a simple path from v_{3i} to $Z(u_0)$ for some copy Z of the clause gadget, then $g_f(v_{3i}, Z(u_0)) = c$ (the corresponding literal is false) and analogically for the simple path from v_{3i+1} to $Z(u_0)$, for which holds $g_f(v_{3i+1}, Z(u_0)) = b$ (the corresponding literal is true). In this case we say that the variable corresponding to X is false.

If $g_f(v_0, v_1) = a$ then we use a similar idea as in the previous paragraph, but we argue in the counterclockwise order ($g_f(v_0, v_{3d-1})$ must be c, etc.) and analogically we get, that if appropriate simple paths exists then $g_f(v_{3i}, Z(u_0)) = b$ (the corresponding literal is true), resp. $g_f(v_{3i+1}, Z(u_0)) = c$ (the corresponding literal is false). In this case we say that the variable corresponding to X is true.

We claim that in this evaluation every clause of ϕ is satisfied. If not, then there exists a copy of the clause gadget Z corresponding to some clause and $g_f(u_1, u_0) = g_f(u_2, u_0) = g_f(u_3, u_0) = c$ in Z. Since there is a copy of the (a,b,c)-blocking gadget at vertex u_0, without loss of generality we suppose that $g_f(u_0, u_1) = c$. Thus the simple path u_0u_1 of length $a+b+c$ allows decomposition $c - c$. But this is not possible because $0 < a + b < c + a$ and $a + b \neq c$, a contradiction.

On the other side, if $\mathcal{T}$ is b-positive and formula ϕ is satisfiable, then we show that there exists a locally injective homomorphism $f : G_\phi \to \mathcal{T}$. Suppose that $e : \{s_1, \ldots, s_r\} \to \{true, false\}$ is a satisfying evaluation of the variables of ϕ and predefine a function $f : G_\phi \to \mathcal{T}$ in the following way. For every $i \in [r]$

let n_i be the number of occurrences of variable s_i in ϕ and let X_i be a copy of the variable gadget α corresponding to s_i, for every $j = 0, \ldots, 3n_i + 2$ define $f(v_j) = A$ and

- if $e(s_i) = true$, then for all $j \in \{0, .., n_i\}$: $g_f(v_{3j}, v_{3j+1}) = a$, $g_f(v_{3j+1}, v_{3j}) = b$, $g_f(v_{3j+1}, v_{3j+2}) = a$, $g_f(v_{3j+2}, v_{3j+1}) = c$, $g_f(v_{3j+2}, v_{3j+3}) = a$, $g_f(v_{3j+3}, v_{3j+2}) = c$,
- if $e(s_i) = false$, then for all $j \in \{0, .., n_i\}$: $g_f(v_{3j}, v_{3j+1}) = b$, $g_f(v_{3j+1}, v_{3j}) = a$, $g_f(v_{3j+1}, v_{3j+2}) = c$, $g_f(v_{3j+2}, v_{3j+1}) = a$, $g_f(v_{3j+2}, v_{3j+3}) = c$, $g_f(v_{3j+3}, v_{3j+2}) = a$.

It is now easy to extend the predefined function f to a locally injective homomorphism from the graph G_ϕ to $\mathcal{T}$.

If $\mathcal{T}$ is not b-positive, the proof is similar to the previous case with the only difference, that we must to prove that in any locally injective homomorphism $f : G_\phi \to \mathcal{T}$, for no copy Z of the clause gadget holds $g_f(u_1, u_0) = g_f(u_2, u_0) = g_f(3, u_0) = b$. If such gadget Z exists, then necessarily $f(u_1) = f(u_2) = f(u_3)$ (because of parity controller Y and construction of variable gadgets). Because of Proposition 5 we have that $f(u_0) = f(u_1)$ and because of (a, b, c)-blocking gadget on vertex u_0 we have, that the simple path of length $a+b+c$ must allows decomposition $b -_k c$. But this is clearly not possible and so in every clause, there exists at least one positive and at least one negative literal. So NAE-3-SAT can be reduced to the $\mathcal{T}$-LIHOM. □

Lemma 2. *Let $\mathcal{T}$ be a Θ-graph such that $a + b \neq c$ and*

$$\mathbf{(a+b)}^{\mathcal{T}}_{\mathbf{a,b,c}} : \ a - b, (a - a)$$
$$\mathbf{(c)}^{\mathcal{T}}_{\mathbf{a,b,c}} : \ b - b, c - c, (a - a)$$

then $\mathcal{T}$-LIHOM is NP-complete.

Lemma 3. *Let $\mathcal{T}$ be a Θ-graph such that $a + b \neq c$ and*

$$\mathbf{(a+b)}^{\mathcal{T}}_{\mathbf{a,b,c}} : \ a - b, (a - a)$$
$$\mathbf{(c)}^{\mathcal{T}}_{\mathbf{a,b,c}} : \ a - b, c - c, (a - a)$$

then $\mathcal{T}$-LIHOM is NP-complete.

While in Lemmas 1, 2 and 3 we reduced 3-SAT, resp. NAE-3-SAT to the $\mathcal{T}$-LIHOM, in the next Lemmas 4, 5, 6 and 7, the NP-complete problem of determining, if there exists a covering projection from a (simple) graph to the weight graph is reduced to the $\mathcal{T}$-LIHOM. The *weight graph* is a multigraph on vertices C and D joined by one edge and one loop at each of them. It is known, that covering projection (or simply *cover*) from a graph $G = (V, E)$ to the weight graph exists if and only if G is cubic and we can split the set of vertices V to two sets V_1 and V_2 such, that every vertex in V_1 has exactly two neighbors in V_1 and every vertex in V_2 has exactly two neighbors in V_2.

Lemma 4. *Let $\mathcal{T}$ be a Θ-graph where*

$$\mathbf{(c)}^{\mathcal{T}}_{\mathbf{a,b,c}} : \ a -_k b, c - c, (b - b) \quad or \quad \mathbf{(c)}^{\mathcal{T}}_{\mathbf{a,b,c}} : \ a -_k b, c - c, (a - a)$$

then $\mathcal{T}$-LIHOM is NP-complete.

Lemma 5. *Let $\mathcal{T}$ be a Θ-graph for which $l \geq 2$. If there exists positive integer p such that*

$$(\boldsymbol{p})^{\mathcal{T}}_{\boldsymbol{a},\boldsymbol{b}}: \ a -_c a, b -_k b$$

then $\mathcal{T}$-LIHom is NP-complete.

Lemma 6. *Let $\mathcal{T}$ be a Θ-graph where*

$$(\boldsymbol{a}+\boldsymbol{c})^{\mathcal{T}}_{\boldsymbol{a},\boldsymbol{b},\boldsymbol{c}}: \ a - c, b -_c b, (a - a), (a - b)$$

then $\mathcal{T}$-LIHom is NP-complete.

Lemma 7. *Let $\mathcal{T}$ be a Θ-graph where*

$$(\boldsymbol{c})^{\mathcal{T}}_{\boldsymbol{a},\boldsymbol{b},\boldsymbol{c}}: \ a -_k a, a -_k b, b -_k b, c - c$$

then $\mathcal{T}$-LIHom is NP-complete.

It is well known, that we can color edges of every cubic bipartite graph with 3 colors in such a way, that all edges incident with one vertex have distinct colors, while determine, if such an edge 3-coloring exists for general cubic graphs is NP-complete problem. However, deciding if a given precoloring of a cubic bipartite graph can be extended to the proper edge 3-coloring of the whole graph is also NP-complete [3]. We prove Lemmas 8, 9 and 10 by reducing this problem to $\mathcal{T}$-LIHom.

Lemma 8. *Let $\mathcal{T}$ be a Θ-graph where*

$$(\boldsymbol{c})^{\mathcal{T}}_{\boldsymbol{a},\boldsymbol{b},\boldsymbol{c}}: \ a -_c b, c - c, (b - b)$$

then $\mathcal{T}$-LIHom is NP-complete.

Lemma 9. *Let $\mathcal{T}$ be a Θ-graph where*

$$(\boldsymbol{a}+\boldsymbol{c})^{\mathcal{T}}_{\boldsymbol{a},\boldsymbol{b},\boldsymbol{c}}: \ a - c, b -_k b, (a - a), (a - b), (b -_c b)$$

then $\mathcal{T}$-LIHom is NP-complete.

Lemma 10. *Let $\mathcal{T}$ be a Θ-graph where $k = 1$ and*

$$(\boldsymbol{c})^{\mathcal{T}}_{\boldsymbol{a},\boldsymbol{b},\boldsymbol{c}}: \ a -_c a, a -_c b, b -_c b, c - c$$

then $\mathcal{T}$-LIHom is NP-complete.

The lemmas are main tools for proving the following two theorems. They clearly cover all Theta graphs with simple paths of at least three different lengths and hence imply Theorem 1. Recall that k is the multiplicity of the shortest simple path a in $\mathcal{T}$.

Theorem 2. *Let $\mathcal{T}$ be a Θ-graph where $k = 1$. Then $\mathcal{T}$-LIHom is NP-complete.*

Theorem 3. *Let $\mathcal{T}$ be a Θ-graph where $k \geq 2$. Then $\mathcal{T}$-LIHom is NP-complete.*

Acknowledgment

We would like to thank Jiří Fiala and Jan Kratochvíl for interesting discussions and to anonymous referees for comments.

References

1. Abello, J., Fellows, M.R., Stillwell, J.C.: On the complexity and combinatorics of covering finite complexes. Australian Journal of Combinatorics 4, 103–112 (1991)
2. Bodlaender, H.L.: The classification of coverings of processor networks. Journal of Parallel Distributed Computing 6, 166–182 (1989)
3. Fiala, J.: NP completeness of the edge precoloring extension problem on bipartite graphs. Journal of Graph Theory 43, 156–160 (2003)
4. Fiala, J., Kratochvíl, J.: Complexity of partial covers of graphs. In: Eades, P., Takaoka, T. (eds.) ISAAC 2001. LNCS, vol. 2223, pp. 537–549. Springer, Heidelberg (2001)
5. Fiala, J., Kratochvíl, J.: Partial covers of graphs. Discussiones Mathematicae Graph Theory 22, 89–99 (2002)
6. Fiala, J., Kratochvíl, J.: Locally injective graph homomorphism: Lists guarantee dichotomy. In: Fomin, F.V. (ed.) WG 2006. LNCS, vol. 4271, pp. 15–26. Springer, Heidelberg (2006)
7. Fiala, J., Kratochvíl, J., Pór, A.: On the computational complexity of partial covers of Theta graphs. Discrete Applied Mathematics 156, 1143–1149 (2008)
8. Fiala, J., Paulusma, D.: The computational complexity of the role assignment problem. In: Baeten, J.C.M., Lenstra, J.K., Parrow, J., Woeginger, G.J. (eds.) ICALP 2003. LNCS, vol. 2719, pp. 817–828. Springer, Heidelberg (2003)
9. Hell, P., Nešetřil, J.: On the complexity of H-colouring. Journal of Combinatorial Theory, Series B 48, 92–110 (1990)
10. Kratochvíl, J., Proskurowski, A., Telle, J.A.: Covering regular graphs. Journal of Combinatorial Theory B 71, 1–16 (1997)
11. Kratochvíl, J., Proskurowski, A., Telle, J.A.: Covering directed multigraphs I. colored directed multigraphs. In: Möhring, R.H. (ed.) WG 1997. LNCS, vol. 1335, pp. 242–257. Springer, Heidelberg (1997)
12. Kratochvíl, J., Proskurowski, A., Telle, J.A.: Complexity of graph covering problems. Nordic Journal of Computing 5, 173–195 (1998)
13. Kristiansen, P., Telle, J.A.: Generalized H-coloring of graphs. In: Lee, D.T., Teng, S.-H. (eds.) ISAAC 2000. LNCS, vol. 1969, pp. 456–466. Springer, Heidelberg (2000)

Ranking and Drawing in Subexponential Time

Henning Fernau[1], Fedor V. Fomin[2], Daniel Lokshtanov[2], Matthias Mnich[3], Geevarghese Philip[4], and Saket Saurabh[4]

[1] Universität Trier FB 4—Abteilung Informatik, 54286 Trier, Germany
fernau@uni-trier.de
[2] Department of Informatics, University of Bergen, 5020 Bergen, Norway
{fomin,daniello}@ii.uib.no
[3] Eindhoven University of Technology, Eindhoven, The Netherlands
m.mnich@tue.nl
[4] The Institute of Mathematical Sciences, Taramani, Chennai 600 113, India
{gphilip,saket}@imsc.res.in

Abstract. In this paper we obtain parameterized subexponential-time algorithms for p-KEMENY AGGREGATION (p-KAGG) — a problem in social choice theory — and for p-ONE-SIDED CROSSING MINIMIZATION (p-OSCM) – a problem in graph drawing (see the introduction for definitions). These algorithms run in time $\mathcal{O}^*(2^{\mathcal{O}(\sqrt{k}\log k)})$, where k is the parameter, and significantly improve the previous best algorithms with running times $\mathcal{O}^*(1.403^k)$ and $\mathcal{O}^*(1.4656^k)$, respectively[1]. We also study natural "above-guarantee" versions of these problems and show them to be fixed parameter tractable. In fact, we show that the above-guarantee versions of these problems are equivalent to a weighted variant of p-DIRECTED FEEDBACK ARC SET. Our results for the above-guarantee version of p-KAGG reveal an interesting contrast. We show that when the number of "votes" in the input to p-KAGG is *odd* the above guarantee version can still be solved in time $O^*(2^{\mathcal{O}(\sqrt{k}\log k)})$, while if it is *even* then the problem cannot have a subexponential time algorithm unless the exponential time hypothesis fails (equivalently, unless FPT=M[1]).

Keywords: Kemeny Aggregation, One-Sided Crossing Minimization, Parameterized Complexity, Subexponential-time Algorithms, Social Choice Theory, Graph Drawing, Directed Feedback Arc Set.

1 Introduction

In this paper we study problems from two different areas of algorithmics: p-KEMENY AGGREGATION (p-KAGG) — a problem in computational social choice theory — and p-ONE-SIDED CROSSING MINIMIZATION (p-OSCM) — a problem in graph drawing — in the realm of parameterized complexity[2].

[1] Karpinski and Schudy [24] have, independently of this work, recently obtained an algorithm for p-KAGG that runs in $\mathcal{O}^*(2^{\mathcal{O}(\sqrt{k})})$ time.

[2] We use the prefix p- to denote parameterized problems.

C.S. Iliopoulos and W.F. Smyth (Eds.): IWOCA 2010, LNCS 6460, pp. 337–348, 2011.

Kemeny Aggregation: Preference lists are extensively used in social science surveys and voting systems to capture information about choice. In many such scenarios there arises the need to combine the data represented by many lists into a single list which reflects the opinion of the surveyed group as much as possible. The p-KAGG problem was introduced by Kemeny [25,26] to abstract out the problem of combining many preference lists into one. This problem appears in a variety of applications, such as in breeding problems in agronomy [21]. In p-KAGG we are given a set of permutations (also called votes) over a set of alternatives (also called candidates), and a positive integer k, and are asked for a permutation π of the set of candidates, called an *optimal aggregation*, such that the sum of the Kendall-Tau distances (KT-distances) of π from all the votes is at most k. The KT-distance between two permutations π_1 and π_2 is the number of pairs of candidates that are ordered differently in the two permutations and is denoted by $KT\text{-}dist(\pi_1, \pi_2)$. The problem is known to be $\mathcal{NP}$-complete [5] and admits polynomial time approximation schemes (PTASs) [27]. Betzler et al. [6] considered this problem from the point of view of parameterized algorithms and obtained an algorithm that runs in time $\mathcal{O}^*(1.53^k)$ [3]. More recently Simjour [31] obtained an algorithm for the problem that runs in time $\mathcal{O}^*(1.403^k)$. Very recently, Karpinski and Schudy [24] obtained an algorithm for p-KAGG that runs in $\mathcal{O}^*(2^{\mathcal{O}(\sqrt{k})})$ time.

One Sided Crossing Minimization: The graph drawing problem that we are interested in is the p-OSCM problem, which is a key ingredient of the well-known "Sugiyama approach" to layered graph drawing [32]. An input to this problem consists of a bipartite graph $G = (V_1, V_2, E)$, a permutation π of V_1, and a positive integer k. The vertices of V_1 are placed on a line, also called a *layer*, in the order induced by π. The objective is to check whether there is a permutation π_m of V_2 such that, when the vertices of V_2 are placed on a second layer parallel to the first one in the order induced by π_m, then drawing a straight-line segment for each edge in E will introduce no more than k pairwise edge crossings. This seemingly simple problem is $\mathcal{NP}$-complete [18], even on sparse graphs [29].

The study of the parameterized algorithmics of graph drawing problems was initiated by Dujmović et al [12], and several new generic results were later obtained by Dujmović and others [13]. Dujmović and Whitesides [16] investigated the p-OSCM problem and obtained an algorithm for this problem which runs in time $\mathcal{O}^*(1.6182^k)$. This was later improved to $\mathcal{O}^*(1.4656^k)$ by Dujmović et al. [15]. There has been a similar race to obtain better approximation algorithms for the problem. To the best of our knowledge, the current best approximation factor for p-OSCM is 1.4664, due to Nagamochi [30].

Our Results. We obtain $\mathcal{O}^*(2^{\mathcal{O}(\sqrt{k}\log k)})$-time algorithms for both p-KAGG and p-OSCM. These significantly improve the previous best algorithms with running times $\mathcal{O}^*(1.403^k)$ and $\mathcal{O}^*(1.4656^k)$, respectively. Both of our algorithms are based on modeling these problems as the p-WEIGHTED DIRECTED FEEDBACK ARC SET (p-WDFAS) problem. In p-WDFAS we are given a directed

[3] The $\mathcal{O}^*$ notation suppresses polynomial terms in the expression.

(multi)graph $D = (V, A)$, a weight function $w : A \to \mathbb{R}^+$ and a positive integer k, and the objective is to find a set of arcs $F \subseteq A$ of total weight at most k such that deleting F from D makes D a directed acyclic graph; such an F is called a *feedback arc set* of D. Both p-KAGG and p-OSCM have been modeled as p-WDFAS in earlier work as well [1,31,32]; the novelty in our modeling is that it allows us to work with p-WDFAS on "tournament–like" structures. We call this specialized problem p-FAST (parameterized Feedback Arc Set on Tournament-like structures). A tournament is a digraph in which between every two vertices there is exactly one arc. By a tournament-like structure, we mean a directed (multi)graph on n vertices that contains a tournament on n vertices as a subgraph. Our modeling allows us to use the chromatic-coding technique recently developed by Alon et al. [3], which they used to obtain the first subexponential time algorithm for p-WDFAS on tournaments.

Very recently, Karpinski and Schudy [24] have obtained a faster algorithm for a special case of p-WDFAS restricted to complete digraphs where, for every two vertices u, v in the digraph, $w(uv) + w(vu) = 1$ (the *probability constraint*). This algorithm runs in $\mathcal{O}^*(2^{\mathcal{O}(\sqrt{k})})$ time. Using essentially the same modeling as we use for p-KAGG, they show that p-KAGG can be solved in $\mathcal{O}^*(2^{\mathcal{O}(\sqrt{k})})$ time. As far as we know, the other problems that we deal with in this paper cannot be modeled as this version of p-WDFAS, and so do not benefit from this improvement in its running time.

We also study natural "above-guarantee" versions of these problems and show them to be fixed parameter tractable. We show that the above-guarantee versions of p-KAGG (A-p-KAGG) and p-OSCM(A-p-OSCM) are both equivalent to p-WDFAS and hence both have algorithms that run in time $\mathcal{O}^*(2^{\mathcal{O}(k \log k)})$ [9]. A finer analysis of A-p-KAGG reveals an interesting contrast in its running time: if the number of votes in the input to p-KAGG is *odd*, then A-p-KAGG can still be solved in time $O^*(2^{\mathcal{O}(\sqrt{k} \log k)})$, while if it is *even*, then the problem cannot have any subexponential-time algorithm unless the exponential time hypothesis (ETH) is false [22], or equivalently [19], unless FPT=M[1].

It is also worth mentioning that our reduction from p-OSCM to p-WDFAS on tournaments implies a PTAS for the graph drawing problem. To summarize, we analyze a common feature of p-KAGG and p-OSCM to provide new insights and findings of interest to both the Graph Drawing community and the Social Choice community.

2 Preliminaries

A *parameterized problem* Π is a subset of $\Gamma^* \times \mathbb{N}$, where Γ is a finite alphabet. An instance of a parameterized problem is a tuple (x, k), where k is called the parameter. A central notion in parameterized complexity is *fixed-parameter tractability (FPT)* which means, for a given instance (x, k), decidability in time $\mathcal{O}(f(k) \cdot p(|x|))$, where f is an arbitrary function of k and p is a polynomial in the input size.

Let Π_1, Π_2 be two parameterized problems. A *parameterized reduction* from Π_1 to Π_2 is an algorithm that takes an instance (x, k) of Π_1 as input, runs in time $\mathcal{O}(f(k) \cdot p(|x|))$, and outputs an instance (y, ℓ) of Π_2 such that ℓ is some function of k alone and (x, k) is a YES instance of Π_1 if and only if (y, ℓ) is a YES instance of Π_2.

A *tournament* is a directed graph in which there is exactly one directed arc between every two vertices. A feedback arc set in a tournament is a set of arcs whose *reversal* results in a DAG. A *tournament-like graph* is a directed (multi)graph on n vertices, for some $n \in \mathbb{N}$, which contains a tournament on n vertices as a subgraph.

3 FPT Algorithms for *p*-KAGG

Let S be a finite set, and let π_1, π_2 be two permutations of S. For $u, v \in S$, we define

$$d_{\pi_2}^{\pi_1}(u, v) = \begin{cases} 0 & \text{if } \pi_1 \text{ and } \pi_2 \text{ rank } u \text{ and } v \text{ in the same order} \\ 1 & \text{otherwise} \end{cases}$$

The Kendall-Tau distance (*KT-distance*) of π_1 and π_2 is defined as: $KT\text{-}dist(\pi_1, \pi_2) = \sum_{\{u,v\} \subseteq S} d_{\pi_2}^{\pi_1}(u, v)$.

Let C be a set of candidates and V a set of votes over C. For any permutation r of C, the *Kemeny Score of r with respect to V* is defined as: $KS(r, V) = \sum_{\pi \in V} KT\text{-}dist(r, \pi)$. Observe that

$$KS(r, V) = \sum_{\pi \in V} KT\text{-}dist(r, \pi) = \sum_{\pi \in V} \sum_{\{u,v\} \subseteq C} d_{\pi}^{r}(u, v) = \sum_{\{u,v\} \subseteq C} \sum_{\pi \in V} d_{\pi}^{r}(u, v) \quad (1)$$

3.1 Parameterized Reduction from *p*-KAGG to *p*-WDFAS

We now describe a parameterized reduction from *p*-KAGG to *p*-WDFAS, briefly mentioned by Betzler et al. [6], which runs in *polynomial* time and takes the parameter from k to k. Let (C, V, k) be an instance of *p*-KAGG. In what follows, we assume without loss of generality that $|V| \geq 1$. We construct a digraph G such that (C, V, k) is a YES instance of *p*-KAGG if and only if G has a feedback arc set of weight at most k. We set the vertex set of G to be the set C of candidates. For each vote $\pi_i \in V$ and for each pair of vertices (u, v) of G, we add a new arc with weight 1 from u to v in G if and only if u appears before v in π_i (equivalently, when u is preferred over v by π_i). This completes the construction; the parameter is k.

Fix a vote $\pi_i \in V$. For each pair of candidates $u, v \in C$, π_i prefers exactly one of these candidates over the other. Thus, for any two vertices u, v of G, each vote contributes exactly one arc between u and v in G. As a consequence, we have:

Observation 1. *Let G be the digraph constructed from an instance (C, V, k) of p-KAGG as described above. For any two vertices u, v of G, let i be the number of arcs in G from u to v, and j the number of arcs from v to u. Then $i + j = |V|$.*

The next two claims show that the reduction is sound.

Claim 1. Let (C, V, k) be a YES instance of p-KAGG, and let G be the digraph constructed from (C, V, k) as described above. Then G has a feedback arc set of weight at most k.

Proof. Since (C, V, k) is a YES instance of p-KAGG, there exists a permutation r of the set C such that $\sum_{\pi \in V} KT\text{-}dist(r, \pi) \leq k$. For $u, v \in V(G)$, let $\overline{r_{uv}}$ be the set of arcs in G between u and v that are oriented *contrary* to the direction implied by r. That is, if u appears before v in r, then $\overline{r_{uv}}$ consists of all arcs from v to u in G; if u appears after v in r, then $\overline{r_{uv}}$ consists of all arcs from u to v in G. Using Equation 1, we get $\sum_{\{u,v\} \subseteq C} \sum_{\pi \in V} d^r_\pi(u, v) \leq k$. By construction, this implies $\sum_{\{u,v\} \subseteq V(G)} |\overline{r_{uv}}| \leq k$.

That is, there are at most k arcs in G, each of weight exactly 1, that are oriented contrary to the directions implied by r. Reversing these arcs, we get a digraph G' in which every arc is oriented according to the direction implied by r. Since r is a permutation of $V(G) = V(G')$, it follows that G' is acyclic. □

Claim 2. Let G be the digraph constructed from an instance (C, V, k) of p-KAGG as described above. If G has a feedback arc set S of weight at most k, then (C, V, k) is a YES instance of p-KAGG.

Proof. Note that since each arc in G has weight exactly 1, S contains exactly k arcs. Consider the DAG G' obtained from G by reversing the arcs in S. Note that this operation preserves the number of arcs between any pair of vertices. From Observation 1, and since G' is a DAG, between each pair u, v of vertices of G' there are exactly $|V|$ arcs, all of which are in the same direction. The arcs of G' thus define a permutation r of C, where for any $u, v \in C$, u appears before v in r if and only if there is an arc (in fact, $|V|$ arcs) from u to v in G'. For $u, v \in V(G)$, let $\overline{r_{uv}}$ be the set of arcs between u and v in G that are oriented *contrary* to the direction implied by r. Then $\cup_{\{u,v\} \subseteq V(G)} \overline{r_{uv}} = S$, $\sum_{\{u,v\} \subseteq V(G)} |\overline{r_{uv}}| = |S| \leq k$, and from this and the construction we get $\sum_{\{u,v\} \subseteq C} \sum_{\pi \in V} d^r_\pi(u, v) \leq k$. From Equation 1 it follows that $KS(r, V) \leq k$, and so (C, V, k) is a YES instance of p-KAGG. □

The above reduction can clearly be done in polynomial time, and the number of vertices in the reduced instance (G, k) is equal to the number of candidates $|C|$ in the input instance (C, V, k). Further, the reduced instance has at least one arc (in fact, exactly $|V|$ arcs) between every pair of vertices. Let H be the edge-weighted digraph obtained from G by replacing parallel arcs with single weighted arcs in the natural way. That is, if there are $i > 0$ arcs from u to v in G, then H contains a single arc of weight i from u to v. It is easy to verify that H has a feedback arc set of weight at most k if and only if G has a feedback arc set of weight at most k. Hence from Claims 1 and 2 we have

Lemma 1. *Given an instance (C, V, k) of p-KAGG, we can construct, in polynomial time, an equivalent instance (G, k) of p-WDFAS where G is a tournament-like graph and $|V(G)| = |C|$.*

3.2 A Subexponential FPT Algorithm for p-KAGG

Our algorithm is based on the observation that the algorithm of Alon et al. [3] for p-WDFAS on tournaments also works for tournament-like graphs. The algorithm presented in [3] starts by preprocessing the instance and obtains an equivalent instance with at most $\mathcal{O}(k^2)$ vertices in polynomial time. That is, given a tournament T and a positive integer k, in polynomial time the preprocessing algorithm either concludes that T does not have a feedback arc set of weight at most k or finds a new tournament T' with $\mathcal{O}(k^2)$ vertices and $k' \leq k$ such that the original tournament T has a feedback arc set of weight at most k, if and only if T' has a feedback arc set of weight at most k'. This preprocessing allows them to assume that the instance where they actually apply the subexponential time algorithm is of size $\mathcal{O}(k^2)$ only, which is integral to their time analysis. Their preprocessing can also be applied to tournament-like graphs by allowing both directed cycles of length two and triangles in the reduction rules proposed in [3, Lemma 1]. So we always first apply these preprocessing rules and obtain a tournament-like graph on $\mathcal{O}(k^2)$ vertices. Let the preprocessed tournament-like graph be $T = (V, A)$.

To obtain our algorithm we also use *universal coloring families* introduced in [3]. For integers m, k and r, a family $\mathcal{F}$ of functions from $[m]$ to $[r]$ is called a universal (m, k, r)-coloring family if for any graph G on the set of vertices $[m]$ with at most k edges, there exists an $f \in \mathcal{F}$ which is a proper vertex coloring of G. The following result gives a bound on the size of universal coloring families.

Proposition 1. [3] *For any $n > 10k^2$ there exists an explicit universal $(n, k, \mathcal{O}(\sqrt{k}))$-coloring family $\mathcal{F}$ of size $|\mathcal{F}| \leq 2^{\mathcal{O}(\sqrt{k}\log k)} \log n$.*

We enumerate each function in the universal coloring family and then color the vertices of T with these functions. Observe that since the number of arcs possible in the solution is at most k, there exists a function $f \in \mathcal{F}$ such that no end-points of the arc in the solution is colored with same color, that is, no arc of the solution is monochromatic. Now using the dynamic programming algorithm proposed in [3, Lemma 3] we can find a feedback arc set of weight at most k of T, if there exists one, in time $\mathcal{O}(2^{\mathcal{O}(\sqrt{k}\log k)})$. This yields the following theorem.

Theorem 1. *The* p-KEMENY AGGREGATION *problem with n candidates can be solved in $2^{\mathcal{O}(\sqrt{k}\log k)} + n^{\mathcal{O}(1)}$ time.*

This is a significant improvement over the previous best known algorithm for p-KEMENY AGGREGATION which runs in $\mathcal{O}^*(1.403^k)$ time [31]. Very recently, Karpinski and Schudy [24] have, by (1) developing an $\mathcal{O}^*(2^{\mathcal{O}(\sqrt{k})})$-time algorithm for a special version of p-WDFAS and (2) reducing p-KAGG to this version of p-WDFAS, obtained an algorithm for p-KAGG that runs in $\mathcal{O}^*(2^{\mathcal{O}(\sqrt{k})})$ time.

3.3 FPT Algorithms for *A*-p-KAGG

Consider an instance of the p-KAGG problem. Let π be *any* permutation of the candidate set C, let V be the set of all votes and let $KS(\pi, V)$ denote the sum of the KT-distances of π from all the votes in the set V. Suppose A and B are two candidates in the input, and let i votes prefer A over B and j votes prefer B over A. Clearly, the pair $\{A, B\}$ contributes at least $\min(i, j)$ to $KS(\pi, V)$. For $\{u, v\} \subseteq C$, let $I(u, v)$ (respectively $J(u, v)$) be the number of votes that rank u before v (respectively v before u), and let $g = \sum_{\{u,v\} \subseteq C} \min\{I(u, v), J(u, v)\}$. Then $KS(\pi, V) \geq g$, and so in the natural above-guarantee version of p-KAGG, which we call *A*-p-KAGG, we ask for a permutation π of C such that $KS(\pi, V) \leq g + k$.

We now describe a reduction from *A*-p-KAGG to p-WDFAS, originally due to Dwork et al. [17]. When the number of votes in the input instance is odd (*A*-p-KAGG(*odd*)), the reduced instance is a tournament with positive integral edge weights. When the number of votes is even (*A*-p-KAGG(*even*)), the reduced instance is not necessarily a tournament. In both cases, the parameter goes from k to k. That is, the reduction takes *A*-p-KAGG(*odd*) to p-WDFAS on tournaments, and *A*-p-KAGG(*even*) to p-WDFAS in general digraphs, in both cases preserving the parameter. Together with the subexponential FPT algorithm of Alon et al. [3] for p-WDFAS on tournaments, this implies a subexponential FPT algorithm for *A*-p-KAGG(*odd*). In the next subsection we describe a parameterized reduction from p-WDFAS to *A*-p-KAGG(*even*) in which the parameter goes from k to $2k$. This implies that *A*-p-KAGG(*even*) does not have a subexponential FPT algorithm unless the exponential time hypothesis is false.

Let (C, V, k) be an instance of *A*-p-KAGG. We construct an instance (H, k) of p-WDFAS in two stages, as follows.

Stage 1. We construct a digraph G exactly as in the previous reduction. We set the vertex set of G to be the set C of candidates. For each vote $\pi_i \in V$ and for each pair of vertices (u, v) of G, we add a new arc of weight 1 from u to v in G if and only if u appears before v in π_i (equivalently, when u is preferred over v by π_i).

Stage 2. We now prune the "above-guarantee" arcs of G. We process every two-vertex subset $\{u, v\}$ of G as follows: Let there be a total of i arcs from u to v and j arcs from v to u in H. Assume without loss of generality that $i \geq j$. We replace all the arcs between u and v by a single arc of weight $i - j$ from u to v. If $i - j = 0$, then we just remove all the arcs between u and v, and do not add any arc to replace them. We repeat this for every 2-subset of vertices of G to obtain a digraph H with integer-weighted arcs. (H, k) is the desired instance of p-WDFAS.

Suppose the number $|V|$ of votes in the input instance (C, V, k) is odd. Then, with the same notation as above, $i + j = |V|$ is odd for each 2-subset $\{u, v\}$ of G (Observation 1), and so $i > j$. Thus there is exactly one arc between every two vertices of H, and so H is a tournament. If $|V|$ is even, then it is possible that $i = j$ for some $\{u, v\} \subseteq V(G)$, and so in H there will not be any arc between u and v. Hence when $|V|$ is odd, H is not necessarily a tournament or a tournament-like graph.

Dwork et al. [17] show that the above reduction is sound; see also Mahajan et al. [28]:

Lemma 2. *[17,28] Let (H, k) be the instance of p-WDFAS obtained from an instance (C, V, k) of A-p-KAGG as described above. Then (H, k) is a* YES *instance of p-WDFAS if and only if (C, V, k) is a* YES *instance of A-p-KAGG.*

The fastest known FPT algorithm for p-WDFAS runs in $\mathcal{O}^*(2^{\mathcal{O}(k \log k)})$ time [9], and the fastest known FPT algorithm for p-WDFAS on tournaments runs in $2^{\mathcal{O}(\sqrt{k} \log k)} + n^{\mathcal{O}(1)}$ time [3]. Hence from Lemma 2 we get

Theorem 2. *The A-p-KAGG problem with n candidates can be solved in $2^{\mathcal{O}(\sqrt{k} \log k)} + n^{\mathcal{O}(1)}$ time when the number of votes is odd, and in $\mathcal{O}^*(2^{\mathcal{O}(k \log k)})$ time when the number of votes is even.*

3.4 A Lower Bound for *A-p*-KAGG(*even*)

We now argue that the A-p-KAGG($even$) problem does not have a subexponential FPT algorithm unless the exponential time hypothesis (ETH) is false. To see this, consider the following sequence of two reductions:

$$\text{Vertex Cover} \to \text{Directed Feedback Arc Set} \to A\text{-}p\text{-KAGG}$$

The first reduction is due to Karp [23], and the second is due to Dwork et al. [17, Theorem 14]. This sequence of reductions take an input instance (G, k) of Vertex Cover where G is a graph on n vertices and m edges and $k \leq n$ is a positive integer, and outputs an instance $(C, V, 2k)$ of A-p-KAGG($even$) where $|C| = 3n + 2m$, $|V| = 4$, and the guarantee is $g = 2(\binom{2n}{2} + \binom{n+2m}{2} + n + 2m)$; see the references for details. Suppose A-p-KAGG($even$) has an algorithm that runs in time $\mathcal{O}^*(2^{o(k)})$. Since $k = \mathcal{O}(n)$ throughout the reduction, we can then use this algorithm to solve Vertex Cover in $\mathcal{O}^*(2^{o(n)})$ time: We first apply the above sequence of reductions and then apply the supposed subexponential FPT algorithm for A-p-KAGG($even$) to the resulting instance. This would in turn imply that ETH is false [22], and so we have

Theorem 3. *The A-p-KAGG problem with an even number of votes cannot be solved in $\mathcal{O}^*(2^{o(k)})$ time unless ETH is false.*

4 FPT Algorithms for *p*-OSCM

Let $(G = (V_1, V_2, E), \pi, k)$ be an instance of p-OSCM. In what follows, we assume without loss of generality that in G, every vertex in V_2 has at least one neighbor in V_1.

4.1 Parameterized Reduction from p-OSCM to p-WDFAS

We now describe a parameterized reduction from p-OSCM to p-WDFAS which runs in *polynomial* time and takes the parameter from k to k. Let $(G=(V_1,V_2,E), \pi, k)$ be an instance of p-OSCM. For two vertices $u,v \in V_2$, let C_{uv} denote the number of crossings of edges incident to u with edges incident to v, when u appears before v in the second layer. It is known [16] that for a given graph G and a fixed ordering π of the vertices of V_1, C_{uv} is a constant and can be computed in polynomial time. We construct a digraph H as follows: H has one vertex for each vertex of V_2. For $\{u,v\} \subseteq V_2$, we draw the arc uv with weight C_{uv} if $C_{uv} > 0$.

Claim 3. Let $(G=(V_1,V_2,E),\pi,k)$ be an instance of p-OSCM, and let H be the digraph obtained from this instance as described above. $(G=(V_1,V_2,E),\pi,k)$ is a YES instance of p-OSCM if and only if H has a feedback arc set of weight at most k.

Proof. Suppose $(G=(V_1,V_2,E),\pi,k)$ is a YES instance of p-OSCM, and let π_m be a permutation of V_2 that witnesses this fact. Place the vertices of H on a line in the order induced by π_m: u is to the left of v if and only if u comes before v in π_m. From the construction it is clear that the sum of the weights of the arcs in H that go from left to right is at most k, and so these arcs together form a feedback arc set of H of weight at most k.

Now suppose S is a *minimal* feedback arc set of H of weight at most k. Let π' be the unique permutation of V_2 such that if we place the vertices of H on a line in the order specified by π', then the arcs that go from left to right are exactly the arcs in S. It is easily verified that if the vertices of V_2 are placed on the second layer in the order specified by π', then the number of crossings will be at most k. □

The above reduction can clearly be done in polynomial time, and the graph H in the reduced instance (H,k) has $|V_2|$ vertices, where the p-OSCM instance is $(G=(V_1,V_2,E),\pi,k)$. Further, it is not difficult to see that the reduced instance has at least one arc between every pair of vertices. Hence from Claim 3 we have

Lemma 3. *Given an instance $(G=(V_1,V_2,E),\pi,k)$ of p-OSCM, we can construct, in polynomial time, an equivalent instance (H,k) of p-WDFAS where H is a tournament-like graph and $|V(H)|=|V_2|$.*

4.2 A Subexponential FPT Algorithm for p-OSCM

From Lemma 3, and using the same argument as in Section 3.2, we get

Theorem 4. *The p-ONE-SIDED CROSSING MINIMIZATION problem can be solved in $2^{\mathcal{O}(\sqrt{k}\log k)} + n^{\mathcal{O}(1)}$ time, where n is the number of vertices in the layer that is not fixed.*

4.3 Lower and Upper Bounds for *A-p*-OSCM

Let $(G = (V_1, V_2, E), \pi, k)$ be an instance of p-OSCM. For two vertices $u, v \in V_2$, let C_{uv} be defined as in Section 4.1. It is known that the minimum possible number of crossings is $g = \sum_{\{u,v\} \subseteq V_2} \min(C_{uv}, C_{vu})$ [16]. So in the natural above-guarantee version of p-OSCM, which we call *A-p*-OSCM, we ask for a permutation π of V_2 such that the number of crossings induced by π is at most $g + k$.

Given an instance $(G = (V_1, V_2, E), \pi, k)$ of p-OSCM, the well-known *penalty graph* construction of Sugiyama et al. [32] constructs a arc-weighted digraph H with V_2 as the vertex set, and there is an arc in H from u to v with weight $C_{vu} - C_{uv}$ if $C_{uv} < C_{vu}$. It is easy to verify that there is a permutation π_m of V_2 such that the number of crossings induced by π_m is at most $g + k$ if and only if H has a feedback arc set of weight at most k. Thus, using the algorithm in [9] we have

Theorem 5. *The A-p-OSCM problem can be solved in* $\mathcal{O}^*(2^{\mathcal{O}(k \log k)})$ *time.*

Muñoz et al. describe a reduction from Directed Feedback Arc Set to p-OSCM that in fact is a parameterized reduction from Directed Feedback Arc Set (where the parameter k is the solution size) to *A-p*-OSCM which takes the parameter from k to $2k$ [29, Proof of Theorem 1]. Hence by a similar argument as in Section 3.4 we have

Theorem 6. *The A-p-OSCM problem cannot be solved in* $\mathcal{O}^*(2^{o(k)})$ *time unless ETH is false.*

5 Conclusion and Future Work

In this paper we modeled two problems, from two different domains, as the weighted feedback arc set problem on tournament-like structures. This allowed us to utilize the recently developed technique of chromatic-coding [3] to obtain subexponential-time algorithms, that is, algorithms that run in time $\mathcal{O}^*(c^{\sqrt{k} \log k})$, for p-Kemeny Aggregation and p-One-Sided Crossing Minimization. The running time of these algorithms might be seen as a breakthrough compared to the hitherto best published algorithms, which had running times of the form roughly $\mathcal{O}^*(1.5^k)$. It is worth mentioning that apart from problems on graphs of bounded genus, only very few problems are known to have running times of the form $\mathcal{O}^*(c^{\sqrt{k}})$ [2,10,11].

Our approach also allowed us to show that the above-guarantee versions of these problems are fixed parameter tractable with algorithms having running times of the form $\mathcal{O}^*(c^{k \log k})$. We also show that the above-guarantee versions of these problems cannot have algorithms that run in $\mathcal{O}^*(2^{o(k)})$ time, unless the well known exponential time hypothesis fails.

We believe that our approach will generalize to other related problems considered in the literature. We cite a few concrete examples in the following.

- Çakiroglu et al. [8] considered drawing graphs with edge weights. If two edges cross, then the crossing receives as a weight the product of both edge weights involved. The overall weight of a crossing is then the sum of all respective crossing weights, and the goal is to minimize this weight.
- Forster [20] considered the so-called *constraint variant* where the ordering of some of the vertices of the free layer is already fixed (as part of the input). This can be clearly modeled by the so-called POSITIVE WEIGHTED COMPLETION OF AN ORDERING (PWCO) as studied in [14]. There, also an FPT result was announced, with a running time of $\mathcal{O}^*(1.52^k)$.
- In radial drawings of graphs, also the restricted ($\mathcal{NP}$-complete) variant called RADIAL ONE-SIDED TWO-LEVEL CROSSING MINIMIZATION has been considered [4].

It also might be interesting to consider the crossing minimization variant of these problems that attempts to minimize the maximum number of crossings per edge as proposed in [7] from the viewpoints of fixed parameter tractability and of approximability.

References

1. Ailon, N., Charikar, M., Newman, A.: Aggregating inconsistent information: Ranking and clustering. J.ACM 55(5), 23:1–23:27 (2008)
2. Alber, J., Fernau, H., Niedermeier, R.: Parameterized complexity: exponential speedup for planar graph problems. Journal of Algorithms 52, 26–56 (2004)
3. Alon, N., Lokshtanov, D., Saurabh, S.: Fast FAST. In: Albers, S., Marchetti-Spaccamela, A., Matias, Y., Nikoletseas, S.E., Thomas, W. (eds.) ICALP 2009. LNCS, vol. 5555, pp. 49–58. Springer, Heidelberg (2009)
4. Bachmaier, C.: A radial adaptation of the Sugiyama framework for visualizing hierarchical information. IEEE Transactions on Visualization and Computer Graphics 13(3), 583–594 (2007)
5. Bartholdi III, J., Tovey, C.A., Trick, M.A.: Voting schemes for which it can be difficult to tell who won the election. Social Choice and Welfare 6, 157–165 (1989)
6. Betzler, N., Fellows, M.R., Guo, J., Niedermeier, R., Rosamond, F.A.: Fixed-parameter algorithms for Kemeny scores. In: Fleischer, R., Xu, J. (eds.) AAIM 2008. LNCS, vol. 5034, pp. 60–71. Springer, Heidelberg (2008)
7. Biedl, T.C., Brandenburg, F.-J., Deng, X.: Crossings and permutations. In: Healy, P., Nikolov, N.S. (eds.) GD 2005. LNCS, vol. 3843, pp. 1–12. Springer, Heidelberg (2006)
8. Çakiroglu, O.A., Erten, C., Karatas, Ö., Sözdinler, M.: Crossing minimization in weighted bipartite graphs. In: Demetrescu, C. (ed.) WEA 2007. LNCS, vol. 4525, pp. 122–135. Springer, Heidelberg (2007)
9. Chen, J., Liu, Y., Lu, S., O'Sullivan, B., Razgon, I.: A fixed-parameter algorithm for the directed feedback vertex set problem. JACM 55(5), 21:1–21:19 (2008)
10. Demaine, E.D., Fomin, F.V., Hajiaghayi, M.T., Thilikos, D.M.: Subexponential parameterized algorithms on graphs of bounded genus and H-minor-free graphs. In: Proceedings of SODA 2004, pp. 823–832. ACM, New York (2004)
11. Dorn, F., Fomin, F.V., Thilikos, D.M.: Subexponential parameterized algorithms. In: Arge, L., Cachin, C., Jurdziński, T., Tarlecki, A. (eds.) ICALP 2007. LNCS, vol. 4596, pp. 15–27. Springer, Heidelberg (2007)

12. Dujmović, V., Fellows, M.R., Hallett, M., Kitching, M., Liotta, G., McCartin, C., Nishimura, N., Ragde, P., Rosamond, F.A., Suderman, M., Whitesides, S., Wood, D.R.: A fixed-parameter approach to 2-layer planarization. Algorithmica 45, 159–182 (2006)
13. Dujmović, V., Fellows, M.R., Kitching, M., Liotta, G., McCartin, C., Nishimura, N., Ragde, P., Rosamond, F.A., Whitesides, S., Wood, D.R.: On the parameterized complexity of layered graph drawing. Algorithmica 52(2), 267–292 (2008)
14. Dujmović, V., Fernau, H., Kaufmann, M.: Fixed parameter algorithms for one-sided crossing minimization revisited. In: Liotta, G. (ed.) GD 2003. LNCS, vol. 2912, pp. 332–344. Springer, Heidelberg (2004)
15. Dujmović, V., Fernau, H., Kaufmann, M.: Fixed parameter algorithms for one-sided crossing minimization revisited. Journal of Discrete Algorithms 6, 313–323 (2008)
16. Dujmović, V., Whitesides, S.: An efficient fixed parameter tractable algorithm for 1-sided crossing minimization. Algorithmica 40, 15–32 (2004)
17. Dwork, C., Kumar, R., Naor, M., Sivakumar, D.: Rank aggregation revisited. Manuscript available at, `http://citeseerx.ist.psu.edu/viewdoc/summary?doi=10.1.1.23.5118`
18. Eades, P., Wormald, N.C.: Edge crossings in drawings of bipartite graphs. Algorithmica 11, 379–403 (1994)
19. Flum, J., Grohe, M.: Parameterized Complexity Theory. Texts in Theoretical Computer Science. An EATCS Series. Springer, Heidelberg (2006)
20. Forster, M.: A fast and simple heuristic for constrained two-level crossing reduction. In: Pach, J. (ed.) GD 2004. LNCS, vol. 3383, pp. 206–216. Springer, Heidelberg (2005)
21. Guénoche, A., Vandeputte-Riboud, B., Denis, J.: Selecting varieties using a series of trials and a combinatorial ordering method. Agronomie 14, 363–375 (1994)
22. Impagliazzo, R., Paturi, R., Zane, F.: Which problems have strongly exponential complexity? Journal of Computer and System Sciences 63(4), 512–530 (2001)
23. Karp, R.M.: Reducibility among combinatorial problems. In: Complexity of Computer Communications, pp. 85–103 (1972)
24. Karpinsky, M., Schudy, W.: Faster algorithms for feedback arc set tournament, kemeny rank aggregation and betweenness tournament. Accepted at ISAAC (2010)
25. Kemeny, J.: Mathematics without numbers. Daedalus 88, 571–591 (1959)
26. Kemeny, J., Snell, J.: Mathematical models in the social sciences. Blaisdell (1962)
27. Kenyon-Mathieu, C., Schudy, W.: How to rank with few errors. In: Proceedings of STOC 2007, pp. 95–103. ACM, New York (2007)
28. Mahajan, M., Raman, V., Sikdar, S.: Parameterizing above or below guaranteed values. Journal of Computer and System Sciences 75(2), 137–153 (2009)
29. Muñoz, X., Unger, W., Vrt'o, I.: One sided crossing minimization is NP-hard for sparse graphs. In: Mutzel, P., Jünger, M., Leipert, S. (eds.) GD 2001. LNCS, vol. 2265, pp. 115–123. Springer, Heidelberg (2002)
30. Nagamochi, H.: An improved bound on the one-sided minimum crossing number in two-layered drawings. Discrete and Computational Geometry 33, 569–591 (2005)
31. Simjour, N.: Improved parameterized algorithms for the Kemeny aggregation problem. In: Chen, J., Fomin, F.V. (eds.) IWPEC 2009. LNCS, vol. 5917, pp. 312–323. Springer, Heidelberg (2009)
32. Sugiyama, K., Tagawa, S., Toda, M.: Methods for visual understanding of hierarchical system structures. IEEE Trans. Systems Man Cybernet. 11(2), 109–125 (1981)

Efficient Reconstruction of RC-Equivalent Strings*

Ferdinando Cicalese[1], Péter L. Erdős[2,**], and Zsuzsanna Lipták[3]

[1] Dipartimento di Informatica ed Applicazioni, University of Salerno, Italy
cicalese@dia.unisa.it

[2] Alfréd Rényi Institute of Mathematics, Budapest, Hungary
elp@renyi.hu

[3] AG Genominformatik, Technische Fakultät, Bielefeld University, Germany
zsuzsa@cebitec.uni-bielefeld.de

Abstract. In the reverse complement (RC) equivalence model, it is not possible to distinguish between a string and its reverse complement. We show that one can still reconstruct a binary string of length n, up to reverse complement, using a linear number of subsequence queries of bounded length. A simple information theoretic lower bound proves the number of queries to be tight. Our result is also optimal w.r.t. the bound on the query length given in [Erdős *et al.*, Ann. of Comb. 2006].

1 Introduction

Reconstructing a string over a finite alphabet Σ from information about its subsequences is a classic string problem, with applications ranging from coding theory to bioinformatics. Because of the confusion in terminology in the literature, we want to give a precise definition right here: Given two strings $\mathbf{s}, \mathbf{t}$ over Σ, $\mathbf{s} = s_1 \ldots s_n$ and $\mathbf{t} = t_1 \ldots t_m$, we say that $\mathbf{t}$ is a *subsequence* (often called *subword*) of $\mathbf{s}$ if there exist $1 \leq i_1 < i_2 < \ldots < i_m \leq n$ such that $\mathbf{t} = s_{i_1} s_{i_2} \ldots s_{i_m}$. It was shown by Simon in 1975 [12] that two strings of length n are equal if their subsequences up to length $\lfloor n/2 \rfloor + 1$ coincide. The proof, as given in Chapter 6 of the classic Lothaire book [11] can be easily adapted to yield an algorithm which reconstructs the string $\mathbf{s}$ of length n, using $O(|\Sigma| n)$ queries of the type "Is $\mathbf{u}$ a subsequence of $\mathbf{s}$?" Here, $\mathbf{u}$ is a string of length at most $\lfloor n/2 \rfloor + 1$.

In this paper, we consider this problem in the RC-equivalence model, which is motivated by reverse complementation of DNA. Our alphabet consists of *pairs* of characters $(a, \bar{a})$, called *complement pairs*, and for every string $\mathbf{s}$ over Σ, $\mathbf{s} = s_1 \ldots s_n$, we define its *reverse complement* as $\tilde{\mathbf{s}} = \bar{s}_n \ldots \bar{s}_1$. Two strings $\mathbf{s}, \mathbf{t}$ are RC-equivalent if $\mathbf{s} = \mathbf{t}$ or $\mathbf{s} = \tilde{\mathbf{t}}$. A string $\mathbf{u}$ is an RC-subsequence of $\mathbf{s}$ if $\mathbf{u}$ or $\tilde{\mathbf{u}}$ is a subsequence of $\mathbf{s}$. Erdős *et al.* showed in [4] that two strings $\mathbf{s}$ and

* This research was carried out in part while F.C. and Zs.L. were visiting with the Alfréd Rényi Institute in Budapest, with support from the Hungarian Bioinformatics MTKD-CT-2006-042794, Marie Curie Host Fellowships for Transfer of Knowledge.

** In part supported by the Hungarian NSF, under contracts NK 78439 and K 68262.

C.S. Iliopoulos and W.F. Smyth (Eds.): IWOCA 2010, LNCS 6460, pp. 349–362, 2011.

$\mathbf{t}$ of length n are equal if and only if all their RC-subsequences up to length $\lceil \frac{2}{3}(n+1) \rceil$ coincide. However, no reconstruction algorithm was given.

Here we present such an algorithm for the case of a binary alphabet, i.e., where the alphabet consists of two complementary characters. Our algorithm reconstructs a string $\mathbf{s}$ of length n, using $O(n)$ queries of the type "Is $\mathbf{u}$ an RC-subsequence of s?" where $\mathbf{u}$ is a string of length at most $\lceil \frac{2}{3}(n+1) \rceil$. We note that our algorithm is optimal both w.r.t. the length of the queries, and w.r.t. the information theoretic lower bound on the number of queries necessary for exact reconstruction. We also give a simple algorithm for arbitrary alphabets, adapted from a paper by Skiena and Sundaram [13], where the length of the queries is not bounded, using $O(n \log |\Sigma|)$ queries.

It should be noted that the problem differs considerably from the classical model. For example, consider the string $\mathbf{s} = \bar{a}\bar{b}a\bar{a}b$. Then aba is not a subsequence of $\mathbf{s}$, but it is an RC-subsequence, because $\bar{a}\bar{b}\bar{a}$ is a subsequence of $\mathbf{s}$.

The RC-equivalence model can be viewed as a special case of erroneous information, where the answers to subsequence queries could be either about the query string or its reverse complement. It is also a special case of a group action on Σ^*, the set of finite strings over Σ. The search in Σ^n is substituted by a search in $\Sigma^n/\sim$, where $\sim$ is the equivalence induced by the group action.

Related work. Most literature deals with the classical, i.e. non-RC, model. In addition to the papers mentioned above, we want to point to the following.

When the multiset of subsequences is known, then much shorter subsequences suffice to uniquely identify a string: A string of length n can be uniquely identified by the multiset of its subsequences of length $\lfloor \frac{16}{7}\sqrt{n} \rfloor + 5$, as shown by Krasikov and Roditty [6]. Dudík and Schulman [3] give asymptotic lower and upper bounds, in terms of k, on the length of strings which can be uniquely determined by the multiset of their subsequences of length k.

Levenshtein [7] investigates the maximal number of common subsequences of length k that two distinct subsequences of length n can have. Here, subsequences are regarded as erroneous versions of the original string. The aim is to find how many times a transmission needs to be repeated, over a channel which allows a constant number of deletions, to make unique recovery of the original message possible.

The case where substrings are considered has also received much attention. Substrings, often called factors, are contiguous subsequences: $\mathbf{t}$ is a substring of $\mathbf{s}$ if there are $1 \leq i \leq j \leq n$ such that $\mathbf{t} = s_i \dots s_j$. The length of substrings of a string $\mathbf{s}$ of length n which are necessary for uniquely determining $\mathbf{s}$ depends on a parameter of $\mathbf{s}$, namely on the maximal length of a repeated substring, as shown by de Luca and Carpi in a series of papers [2,1]. An algorithm for reconstruction was given by Fici *et al.* in [5], while the uniqueness bound for multisets of substrings was recently shown to be $\lfloor \frac{n}{2} \rfloor + 1$ by Piña and Uzcágetui [9].

The problem of reconstructing a string of length n using substring queries has also been extensively studied in the setting of Sequencing by Hybridization (SBH), first suggested by Pevzner [8]. Here, a large number of strings of a certain length are queried in parallel, using a DNA chip, and the resulting answers are then used to reconstruct all or parts of the DNA string. A number of different SBH techniques have been proposed, leading to different string combinatorial questions. (See, for example, [14,10] for some more recent results.)

2 Preliminaries

By a paired alphabet we understand a finite set $\Sigma = \{a_1, \ldots, a_{2\delta}\}$, for some integer $\delta \geq 1$, together with a non-identity involution operation $\bar{} : \Sigma \mapsto \Sigma$, which we call complement. Thus, for each $i = 1, \ldots, 2\delta$, there is a $j \neq i$ such that $a_i = \overline{a_j}$. Notice that by definition, $\overline{\overline{a_i}} = a_i$, for each i.

Let $\mathbf{s} = s_1 \ldots s_n$ be a string (or word) over Σ, i.e., $\mathbf{s} \in \Sigma^* = \bigcup_{i=0}^{\infty} \Sigma^i$, where, following standard notation, $\Sigma^i = \{x_1 \ldots x_i \mid x_k \in \Sigma, \text{ for each } k = 1, \ldots, i\}$, and Σ^0 is the singleton containing only the empty string ϵ. For each $x \in \Sigma$ we also set $x^0 = \epsilon$. The *reverse complement* of $\mathbf{s}$ is defined as $\tilde{\mathbf{s}} = \overline{s_n}\,\overline{s_{n-1}} \ldots \overline{s_1}$. Two strings $\mathbf{s}, \mathbf{t}$ are RC-*equivalent*, denoted $\mathbf{s} \equiv_{\mathrm{RC}} \mathbf{t}$, if either $\mathbf{s} = \mathbf{t}$ or $\mathbf{s} = \tilde{\mathbf{t}}$. For a string $\mathbf{s} = s_1 \ldots s_n$ over the alphabet Σ, we denote by $|\mathbf{s}| = n$ the length of $\mathbf{s}$, and by $|\mathbf{s}|_a = |\{i \mid s_i = a\}|$ the number of a's in $\mathbf{s}$, for $a \in \Sigma$.

Given two strings $\mathbf{s}, \mathbf{t}$ over Σ, $\mathbf{s} = s_1 \ldots s_n$, $\mathbf{t} = t_1 \ldots t_m$, we say that $\mathbf{t}$ is a *subsequence*[1] of $\mathbf{s}$, denoted by $\mathbf{t} \prec \mathbf{s}$, if there exist $1 \leq i_1 < i_2 < \ldots < i_m \leq n$ such that $\mathbf{t} = s_{i_1} s_{i_2} \ldots s_{i_m}$. Further, we define $\mathbf{t}$ to be an *RC-subsequence*, denoted $\mathbf{t} \prec_{\mathrm{RC}} \mathbf{s}$ if and only if $\mathbf{t} \prec \mathbf{s}$ or $\mathbf{t} \prec \tilde{\mathbf{s}}$, i.e., if $\mathbf{t}$ is a subsequence of $\mathbf{s}$ or of its reverse complement. Note that the condition $\mathbf{t} \prec \tilde{\mathbf{s}}$ is equivalent to $\tilde{\mathbf{t}} \prec \mathbf{s}$.

Example 1. Our motivating example is the alphabet of the 4 nucleotides (DNA) $\Sigma = \{\mathrm{A, C, G, T}\}$ where $(\mathrm{A, T})$ and $(\mathrm{G, C})$ are complement pairs. Let $\mathbf{s} =$ ACCGATTAC. Then $\tilde{\mathbf{s}} =$ GTAATCGGT, GTTT $\not\prec \mathbf{s}$ but GTTT $\prec_{\mathrm{RC}} \mathbf{s}$.

We are now ready to state the problem we investigate in the present paper.

The RC-String Identification Problem. Fix a paired alphabet Σ, together with a string $\mathbf{s}$ over Σ, and let $n = |\mathbf{s}|$. For any positive integer $T \leq n$, a T-bounded RC-subsequence query is any $\mathbf{t} \in \bigcup_{i=1}^{T} \Sigma^i$. The answer to such a query is *yes* (or *positive*) if and only if $\mathbf{t} \prec_{\mathrm{RC}} \mathbf{s}$. Otherwise the answer is *no* (or *negative*). Given the alphabet Σ, the size of the string n, and the threshold on the length of the queries $T \leq n$, the RC-String Identification Problem asks for the minimum number of T-bounded RC-subsequence queries which are sufficient to determine the pair $(\mathbf{s}, \tilde{\mathbf{s}})$, for any unknown string $\mathbf{s}$ of size n.

We first present an information theoretic lower bound that holds even in the case of unbounded queries, i.e. if $T = n$.

Proposition 1 (Lower Bound). *Given a string* $\mathbf{s}$ *of size* n *from an alphabet* Σ*. Any deterministic algorithm that identifies* $\mathbf{s}$ *(up to reverse complement) by asking RC-subsequence queries needs at least* $n \log |\Sigma| - 1$ *queries.*

[1] In the literature, the term 'subword' is also common. However, 'subword' is also used to mean a contiguous subsequence. We avoid the term.

Proof. Upon identifying a string with its reverse complement, there are at least $|\Sigma^n|/2$ possible *distinct* strings of length n. Any query $\mathbf{t}$ splits the space of candidate solutions into two parts. Therefore, at least $\log|\Sigma^n|/2 = n\log|\Sigma| - 1$ questions are necessary to identify $\mathbf{s}$.

3 Unbounded Query Size

If $T = n$ (i.e., no constraint is set on the length of a query), then it is easy to reconstruct a string in linear time. We adapt a simple algorithm from [13], originally developed for the classic case (where queries would answer *no* if the subsequence only appears in the reverse complement of the string). Here we give a proof sketch and defer the precise analysis to the full version of the paper.

Theorem 1. *There exists an algorithm for reconstructing a string using* $\Theta(n\log|\Sigma|)$ *RC-subsequence queries of unbounded length.*

Proof. (Sketch.) For the binary case $\Sigma = \{a, b\}$, we first find $A := \max\{|s|_a, |s|_b\}$ by asking queries a^χ for $\chi = 1, 2, 3\ldots$. Clearly, $A = \chi - 1$ for the first χ that gives a *no* answer. Now there are indices $0 \leq i_0, i_1, \ldots, i_A$ s.t. $\mathbf{s} = b^{i_0}ab^{i_1}a\ldots ab^{i_{A-1}}ab^{i_A}$. We find i_0 by asking ba^A, b^2a^A, b^3a^A, etc., then find i_1 by asking $b^{i_0}aba^{A-1}, b^{i_0}ab^2a^{A-1}$ etc. The total number of queries is at most $\frac{3}{2}n + 2$.

Now let $\Sigma = \{a_1, \overline{a_1}, \ldots, a_\delta, \overline{a_\delta}\}$. For each complement pair $a_i, \bar{a}_i$, we first determine $\mathbf{s}_{|i}$, the longest subsequence of $\mathbf{s}$ which consists only of a_i's and $\bar{a}_i$'s. This can be done by using the algorithm for the binary case sketched above. Now we iteratively interleave the projections: first $\mathbf{s}_{|1}$ with $\mathbf{s}_{|2}$, yielding $\mathbf{s}_{|1,2}$, then $\mathbf{s}_{|1,2}$ with $\mathbf{s}_{|3,4}$ etc. Interleaving two strings $\mathbf{u}, \mathbf{v}$ that only contain characters from different complement pairs can be done with $2(|\mathbf{u}| + |\mathbf{v}| + 1)$ queries using the same idea as for the binary case, with the following small alteration: Since either $\mathbf{u}$ and $\mathbf{v}$, or $\mathbf{u}$ and $\tilde{\mathbf{v}}$ have to be interleaved, we start with $\mathbf{u}$ and $\mathbf{v}$, and if we get a contradictory answer at some point, then we start over with $\mathbf{u}$ and $\tilde{\mathbf{v}}$ (hence the factor 2). So the total number of queries for interleaving the projections $\mathbf{s}_{|i}$ is at most $2n\log\delta + 2(\delta - 1)$. The number of queries of the first phase is at most $\sum_{i=1}^{\delta}(\frac{3}{2}A_i + 2)$, where $A_i = |s|_{a_i} + |s|_{\bar{a}_i}$, yielding $O(n\log|\Sigma|)$ questions in total, using the (natural) assumption that $|\Sigma| = O(n)$.

4 Bounded Query Size (Binary Alphabet)

We now turn to subsequence queries whose length is bounded by a threshold T. In the following, the alphabet is binary, i.e., $\Sigma = \{a, b\}$, with $b = \overline{a}$. The following result shows that string identification by T-bounded subsequence queries cannot be attained in general if the threshold T on the size of the subsequence queries is set below $\lceil \frac{2}{3}n \rceil$.

Fact 1 (Erdős et al., 2006 [4]). *For any* $n \geq 4$ *there exist two distinct strings of size* n *with exactly the same set of subsequences of length up to* $\lceil \frac{2n}{3} \rceil - 1$.

This implies that if we are looking for algorithms which are able to reconstruct *any* binary string of size n, we must allow queries of size $\geq \lceil 2n/3 \rceil$.

Any string $\mathbf{s}$ over Σ can be written uniquely in its runlength encoded form:

$$\mathbf{s} = a^{x_1} b^{y_1} a^{x_2} b^{y_2} \ldots a^{x_{\rho-1}} b^{y_{\rho-1}} a^{x_\rho} b^{y_\rho}, \tag{1}$$

with x_1 and y_ρ possibly 0, and all other $x_i, y_i > 0$. The number of non-zero x_i, y_i is the number of runs of $\mathbf{s}$. We denote by $A = |\mathbf{s}|_a$ the number of a's and by $B = |\mathbf{s}|_b$ the number of b's in $\mathbf{s}$. In the following we assume that $A \geq B$. This is without loss of generality since otherwise, swap $\mathbf{s}$ and $\tilde{\mathbf{s}}$. We will denote by ρ_a the number of a-runs, and by ρ_b the number of b-runs of $\mathbf{s}$. Note that both have value either ρ or $\rho - 1$. (We have $\rho_a = \rho_b = \rho - 1$ if and only if the string starts with a b and ends with an a.)

In this section we prove the main result of the paper, which is given in the following theorem:

Theorem 2. *There is an algorithm which reconstructs a binary string* $\mathbf{s}$ *of length* n *using* $O(n)$ *many RC-subsequence queries of length at most* $\lceil \frac{2}{3}(n+1) \rceil$.

Notice that this is tight w.r.t. the lower bound of Fact 1 in all cases except where n is a multiple of 3. Even for these n, a gap of 1 unit is only necessary in the special case $A = \frac{2}{3}n$. In all other cases, our analysis resists the stricter bound of $T = \lceil \frac{2}{3}n \rceil$.

The proof of the theorem is by examining four cases separately. Recall that $A = |\mathbf{s}|_a, B = |\mathbf{s}|_b$, and $T = \frac{2}{3}(n+1)$. The four cases are: *1.* $A \geq T$, *2.* $T > A > B$, *3.* $A = B$ and $s_1 = s_n$, and *4.* $A = B$ and $s_1 \neq s_n$. The following simple lemma will be used to distinguish these cases.

Lemma 1. *Let* $\mathbf{s}$ *be a string of length at least* 8 *over* $\{a, b\}$, $T = \lceil \frac{2}{3}(n+1) \rceil$, *and* $A = |\mathbf{s}|_a \geq |\mathbf{s}|_b$. *Then,*

1. *using* $O(\log n)$ *RC-subsequence queries of length at most* T, *it is possible to determine the exact value of* $A = |\mathbf{s}|_a$ *if* $A < T$, *or to establish the fact that* $A \geq T$.
2. *Moreover, if* $A < T$, *then it can be determined whether* $\mathbf{s}$ *starts and ends with the same character; furthermore, unless* $A = \frac{n}{2}$ *and* $s_1 = s_n$, *we can determine* s_1 *and* s_n. *Altogether we require at most* 3 *additional RC-subsequence queries of length at most* T.

Proof. 1. Binary search for A, using queries of the form a^χ, for $\chi \in [\frac{n}{2}, T]$, will either return the exact value of A (if $A < T$), or will exit with the maximum size query $a^T \prec_{\text{RC}} \mathbf{s}$, thus showing that $A \geq T$.

2. Notice that if $A = B = \frac{n}{2}$, then the query $\mathbf{t} = ab^{\frac{n}{2}}a$ will return *yes* if and only if $s_1 = s_n$. If $s_1 = s_n$, then, due to the complete symmetry, we cannot determine the exact nature of s_1 and s_n. Otherwise, either $T > A = B$ and $s_1 \neq s_n$, or $T > A > B$. In either case, the query ba^A has length at most T and will answer positively if and only if $s_1 = b$. Likewise, the query $a^A b$ will answer positively if and only if $s_n = b$.

Example 2. Let $\mathbf{s}_1 = aababbba$. Then $\tilde{\mathbf{s}}_1 = baaababb$. The query ab^4a will return *yes* and we can only determine that the first and last characters are equal, but not what they are. Instead, for $\mathbf{s}_2 = aababbab$, we have $\tilde{\mathbf{s}}_2 = abaababb$, the query ab^4a will return *no*, and since the query ab^4 is positively answered, we know that the first character is a (and thus the last character is b).

Given a string $\mathbf{s}$ and a subsequence $\mathbf{t}$ of $\mathbf{s}$, we say that $\mathbf{t}$ *fixes the direction of* $\mathbf{s}$ if $\mathbf{t} \not\prec \tilde{\mathbf{s}}$. If $\mathbf{t}$ fixes the direction of $\mathbf{s}$ then for any $\mathbf{t}' \prec \mathbf{s}$, such that $\mathbf{t} \prec \mathbf{t}'$ we also have that $\mathbf{t}'$ fixes the direction of $\mathbf{s}$. In general, we shall try to identify $\mathbf{s}$ by first finding some sequence $\mathbf{t}$ which fixes the direction of $\mathbf{s}$ or $\tilde{\mathbf{s}}$ and then extending this $\mathbf{t}$. The importance of "direction-fixing" is that once we have found $\mathbf{t}$ which fixes the direction of $\mathbf{s}$, by asking queries about super-sequences of $\mathbf{t}$ we are sure that the answers to our queries are only about $\mathbf{s}$ and not its reverse complement.

The following two statements formalize two simple facts which will be used repeatedly in the following, thus, for the sake of completeness, we formally state and prove them here. Let $\mathbf{s}$ be fixed for the rest of this section.

Lemma 2. *Let* $\mathbf{t} = t_1 \ldots t_m$ *be a sequence which fixes the direction of* $\mathbf{s}$. *Fix a character* $c \in \Sigma$. *For each* $i = 1, \ldots, m+1$, *let* $\gamma_i = \min\{\max\{j \mid t_1 \ldots t_{i-1} c^j t_i \ldots t_m \prec \mathbf{s}\}, T - m\}$. *Then, for each* $i = 1, \ldots, m+1$, *we can determine* γ_i *using* $2\log\gamma_i + 1$ *queries, or alternatively, using* $\gamma_i + 1$ *queries. In particular, we can determine* all γ_i *using at most* $m + 1 + \sum_{i=1}^{m+1} \gamma_i$ *queries.*

Proof. We can determine all the values γ_i either with one-sided binary search, using $2\log\gamma_i + 1$ queries, or with linear search, using $\gamma_i + 1$ queries.

Example 3. Note that the lemma only assumes that $\mathbf{t}$ fixes the direction of $\mathbf{s}$, but not that the *positions* in $\mathbf{s}$ to which the characters of $\mathbf{t}$ are matched are also fixed. Consider the following example. Let $\mathbf{s} = a^{10}ba^{10}ba^{10}$. Then $\mathbf{t} = aaa$ fixes the direction of $\mathbf{s}$. For $c = b$, we get $\gamma_1 = \gamma_2 = \gamma_3 = \gamma_4 = 2$. For $c = a$, we have $\gamma_i = \min(27, 22) = 22$ for all i (since $T = 22$).

The next lemma says that if there are large a-runs or large b-runs, then there cannot be many runs.

Lemma 3. *Let* $\mathbf{s} = a^{x_1}b^{y_1} \ldots a^{x_\rho}b^{y_\rho}$. *Assume that there are* $1 \le i_1 < i_2 < \cdots < i_q \le \rho$ *and* $k \ge 0$, *such that* $x_{i_j} \ge T - B - k$ *(resp.* $y_{i_j} \ge T - A - k$*) for each* $j = 1, \ldots, q$, *and for at least one value of* j *it holds that* $x_{i_j} > T - B - k$ *(resp.* $y_{i_j} > T - A - k$*). Then*

$$\rho_a \le n - B - q(T - B - k - 1) - 1 \qquad (\textit{resp. } \rho_b \le n - A - q(T - A - k - 1) - 1).$$

Proof. We limit ourselves to showing the argument for ρ_a, the number of non-empty a-runs. Since each run counted by ρ_a has at least one a, we have the desired inequality:

$$n - B = A \ge \sum_{j=1}^{q} x_{i_j} + \rho_a - q \ge q(T - B - k) + 1 + \rho_a - q.$$

4.1 The Case Where $A \geq T$

Since $A \geq T = \lceil \frac{2}{3}(n+1) \rceil$, we have that $B \leq \frac{n}{3} - \frac{2}{3}$, so $2B+1 = 2(n-A)+1 \leq \frac{2}{3}(n+1) \leq T$. This implies that we can ask queries which include $B+1$ many a's and up to B many b's. Let $\beta = \lceil \frac{n}{3} - \frac{2}{3} \rceil$, and $\mathbf{t} = a^{\beta+1}$. We have $B \leq \beta$ and, therefore, $\mathbf{t}$ fixes the direction of $\mathbf{s}$. Notice also that $B + \beta + 1 \leq T$.

By Lemma 2, with $\mathbf{t} = a^{\beta+1}$, we can find $L = \max\{j \mid b^j a^{\beta+1} \prec \mathbf{s}\}$ with $O(\log L)$ queries. Likewise, with $\mathbf{t} = b^L a^{\beta+1}$ we can find $R = \max\{j \mid b^L a^{\beta+1} b^j \prec \mathbf{s}\}$, with $O(\log R)$ queries.

Notice that in $\mathbf{s}$, between the left-most L many b's and the right-most R many b's, there may be more than $\beta+1$ many a's. More precisely, with reference to (1), the previous queries guarantee that there are $1 \leq i \leq j \leq \rho$ such that $\sum_{k=1}^{i-1} y_k = L$ and $\sum_{k=j}^{\rho} y_k = R$. Let $\mathbf{w}$ be the substring of $\mathbf{s}$ between the L left-most and the R right-most b's, i.e., $\mathbf{w} = a^{x_i} b^{y_i} \cdots a^{x_j}$. Moreover, let $\mathbf{s}_{\text{left}}$ and $\mathbf{s}_{\text{right}}$ be such that $\mathbf{s} = \mathbf{s}_{\text{left}} \mathbf{w} \mathbf{s}_{\text{right}}$. We know that $|\mathbf{s}_{\text{left}}|_b = L$, $|\mathbf{s}_{\text{right}}|_b = R$, and $|\mathbf{w}|_a \geq \beta+1$. We will first determine all but the first a-run of $\mathbf{w}$ and all of its b-runs, in particular yielding the exact value of B. Then we determine $\mathbf{s}_{\text{left}}$ and $\mathbf{s}_{\text{right}}$. For any a-runs that have length at least $T-B$, their exact value will be determined during the final stage.

We have $a^{\beta+1} \prec \mathbf{w}$, and by the definition of L and R we also have that $\sum_{k=i+1}^{\rho} x_k \leq \beta$ and $x_i > \sum_{k=j+1}^{\rho} x_k$. It follows that, for $\chi = 1, 2, 3 \ldots, \beta$, the query $b^L a^{\chi} b a^{\beta+1-\chi} b^R$ answers negatively as long as $\sum_{k=i+1}^{j} x_k < \beta + 1 - \chi$. Let χ^* be the first value for which the answer to this query is *yes*, and $\chi^* = \beta+1$ if the answer is *no* for all values of χ. It is easy to see that $\chi^* = \beta + 1 - \sum_{k=i+1}^{j} x_k$. In particular, $\chi^* = \beta+1$ if and only if $\mathbf{w}$ does not contain any b's. In this case, set $\mathbf{w}' = a^{\beta+1}$.

If $\chi^* \leq \beta$, define $\mathbf{t} = b^L a^{\chi^*} b a^{\beta+1-\chi^*} b^R$. By Lemma 2, with $\mathbf{t}$ we can find the value of y_k for each $k = i, \ldots, j-1$. As a side effect, we also determine the value of x_k for $k = i+1, \ldots, j$. Now we know that $b^L \mathbf{w}' b^R \prec \mathbf{s}$, where $\mathbf{w}' = a^{\chi^*} b^{y_i} a^{x_{i+1}} \ldots b^{y_{j-1}} a^{x_j}$. In other words, we know $\mathbf{w}$ except for its first a-run, which may be longer than χ^*. We also know B, the number of b's of $\mathbf{s}$.

Now we turn to $\mathbf{s}_{\text{right}}$. Let us denote by $\mathbf{w}' - a^{\ell}$ an arbitrary sequence obtained by removing exactly ℓ many a's from $\mathbf{w}'$ and leaving the rest as it is. Now we can use queries of the form $b^L(\mathbf{w}' - a^{\ell}) b^r a^{\ell} b^{R-r}$ with $r = 1, \ldots, R$ and $\ell = 1, 2, 3 \ldots$, in order to determine the values of x_k, for each $k = j+1, \ldots, \rho$. To see this, it is enough to notice that each such query contains $\beta+1$ many a's, therefore it can only be a subsequence of $\mathbf{s}$ and not of $\tilde{\mathbf{s}}$. Moreover, we notice that in order to determine x_k we need to receive a positive answer to the query $b^L(\mathbf{w}' - a^{x_k}) b^{\sum_{\ell=j}^{k-1} y_\ell} a^{x_k} b^{R - \sum_{\ell=j}^{k-1} y_\ell}$ and a negative answer to the query $b^L(\mathbf{w}' - a^{x_k - 1}) b^{\sum_{\ell=j}^{k-1} y_\ell} a^{x_k+1} b^{R - \sum_{\ell=j}^{k-1} y_\ell}$. Because of $\sum_{k=j+1}^{\rho} x_k < \beta + 1$, both these queries have length not larger than T. Again, by determining x_k for each $k = j+1, \ldots, \rho$, we also determine y_k for each $k = j+1, \ldots, \rho$.

By an analogous procedure, we can determine $\mathbf{s}_{\text{left}}$ and the first a-run of $\mathbf{w}$, i.e. all the values x_k, for $k = 1, \ldots, i$, where $x_k \leq T - B$. Again, in this process, we also determine the size of the runs of b's, i.e., the y_k, for each $k = 1, \ldots, i-1$.

Finally, we compute the size of the a-runs in $\mathbf{s}$ that are larger than $T-B$. Notice that for at most two indices we can have $x_k \geq T-B$, for otherwise their total sum would be larger than n, the total length of the string. If there is exactly one such x_k, then we can compute it as $x_k = n - B - \sum_{\ell \neq k} x_\ell$. Otherwise, let $1 \leq i_1 < i_2 \leq i$ be such that $x_{i_1}, x_{i_2} \geq T - B$. Then it must hold that $n - B - \sum_{\ell \neq i_1, i_2} x_\ell = 2(T-B)$, and thus, $x_{i_1}, x_{i_2} = T - B$. Otherwise, we would have that $x_1 + x_2 > 2(T-B)$, and using Lemma 3, with $k = 0$, we can then conclude that $\rho_a \leq n - 2T + B + 1 \leq -1$, a contradiction.

Notice that we use at most one query per character of $\mathbf{s}$ plus at most one query for each run of $\mathbf{s}$. Therefore, in total we have $O(\sum_i (x_i + 1) + \sum_i (y_i + 1)) = O(n)$.

4.2 The Case $T > A > B$

By Lemma 2 with $\mathbf{t} = a^A$ and $c = b$, with $O(n)$ queries we can determine exactly y_k for each k such that $y_k < T - A$. In the process, we also find out exactly x_k for each $k = 1, \ldots, \rho_a$. The only problem now is to determine those runs of b's which have length at least $T - A$.

Let $i_1, \ldots, i_q$ be the q distinct indices of the runs of b's such that $y_{i_j} \geq T - A$, so we have not yet been able to determine their exact value. Clearly, if $q = 1$, we can compute $y_{i_1} = B - \sum_{\ell \neq i_1} y_\ell$. Likewise, if $B - \sum_{\ell \neq i_1, \ldots, i_q} y_\ell = q(T-A)$, then we know $y_{i_j} = T - A$ for all i_j. Otherwise, it must hold that $\sum_{j=1}^{q} y_{i_j} > q(T-A)$. Let $y_{i_1} \geq y_{i_2} \geq \cdots \geq y_{i_q}$ and $\alpha > 0$ such that $y_{i_1} = T - A + \alpha$. We have

$$\rho_b \leq B - (y_{i_1} + y_{i_2}) + 2 = n - A - (T - A + \alpha) - y_{i_2} + 2 \leq \frac{n}{3} + \frac{4}{3} - \alpha - y_{i_2}.$$

Now, consider the sequence $\mathbf{t}_\chi = (ab)^{i_2 - 1} a b^\chi (ab)^{\rho_b - i_2}$. For each $\chi = T - A + 1, T - A + 2, \ldots, y_{i_2} + 1$, such a string has length at most T, since we have

$$|\mathbf{t}_\chi| = 2\rho_b + \chi - 1 \leq 2\rho_b + y_{i_2} \leq \frac{2n}{3} + \frac{8}{3} - 2\alpha - 2y_{i_2} + y_{i_2} = \frac{2n}{3} + \frac{8}{3} - 2\alpha - y_{i_2} \leq T - 1, \tag{2}$$

where the last inequality follows from the fact that $\alpha, y_{i_2} \geq 1$.

We will finish the proof for the case $T > A > B$ by distinguishing four cases according to whether $s_1 = s_n$ and whether $s_1 = a$ or $s_1 = b$. (Note that due to the assumption $A > B$ we cannot assume w.l.o.g. the identity of the first character.)

Case 1. If $s_1 = s_n = b$, then $\rho_b = \rho$, we can remove the first a from $\mathbf{t}_\chi$, and the new query fixes the direction of $\mathbf{s}$. This query has length at most T, so we can identify y_{i_2}. By the same argument, we can also identify y_{i_j}, for each $j = 3, \ldots, q$, since $y_{i_j} \leq y_{i_2}$, for each such j. Finally we can determine y_{i_1} by subtraction.

Case 2. If $s_1 = s_n = a$, then $\rho_b = \rho - 1$. Now we have to add an a at the end to get a query which fixes the direction of $\mathbf{s}$, and its length is again at most T. The argument is then analogous to Case 1.

Case 3. Let $s_1 \neq s_n$ and $s_1 = b$. This case is analogous to Case 4. below, replacing $\mathbf{t}_\chi$ by $\mathbf{u}_\chi = (ba)^{i_2 - 1} b^\chi a (ba)^{\rho_b - i_2}$ and all following sequences accordingly.

Case 4. As the final case we have $s_1 \neq s_n$ and $s_1 = a$, so $\rho_b = \rho$. We will now look at the value of $X := x_{\rho-i_2+1}$. Note that any query $\mathbf{t}_\chi$ with $\chi \leq X$ would answer *yes* because it would be interpreted as $\tilde{\mathbf{t}}_\chi$. Notice that we know the value of X. If $X < T - 2\rho$, then we ask query $\mathbf{t}_\chi$ for $\chi = X + 1$. If the answer is *yes*, we continue with $X+2, X+3\ldots$ until we receive the first *no*, and we are done, since the last χ where $\mathbf{t}_\chi$ answered positively was equal to y_{i_2}. By (2), these queries do not exceed the threshold.

Otherwise, if the query $\mathbf{t}_{X+1}$ answered *no* or if $X > T - 2\rho$, then we know that $y_{i_2} \leq X$. In this case, we use the following queries to determine y_{i_2}.

Let w.l.o.g. $i_2 \leq \rho - i_2 + 1$ (otherwise exchange the roles of i_2 and $\rho - i_2 + 1$ in the formulas below). Define $\mathbf{t}'_\xi = (ab)^{i_2-1}a^\xi(ba)^{\rho-2i_2+1}b^{\xi+1}(ab)^{i_2-1}$. One can verify that for each $\xi = T - A, T - A + 1, \ldots, y_{i_2}$, we have

$$|\mathbf{t}'_\xi| \leq 2\rho_b + 2y_{i_2} - 1 \leq \frac{2n}{3} + \frac{2}{3} - 2\alpha + 1 \leq T + 1 - 2\alpha \leq T - 1, \qquad (3)$$

where the last inequality follows from the fact that $\alpha \geq 1$.

We can ask queries $\mathbf{t}'_\xi$ until either we receive a negative answer or we cannot enlarge it further because it would violate the bound T. The largest value of ξ for which we receive a positive answer to query $\mathbf{t}'_\xi$ correctly gives the value of y_{i_2}. Clearly this is true if we also receive a negative answer, for the next larger value. If, instead, we had to stop because of the bound T, we can be sure that $\xi = y_{i_2}$, because if $y_{i_2} > \xi$, then this would contradict the inequality (3).

We ask at most one query per character plus one query per run, except for Case 4, where we might use two queries per character of the y_{i_2}'th run of b's. Altogether, we have that the total number of queries is $O(n)$.

4.3 The Case $T > A = B = \frac{n}{2}$, $s_1 = s_n$

We assume w.l.o.g. that the string starts and ends in a. Therefore, with reference to (1), in this section we have $y_\rho = 0$ and our string looks like this:

$$\mathbf{s} = a^{x_1}b^{y_1}a^{x_2}b^{y_2}\ldots a^{x_{\rho-1}}b^{y_{\rho-1}}a^{x_\rho},$$

with all $x_i, y_i > 0$, i.e., it includes $\rho = \rho_a$ runs of a's and $\rho - 1 = \rho_b$ runs of b's.

By Lemma 2 with $\mathbf{t} = ab^{\frac{n}{2}}$ we can exactly determine x_k (run of a's) for each k, such that $x_k < T - \frac{n}{2} - 1 \leq \frac{n}{6} - \frac{1}{3}$. In this process, we determine exactly y_k, for each $k = 1, \ldots, \rho_b$.

Let $1 \leq i_1 < i_2 < \cdots < i_q \leq \rho_a$ be all the indices of the runs of a's whose length we have not been able to determine exactly, i.e., such that $x_{i_j} \geq T - \frac{n}{2} - 1$. By $A = \frac{n}{2}$, we have that $q \leq 3$. In fact, the only interesting cases are $q = 2$ and $q = 3$, since, for $q = 1$ we can determine the only missing x_{i_1}, as the difference between A and the sum of the remaining x_k's.

For $q = 3$, by Lemma 3, we have $\rho_a \leq 3$, thus it follows that $\rho_a = 3$. Let $\mathbf{t} = ababa$, and $c = a$. By Lemma 2 we can determine each x_k, such that $x_k \leq T - 5$. Suppose that for all $k = 1, 2, 3$, it holds that $x_k \geq T - 5$. Since there must exist one run of a's of length $\leq \frac{n}{6}$, we have that $n \leq 9$, whence $A \leq 4$, implying that the only possible case is to have two runs of a's of length 1 and one run of a's of length 2. Direct inspection shows that in this case we can easily reconstruct the whole string with T-bounded queries.

Finally, if $q = 2$, by Lemma 3, we have $\rho_a \leq \frac{n}{6} + \frac{5}{3}$. We can now use query $\mathbf{t}_1 = (ab)^{i_1-1}a^{T-\frac{n}{2}-1+\chi}(ba)^{\rho-i_1}$, for $\chi = 1,2,3,\ldots$, until we receive a negative answer, then $x_{i_1} = T - \frac{n}{2} - 1 + \chi - 1$. If we never receive a negative answer and the query becomes of length T, we can resort to the query $\mathbf{t}_2 = (ab)^{i_2-1}a^{T-\frac{n}{2}-1+\chi}(ba)^{\rho-i_2}$, for $\chi = 1,2,\ldots$, and proceed analogously. It is easy to see that we cannot have that both $\mathbf{t}_1$ and $\mathbf{t}_2$ exceed the threshold T; the other value can then be determined by difference.

We have used $O(\log n + A) = O(n)$ many queries.

4.4 The Case $T > A = B = \frac{n}{2}$, $s_1 \neq s_n$

Recall that by Lemma 1, in this case we can exactly determine s_1 and s_n. Let us assume w.l.o.g. that $s_1 = a$ and $s_n = b$. (Otherwise, rename the characters.) Then the string $\mathbf{s}$ has the following shape

$$\mathbf{s} = a^{x_1}b^{y_1}a^{x_2}b^{y_2}\ldots a^{x_{\rho-1}}b^{y_{\rho-1}}a^{x_\rho}b^{y_\rho}.$$

In particular, it starts with a run of a's and ends with a run of b's.

We need some more notation. For each $i = 1,2,\ldots,2\rho$, we use r_i to denote the size of the i'th run in $\mathbf{s}$ starting from the left. I.e., we have $x_i = r_{2i-1}$ and $y_i = r_{2i}$ for each $i = 1,\ldots,\rho$. Also we denote by $m_i = \min\{r_i, r_{2\rho-i+1}\}$ and by $M_i = \max\{r_i, r_{2\rho-i+1}\}$. We use the following technical lemma.

Lemma 4. *Fix $i < \rho$ and assume that for each $k = 1,\ldots,i-1$, we know r_k and $r_{2\rho-k+1}$ and it holds that $r_k = r_{2\rho-k+1} < T - \frac{n}{2}$. Then we can determine m_i and $\min\{M_i, T - \frac{n}{2}\}$, asking at most $\max\{m_i, \min\{M_i, T - \frac{n}{2}\}\}$ queries.*

Proof. For each odd i (i.e., r_i denotes the length of a run of a's) we have

$$m_i = \min\left\{\chi = 1,2,3,\ldots \mid \mathbf{t}_\chi = a^{x_1+\cdots+x_{i-1}+\chi}ba^{\frac{n}{2}-(x_1+\cdots+x_{i-1}+\chi)} \prec_{\mathrm{RC}} \mathbf{s}\right\},$$

$$\min\left\{M_i, T - \frac{n}{2}\right\} = \max\left\{\chi = m_i, m_i+1,\ldots,T-\frac{n}{2} \mid \right.$$
$$\left. \mathbf{q}_\chi = a^{\frac{n}{2}-(y_1+\cdots+y_{i-1})}b^\chi a^{y_1+\cdots+y_{i-1}} \prec_{\mathrm{RC}} \mathbf{s}\right\}.$$

Using the above equalities, one can determine the value m_i (resp. M_i) by asking the query $\mathbf{t}_\chi$ (resp. $\mathbf{q}_\chi$) for increasing values of χ, until the first positive (resp. negative) answer. This settles the case of i odd.

It is not hard to see that exactly the same argument holds for even i, using the following:

$$m_i = \min\left\{\chi = 1,2,\ldots \mid \mathbf{t}_\chi = b^{y_1+\cdots+y_{i-1}+\chi}ab^{\frac{n}{2}-(y_1+\cdots+y_{i-1}+\chi)} \prec_{\mathrm{RC}} \mathbf{s}\right\},$$

$$\min\left\{M_i, T-\frac{n}{2}\right\} = \max\left\{\chi = m_i, m_i+1, \ldots, T-\frac{n}{2} \mid \right.$$
$$\left. \mathbf{q}_\chi = b^{\frac{n}{2}-(x_1+\cdots+x_{i-1})} a^\chi b^{x_1+\cdots+x_{i-1}} \prec_{\mathrm{RC}} \mathbf{s}\right\}.$$

This completes the proof of the lemma.

Now, let us consider the largest $k \geq 1$ such that $r_j = r_{2\rho-j+1} < T - \frac{n}{2}$ for each $j < k$. Note that by repeated application of Lemma 4, we can determine all these r_j's. Assume w.l.o.g. that k is odd and let $i = \lceil k/2 \rceil$. Then we can write:

$$\mathbf{s} = \mathbf{u} a^{x_i} \mathbf{s}' b^{y_{\rho-i+1}} \tilde{\mathbf{u}}, \tag{4}$$

where $\mathbf{u} = a^{x_1}b^{y_1}\ldots a^{x_{i-1}}b^{y_{i-1}}$ is known, and the string $\mathbf{s}'$ is still unknown. Note that also the two values $\min\{x_i, y_{\rho-i+1}\}$ and $\min\{\max\{x_i, y_{\rho-i+1}\}, T-\frac{n}{2}\}$ are known (again by application of Lemma 4). Moreover, for determining these two values and string $\mathbf{u}$, we have used a number of queries linear in $2|\mathbf{u}| + \min\{\max\{x_i, y_{\rho-i+1}\}, T-\frac{n}{2}\}$.

According to the magnitude of x_i and $y_{\rho-i+1}$, we will enter one of the following three cases, where we will assume, w.l.o.g., that $x_i \leq y_{\rho-i+1}$. (The case where $y_{\rho-i+1} < x_i$ is symmetric.) We illustrate the situation in Figure 1.

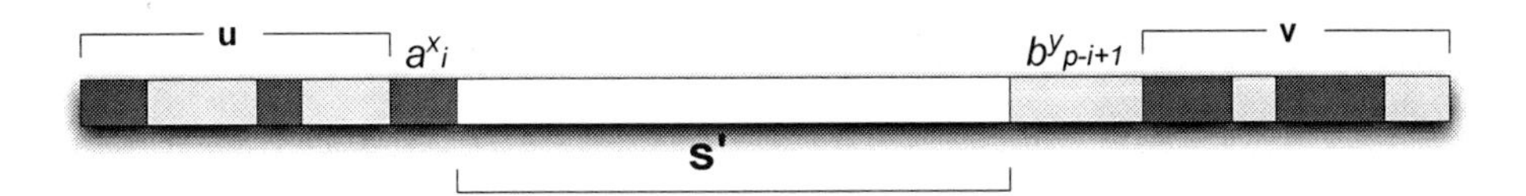

Fig. 1. The case where $|\mathbf{s}|_a = |\mathbf{s}|_b$ and $s_1 \neq s_n$. We determine $\mathbf{s}$ by first finding the first assymetry in $\mathbf{s}$ ($x_i \neq y_{\rho-i+1}$), and then extending queries for $\mathbf{s}'$, which has fewer b's than a's. Note that up to index i, string $\mathbf{s}$ is perfectly symmetric, i.e. we have $\mathbf{v} = \tilde{\mathbf{u}}$.

The case $A = B = \frac{n}{2}$, $s_1 \neq s_n$, $x_i, y_{\rho-i+1} < T - \frac{n}{2}$. With reference to (4), we can use a recursive argument to show how to determine $\mathbf{s}'$. Let $n' = |\mathbf{s}'|$. Note that $|\mathbf{s}'|_a > |\mathbf{s}'|_b$ and that $\mathbf{s}'$ starts with a b and ends with an a.

Let $\mathbf{t}'$ be a query for $\mathbf{s}'$: Since $|\mathbf{s}'|_a > |\mathbf{s}'|_b$, such queries were defined by one of the previous cases (Section 4.1 or 4.2). Let $\mathbf{t}'_+$ be the query obtained by adding to $\mathbf{t}'$ an initial b, if $\mathbf{t}'$ does not begin with b, and a final a, if $\mathbf{t}'$ does not end with an a. Define a query $\mathbf{t}$ for $\mathbf{s}$ in the following way:

$$\mathbf{t} = a^{|\mathbf{u}|_a} \mathbf{t}'_+ a^{|\mathbf{u}|_b} \tag{5}$$

Lemma 5. *Let* $\mathbf{t}$ *be defined as in Eq.* (5)*. Then, it holds that*

1. $\mathbf{t} \prec_{\mathrm{RC}} \mathbf{s}$ *if and only if* $\mathbf{t}' \prec_{\mathrm{RC}} \mathbf{s}'$.
2. *If* $|\mathbf{t}'| \leq \frac{2(n'+1)}{3}$, *then* $|\mathbf{t}| \leq \frac{2(n+1)}{3}$.

Proof. *1.* Let $\mathbf{t} \prec_{\mathrm{RC}} \mathbf{s}$. First assume that $\mathbf{t} \prec \mathbf{s}$. Notice that $\mathbf{t}'_+$ starts with a b and ends with an a, and that $\mathbf{t} = a^{|\mathbf{u}|_a}\mathbf{t}'_+ a^{|\tilde{\mathbf{u}}|_a}$, i.e., the number of a's in $\mathbf{t}$ following

$\mathbf{t}'_+$ equals the number of a's in $\tilde{\mathbf{u}}$. Because of the $|\mathbf{u}|_a$ many a's at the beginning of $\mathbf{t}$, the fact that $\mathbf{t}$ is a subsequence of $\mathbf{s}$ implies $\mathbf{t}'_+ a^{|\tilde{\mathbf{u}}|} \prec a^{x_i}\mathbf{s}'b^{y_{\rho-i+1}}\tilde{\mathbf{u}}$, and because $\mathbf{t}'_+$ starts with a b, we also have $\mathbf{t}'_+ a^{|\tilde{\mathbf{u}}|} \prec \mathbf{s}'b^{y_{\rho-i+1}}\tilde{\mathbf{u}}$. This again implies that $\mathbf{t}'_+ \prec \mathbf{s}'b^{y_{\rho-i+1}}$, and because $\mathbf{t}'_+$ ends with an a, also $\mathbf{t}'_+ \prec \mathbf{s}'$, and thus, $\mathbf{t} \prec \mathbf{s}'$.

Now let $\mathbf{t} \prec \tilde{\mathbf{s}}$, or, equivalently, $\tilde{\mathbf{t}} \prec \mathbf{s}$. We have $\tilde{\mathbf{t}} = b^{|\mathbf{u}|_b}\tilde{\mathbf{t}}'_+ b^{|\mathbf{u}|_a} = b^{|\mathbf{u}|_b}\tilde{\mathbf{t}}'_+ b^{|\tilde{\mathbf{u}}|_b}$, and $\tilde{\mathbf{t}}'_+$ starts with an a and ends with a b. Thus, because of the $|\mathbf{u}|_b$ many b's at the beginning of $\tilde{\mathbf{t}}$ and the fact that $\tilde{\mathbf{t}}'_+$ starts with an a, we have $\tilde{\mathbf{t}}'_+ \prec \mathbf{s}'b^{y_{\rho-i+1}}\tilde{\mathbf{u}}$. Further, because of the $|\tilde{\mathbf{u}}|_b$ many b's at the end and the fact that $\tilde{\mathbf{t}}'_+$ ends with a b, this implies $\tilde{\mathbf{t}}'_+ \prec \mathbf{s}'$. It follows that $\mathbf{t}' \prec \mathbf{s}'$.

Conversely, if $\mathbf{t}'$ (resp. $\tilde{\mathbf{t}'}$) is a subsequence of $\mathbf{s}'$, then clearly, $\mathbf{t}$ (resp. $\tilde{\mathbf{t}}$) is a subsequence of $\mathbf{s}$.

2. The length of $\mathbf{t}$ is $|\mathbf{t}| \leq |\mathbf{u}| + 2 + |\mathbf{t}'|$, where $|\mathbf{t}'| \leq \frac{2}{3}(n'+1)$ and $n' = n - 2|\mathbf{u}| - x_i - y_{\rho-i+1}$, and $y_{\rho-i+1} > x_i \geq 1$. This implies $x_i + y_{\rho-i+1} \geq 3$. Thus,

$$\begin{aligned}|\mathbf{t}| &\leq |\mathbf{u}| + 2 + \frac{2}{3}(n - 2|\mathbf{u}| - x_i - y_{\rho-i+1} + 1)\\ &= \frac{2}{3}(n+1) + 2 - \frac{1}{3}|\mathbf{u}| - \frac{2}{3}(x_i + y_{\rho-i+1})\\ &\leq \frac{2}{3}(n+1) + 2 - \frac{1}{3}|\mathbf{u}| - 2 \leq \frac{2}{3}(n+1).\end{aligned}$$

Thus it follows that we can use the analysis of the previous sections to prepare a sequence of queries on $\mathbf{s}$ which is *(i)* linear in $|\mathbf{s}'|$ and *(ii)* allows us to determine the substring $\mathbf{s}'$ of $\mathbf{s}$. Once this is accomplished, the whole $\mathbf{s}$ can be fully determined (up to reverse complement).

The case $A = B = \frac{n}{2}$, $s_1 \neq s_n$, $x_i, y_{\rho-i+1} \geq T - \frac{n}{2}$. Notice that, because of the assumption $n \geq 8$ and $T - \frac{n}{2} \leq x_i, y_{\rho-i+1}$, it follows that $x_i + y_{\rho-i+1} \geq 4$. We have $|\mathbf{s}'| = n - 2|\mathbf{u}| - x_i - y_{\rho-i+1}$. This implies

$$|\mathbf{s}'| + |\mathbf{u}| + 2 \leq n - x_i - y_{\rho-i+1} + 2 + |\mathbf{u}| \leq 2n - 2T + 2 - |\mathbf{u}| \leq \frac{2n}{3} + \frac{2}{3} - |\mathbf{u}| \leq T. \quad (6)$$

Thus we can adapt the strategy we described in Section 3 for unbounded RC-reconstruction to determine $\mathbf{s}'$ and then, by subtraction, also x_i and $y_{\rho-i+1}$. We proceed as follows: Suppose that in the strategy for reconstructing $\mathbf{s}'$, in the unbounded-query case, we ask a question $\mathbf{t}'$, starting with b and ending with a. Then we will ask query $\mathbf{t} = a^{|\mathbf{u}|_a+1}\mathbf{t}'ba^{|\mathbf{u}|_b}$. It is not hard to see that such $\mathbf{t}$ answers positively on $\mathbf{s}$ if and only if $\mathbf{t}'$ answers positively on $\mathbf{s}'$. By (6), $|\mathbf{t}| = |\mathbf{t}'| + 2 + |\mathbf{u}| \leq T$.

The only requirement is that $\mathbf{t}'$ begin with b and end with a. However, the strategy in Section 3 can be easily adapted to this case, under the assumption that the string to be reconstructed begins with b and ends with a, a condition that holds for $\mathbf{s}'$. (Notice, in fact, that because the query size is unbounded, any query in the strategy in Section 3 can be safely extended by an arbitrary prefix and/or suffix of the string we are trying to reconstruct.)

Finally, once we have reconstructed $\mathbf{s}'$ we can determine $\max\{x_i, y_{\rho-i+1}\}$ as $\frac{n}{2} - |\mathbf{s}'|_b - |\mathbf{u}|$. (Recall that we have assumed w.l.o.g. that $x_i \leq y_{\rho-i+1}$; in fact, now that we know $\mathbf{s}'$, we can determine whether this is the case: we have $x_i \leq y_{\rho-i+1}$ if and only if $|\mathbf{s}'|_a \geq |\mathbf{s}'|_b$.)

The case $A = B = \frac{n}{2}$, $s_1 \neq s_n$, $x_i < T - \frac{n}{2}, y_{\rho-i+1} \geq T - \frac{n}{2}$. In order to determine ρ and $x_{i+1}, \ldots, x_{\rho-i+1}$, we can use the query

$$\mathbf{t}_\chi = a^{|\mathbf{u}|_a + x_i} b a^\chi b a^{\frac{n}{2} - |\mathbf{u}|_a - x_i - \chi} \tag{7}$$

as follows. Under the standing hypothesis, we have $x_i < \frac{2(n+1)}{3} - \frac{n}{2} \leq y_{\rho-i+1}$. The above query $\mathbf{t}_\chi$ has size $\frac{n}{2} + 2 \leq T$, for any $n \geq 8$. Moreover, the fact that $x_i < y_{\rho-i+1}$ guarantees that if $\mathbf{t}_\chi \prec \mathbf{s}$ then $\mathbf{t}$ fixes the direction of $\mathbf{s}$. To see this, with reference to (4), it is enough to observe that in this case, in $\mathbf{s}$ there are more a's following the first b of $\mathbf{s}'$ than there are b's preceeding the last a of $\mathbf{s}'$.

We use the query $\mathbf{t}_\chi$ as follows: We ask $\mathbf{t}_\chi$ for each $\chi = 1, 2, 3 \ldots$, until we get the first positive answer. Let χ_1 be the minimum value of χ for which the answer is positive. It is not hard to see that this implies $x_{i+1} = \chi_1$. We now continue asking query $\mathbf{t}_\chi$ for each $\chi = \chi_1 + 1, \chi_1 + 2, \ldots$. Let χ_2 be the minimum value of χ for which we get a new positive answer. Again, this implies that $x_{i+2} = \chi_2 - \chi_1$. More generally, for each $j = 1, \ldots, \rho - i + 1$, let χ_j be the value of χ when we receive the ith positive answer. Then, we have $x_{i+j} = \chi_j - \chi_{j-1}$ (where we set $\chi_0 = 0$ for sake of definiteness).

Note, however, that at this point we do not know ρ. We continue asking $\mathbf{t}_\chi$ as long as $\frac{n}{2} - |\mathbf{u}|_a - x_i - \chi > |\mathbf{u}|_b$, or equivalently, $\chi < \frac{n}{2} - |\mathbf{u}| - x_i$. This way we determine x_j, for $j = i+1, \ldots, \rho - i + 1$ and, in particular, we determine ρ.

Now by Lemma 2, with $\mathbf{t} = a^{|\mathbf{u}|_a + x_i} b a^{\frac{n}{2} - x_i - |\mathbf{u}|_a}$, we can determine exactly y_j, for each $j = i, \ldots, \rho - i$ such that $y_j < T - \frac{n}{2}$, or, otherwise, establish the fact that $y_j \geq T - \frac{n}{2}$. As in the previous cases, it now remains to determine the exact values of those runs with length at least $T - \frac{n}{2}$.

Let $i_1, \ldots, i_q$, be such that $y_{i_j} \geq T - \frac{n}{2}$, for each $j = 1, \ldots, q$. We can also assume that for at least one $1 \leq j \leq q$ it holds that $y_{i_j} > T - \frac{n}{2}$, for otherwise we can identify this situation by the fact that $n - \sum_{\ell \notin \{i_1, \ldots, i_q\}} y_\ell = q(T - \frac{n}{2})$, whence we have $y_{i_j} = T - \frac{n}{2}$, for each j.

For each j such that $y_{i_j} \geq T - \frac{n}{2}$ and whose value is not determined yet, we use a query of the form:

$$\mathbf{t}_\chi = (ab)^{i_j - 1} a b^\chi (ab)^{\rho - i - i_j} a b^{x_i + 1} (ab)^{i-1},$$

increasing χ until we get the first positive answer. It remains to show that each of these queries has length smaller or equal to T.

We have that $|\mathbf{t}_\chi| = 2\rho + x_i + \chi - 1$. To see that this is smaller or equal to T for each $\chi \leq y_{i_j}$, notice that $y_{i_j} \geq \frac{2(n+1)}{3} - \frac{n}{2} = \frac{n}{6} + \frac{2}{3}$. Further, by assumption, we have $y_{\rho-i+1}, y_{i_j} \geq \frac{n}{6} + \frac{2}{3}$, implying $t \leq \frac{n}{2} - \frac{2n}{6} + \frac{2}{3} = \frac{n}{6} + \frac{2}{3} \leq y_{\rho-i+1}$. Thus, we have $\rho \leq y_{\rho-i+1}$. Moreover, recall that $\frac{n}{2} = B \geq y_{\rho-i+1} + \rho + y_{i_j} - 2$. Putting it all together, we get

$$\mathbf{t}_{y_{i_j}} = 2\rho + x_i + y_{i_j} - 1 \leq \underbrace{y_{\rho-i+1} + \rho + y_{i_j} - 1}_{\leq B+1=\frac{n}{2}+1} + x_i \leq \frac{n}{2} + \underbrace{x_i + 1}_{\leq \frac{n}{6}+\frac{2}{3}} < \frac{2(n+1)}{3} \leq T.$$

As can be readily seen, in all three subcases we use $O(|\mathbf{s}'|)$ queries to determine $\mathbf{s}'$, hence, altogether $O(|\mathbf{s}|)$ queries to complete the reconstruction.

References

1. Carpi, A., de Luca, A.: Words and special factors. Theor. Comput. Sci. 259(1-2), 145–182 (2001)
2. de Luca, A.: On the combinatorics of finite words. Theor. Comput. Sci. 218(1), 13–39 (1999)
3. Dudík, M., Schulman, L.J.: Reconstruction from subsequences. J. Comb. Theory, Ser. A 103(2), 337–348 (2003)
4. Erdős, P.L., Ligeti, P., Sziklai, P., Torney, D.C.: Subwords in reverse-complement order. Annals of Combinatorics 10, 415–430 (2006)
5. Fici, G., Mignosi, F., Restivo, A., Sciortino, M.: Word assembly through minimal forbidden words. Theor. Comput. Sci. 359(1-3), 214–230 (2006)
6. Krasikov, I., Roditty, Y.: On a reconstruction problem for sequences. J. Comb. Theory, Ser. A 77(2), 344–348 (1997)
7. Levenshtein, V.I.: Efficient reconstruction of sequences. IEEE Transactions on Information Theory 47(1), 2–22 (2001)
8. Pevzner, P.: l-tuple DNA sequencing: Computer analysis. Journal of Biomolecular Structure and Dynamics 7, 63–73 (1989)
9. Piña, C., Uzcátegui, C.: Reconstruction of a word from a multiset of its factors. Theor. Comput. Sci. 400(1-3), 70–83 (2008)
10. Preparata, F.P.: Sequencing-by-hybridization revisited: The analog-spectrum proposal. IEEE/ACM Trans. Comput. Biology Bioinform. 1(1), 46–52 (2004)
11. Schützenberger, M.-P., Simon, I.: Combinatorics on Words, by M. Lothaire. Subwords. ch. 6. Cambridge University Press, Cambridge (1983)
12. Simon, I.: Piecewise testable events. In: Brakhage, H. (ed.) Automata Theory and Formal Languages. LNCS, vol. 33, pp. 214–222. Springer, Heidelberg (1975)
13. Skiena, S., Sundaram, G.: Reconstructing strings from substrings. Journal of Computational Biology 2(2), 333–353 (1995)
14. Tsur, D.: Tight bounds for string reconstruction using substring queries. In: Chekuri, C., Jansen, K., Rolim, J.D.P., Trevisan, L. (eds.) APPROX 2005 and RANDOM 2005. LNCS, vol. 3624, pp. 448–459. Springer, Heidelberg (2005)

Improved Points Approximation Algorithms Based on Simplicial Thickness Data Structures*

Danny Z. Chen and Haitao Wang**

Department of Computer Science and Engineering
University of Notre Dame, Notre Dame, IN 46556, USA
{dchen,hwang6}@nd.edu

Abstract. Given a real $\epsilon > 0$, an integer $g \geq 0$ and a set of points in the plane, we study the problem of computing a piecewise linear functional curve with minimum number of line segments to approximate all points after removing g outliers such that the approximation error is at most ϵ. We give an improved algorithm over the previous work. The algorithm is based on two dynamic data structures developed in this paper for the simplicial thickness queries, which are of independent interest. For a set S of simplices in the d-D space E^d ($d \geq 2$ is a constant), the *simplicial thickness* of a point p is defined as the number of simplices in S that contain p. Given a set P of n points in E^d, we develop two linear-space dynamic data structures to support the following operations. (1) Simplex insertion: Insert a simplex into S. (2) Simplex deletion: Delete a simplex from S. (3) Simplicial thickness query: Given a query simplex σ, compute the minimum simplicial thickness among all points in $\sigma \cap P$. The first data structure supports each operation in $O(n^{1-1/d})$ time with $O(n^{1+\delta})$ time preprocessing, for any constant $\delta > 0$; the second one supports each operation in $O(n^{1-1/d}(\log n)^{O(1)})$ time with $O(n \log n)$ time preprocessing. These data structures may also find other applications.

1 Introduction

In this paper, we study the following points approximation problem. Let $P = \{p_i = (x_i, y_i) \mid 1 \leq i \leq n\}$ be an input point set in the plane with $x_1 < \cdots < x_n$. Each point p_i has a weight $u_i > 0$. Let f be a piecewise linear functional curve for approximating the points in P. The *vertical distance* between each point p_i and f is defined as $d(p_i, f) = u_i \cdot |y_i - f(x_i)|$. Denote by $P' \subseteq P$ an outlier set for f. Then, the approximation error of f with respect to P' is defined as $e(f, P, P') = \max_{p_i \in P \setminus P'} d(p_i, f)$. Given $\epsilon > 0$ and an integer $g \geq 0$, to approximate the points in P with at most g outliers, we seek a piecewise linear function f in which any two consecutive line segments need not to be joined

* This research was supported in part by NSF under Grants CCF-0515203 and CCF-0916606.

** Corresponding author. The work of this author was also supported in part by a graduate fellowship from the Center for Applied Mathematics, University of Notre Dame.

C.S. Iliopoulos and W.F. Smyth (Eds.): IWOCA 2010, LNCS 6460, pp. 363–376, 2011.

(see Fig. 1(a)) and an outlier set P' with $|P'| \leq g$, such that the number of segments in f is minimized and $e(f, P, P') \leq \epsilon$. We call the problem *the piecewise linear approximation with outliers*. It was first studied in [7], where it was called VWPF *min-#*. In this paper, we use VWPF to denote it. An $O(ng^4 \log^2 n)$ time algorithm was given in [7] for solving VWPF. In this paper, we derive an improved algorithm with running time $O(n^2 g)$ when $g = \Omega((\frac{n}{\log^2 n})^{\frac{1}{3}})$. The problem is motivated by the desire to obtain an approximating function with the smallest possible size while maintaining a certain level of approximation accuracy after the outliers are detected and removed.

If the approximating function f is required to be a step function (which is a special case of the piecewise linear function), then the problem is solvable in $O(ng^2)$ time [7]. If $g = 0$, i.e., no outliers are allowed, then both the step function case and the piecewise linear function case are solvable in $O(n)$ time [9,12]. These problems are usually referred to as the *min-#* versions. For each problem, there is a corresponding *min-ϵ* version, where an integer k is given as input and the objective is to find an approximating function f (under certain constraints) minimizing the approximation error such that the number of segments in f is at most k. The *min-ϵ* versions with outliers have also been considered. Refer to [6,7] for the *min-ϵ* results and more discussions.

Our algorithm for VWPF is based on dynamic data structures for a simplex-related problem, which we call *simplicial thickness queries*, as follows. For a set S of simplices in the d-D space E^d ($d \geq 2$ is a constant), the *simplicial thickness* of a point p is defined as the number of simplices in S that contain p, denoted by $\beta(p, S)$. Given a set of n points in E^d, $P = \{p_1, p_2, \ldots, p_n\}$, for a simplex σ, define the *simplicial thickness* of σ as the minimum simplicial thickness among all points in $\sigma \cap P$, and denote it by $\beta(\sigma, S)$. We seek dynamic data structures to support the following operations ($S = \emptyset$ initially). (1) Simplex insertion: Insert a simplex into S. (2) Simplex deletion: Delete a simplex from S. (3) Simplicial thickness query: Given a query simplex σ, report $\beta(\sigma, S)$ and the corresponding point whose simplicial thickness is $\beta(\sigma, S)$. Denote the above problem by STQ.

We develop two data structures for STQ by modifying extensively the simplex range searching data structures in [10,11]. The *simplex range searching* problem is to preprocess a set P of n points in E^d such that given any query simplex σ, the number of points in $P \cap \sigma$ can be reported efficiently [4,5,10,11,14]. A best result is due to Matoušek [11]: An $O(n)$ space data structure with $O(n^{1+\delta})$ time ($\delta > 0$ is an arbitrary constant) preprocessing and $O(n^{1-1/d})$ time query. The above bounds match the lower bound given in [3] for $d = 2$ and quite likely for $d > 2$. Further, an $O(n)$ space data structure is given in [10] with a slower $O(n^{1-1/d} \log^{O(1)} n)$ time query but faster $O(n \log n)$ time preprocessing. Note that the above two data structures are static. By using some standard techniques [1,13], Matoušek in [10] also gave results for the dynamic version of the simplex range searching that allows point insertions and deletions (note that our STQ problem considers simplex insertions and deletions).

By modifying the data structure in [11], we build a STQ data structure in $O(n^{1+\delta})$ time and $O(n)$ space, which can support the three operations each in $O(n^{1-1/d})$ time. By modifying the one in [10], we build another STQ data structure in $O(n)$ space and $O(n \log n)$ time, which can support each simplex operation in $O(n^{1-1/d}(\log n)^{O(1)})$ time. Both data structures match the performances of those in [10,11]. These data structures are of independent interest.

Our two STQ data structures can be easily extended with the same performances to solve the *weighted STQ*, where each simplex of S has a weight and the *weighted simplicial thickness* of a point p is defined as the weight sum of all simplices in S containing p. Further, if the simplicial thickness of each query simplex σ is defined as the *maximum* simplicial thickness of all points in $\sigma \cap P$, then the problem (denoted by *max-STQ*) can also be solved with the same performances.

As by-products, in Section 2, we also solve some other interesting query problems, e.g., 3-D q-level lowest point queries and 2-D minimum point weight queries. Due to the space limit, the proofs of all lemmas in the paper are omitted.

In the following paper, we first present our algorithm for VWPF by assuming that we already have the STQ data structures.

2 An Improved Algorithm for VWPF

This section presents our improved algorithm for VWPF. by assuming we already have the STQ data structures in Theorems 2 and 3. The VWPF problem was studied in [7]. Our new algorithm follows the high-level framework of the algorithm in [7] but uses a better approach to handle the low-level computations. We first reduce a sub-problem in the VWPF algorithm [7] to a new problem, called *3-D q-level lowest point queries*, and we further model it as the *2-D minimum point weight queries*, which can be solved by using our STQ data structures. Consequently, the previous VWPF result in [7] can be improved.

Let $P = \{p_i = (x_i, y_i) \mid 1 \leq i \leq n\}$ be an input point set in the plane with $x_1 < \cdots < x_n$. Each point p_i has a positive weight u_i. Let $\epsilon > 0$ be the error tolerance and $g \geq 0$ be the number of allowed outliers. A sub-problem (call it *SUB*) in the VWPF algorithm in [7] is as follows. Given $P' \subseteq P$ and q $(0 \leq q \leq g)$ with $P' = \emptyset$ initially, derive a data structure for maintaining P', to support the following operations. (1) Point insertion: Insert a point from $P \setminus P'$ to P'; (2) point deletion: Delete a point from P'; (3) feasibility test: Determine whether there exists a line segment for approximating the points in the current P' with at most q outliers such that the approximation error of the segment is at most the given ϵ. The *approximation error* of the segment is the maximum vertical distance between the segment and all non-outlier points in P'. The lemma below is given in [7],

Lemma 1. *[7] If one can build a data structure in $O(T)$ time, which can support point insertion, point deletion, and feasibility test in $O(I)$, $O(D)$, and $O(F)$ time each, respectively, for the SUB problem, then the VWPF problem can be solved in $O(T + ng(g + I + D + F))$ time.*

Thus, our goal is to develop an efficient data structure for SUB. We first reduce SUB to the 3-D q-level lowest point query as follows. We assume $|P'| > q$ since otherwise the answer to the feasibility test is always true. For each point $p_t = (x_t, y_t)$ with a weight u_t in P', we define two (upper) half-spaces $u_t(ax_t - b - y_t) \leq \xi$ and $-u_t(ax_t - b - y_t) \leq \xi$ in the 3-D coordinate system $\mathcal{P}$ with a as the x-axis, b as the y-axis, and ξ as the z-axis. Denote by H_s the set of all these $2|P'|$ half-spaces. Note that given two half-spaces defined by a point in P', for any point p in $\mathcal{P}$, if p is above or on the plane $\xi = 0$, then p must be in at least one of the two half-spaces. An important observation is that for any point $p = (a', b', \xi')$ in $\mathcal{P}$, suppose we use the line $y = a'x - b'$ to approximate P', then for any point p_t in P', the vertical distance between p_t and the approximating line is at most ξ' if and only if p is contained in both half-spaces defined by p_t. Let H be the set of planes in $\mathcal{P}$ bounding the half-spaces in H_s. Let $\mathcal{A}$ denote the arrangement of the planes in H. We define the *q-level* of $\mathcal{A}$, denoted by $\mathcal{A}^q$, as the closure of the set of points that lie on the planes of H and have exactly q planes in H above them. Due to $|P'| > q$, no point on A^q can be lower than the plane $\xi = 0$. Thus, we have the following lemma.

Lemma 2. *If $p^* = (a^*, b^*, \xi^*)$ is the lowest point on $\mathcal{A}^q$, then the answer to the feasibility test is true if and only if $\xi^* \leq \epsilon$; further, if $\xi^* \leq \epsilon$, the function $y = a^*x - b^*$ can be used to approximate the points in P' such that the approximation error is at most ϵ and the number of outliers is at most q.*

Let h_ϵ be the plane $\xi = \epsilon$ in $\mathcal{P}$. By the preceding lemma, inserting (resp., deleting) a point into (resp., from) P' is equivalent to inserting (resp., deleting) two planes into (resp., from) H, and the feasibility test for P' can be done by checking whether the lowest point on $\mathcal{A}^q$ is above the plane h_ϵ. Fig. 1(b) shows a 2-D example. Therefore, SUB is reduced to the following *3-D q-level lowest point query* problem. Denote by H^* the planes defined by all points in P. Given $H \subseteq H^*$ ($H = \emptyset$ initially), design a data structure to support plane insertion (i.e., inserting a plane from $H^* \setminus H$ into H), plane deletion (i.e., deleting a plane from H), and q-level lowest point query (i.e., determining whether the lowest point on the q-level of the arrangement of H is above h_ϵ).

To solve it, we model the q-level lowest point query as follows. For each $h_i \in H$, denote by l_i the intersection line of h_ϵ and h_i. Note that every h_i intersects h_ϵ due to $u_i > 0$ and no plane in H is perpendicular to h_ϵ. Let h_ϵ^i be the *open* half-plane of h_ϵ that is bounded by l_i (but not including l_i) and is below the plane h_i. Let $L_H = \{l_i \mid h_i \in H\}$ and $H_\epsilon = \{h_\epsilon^i \mid h_i \in H\}$. Clearly, all lines in L_H lie on h_ϵ. Let $\mathcal{A}_{L_H}$ be the line arrangement of L_H. For each cell c_j of $\mathcal{A}_{L_{H^*}}$, we pick an arbitrary point $\hat{p}_j$ in its *interior*. Let $\hat{P}$ be the set of all such points and all vertices of $\mathcal{A}_{L_{H^*}}$. Then $|\hat{P}| = O(n^2)$.

Lemma 3. *For any $H \subseteq H^*$, let $\mathcal{A}$ be the arrangement of H. Then the lowest point p^* on $\mathcal{A}^q$ is below or on h_ϵ if and only if there is a point in $\hat{P}$ which is in at most q half-planes of H_ϵ.*

Given H_ϵ, for each $\hat{p}_j \in \hat{P}$, define its *weight* as the number of half-planes in H_ϵ which contain $\hat{p}_j$. Based on Lemma 3, we model the 3-D q-level lowest point query as the following 2-D *minimum point weight query* (MPWQ for short) problem: Given a half-plane set H_ϵ with $H_\epsilon = \emptyset$ initially, design a data structure for maintaining H_ϵ, to support half-plane insertion (i.e., insert a half-plane to H_ϵ), half-plane deletion (i.e., delete a half-plane from H_ϵ), and the *minimum point weight query*, i.e., finding in $\hat{P}$ the point with minimum weight.

By the STQ data structures in Theorems 2 and 3, we can easily handle the MPWQ problem. Note that an open half-plane can be considered as a special 2-D simplex. Let l be an arbitrary line on the plane. Denote by h_l^1 and h_l^2 the two (closed) half-planes bounded by l. Then, for each MPWQ query, we first apply two simplicial thickness queries on $\hat{P}$ (with respect to H_ϵ) using h_l^1 and h_l^2, respectively, and then report the minimum value of these two queries. In this way, we can obtain similar performance results for MPWQ as those in Theorems 3 and 2, and each MPWQ query takes roughly $O(\sqrt{n})$ time. By careful examining our STQ data structures, we can actually handle each MPWQ query in $O(1)$ time, as summarized in the following lemma.

Lemma 4. *Given n points, a data structure of $O(n)$ space can be built in $O(n^{1+\delta})$ time that supports the half-plane insertion, deletion and the MPWQ query each in $O(\sqrt{n})$, $O(\sqrt{n})$ and $O(1)$ time, respectively; another data structure of $O(n)$ space can be built in $O(n \log n)$ time that supports the three operations each in $O(\sqrt{n}(\log n)^{O(1)})$, $O(\sqrt{n}(\log n)^{O(1)})$ and $O(1)$ time, respectively.*

Since the 3-D q-level lowest point query problem can be modeled as an MPWQ problem instance in $O(n^2)$ time with a point set of size $O(n^2)$, each point insertion, deletion, and feasibility test in Lemma 1 can be handled in $O(n)$, $O(n)$, and $O(1)$ time, respectively, with an $O(n^{2+\delta})$ preprocessing time, or the three operations can be handled each in $O(n(\log n)^{O(1)})$, $O(n(\log n)^{O(1)})$, and $O(1)$ time, respectively, with an $O(n^2 \log n)$ preprocessing time. By Lemma 1, we have the following result.

Theorem 1. *VWPF is solvable in $O(\min\{n^{2+\delta} + n^2 g, n^2 g \log^{O(1)} n\})$ time.*

When $g = \Omega((\frac{n}{\log^2 n})^{\frac{1}{3}})$, the time bound of the above algorithm is $O(n^2 g)$, which is faster than the previously best-known $O(ng^4 \log^2 n)$ algorithm in [7].

3 Data Structure Based on Hierarchical Cuttings

In this section, we give an STQ data structure by modifying the data structure in [11], which is based on hierarchical cuttings [2]. The main result is given below.

Theorem 2. *Given a set P of n points in E^d, we can build a data structure in $O(n)$ space and $O(n^{1+\delta})$ time, which can support the simplex insertion, simplex deletion, and simplicial thickness query each in $O(n^{1-1/d})$ time for STQ.*

We first present a main data structure with performance bounds as stated in the above theorem for the case with $d > 2$. For the 2-D case, the main data structure still works but supports each operation in $O(\sqrt{n} \log n)$ time. To get rid of the $\log n$ factor, we derive an auxiliary data structure and incorporate it into the main data structure, which is the most challenging work in this section.

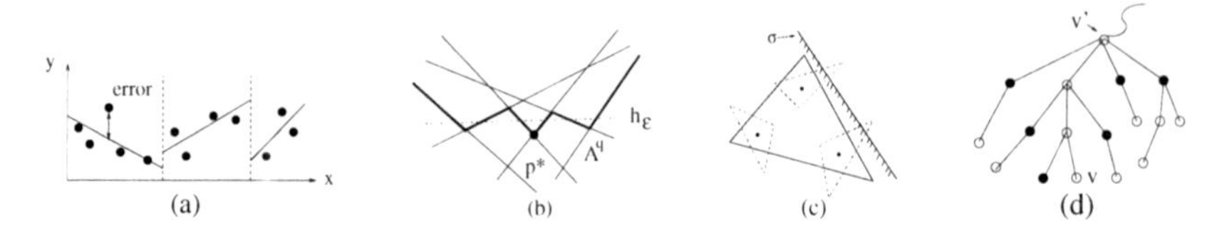

Fig. 1. (a) Illustrating a piecewise linear function for approximating points in VWPF. (b) The lowest point on $\mathcal{A}^q$ (the bold curve) is p^*. (c) The solid simplex (triangle) is a node v in Υ, the three points are super-points in $\hat{P}_v$, and the dashed (red) triangles are primary simplices. (d) Illustrating a simplex insertion. V' consists of the four nodes on the v-to-v' path plus $\hat{p}_i$ (not shown). The black nodes are siblings of the nodes in V'.

3.1 High Dimensional Space ($d > 2$)

To make the paper self-contained, we first sketch the data structure in [11]. Its main idea is to construct a partitioning scheme in the following manner: Given n points in E^d, divide the space into several regions, such that each region contains at least a constant fraction of the points and the boundary (i.e., a hyper-plane) of any half-space misses one (or several) of the regions. Given a query half-space hs, one can then treat the points in the regions missed by the boundary of hs very efficiently: Either they are all inside hs or are all outside. Thus, it remains to handle the points in the regions intersected by the boundary of hs. To this end, the partition scheme is applied recursively to the subsets of the points in each of these regions, until trivially small subsets (of $O(1)$ size) are reached. A data structure capturing this recursive partition of the point set in a suitable way is usually called a *partition tree*. In the query answering process, the regions missed by the boundary of the query half-space are handled directly, and the regions intersected by the boundary are processed recursively down the tree.

For a point set P of size n in E^d, the linear-space data structure [11], which supports each simplex range searching query in $O(n^{1-1/d})$ time, can be built in $O(n^{1+\delta})$ time. Using a similar presentation as in [11], we first describe a way of constructing a certain subset $P' \subseteq P$ with $|P'| \geq |P|/2$, and a simplex range searching data structure for P' with performance bounds as stated above. To obtain the data structure for the whole set P, first the construction for P' is performed, and then for $P \setminus P'$, in the same manner, a logarithmic number of data structures with geometrically decreasing sizes are produced. The space and construction time of the data structure for the whole set P are the same as stated above. Any query for P can be answered by querying each of the logarithmic number of data structures, with a total query time of $O(n^{1-1/d})$.

We now describe the construction of P' and the corresponding data structure Ψ for it. Ψ has a set of possibly unbounded (not necessarily disjoint) simplices, $\Psi_0 = \{s_1, \ldots, s_t\}$, with $t = n^{1/d} \log n$. For every $1 \leq i \leq t$, there is a set $P_i \subset P'$ of size at most $\frac{n}{2t}$, with $P_i \subset s_i$ (i.e., all points of P_i are contained in s_i). The sets P_i are disjoint and together form the set P'. For each s_i, there is a rooted tree T_i whose nodes are simplices, with s_i as the root. Each non-leaf simplex of T_i has $O(1)$ children, which are simplices with disjoint interior that together cover their parent. Each point of P_i is contained in exactly one leaf simplex of T_i. The depth of each tree T_i is at most $l = O(\log n)$. Hence, Ψ is essentially a forest of t trees, with each tree node corresponding to a possibly unbounded simplex. Let Ψ_b be the collection of all simplices that lie at distance b from the roots of their trees. For a simplex $s \in \Psi_b$, let $P(s) = P_i \cap s$, where T_i is the tree containing s. For any hyper-plane h, let $K_b(h)$ be the set of simplices in Ψ_b intersected by h, and $L_b(h)$ be the set of leaf simplices in $K_b(h)$. Let $K(h) = \cup_{b=0}^{l} K_b(h)$ and similarly for $L(h)$. It was shown in [11] that $\sum_{b=0}^{l} |\Psi_b| = O(n)$, and for any hyper-plane h, $|K(h)| = O(n^{1-1/d})$ and $\sum_{s \in L(h)} |P(s)| = O(n^{1-1/d})$. In the forest Ψ, each node s stores the *cardinality* of $P(s)$ (i.e., the number of points in $P(s)$), and each leaf simplex s also stores the list of points in $P(s)$. Ψ can be constructed in $O(n)$ space and $O(n^{1+\delta})$ time.

For a simplex range query σ, the number of points in $P' \cap \sigma$ is computed as follows. (1) Compute the total cardinality of the point sets P_i whose simplices $s_i \in \Psi_0$ are completely contained in σ; compute the set K' of all simplices in Ψ_0 intersected by the boundary of σ. (2) Repeat the following until $K' = \emptyset$: Remove a simplex s from K'; if s is a leaf simplex, test directly the membership of each point of $P(s)$ in σ; if s is a non-leaf simplex, determine the position of each child of s with respect to σ, add to K' those children poked by σ, and count the cardinalities of the point sets in the children of s that are completely contained in σ. To analyze the running time, since $|K(h)| = O(n^{1-1/d})$ and $\sum_{s \in L(h)} |P(s)| = O(n^{1-1/d})$ for any hyper-plane h, Step (2) takes $O(n^{1-1/d})$ time. For Step (1), if all simplices of Ψ_0 are inspected straightforwardly, it needs $O(n^{1/d} \log n)$ time. For $d > 2$, this is $o(n^{1-1/d})$; for $d = 2$, an auxiliary data structure in [10] can be utilized to reduce the time to $O(\sqrt{n})$.

Note that for the simplex range searching problem, Chazelle [3] gave some lower bounds: Under reasonable assumptions on the computation model, a data structure with $O(m)$ space can support a simplex range query in no better than $\Omega(n/m^{1/d} \log n)$ (resp., $\Omega(n/\sqrt{m})$ for $d = 2$) time. In view of these lower bounds, the data structure Ψ is optimal for $d = 2$, and quite likely for $d > 2$ as well.

Based on the above data structure, our STQ data structure is built as follows. Similarly, we first describe a construction of a subset $P' \subseteq P$ with $|P'| \geq |P|/2$, and a data structure for P' with performance bounds as stated in Theorem 2. To obtain the data structure for the whole set P, we first perform the construction for P', and then for $P \setminus P'$, produce a logarithmic number of data structures with geometrically decreasing sizes. The space and construction time of the data structure for P are as stated in Theorem 2. Every simplex insertion, deletion, or simplicial thickness query for P is processed by performing it on each of the logarithmic number of data structures, with performance as in Theorem 2.

For P and a dynamic set S of simplices, the construction of $P' \subseteq P$ and the corresponding STQ data structure Ψ' is the same as Ψ in [11], except that for each simplex $s \in \Psi'$, we store two key values $k_1(s)$ and $k_2(s)$. To define them, we need some new definitions. We say that a simplex s' *pokes* another simplex s'' if s'' is intersected by the boundary of s'. In other words, s' pokes s'' if and only if their boundaries intersect each other or s' is properly contained in s''. Note that s' poking s'' does not necessarily imply s'' also poking s'. Ψ' is actually a forest with t trees, $T_1, T_2, \ldots, T_t$. Let $parent(s)$ denote the parent simplex of a simplex $s \in \Psi'$. For each simplex s in a tree T_i, if s is the root s_i, then define $IS(parent(s))$ to be S; otherwise, define $IS(s)$ to be the set of simplices in $IS(parent(s))$ that poke s. Note that if s is an ancestor of s', then $IS(s') \subseteq IS(s)$. We define $k_1(s)$ to be the number of simplices of S in $IS(parent(s))$ that contain s completely; define $k_2(s)$ to be the minimum simplicial thickness among all points in $P(s)$ ($P(s) = P_i \cap s$, where $s \in T_i$) with respect to the simplices in $IS(s)$, i.e., $k_2(s) = \min_{p_i \in P(s)} \beta(p_i, IS(s))$. Further, for each point $p_i \in P(s)$ when s is a leaf simplex, we store $k_1(p_i)$ as the number of simplices in $IS(s)$ containing p_i, i.e., $k_1(p_i) = \beta(p_i, IS(s))$. Suppose the k_1 values of all simplices and points in Ψ' have been set correctly. Then the k_2 values of all simplices in Ψ' can be determined by the following lemma. For each non-leaf simplex $s \in \Psi'$, denote by $C(s)$ the set of its children.

Lemma 5. *For each non-leaf simplex* $s \in \Psi'$, $k_2(s) = \min_{s' \in C(s)}\{k_1(s') + k_2(s')\}$*; for each leaf simplex* $s \in \Psi'$, $k_2(s) = \min_{p_i \in P(s)}\{k_1(p_i)\}$.

Comparing with Ψ, since we add only $O(1)$ space to each node (for storing the k_1 and k_2 values), Ψ' uses $O(n)$ space. ($IS(s)$ is only a concept for analysis which is not computed explicitly.) As $S = \emptyset$ initially, all key values are initially zero; thus the construction time of Ψ' is the same as Ψ.

To insert a simplex σ, we do the following. For simplicity of discussion, if a point p_i is in $P(s)$ where s is a leaf simplex, we view p_i as a child of s. We call a simplex s (resp., a point p_i) in Ψ' an *ending node* with respect to σ if σ contains s (resp., p_i) completely but pokes the simplices of all its proper ancestors. In the simplex range query answering procedure on σ, a simplex or a point is an ending node if and only if all its ancestors will be visited and the procedure will not proceed onto its children. To insert the simplex σ, for each ending node in Ψ', we increase its k_1 value by 1. After the k_1 values of all ending nodes have been updated, we update the k_2 values of all involved simplices in Ψ' based on Lemma 5 in a bottom-up manner. Precisely, if s' is an ending node whose k_1 value has been increased, then the k_2 values of all its ancestors are updated accordingly by Lemma 5. To analyze the running time, recall that for any hyper-plane h, we have $|K(h)| = O(n^{1-1/d})$ and $\sum_{s \in L(h)} |P(s)| = O(n^{1-1/d})$. If a simplex s in Ψ' is in $K(h)$, then all its ancestors in Ψ' are in $K(h)$ as well. Further, since each non-leaf simplex in Ψ' has $O(1)$ children, the running time for updating the k_1 and k_2 values of all involved simplices in Ψ' is $O(n^{1-1/d})$ for a simplex insertion if we know all the ending nodes. The ending nodes can be obtained by the procedure for answering the simplex range query σ in Ψ. However, as we discussed above, in that procedure, since $|\Psi_0| = n^{1/d} \log n$, if we inspect straightforwardly all simplices in Ψ_0, it takes $O(n^{1/d} \log n)$ time. For $d > 2$, this is $o(n^{1-1/d})$; but for $d = 2$, it is $O(\sqrt{n} \log n)$. We will derive an auxiliary data structure Υ below to reduce the running time to $O(\sqrt{n})$ time for the 2-D case.

To delete a simplex, we decrease the k_1 value of each ending node by 1, and update the corresponding k_2 values. The running time is also $O(n^{1-1/d})$ for

$d > 2$ and $O(\sqrt{n}\log n)$ for $d = 2$. The simplicial thickness query σ is to compute $\beta(\sigma, S)$. Let $V(\sigma)$ (resp., $P(\sigma)$) denote the set of simplices (resp., points) in Ψ' that are ending nodes with respect to σ. We have the following lemma.

Lemma 6. *For each $s \in V(\sigma)$, denote by $A(s)$ the set of all ancestors of s in Ψ' including s, i.e., all nodes in the path from s to the root of the corresponding tree. For each $p_i \in P(\sigma)$, denote by $s(p_i)$ the leaf simplex such that $p_i \in P(s(p_i))$. Then the answer to the simplicial thickness query, i.e., $\beta(\sigma, S)$, is $\min\{\min_{s\in V(\sigma)}\{k_2(s)+\sum_{s'\in A(s)} k_1(s')\}, \min_{p_i\in P(\sigma)}\{k_1(p_i)+\sum_{s'\in A(s(p_i))} k_1(s')\}\}$.*

By Lemma 6, a simplicial thickness query σ on Ψ' can be handled as follows. We apply a similar procedure as for the simplex insertion, in which we accumulate the k_1 values from the root of the corresponding tree in Ψ' down to the simplices on each search path. As we encounter an ending node, if it is a simplex s, we compute the value $k_2(s)+\sum_{s'\in A(s)} k_1(s')$, and call it the *thickness-query value* of s. Since we have the value $\sum_{s'\in A(s)} k_1(s')$ available, the thickness-query value of s can be obtained in $O(1)$ time. If the ending node is a point p_i, we also compute its *thickness-query value* $k_1(p_i) + \sum_{s'\in A(s(p_i))} k_1(s')$ in $O(1)$ time. Further, we maintain the minimum thickness-query value among those that have already been computed. When the procedure stops, report the minimum thickness-query value as the answer to the simplicial thickness query. To analyze the running time, it takes $O(1)$ time at each node for either accumulating the k_1 values or computing the thickness-query value and maintaining the minimum one. Thus, the running time of a query is the same as a simplex insertion, i.e., $O(n^{1-1/d})$ for $d > 2$ and $O(\sqrt{n}\log n)$ for $d = 2$.

3.2 Auxiliary Data Structure for the Planar Case

We design an auxiliary data structure Υ for the 2-D case to organize the $t = \sqrt{n}\log n$ simplices in Ψ_0, so that all three operations (i.e., simplex insertion, simplex deletion, and simplicial thickness query) take $O(\sqrt{n})$ time each. Υ is built based on the partition tree given in [8]. Note that the auxiliary data structure in [11] is also a partition tree that was given in [10], but this particular partition tree does not work well for our STQ problem.

To make Υ work, given a simplex σ for any operation, we need to handled two questions efficiently. Q_1: Which simplices in Ψ_0 are poked by σ? Q_2: How should the simplices in Ψ_0 which are completely contained in σ be handled? Q_1 can be answered in $O(\sqrt{n})$ time by another data structure in [11]. Consequently, those simplices in Ψ_0 which are poked by σ can be handled by Ψ' in $O(\sqrt{n})$ time as before. Υ is then used to solve Q_2. Our goal is to construct Υ in $O(n^{1+\delta})$ time and $O(n)$ space to handle the three operations each in $O(\sqrt{n})$ time.

We pick an arbitrary point $\hat{p}_i$ in each simplex s_i of Ψ_0 as a *representing point* for s_i, and call $\hat{p}_i$ a *super-point*. Recall that each s_i corresponds to the point set P_i. Below, we call each simplex s_i in Ψ_0 a *primary simplex*. Denote by $\hat{P}$ the set of all t super-points. For $\hat{P}$, we build a partition tree Υ as in [8], which is built by applying the simplicial partition in [10] recursively. Each node $v \in \Upsilon$

corresponds to a simplex Δ_v and a super-point subset $\hat{P}_v$ with Δ_v containing $\hat{P}_v$. The simplex Δ_v may also contain other points in $\hat{P}$ than those in $\hat{P}_v$. Each internal node v in Υ has $O(1)$ children. The super-point subsets corresponding to the children of v are pairwise disjoint and form a partition of $\hat{P}_v$. If v is a leaf node, $\hat{P}_v$ has $O(1)$ super-points. The height of Υ is $O(\log t)$. It was shown in [8] that for any constant $\epsilon > 0$, Υ can be built in $O(t^{1+\epsilon})$ time and $O(t)$ space, and each simplex range query on $\hat{P}$ can be answered in $O(t^{1/2+\epsilon})$ time.

For each node v in Υ, we store two key values $k_1(v)$ and $k_2(v)$. To define $k_1(v)$ and $k_2(v)$, we need some new concepts. Denote by $\psi(v)$ the union of Δ_v and all primary simplices represented by the super-points in $\hat{P}_v$, i.e., $\psi(v) = \cup_{\hat{p}_i \in \hat{P}_v} s_i \cup \Delta_v$. We say that a simplex s' *pokes* $\psi(v)$ if $\psi(v)$ is intersected by the boundary of s' (i.e., the boundaries of s' and $\psi(v)$ intersect or s' is properly contained in $\psi(v)$). It should be noted that the concept of $\psi(v)$ is used only for analysis and is never computed in our algorithm. For each $v \in \Upsilon$, denote by $\pi(v)$ the union of all point sets corresponding to the primary simplices represented by the super-points in $\hat{P}_v$, i.e., $\pi(v) = \cup_{\hat{p}_i \in \hat{P}_v} P_i$. For each primary simplex $s_i \in \Psi_0$, denote by S_i the set of simplices in S that poke s_i. For each $v \in \Upsilon$, if v is the root, define $IS(v)$ to be S; otherwise, define $IS(v)$ to be the set of simplices of S in $IS(parent(v))$ that poke $\psi(v)$. Note that for any $\hat{p}_i \in \hat{P}$, if $\hat{p}_i$ is in $\hat{P}_v$, it must be $S_i \subseteq IS(v)$.

For each $v \in \Upsilon$, $k_1(v)$ is defined to be the number of simplices of S in $IS(parent(v))$ that contain $\psi(v)$ completely (if v is the root, let $k_1(v) = 0$); $k_2(v)$ is defined to be the minimum simplicial thickness among all points in $\pi(v)$ with respect to the simplex set $IS(v)$, i.e., $k_2(v) = \min_{p_j \in \pi(v)} \beta(p_j, IS(v))$. Further, for each super-point $\hat{p}_i$ in a leaf node v (i.e., $\hat{p}_i \in \hat{P}_v$), we store $k_1(\hat{p}_i)$ as the number of simplices in $IS(v)$ that contain s_i completely, and store $k_2(\hat{p}_i)$ as the minimum simplicial thickness among all points in P_i with respect to the simplex set S_i, i.e., $k_2(\hat{p}_i) = \min_{p_j \in P_i} \beta(p_j, S_i)$. For each tree T_i, let its root be $root(T_i)$. Also, for each tree T_i in Ψ', let $IS(root(T_i))$ be S_i and the key values in Ψ' be defined as above. Thus, for each super-point $\hat{p}_i$, $k_2(\hat{p}_i) = k_2(root(T_i))$. For simplicity of discussion, let $k_1(root(T_i)) = 0$ for each T_i. If the k_1 values of all nodes in Υ and the k_1 and k_2 values of all super-points of $\hat{P}$ have been set correctly, the k_2 values of all nodes in Υ can be obtained by the lemma below.

Lemma 7. *For each internal node $v \in \Upsilon$, $k_2(v) = \min_{v' \in C(v)}\{k_1(v') + k_2(v')\}$; for each leaf node $v \in \Upsilon$, $k_2(v) = \min_{\hat{p}_i \in \hat{P}_v}\{k_1(\hat{p}_i) + k_2(\hat{p}_i)\}$.*

Since $S = \emptyset$ initially, all k_1 and k_2 values in Υ are initially zero. Thus, as in [8], Υ can be built in $O(t^{1+\epsilon})$ time and $O(t)$ space, with $t = \sqrt{n} \log n$. If we set $\epsilon = 1/3$, the construction time for Υ is $O(n)$ (and each simplex range query can then be answered in $O(\sqrt{n})$ time). Next, we discuss how to handle the simplex insertion, simplex deletion, and simplicial thickness query in Υ and Ψ'.

Given a simplex σ for an operation (e.g., inserting σ), there are three types of primary simplices in Ψ_0 with respect to σ: (i) Those completely contained in σ, (ii) those disjoint from σ, and (iii) those poked by σ. The primary simplices of types (i) and (ii) are mainly handled by Υ, and those of type (iii) are mainly handled by Ψ'. We also need to determine which primary simplices (i.e., their types) should be handled in Υ, and which should be done in Ψ'.

To insert a simplex σ, we perform four steps. (1) Clearly, we need to update the k_1 values of some nodes due to the insertion of σ. One way is that for each node $v \in \Upsilon$ that we need to check, we first compute $\psi(v)$ and then determine whether σ pokes $\psi(v)$. However, since computing $\psi(v)$ is time-consuming, this method takes too much time. Instead, we use a faster procedure to update the k_1 values of the involved nodes in Υ but may make some errors, and then correct the errors in a way without computing $\psi(v)$ explicitly. Precisely, we update the key values of Υ in the same way as we did on Ψ', as follows. We increase $k_1(v)$ by 1 if σ contains Δ_v completely but pokes all its proper ancestor simplices in Υ, and similarly, increase $k_1(\hat{p}_i)$ by 1 if σ contains $\hat{p}_i$ but pokes all its ancestor simplices in Υ. We then update the corresponding k_2 values by Lemma 7. But, after this step, some key values may be incorrect. For example, in Fig. 1(c), suppose the simplex with solid edges is Δ_v such that σ contains Δ_v completely but pokes all its proper ancestor simplices in Υ. Thus, $k_1(v)$ is increased by 1 in Step (1). But, since σ actually pokes $\psi(v)$, by the definition of $k_1(v)$ in Υ, $k_1(v)$ should not be increased. This error occurs as we did not use the simplex union $\psi(v)$ in checking the poking relation. We will fix all errors in the later steps.

(2) Using the data structure in [10] (Lemma 5.3), find the set K' of all primary simplices in Ψ_0 that are poked by σ. (3) This step corrects all errors made in Step (1). We first discuss where these errors can occur. Clearly, all super-points of $\hat{P}_v$ for each $v \in \Upsilon$ are contained in Δ_v. Thus, if for every super-point $\hat{p}_i \in \sigma$, its simplex s_i is contained entirely in σ (i.e., $s_i \notin K'$), then all k_1 values in Υ have been set correctly. Suppose a primary simplex s_i is poked by σ (i.e., $s_i \in K'$) and $\hat{p}_i \in \sigma$. Let v be the leaf simplex in Υ such that $\hat{p}_i \in \hat{P}_v$, and V be the set of nodes in the path from v to $root(\Upsilon)$ plus $\hat{p}_i$ ($\hat{p}_i$ may be viewed as a node in V). If the k_1 value of any node in V is increased in Step (1), this is an error. Since σ contains $\hat{p}_i$, there must be one and only one node in V whose k_1 value is increased in Step (1). That is, after Step (1), the observation below holds.

Observation 1. *For each super-point $\hat{p}_i \in \Upsilon$, suppose $\hat{p}_i$ is in a leaf node v. If $\hat{p}_i$ is contained in the simplex σ, then there is one and only one node in V whose k_1 value is increased by 1 due to the insertion of σ, where V is the set of nodes in the path from v to $root(\Upsilon)$ plus the super-point $\hat{p}_i$.*

Thus, the errors in Step (1) are caused by the primary simplices in K' whose representing super-points are contained in σ. Denote by K'' the set of primary simplices in K' whose representing super-points are contained in σ.

We fix the errors as follows. Consider a primary simplex $s_i \in K''$ (i.e., $\hat{p}_i \in \sigma$), and let $\hat{p}_i$ be stored in a leaf node v of Υ. By following the path from v to $root(\Upsilon)$, we can find the only node v' whose k_1 value gets increased in Step (1) due to the insertion of σ. Denote by V' the set of nodes in the path from v to v' plus $\hat{p}_i$ (see Fig. 1(d)). Since σ pokes s_i, σ also pokes $\psi(v'')$ for each $v'' \in V$, implying that the k_1 values of all nodes in V should not be increased. But the value of $k_1(v')$ was actually increased. To correct this error while maintaining the validity of other key values in Υ, and in particular the validity of Observation 1 for those super-points whose primary simplices are not in K'', we do the following. (a) Decrease $k_1(v')$ by 1; (b) for every node in $V' \setminus \{v'\}$, increase the k_1 value of each of its siblings by 1 (see Fig. 1(d)) (for $\hat{p}_i$, we consider all other super-points in $\hat{P}_v$ as its siblings); (c) for each node in V, update its k_2 value by Lemma 7. After the above process, the error incurred due to σ poking the simplex s_i is corrected while the property of Observation 1 is still maintained (for other super-points minus $\hat{p}_i$). By performing this process for each simplex in K'', we correct all errors made in Step (1), i.e., the k_1 values of all nodes and super-points in Υ are then set correctly for the insertion of σ. Note that once the error on a $\hat{p}_i$-to-$root(\Upsilon)$ path gets corrected, the subsequent error correction process (for other simplices in K'') does not increase (wrongly) the k_1 value of any node on this path again. The reason is that the correction process always "pushes" the value increasing in the downward fashion in Υ and away from each such path. Further, based on the current k_2 value of each super-point, the k_2 values of all nodes in Υ are also set correctly.

(4) For each $s_i \in K'$, apply the insertion of σ on T_i, and at the end of this insertion, if the k_2 value of $root(T_i)$ is updated, then after setting $k_2(\hat{p}_i)$ to the new value of $k_2(root(T_i))$, update the k_2 value of each node in the path from v to the root of Υ by Lemma 7, where v is the leaf node with $\hat{p}_i \in \hat{P}_v$. After this step, for the insertion of σ, the key values of all nodes and super-points in Υ are set correctly. This finishes the simplex insertion operation for σ.

For Step (1), its running time is $O(\sqrt{n})$. For Step (2), with the data structure in [10] (whose preprocessing time is $O(n)$ due to $t = \sqrt{n}\log n$), K' can be computed in $O(\sqrt{n}+|K'|)$ time. As shown in [11], for any hyper-plane h, $K_j(h) = O(\sqrt{n} \cdot 4^{-(q-j)})$ for $0 \le j \le q$, where $q = \theta(\log n)$ [1]. Thus, we have $|K'| = |K_0(h)| = O(\sqrt{n} \cdot 4^{-q})$. Hence, Step (2) takes $O(\sqrt{n})$ time. For Step (3), since the height of Υ is $O(\log t)$ and each node in Υ has $O(1)$ children, for every $s_i \in K''$, it takes $O(\log t)$ time to update the corresponding k_1 values in Υ, and updating the corresponding k_2 values of all involved nodes also takes $O(\log t)$ time. Since $|K'| = O(\sqrt{n} \cdot 4^{-q})$, $|K'| \cdot \log t = O(\sqrt{n})$, and $K'' \subseteq K'$, this step takes $O(\sqrt{n})$ time. For Step (4), as analyzed before, the computation in all the trees of Ψ' takes $O(\sqrt{n})$ time. For each simplex in K', the procedure of updating the k_2 values of the corresponding nodes in Υ takes $O(\log t)$ time since each its node has $O(1)$ children. Thus, this step takes $O(\sqrt{n})$ time due to $|K'| \cdot \log t = O(\sqrt{n})$. In summary, the simplex insertion takes $O(\sqrt{n})$ time.

The simplex deletion proceeds similarly, except changing the key value increasing (resp., decreasing) to decreasing (resp., increasing), which takes $O(\sqrt{n})$ time.

For each simplicial thickness query σ, suppose we apply σ as for a simplex insertion on Υ and Ψ', and denote by $V(\sigma)$ the set of nodes in Υ whose k_1 values are increased, and by $\hat{P}(\sigma)$ the set of super-points in Υ whose k_1 values are increased. One may view the nodes in $V(\sigma)$ and the super-points in $\hat{P}(\sigma)$ as the *ending nodes* with respect to σ. Denote by K' the set of primary simplices in

[1] It is shown $q = O(\log n)$ explicitly in [11]. By a careful analysis, we can also obtain $q = \Omega(\log n)$.

Ψ_0 that are poked by σ. To compute $\beta(\sigma, S)$, we use the lemma below. For each $s_i \in K'$, denote by P_i' the set of points in P_i that are contained in σ.

Lemma 8. *For each $v \in V(\sigma)$, denote by $A(v)$ the set of all ancestors of v in Υ and $A(v)$ also contains v. For each $\hat{p}_i \in \hat{P}(\sigma)$, denote by $v(\hat{p}_i)$ the leaf node of Υ such that $\hat{p}_i \in \hat{P}_{v(\hat{p}_i)}$. Then the answer to the simplicial thickness query, i.e., $\beta(\sigma, S)$, is $\min\{\beta_1, \beta_2, \beta_3\}$, where $\beta_1 = \min_{v \in V(\sigma)}\{k_2(v) + \sum_{v' \in A(v)} k_1(v')\}$, $\beta_2 = \min_{\hat{p}_i \in \hat{P}(\sigma)}\{k_1(\hat{p}_i) + k_2(\hat{p}_i) + \sum_{v' \in A(v(\hat{p}_i))} k_1(v')\}$, and $\beta_3 = \min_{s_i \in K'} \min_{p_j \in P_i'} \beta(p_j, S)$.*

By the above lemma, for each simplicial thickness query on σ, we apply a procedure as for simplex insertion and accumulate the sum of the k_1 values from the root of Υ down to the current node in Υ visited by the procedure. In this way, β_1 and β_2 can be obtained in the same time as the simplex insertion operation. For β_3, it can be obtained in the same way as in Lemma 6. Thus, it takes $O(\sqrt{n})$ time to answer each simplicial thickness query. Theorem 2 thus follows.

4 Data Structure Based on Simplicial Partitions

Our second STQ data structure is based on the simplicial partitions in [10]. A *simplicial partition* of a point set P is a collection $\Pi = \{(P_1, \Delta_1), \ldots, (P_m, \Delta_m)\}$, where the P_i's are pairwise disjoint subsets (called the *classes* of Π) forming a partition of P, and each Δ_i is a possibly open simplex containing the set P_i. The *size* of Π is m. The simplex Δ_i may also contain other points in P than those in P_i. A simplicial partition is called *special* if $\max_{1 \leq i \leq m}\{|P_i|\} < 2 \min_{1 \leq i \leq m}\{|P_i|\}$, i.e., all the classes are of roughly the same size.

The data structure in [10] is a partition tree, denoted by T, based on constructing special simplicial partitions on P recursively. Given n points in E^d for any $d \geq 2$, T can be built in $O(n)$ space and $O(n \log n)$ time, and can answer every simplex range query in $O(n^{1-1/d}(\log n)^{O(1)})$ time [10]. Our modifications on T are similar as before. Namely, we define the k_1 and k_2 values on each node in T in the same way as in Section 3 but the auxiliary data structure is not needed. Due to the space limit, we omit all the details and only give the following result.

Theorem 3. *Given n points in E^d, we can build a data structure in $O(n)$ space and $O(n \log n)$ time, which can support the simplex insertion, simplex deletion, and simplicial thickness query each in $O(n^{1-1/d}(\log n)^{O(1)})$ time for STQ.*

5 Handling the Weighted STQ and Max-STQ

It is straightforward to extend our STQ data structures to the weighted STQ problem. Precisely, we change the definitions of the k_1 and k_2 values by taking into account the simplex weights. For example, in the data structure Ψ' in Section 3, to solve the weighted case, we define $k_1(s)$ as the weight sum of the simplices of S in $IS(parent(s))$ that contain the simplex s completely and define $k_2(s)$ as the minimum weighted simplicial thickness among all points in $P(s)$ with respect to the simplex set $IS(s)$. The computation of the k_2 values still follows Lemma 5. To insert a simplex σ, whenever the k_1 value of a simplex s needs to be increased, instead of increasing it by 1, we increase it by the weight of σ. The simplex deletion and weighted simplicial thickness query are handled similarly. Thus, the results in Theorem 2 still apply to the weighted case. By a similar extension, the results in Theorem 3 also apply to the weighted STQ.

For the max-STQ, the same results in Theorems 2 and 3 can also be obtained by replacing the *min* definitions and operations in the data structures and algorithms by the corresponding *max* definitions and operations.

References

1. Bentley, J.L.: Decomposable searching problems. Information Processing Letters 8, 244–251 (1979)
2. Chazelle, B.: Cutting hyperplanes for divide-and-conquer. Discrete & Computational Geometry 9(2), 145–158 (1993)
3. Chazelle, B.: Lower bounds on the complexity of polytope range searching. J. Amer. Math. Soc. 2(4), 637–666 (1999)
4. Chazelle, B., Sharir, M., Welzl, E.: Quasi-optimal upper bounds for simplex range searching and new zone theorems. In: Proc. of the 6th ACM Symposium on Computational Geometry, pp. 23–33 (1990)
5. Chazelle, B., Welzl, E.: Quasi-optimal range searching in spaces of finite vc-dimension. Discrete & Computational Geometry 4(5), 467–489 (1989)
6. Chen, D.Z., Wang, H.: Approximating points by a piecewise linear function: I. In: Dong, Y., Du, D.-Z., Ibarra, O. (eds.) ISAAC 2009. LNCS, vol. 5878, pp. 224–233. Springer, Heidelberg (2009)
7. Chen, D.Z., Wang, H.: Approximating points by a piecewise linear function: II. Dealing with outliers. In: Dong, Y., Du, D.-Z., Ibarra, O. (eds.) ISAAC 2009. LNCS, vol. 5878, pp. 234–243. Springer, Heidelberg (2009)
8. de Berg, M., van Kreveld, M., Overmars, M., Schwarzkopf, O.: Computational Geometry – Algorithms and Applications, 1st edn. Springer, Berlin (1997)
9. Karras, P., Sacharidis, D., Mamoulis, N.: Exploiting duality in summarization with deterministic guarantees. In: Proc. of the 13th International Conference on Knowledge Discovery and Data Mining, pp. 380–389 (2007)
10. Matoušek, J.: Efficient partition trees. Discrete and Computational Geometry 8(3), 315–334 (1992)
11. Matoušek, J.: Range searching with efficient hierarchical cuttings. Discrete & Computational Geometry 10(1), 157–182 (1993)
12. O'Rourke, J.: An on-line algorithm for fitting straight lines between data ranges. Commun. of ACM 24, 574–578 (1981)
13. Overmars, M.H.: The design of dynamic data structures. Springer, Berlin (1983)
14. Willard, D.E.: Polygon retrieval. SIAM Journal on Computing 11, 149–165 (1982)

The Cover Time of Cartesian Product Graphs

Mohammed Abdullah, Colin Cooper, and Tomasz Radzik

Department of Computer Science, King's College London
mohammed.abdullah,colin.cooper,tomasz.Radzik@kcl.ac.uk

Abstract. Let $P = G\Box H$ be the cartesian product of graphs G, H. We relate the cover time $\mathbf{COV}[P]$ of P to the cover times of its factors. When one of the factors is in some sense larger than the other, its cover time dominates, and can become of the same order as the cover time of the product as a whole. Our main theorem effectively gives conditions for when this holds. The probabilistic technique which we introduce, based on the blanket time, is more general and may be of independent interest, as might some of our lemmas.

Keywords: Random walks, cover time, blanket time, effective resistance, cartesian product graphs.

1 Introduction

For a connected graph Let G, denote by $V(G)$ and $E(G)$ the vertex and edge set respectively. The *vertex cover time* $\mathbf{COV}[G]$ of G is defined as the expected time it takes a random walk to visit all vertices of the graph, maximised over all possible starting vertices. This quantity is a fundamental area in the study of random walks has been extensively studied giving rise to a large body of theory and application. Let $n = |V(G)|$ and $m = |E(G)|$. It is a classic result of Aleliunas, Karp, Lipton, Lovász and Rackoff [1] that $\mathbf{COV}[G] \leq 2m(n-1)$. It was shown by Feige [7], [8], that for any connected graph G, the cover time satisfies $(1-o(1))n\log n \leq \mathbf{COV}[G] \leq (1+o(1))\frac{4}{27}n^3$. Between these two extremal examples, the cover time, both exact and asymptotic, has been determined for a number of different classes of graphs.

In this work, we study the cover time of the *cartesian product* P of two graphs G, H defined as follows:

Definition 1. *The* cartesian product $P = G\Box H$ *of finite connected graphs G, H, is the graph such that*

- $V(P) = V(G) \times V(H)$
- $((a,x),(b,y)) \in E(P)$ *if and only if either*
  - $(a,b) \in E(G)$ *and* $x = y$, *or*
  - $a = b$ *and* $(x,y) \in E(H)$

For a natural number d, we denote by G^d the d'th cartesian power, that is, $G^d = G$ when $d = 1$ and $G^d = G^{d-1}\Box G$ when $d > 1$. We can think of $P = G\Box H$

C.S. Iliopoulos and W.F. Smyth (Eds.): IWOCA 2010, LNCS 6460, pp. 377–389, 2011.

in terms of the following construction: We make a copy of one of the graphs, say G, once for each vertex of the other, H. For the copy of G corresponding to vertex $x \in V(H)$, G_x, and a vertex $a \in V(G)$, we add an edge from $a \in G_x$ to $a \in G_y$ for all vertices $y \in V(H)$ such that $(x, y) \in E(H)$.

In the following, if z is a parameter, let z_G represent that parameter for a graph G. We have the following : n the number of vertices; m the number of edges; δ the minimum degree; θ average degree; Δ the maximum degree; D diameter.

In this paper we prove the following

Theorem 1. *Let $P = G \Box H$ where G, H are any connected, finite graphs. We have*

$$\max\left\{\left(\frac{\delta_G}{\Delta_H}+1\right)\mathbf{COV}[H], \left(\frac{\delta_H}{\Delta_G}+1\right)\mathbf{COV}[G]\right\} \leq \mathbf{COV}[P] \tag{1}$$

and

$$\mathbf{COV}[P] \leq K \min\left\{\left(1+\frac{\Delta_G}{\delta_H}\right)\mathbf{COV}[H] + \frac{M m_G m_H n_H l^2}{\mathbf{COV}[H] D_G},\right.$$
$$\left.\left(1+\frac{\Delta_H}{\delta_G}\right)\mathbf{COV}[G] + \frac{M m_G m_H n_G l^2}{\mathbf{COV}[G] D_H}\right\} \tag{2}$$

Where $M = |E(P)|$, $l = \log D_G \log(n_G D_G)$ and K is a universal constant.

This extends much work done on the particular case of the two-dimensional toroid on n^2 vertices, i.e., $\mathbb{Z}_n \Box \mathbb{Z}_n$ where $\mathbb{Z}_n$ is the n-vertex cycle, and on powers G^d done by [9]. To prove Theorem 1, we present a framework to analyse the cover time of a random walk on a graph which works by dividing the graph up into (possibly overlapping) regions, analysing the behaviour of the walk when locally observed on those regions, and then composing the analyses of all the regions over the whole graph. The technique facilitates the analysis of the local observation on a region by relating it to a walk on a graph derived from that region. Thus the analysis of the whole graph is reduced to analysis of outcomes on local regions and subsequent compositions of those outcomes. This framework can be applied more generally than cartesian products. Some of the lemmas we use may be of independent interest. In particular, Lemmas 7 and 8 provided bounds on effective resistances of graph products that extend well-known and commonly used bounds for the $n \times n$ lattice graph.

Our paper uses the very recently proved conjecture that the blanket time of a graph is within a universal constant factor of the cover time. The *blanket time* $\mathbf{B}[G]$ of a graph G, introduced in [12], is the expected time of the random walk on G not only to visit every vertex, but to visit all vertices moreless uniformly (the exact definition given in 2.2). Our analysis is an example of how to exploit the relation $\mathbf{B}[G] = O(\mathbf{COV}[G])$. The lower bound in Theorem 1 implies that $\mathbf{COV}[G \Box H] \geq \mathbf{COV}[H]$, and the upper bound can be viewed as providing conditions sufficient for $\mathbf{COV}[G \Box H] = O(\mathbf{COV}[H])$. For example,

$\mathbf{COV}[\mathbb{Z}_p \Box \mathbb{Z}_q] = \Theta(\mathbf{COV}[\mathbb{Z}_q]) = \Theta(q^2)$ subject to the condition $p \log^2 p = O(q)$. Thus for this example, the lower and upper bounds in Theorem 1 are within a constant factor.

2 Preliminaries

2.1 Some Notation

We make use of the following notation: For a graph G let $V(G)$ and $E(G)$ denote the vertex and edge set of G respectively. For a random variable A representing a function of a walk, and a vertex $u \in V(G)$ let $\mathbf{E}_u[A]$ represent the expectation of A when the walk starts at u. Let $\tau(u)$ be a random variable representing the first time that u is visited by the walk and $\kappa(G)$ the first time every vertex in G has been visited by the walk. $\mathbf{H}[u,v] = \mathbf{E}_u[\tau(v)]$ is the hitting time from u to v, $\mathbf{COM}[u,v] = \mathbf{H}[u,v] + \mathbf{H}[v,u]$ is the commute time between u and v. $\mathbf{COV}[G] = \max_{u \in V(G)} \mathbf{E}_u[\kappa(G)]$ is the cover time of a graph G. Let $\mathbf{H}^+[G] = \max_{u,v} \mathbf{H}[u,v]$. The function $d(u)$ gives the degree of vertex u. For clarity, and because a vertex u may be considered in two different graphs, we may use $d_G(u)$ to explicitly denote the degree of u in graph G.

L_n denotes the n'th harmonic number, that is, $L_n = \sum_{i=1}^{n} 1/i$. Note $L_n = \log n + \gamma + O(1/n)$ Where $\gamma \approx 0.577$. In this paper all logarithms are base-e.

2.2 Blanket Time

Definition 2 ([12]). *For a graph G, and $\delta \in [0,1)$ define the random variable $B_\delta(G) = \min\{t : (\forall v) N_v(t) > \delta \pi_v t\}$ where $N_v(t)$ is the number of times v has been visited by time t and π_v is the stationary probability of vertex v. The* blanket time $\mathbf{B}_\delta[G] = \max_{v \in V(G)} \mathbf{E}_v[B_\delta(G)]$.

The following was very recently proved.

Theorem 2 ([5]). *For any graph G, and any $\delta \in (0,1)$, we have*

$$\mathbf{B}_\delta[G] \leq c(\delta)\mathbf{COV}[G] \tag{3}$$

Where the constant $c(\delta)$ depend only on δ.

As stated in [12], this is equivalent to saying that the expected time until each vertex v is visited $\pi_v \mathbf{COV}[G]$ times - which we shall refer to as the *blanket-cover criterion* - is $O(\mathbf{COV}[G])$.

2.3 Random Walks and Electrical Networks

We give some key facts and ideas relevant to this work, drawing on [10], which discusses electrical network theory in the wider context of Markov chains. [6] is the classical treatment. Consider a finite, connected graph $G = (V,E)$ with edge weights $\{c(e) : e \in E\}$. For a vertex u, let $c(u) = \sum_{v:(u,v) \in E} c(u,v)$ with each loop counted once, and let $c(G) = \sum_{u \in V} c(u)$. In the language of electrical network theory, the weight $c(u,v)$ is known as the *conductance* of the edge (u,v), and the *resistance* $r(u,v) = 1/c(u,v)$. The random walk on G defined by the transition matrix $P(u,v) = c(u,v)/c(u)$ defines a reversible Markov chain with the vertices of G as states and the transition matrix P. The stationary distribution is $\pi(u) = c(u)/c(G)$. Conversely, every reversible Markov chain can be shown to be a network. Thus the two are equivalent. A *flow* f is an asymmetric function on oriented edges, i.e., for $(u,v) \in E(G)$, $f(u,v) = -f(v,u)$ and *net flow* $f(u)$ at a vertex u is $\sum_{v:(u,v)\in E(G)} f(u,v)$. We note $\sum_{u\in V(G)} f(u) = 0$. For vertices a, z, a *flow from* a *to* z f is a flow with the additional properties that (i) $f(u) = 0$ for all $u \in V(G)\backslash\{a,z\}$ and (ii)the *strength* of the flow $f(a) \geq 0$. The *energy* $\mathcal{E}(f)$ of a flow from a to z, f is defined as $\mathcal{E}(f) = \sum_{e\in E} f^2(e)r(e)$ where the sum is over unoriented edges (i.e., each edge is considered once). We have the following

Lemma 1 (Thomson's principle). *For any finite connected graph, the* effective resistance *$R(a,z)$ between a and z is such that*

$$R(a,z) = \min\{\mathcal{E}(f) : f \text{ is a unit flow from } a \text{ to } z\}. \tag{4}$$

There is unique minimiser in the above known as the current flow.

This allows us to say that the energy of any unit flow we contruct is an upperbound on effective resistance. The following facts are useful. **Series law** Edges $(a,b), (b,c)$ with can be replaced by a single edge (a,c) with $r(a,c) = r(a,b) + r(b,c)$ if there are no other edge incident on b. **Parallel law** Parallel edges $(a,b)_1, (a,b)_2$ can be replaced by a single edge (a,b) with $c(a,b) = c((a,b)_1) + c((a,b)_2)$. **Shorting law** Adding an edge of zero resistance between two vertices is equivalent to merging them into one vertex, and cannot increase effective resistance anywhere in the network. **Cutting law** Removing an edge with positive conductance cannot decrease effective resistance anywhere in the network. **Monotonicity law** The effective resistance between two given vertices is monotonic in the resistances of the edges in the whole network.

The $k \times k$ lattice graph P_k^2, where P_k is the k-path, plays an important role in our work. We shall analyse random walks on subgraphs isomorphic to this structure. It is well known in the literature (see, e.g. [10]) that for any pair of vertices $u, v \in V(P_k^2)$, we have $R(u,v) \leq C \log k$ where C is some universal constant. We shall quote part of [9] Lemma 3.1 in our notation and refer the reader to the proof there.

Lemma 2 ([9], Lemma 3.1). *(a) Let u and v be any two vertices of P_k^2. Then $R(u,v) < 8L_k$, where L_k is the k'th harmonic number.*

The following important lemmas are widely used in the field

Lemma 3 ([3]). *For vertices $u, v \in V(G)$*

$$\mathbf{COM}[u,v] = c(G)R(u,v) \tag{5}$$

Lemma 4 ([11]). *For a finite connected graph G, (a)* $\mathbf{COV}[G] \leq \mathbf{H}^+[G]L_n$.

3 Related Work

A d-demensional torus on $N = n^d$ vertices is the d'th power of an n-cycle, $\mathbb{Z}_n^d$. The behaviour of random walks on this structure is well studied. It is well-known (see, e.g., [10]) that $\mathbf{COV}[\mathbb{Z}_n^d] = \Theta(N(\log N)^2)$ when $d = 2$ and $\mathbf{COV}[\mathbb{Z}_n^d] = \Theta(N \log N)$ when $d \geq 3$. [4] gives $\mathbf{COV}[\mathbb{Z}_n^2] \sim \frac{1}{\pi} N(\log N)^2$. [9] extends the study of graph powers giving the following theorem

Theorem 3 ([9], Theorem 1.2). *Let $G = (V, E)$ be any connected graph on n vertices with $\theta_G = 2|E|/n$. Let $d \geq 2$ be an integer and let $N = n^d$. For $d = 2$, $\mathbf{COV}[G^d] = O(\theta_G N(\log N)^2)$ and for $d \geq 3$, $\mathbf{COV}[G^d] = O(\theta_G N \log N)$. These bounds are tight.*

[2] gives a number of theorems related to random walks and effective resistance between pairs of vertices in graph products. To give the reader a flavour we quote Theorem 1 of that paper, which is useful as a lemma implicitly in this paper and in the proof of [9] Theorem 1.2 to justify the intuition that the effective resistance is maximised between opposite corners of the square lattice.

Lemma 5 ([2], Theorem 1). *Let P_n be an n-vertex path with endpoints x and y. Let G be a graph and let a and b be any two distinct vertices of G. Consider the graph $G \times P_n$. The effective resistance $R((a,x),(b,v))$ is maximised over vertices v of P_n at $v = y$.*

For P_n^2 this is used twice: $R((0,0),(r,s)) \leq R((0,0),(n-1,s)) \leq R((0,0),(n-1,n-1))$.

4 Locally Observed Random Walk

Let $G = (V, E)$ be a connected, unweighted (equiv., uniformly weighted) graph. Let $S \subset V$ and let $G[S]$ be the subgraph of G induced by S. Let $B = \{v \in S : \exists x \notin S, (v, x) \in E\}$. Call B the *boundary* of S, and the vertices of $V \backslash S$ *exterior vertices.* If $v \in S$ then $d_G(v)$ (the degree of v in G) is partitioned into $d(v, in) = |N(v, in)| = |N(v) \cap S|$ and $d(v, out) = |N(v, out)| = |N(v) \cap (V - S)|$, (*inside* and *outside* degree). Here $N(v)$ denotes the neighbour set of v.

Let $u, v \in B$. Say that u, v are *exterior-connected* if there is a (u, v)-path $u, x_1, ...x_k, v$ where $x_i \in V \backslash S$, $k \geq 1$. Thus all vertices of the path except u, v are exterior, and the path contains at least one exterior vertex. Let $A(B) = \{(u, v) :$ u, v are exterior-connected $\}$. Note $A(B)$ may include self-loops.

Call edges of $G[S]$ *interior*, edges of $A(B)$ *exterior*. We say that a walk $\omega = (u, x_1, ...x_k, v)$ on G is an exterior walk if $u, v \in S$ and $x_i \notin S$, $1 \leq i \leq k$.

We derive a weighted multi-graph H from G and S as follows: $V(H) = S$, $E(H) = E(G[S]) \cup A(B)$. Note if $u, v \in B$ and $(u, v) \in E$ then $(u, v) \in E(G[S])$, and if, furthermore, u, v are exterior connected, then $(u, v) \in A(B)$ *and these edges are distinct*, hence, H may not only have self-loops but also parallel edges, ie., $E(H)$ is a multiset.

Associate with an orientation $(\boldsymbol{u}, \boldsymbol{v})$ of an edge $(u, v) \in A(B)$ the set of all exterior walks $\omega = (u, x_1, ...x_k, v)$, $k \geq 1$ that start at u and end at v, and associate with each such a walk the value $p(\omega) = 1/(d_G(u)d_G(x_1)...d_G(x_k))$ (note, the $d(x_i)$ is not ambiguous, since $x_i \notin E(H)$, but we leave the 'G' subscript in for clarity). This is precisely the probability that the walk ω is taken by a simple random walk on G starting at u. Let

$$p_H(\boldsymbol{u}, \boldsymbol{v}) = \sum_{k \geq 1} \sum_{\omega=(u,x_1...x_k,v)} p(\omega), \tag{6}$$

where the sum is over all exterior walks ω.

We set the edge conductances (weights) of H as follows: If e is an interior edge, $c(e) = 1$. If it is an exterior edge $e = (u, v)$ define $c(e)$ as

$$c(e) = d_G(u)p_H(\boldsymbol{u}, \boldsymbol{v}) = \sum_{k \geq 1} \sum_{\omega=(u,x_1...x_k,v)} \frac{1}{d_G(x_1)...d_G(x_k)} = d_G(v)p_H(\boldsymbol{v}, \boldsymbol{u}) \tag{7}$$

Thus the edge weight is consistent. A weighted random walk on H is thus a finite reversible Markov chain with all the associated properties that this entails.

Definition 3. *The weighted graph H derived from (G, S) is termed* the local observation of G at S, *or* G locally observed at S. *We shall denote it as $H = Loc(G, S)$.*

The intuition in the above is that we wish to observe a random walk $\mathcal{W}(G)$ on a subset S of the vertices. When $\mathcal{W}(G)$ makes an external transition at the border B, we cease observing and resume observing if/when it returns to the border. It will thus appear to have transitioned a virtual edge between the vertex it left off and the one it returned on. It will therefore appear to be a weighted random walk on H. This equivalence is formalised thus

Definition 4. *Let G be a graph and $S \subset V(G)$. For a (unweighted) random walk $\mathcal{W}(G)$ on G starting at $x_0 \in S$, derive the Markov chain $\mathcal{M}(G, S)$ on the states of S as follows: (i) $\mathcal{M}(G, S)$ starts on x_0 (ii) If $\mathcal{W}(G)$ makes a transition through an internal edge (u, v) then so does $\mathcal{M}(G, S)$ (iii)If $\mathcal{W}(G)$ takes an exterior walk $\omega = (u, x_1...x_k, v)$ then $\mathcal{M}(G, S)$ remains at u until the walk is complete and subsequently transitions to v. We call $\mathcal{M}(G, S)$* the local observation of $\mathcal{W}(G)$ at S, *or* $\mathcal{W}(G)$ locally observed at S.

Lemma 6. *For a walk $\mathcal{W}(G)$ and a set $S \subset V(G)$, the local observation of $\mathcal{W}(G)$ at S, $\mathcal{M}(G, S)$ is equivalent to the weighted random walk $\mathcal{W}(H)$ where $H = Loc(G, S)$.*

Proof. See Appendix.

5 A General Bound

We give **COV**$[P]$ bounds in terms of H, and by symmetry, Theorem 1 can be inferred. The lower bound is easy: It is clear that the H dimension needs to be covered - that is, each copy of G needs to be visited at least once. The probability of moving through the H dimension is at least $\frac{\Delta_H}{\Delta_H+\delta_G}$, and the lower bound follows.

For the upper bound, we first require the following lemmas. Denote by $R_{max}(G)$ the maximum effective resistance between any pair of vertices in a graph G.

Lemma 7. *For a graph G and tree T, $R_{max}(G\Box T) < 4R_{max}(G\Box P_r)$ where $|V(T)| \leq r \leq 2|V(T)|$ and P_r is the path on r vertices.*

Proof. Note first the following: (i) By the parallel law, an edge (a, b) of unit resistance can be replaced with two parallel edges between a, b, each of resistance 2. (ii) By the shorting law, a vertex a can be replaced with two vetices a_1, a_2 with a zero-resistance edge between them and the ends of edges incident on a disributed arbitrarily between a_1 and a_2. These transformations preserve electrical properties of a network.

Let $F = G\Box T$. Starting from some vertex v in T, perform a depth-first search (DFS) of T stopping when all vertices in T have been visited. Each edge of T is traversed at most twice; once in each orientation (though a particular vertex x will be visited up to $d(x)$ times). Let (e_i) be the sequence of oriented edges generated by the search. The idea is to use (e_i) to construct a transformation from $F = G\Box T$ to $G\Box P_r$. From (e_i), we derive another sequence (a_i), which is generated by following (e_i) and if we have edges e_i, e_{i+1} with $e_i = (a, b)$, $e_{i+1} = (b, c)$ such that it is neither the first time nor the last time b is visited in the DFS, then we replace e_i, e_{i+1} with (a, c). We term such an operation an *aggregation.* Consider F; by (i) above we can replace each unit-resistance edge by a pair of parallel edges each of resistance 2. For a pair of parallel edges in the T dimension, arbitrarily label one of them with an orientation, and label the other with the opposite orientation. Note, orientations are only an aid to the proof, and are not a flow restriction. We therefore see that (e_i) can be interpreted as a sequence of these parallel oriented edges. Now we modify F using (a_i): If $(a, b), (b, c)$ was aggregated to (a, c), then replace each pair of **oriented** edges $((x.a), (x, b))$ and $((x, b)(x, c))$ in F with an oriented edge $((x, a), (x, c))$ and set the resistance of it the sum of the resistances of the replaced edges. This operation is the same as restricting flow through $((x.a), (x, b))$ and $((x, b)(x, c))$ to only going from one to the other at vertex (x, b), without the possiblity of going through other edges, The infimum of this subset of flows is at least the infimum of the previous set and so by Thomson's principle, the effective resistance cannot be decreased by this operation.

For each copy of G in F excluding those that do not correspond to a leaf of T, by(ii), we can do the following: Create a "twin" copy by associating with each vertex $x \in V(F)$ (except those excluded) a twin vertex x', putting a zero-resistance edge between x and x'. We then (a) redistribute the parallel edges in the G dimension so as to preserve structural isomorphism between each copy and G and (b) redistribibute edges in the T dimension so as to respect the sequence (a_i). This means that when we trace (a_i) via any vertex $x \in V(G)$, then if we have $(a, b), (b, c)$ in (a_i), we must have the corresponding path of oriented edges $((x, a), (x, b)), ((x, b), (x, c))$. We then remove the zero-resistance edges between each pair of twin vertices, and by Rayleigh's cutting law, this cannot decrease the effective resistance. Using the sequence (a_i) to trace a path of copies of G along the T dimension, we see that the resulting structure is isomorphic to $G\Box P_r$. Since the aggregation process only aggregates edges that pass through a previously seen vertex, r is at least k. Also, because each edge is traversed at most once in each direction, r is at most $2k$. Each edge has resistance at most 4, and so the theorem follows.

Lemma 8. *For graphs G, H with with $D_G = k$ and $k \leq n_H \leq Rk$, $R_{max}(G\Box H) < 32(3+R)L_k \leq \zeta R \log D_G$ where L_k is the k'th harmonic number and ζ is some universal constant.*

Proof. Let $(a, x), (b, y)$ be any two vertices in $G\Box H$. Let D be some diametric path of G. Let $\langle a, D\rangle$ represent the shortest path from a to D in G (which may trivially be a if it is on D). Similarly with $\langle b, D\rangle$. Let $T_D = D \cup \langle a, D\rangle \cup \langle b, D\rangle$. Note $k \leq |V(T_D)| \leq 3k$. Now let T_H be any spanning tree of H. Applying Lemma 7 twice we have

$$R_{max}(T_D\Box T_H) < 4R_{max}(T_D\Box P_r) < 16R_{max}(P_r\Box P_s) \tag{8}$$

where $k \leq r \leq 6k$ and $k \leq s \leq 2Rk$. Considering a series of connected P_k^2 subgraphs and using the triangle inequality for effective resistance, we have $R_{max}(P_r\Box P_s) \leq 32(3+R)L_k$ and the theorem follows.

Lemma 8 gives us an upperbound of $\zeta \log D_G$ for the effective resistance in a block (definition below), which in turn allows us to bound the maximum hitting time within a block, and therefore the cover time via Matthews' technique.

The following proves the upperbound in Theorem 1.

Theorem 4. *Let $P = G\Box H$ where G, H are any connected, finite graphs. We have*

$$\mathbf{COV}[P] \leq K\left(\left(1+\frac{\Delta_G}{\delta_H}\right)\mathbf{COV}[H] + \frac{Mm_G m_H n_H l^2}{\mathbf{COV}[H] D_G}\right) \tag{9}$$

where $l = \log D_G \log(n_G D_G)$ and K is some universal constant.

Proof. We group the vertices of H into sets such that for any set S and the subgraph of H induced by S, $H[S]$: (i)$|S| \geq D_G$, (ii)$H[S]$ is connected, (iii) The diameter of $H[S]$ is at most $4D_G$. We demonstrate this grouping is possible through the following algorithm on H: Choose some arbitrary vertex v as the root, and using a breadth-first search (BFS), descend from v at most distance D_G. The resulting tree $T(v)$ will have diameter at most $2D_G$. For each leaf l of $T(v)$, continue the BFS using l as a root. If $T(l)$ has fewer than D_G vertices, append it to $T(v)$. If not, recurse on the leaves of $T(l)$. Each tree then forms a group that satisfies the three conditions above. The root is part of a new group, unless it has been appended to another tree.

In the product P we refer to copies of G as *columns*. In P we have a natural association of each column with the set $S \subseteq V(H)$ defined above. We denote by $Block[S]$ the set of columns in P associated with S union with all edges incident on vertices in those columns. Therefore $Block[S] = (G\Box H[S]) \cup \{(u,v) \in E(P) : u \in V(G\Box H[S])\}$ (note, this is not a graph since it contains edges with free ends). For any two vertices $(.a), (.,b) \in Block[S]$ we can find a connected subgraph $T(a,b)$ of the tree T that generated S such that a and b are connected in $T(a,b)$ and $D_G \leq |V(T(a,b))| \leq 4D_G$. Then using Lemma 8, we can bound the effective resistance $R((.a),(.,b)) \leq 4\zeta \log D_G$.

Hence

$$R_{max}(Block[S]) \leq 4\zeta \log D_G \tag{10}$$

Similarly, $G\Box H[S] \subseteq Loc(P, V(Block[S]))$, and so

$$R_{max}(Loc(P, V(Block[S]))) \leq 4\zeta \log D_G \tag{11}$$

It is envisaged that the following is used with the idea in mind that G is small relative to H, and so the cover time of the product is essentially dominated by the cover time of H.

We use the following two-phase approach

Phase 1. Perform a random walk on $\mathcal{W}(P)$ until the blanket-cover criterion is satisfied for the H dimension.

Phase 2. Starting from the end of phase 1, perform a random walk on P until all vertices of P not visited in phase 1 are visited.

Phase 1 can be thought of in the following way: We couple $\mathcal{W}(P)$ with a walk $\mathcal{W}(H)$ such that (i)if $\mathcal{W}(P)$ starts at $(.,x)$, then $\mathcal{W}(H)$ starts at x, and (ii) $\mathcal{W}(H)$ moves to a new vertex y from a vertex x when and only when $\mathcal{W}(P)$ moves from $(.,x)$ to $(.,y)$ for the first time. This coupled process runs until $\mathcal{W}(H)$ satisfies the blanket-cover criteria for H, ie, when each vertex $u \in V(H)$ has been visited at least $\pi(u)\mathbf{COV}[H]$ times.

Having grouped P into blocks, we analyse the outcome of phase 1 by relating $\mathcal{W}(P)$ to the local observation on each block. A particular block B will have some vertices unvisited by $\mathcal{W}(P)$ if and only if $\mathcal{W}(P)$ locally observed on B fails to visit all vertices. We refer to such a block as *failed.* Consider the weighted random walk $\mathcal{W}(B')$ on $B' = Loc(P, V(B))$. This has the same law as $\mathcal{W}(P)$ locally observed on B. Hence, we bound the probability of $\mathcal{W}(P)$ failing to cover B by bounding the probability that $\mathcal{W}(B')$ fails to cover B'. Done for all blocks, we can bound the expected time it takes phase 2 to cover the failed blocks. We think of phase 1 as doing most of the "work", and phase 2 as a "mopping up" phase. Mopping up a block in phase 2 is costly, but if there are few of them, the overall cost is within a small factor of phase 1.

We bound $\mathbf{Pr}(\mathcal{W}(B')$ fails) by exploiting the fact that $\mathcal{W}(B')$ will have made some minimal number of transitions t. This is guaranteed because phase 1 terminates only when $\mathcal{W}(H)$ has statisfied the blanket-cover criterion on H, that is, each vertex $u \in V(H)$ has been visited at least $\pi(u)\mathbf{COV}[H]$ times, so each column G_u in P will have been visited at least that many times. If κ counts the number of steps of a walk $\mathcal{W}(B')$ until B' is covered, then $\mathbf{Pr}(\mathcal{W}(B')$ fails to cover $B') = \mathbf{Pr}(\kappa > t) \leq \mathbf{E}[\kappa]/t$ by Markov's inequality.

Definition 5. *For graphs $I = J \Box K$, and $S \subset I$, denote by $S.J$ the* projection of S on to J, *that is,* $S.J = \{u \in J : (u, .) \in S\}$.

For a weighted graph G recall $c(G)$ is the total conductance (weight) of all edges of G.

Let B be a block and let $B' = Loc(P, V(B))$. By Section 4 $c(B') = c(B)$, given by the following

$$c(B) \leq m_G |\{u \in V(B).H\}| + n_G \sum_{u \in V(B).H} d(u) \tag{12}$$

Using 11 and Lemma 3 we therefore have for any $u, v \in V(B')$, $\mathbf{COM}[u, v] \leq Kc(B') \log D_G$ for some universal constant K. (In what follows K will change, but we shall keep the same symbol, with an understanding that what we finish with is a univeral constant). Hence, by Lemma 4

$$\mathbf{COV}[B'] \leq Kc(B') \log D_G \log(|V(B')|) = Kc(B) l_B \tag{13}$$

where $l_B = \log D_G \log(|V(B)|)$

For a block B, the number of transitions on the H dimension - and therefore the number of transitions on B - as demanded by the blanket-cover criterion is at least

$$\sum_{u \in V(B).H} \mathbf{COV}[H] \pi(u) = \frac{\mathbf{COV}[H]}{2m_H} \sum_{u \in V(B).H} d(u) \tag{14}$$

Now

$$\mathbf{Pr}(\mathcal{W}(P) \text{ fails on } B) = \mathbf{Pr}(\mathcal{W}(B') \text{ fails on } B') \tag{15}$$

$$\leq Kc(B) l_B \left(\frac{\mathbf{COV}[H]}{2m_H} \sum_{u \in V(B).H} d(u) \right)^{-1} . \tag{16}$$

The second equality by Markov's inequality.

Phase 2 consists of two components: movement between failed blocks, and covering a failed block it has arrived at. The total block-to-block movement is upperbounded by the time is takes to cover the H dimension of P (in other words, for each column to have been visited at least once). We denote this by $\mathbf{COV}_P[P.H]$. Denoting the covertime of a block B by the walk $\mathcal{W}(P)$ by $\mathbf{COV}_P[B]$,

$$\mathbf{E}[Ph2] \leq \mathbf{COV}_P[P.H] + \sum_{B \in P} \mathbf{Pr}(\mathcal{W}(P) \text{ fails on } B) \mathbf{COV}_P[B] \tag{17}$$

For $\mathcal{W}(H)$, the r.v. $\beta_H = \min\{t : (\forall v) N_v(t) > \pi(v) \mathbf{COV}[H]\}$ counts the time it takes to satisfy the blanket-cover criterion on H.

The expected number of movements on P per movement on the H dimension is at most $\frac{\Delta_G+\delta_H}{\delta_H}$. Therefore $\mathbf{E}[Ph1] \leq \frac{\Delta_G+\delta_H}{\delta_H}\mathbf{E}[\beta_H]$. Similarly, $\mathbf{COV}_P[P.H] \leq \frac{\Delta_G+\delta_H}{\delta_H}\mathbf{COV}[H]$.

Using 10 and Lemmas 3 and 4 again, we have $\mathbf{COV}_P[B] \leq K'c(P)l_B$ where $c(P) = |E(P)| = M$. Theorem 2 gives us $\mathbf{E}\beta_H \leq K\mathbf{COV}[H]$, for some universal constant K and so

$$\mathbf{COV}[P] \leq \mathbf{E}[Ph1] + \mathbf{E}[Ph2] \tag{18}$$

$$\leq K\frac{\Delta_G + \delta_H}{\delta_H}\mathbf{COV}[H] + \sum_{B\in P} \mathbf{Pr}(\mathcal{W}(P) \text{ fails on } B)\mathbf{COV}_P[B]. \tag{19}$$

We have, using 16

$$\sum_{B\in P} \mathbf{Pr}(\mathcal{W}(P) \text{ fails on } B)\mathbf{COV}_P[B] \leq K\frac{Mm_H}{\mathbf{COV}[H]}\sum_{B\in P}\frac{c(B)l_B^2}{\sum_{u\in V(B).H} d(u)} \tag{20}$$

and

$$\sum_{B\in P}\frac{c(B)l_B^2}{\sum_{u\in V(B).H} d(u)} \leq \sum_{B\in P}\left(n_G + \frac{m_G|\{u \in V(B).H\}|}{\sum_{u\in V(B).H} d(u)}\right) l_B^2 \tag{21}$$

Since $\sum_{u\in V(B).H} d(u) \geq |\{u \in V(B).H\}|$, we have

$$\sum_{B\in P} \mathbf{Pr}(\mathcal{W}(P) \text{ fails on } B)\mathbf{COV}_P[B] \leq K\frac{Mm_Gm_H}{\mathbf{COV}[H]}\sum_{B\in P} l_B^2 \tag{22}$$

$$\leq K\frac{Mm_Gm_Hn_Hl^2}{\mathbf{COV}[H]D_G} \tag{23}$$

where $l = \log D_G \log(n_G D_G)$

References

1. Aleliunas, R., Karp, R.M., Lipton, R.J., Lovász, L., Rackoff, C.: Random Walks, Universal Traversal Sequences, and the Complexity of Maze Problems. In: Proceedings of the 20th Annual IEEE Symposium on Foundations of Computer Science, pp. 218–223 (1979)
2. Bollobas, B., Brightwell, G.: Random walks and electrical resistances in product graphs. Discrete Applied Mathematics 73, 69–79 (1997)
3. Chandra, A.K., Raghavan, P., Ruzzo, W.L., Smolensky, R., Tiwari, P.: The electrical resistance of a graph captures its commute and cover times. Computational Complexity 6, 312–340 (1997)
4. Dembo, A., Peres, Y., Rosen, J., Zeitouni, O.: Cover times for Brownian motion and random walks in two dimensions. Ann. Math. 160, 433–464 (2004)
5. Ding, J., Lee, J.R., Peres, Y.: Cover times, blanket times and majorizing measures (2010) (manuscript)

6. Doyle, P.G., Laurie Snell, J.: Random walks and electrical networks (2006)
7. Feige, U.: A tight upper bound for the cover time of random walks on graphs. Random Structures and Algorithms 6, 51–54 (1995)
8. Feige, U.: A tight lower bound for the cover time of random walks on graphs. Random Structures and Algorithms 6, 433–438 (1995)
9. Jonasson, J.: An upper Bound on the Cover Time for Powers of Graphs. Discrete Mathematics 222, 181–190 (2000)
10. Levin, D.A., Peres, Y., Wilmer, E.L.: Markov Chains and Mixing Times (2009)
11. Matthews, P.: Covering problems for Brownian motion on spheres. Ann. Prob. 16, 189–199 (1988); Nash-Williams, C.S.J.A.: Random walk and electric currents in networks. Proc. Camb. Phil. Soc. 55, 181–194 (1959)
12. Winkler, P., Zuckerman, D.: Multiple Cover Time. Random Structures and Algorithms 9, 403–411 (1996)

A Appendix

A.1 Proof of Lemma 6

The states are clearly the same so it remains to show that the transition probability $P_{\mathcal{M}}(u,v)$ from u to v in $\mathcal{M}(G,S)$ is the same as $P_{\mathcal{W}(H)}(u,v)$ in $\mathcal{W}(H)$. Recall that B is the border of the induced subgraph $G[S]$. If $u \notin B$ then an edge $(u,v) \in E(H)$ is internal and so has unit conductance in H, as it does in G. Furthermore, for an internal edge e, $e \in E(H)$ if and only if $e \in E(G)$, thus $d_H(u) = d_G(u)$ when $u \notin B$. Therefore $P_{\mathcal{W}(H)}(u,v) = 1/d_H(u) = 1/d_G(u) = P_{\mathcal{M}}(u,v)$.

Now suppose $u \in B$. Let $E(u)$ denote the set of all edges incident with u in H and recall $A(B)$ above is the set of exterior edges. The total conductance (weight) of the exterior edges at u is

$$\begin{aligned}\sum_{e\in E(u)\cap A(B)} c_H(e) &= \sum_{x\in N(u,out)}\sum_{v\in B}\mathbf{Pr}(\text{walk from } x \text{ returns to } B \text{ at } v)\\ &= \sum_{x\in N(u,out)} 1\\ &= d(u,out).\end{aligned}$$

(Note the 'H' subscript in $c_H(e)$ above is redundant since exterior edges are only defined for H, but we leave it for clarity).

Thus for $u \in B$

$$\begin{aligned}c_H(u) = \sum_{e\in E(u)} c_H(e) &= \sum_{e\in E(u)\cap G[S]} 1 + \sum_{e\in E(u)\cap A(B)} c_H(e)\\ &= d(u,in) + d(u,out)\\ &= d_G(u)\end{aligned}$$

Now

$$P_{\mathcal{M}}(u,v) = \mathbf{1}_{\{(u,v)\in G[S]\}}\frac{1}{d_G(u)} + \sum_{k\geq 1}\sum_{\omega=(u,x_1\dots x_k,v)}\frac{1}{d_G(u)d_G(x_1)\dots d_G(x_k)} \quad (24)$$

where the sum is over all exterior walks ω. Thus

$$P_{\mathcal{M}}(u,v) = \mathbf{1}_{\{(u,v)\in G[S]\}}\frac{1}{d_G(u)} + p_H(\boldsymbol{v},\boldsymbol{u}) \tag{25}$$

$$\begin{aligned}
P_{\mathcal{W}(H)}(u,v) &= \frac{1}{c_H(u)}\left[\mathbf{1}_{\{(u,v)\in G[S]\}} + \mathbf{1}_{\{(u,v)\in A(S)\}}c_H(u,v)\right] &(26)\\
&= \frac{1}{d_G(u)}\left[\mathbf{1}_{\{(u,v)\in G[S]\}} + \mathbf{1}_{\{(u,v)\in A(S)\}}d_G(u)p_H(\boldsymbol{v},\boldsymbol{u})\right] &(27)\\
&= \mathbf{1}_{\{(u,v)\in G[S]\}}\frac{1}{d_G(u)} + \mathbf{1}_{\{(u,v)\in A(S)\}}p_H(\boldsymbol{v},\boldsymbol{u}) &(28)\\
&= P_{\mathcal{M}}(u,v) &(29)
\end{aligned}$$

Dictionary-Symbolwise Flexible Parsing

Maxime Crochemore[1,4], Laura Giambruno[2], Alessio Langiu[2,4], Filippo Mignosi[3], and Antonio Restivo[2]

[1] Dept. of Computer Science, King's College London, London WC2R 2LS, UK
maxime.crochemore@kcl.ac.uk
[2] Dipartimento di Matematica e Informatica, Università di Palermo, Palermo, Italy
{lgiambr,alangiu,restivo}@math.unipa.it
[3] Dipartimento di Informatica, Università dell'Aquila, L'Aquila, Italy
filippo.mignosi@di.univaq.it
[4] Institut Gaspard-Monge,Université Paris-Est, France

Abstract. Linear time optimal parsing algorithms are very rare in the dictionary based branch of the data compression theory. The most recent is the *Flexible Parsing* algorithm of Mathias and Shainalp that works when the dictionary is prefix closed and the encoding of dictionary pointers has a constant cost. We present the *Dictionary-Symbolwise Flexible Parsing* algorithm that is optimal for prefix-closed dictionaries and any symbolwise compressor under some natural hypothesis. In the case of LZ78-alike algorithms with *variable costs* and any, linear as usual, symbolwise compressor can be implemented in linear time. In the case of LZ77-alike dictionaries and any symbolwise compressor it can be implemented in $O(n \log(n))$ time. We further present some experimental results that show the effectiveness of the dictionary-symbolwise approach.

1 Introduction

In [16] Mathias and Shainalp gave a linear time optimal parsing algorithm in the case of dictionary compression where the dictionary is prefix closed *and* the cost of encoding dictionary pointer is constant. They called their parsing algorithm *Flexible Parsing*. The basic idea of *one-step-lookahead parsing* that is at the base of flexible parsing was firstly used to our best knowledge in [6] in the case of dictionary compression where the dictionary is prefix closed *and* the cost of encoding dictionary pointer is constant *and* the dictionary is static. A first intuition, not fully exploited, that this idea could be successful used in the case of dynamic dictionaries, was given in [7] and also in [11], where it was called MTPL parsing (maximum two-phrase-length parsing).

Optimal parsing algorithms are rare and linear time optimal parsing results are rather rare. We can only also cite the fact that greedy parsing is optimal and linear for LZ77-alike dictionaries and constant cost dictionary pointers (see [19]) and its generalization to suffix closed dictionaries and constant cost dictionary pointers (see [2]) later used also in [11].

C.S. Iliopoulos and W.F. Smyth (Eds.): IWOCA 2010, LNCS 6460, pp. 390–403, 2011.

In this paper we consider the case of a free mixture of a dictionary compressor and a symbolwise compressor and we extend, in a non obvious way, the result of Mathias and Shainalp. We have indeed an optimal parsing algorithm in the case of dictionary-symbolwise compression where the dictionary is prefix closed and the cost of encoding dictionary pointer is *variable* and the symbolwise is any classical one that works in linear time. Our algorithm works under the assumption that a special graph that will be described in next section is well defined. Even in the case where this condition is not satisfied it is possible to use the same method to obtain almost optimal parses. In particular, when the dictionary is LZ78-alike our algorithm can be implemented in linear time and when the dictionary is LZ77-alike our algorithm can be implemented in time $O(n \log(n))$.

The study of free mixtures of two compressor is quite involved and it represents a new theoretical challenge. Free mixture has been implicitly or explicitly using for a long time in many fast and effective compressors such as gzip (see [5]), PkZip (see [10]) and Rolz algorithms (see [14]). For a quick look to compression performance on texts see Mahony challenge's page (see [13]).

So, why linear time optimal parsing algorithms are rather rare? Classically (see for example [18]), for static dictionaries it is possible to associate to any dictionary algorithm $\mathcal{A}$ and to any text T a weighted graph $G_{\mathcal{A},T}$ such that there is a bijection between optimal parsings and minimal paths in this graph. The extension of this approach to dynamical dictionaries has been firstly studied, to our best knowledge, in [17] and it has also been later used in [4]. More details will be given in next sections.

The graph $G_{\mathcal{A},T}$ is a Directed Acyclic Graph and it is possible to find a minimal path in linear time with respect to the size of it (see [3]). Unfortunately the size of the graph can be quadratic in the size of the text and this approach was not recommended in [18], because it is too time consuming. From a philosophical point of view, the graph $G_{\mathcal{A},T}$ represents a mathematical modelling of the optimal parsing problem. Thus, finding an optimal parsing in linear time corresponds to discovering a strategy for using only a subgraph of linear size. Indeed, in order to get over the quadratic worst case problem, there are many different approaches and many papers deal with optimal parsing in dictionary compressions. For instance the reader can see [1,2,4,6,8,9,11,12,15,16,19,20]. Among them, we stress [4] where it is shown that a minimal path can be obtained by using a subgraph of $G_{\mathcal{A},T}$ of size $O(n\ log(n))$, in the LZ77 case under some natural assumptions on the cost function, by exploiting the discreteness of the cost functions.

In this paper we use a similar strategy, i.e. we consider static or dynamical dictionaries, following the approach of [17] and we discover a "small" subgraph of $G_{\mathcal{A},T}$ that is linear in the size of the text for LZ78-alike dictionaries and $O(n \log(n))$ for LZ77-alike dictionaries. This "small" subgraph is such that any minimal path in it is also a minimal path in $G_{\mathcal{A},T}$.

In Sect. 2 we recall some literature notions about dictionary and dictionary-symbolwise compression algorithms and we define the graph $G_{\mathcal{A},T}$. In Sect. 3 we formalize the definition of optimal algorithm and optimal parsing and extend them to the dictionary-symbolwise domain. In Sect. 4 we present the *Dictionary-Symbolwise Flexible Parsing*, a parsing algorithm that extends in some sense the *Flexible Parsing* (see [16]). We prove its optimality by showing that it corresponds to a shortest path in the full graph, and in Sect. 5 we describe some data structures that can be used for our algorithm in the two main cases of LZ78-alike and LZ77-alike dictionaries together the time analysis. Section 6 reports some experiments, open problems and our conclusions.

2 Preliminaries

Analogously to what stated in [17] and extending the approach introduced in [18] to the dynamical dictionary case, we show how it is possible to associate a directed weighted graph $G_{\mathcal{A},T} = (V, E, L)$ to any dictionary compression algorithm $\mathcal{A}$, any text $T = a_1 a_2 a_3 \cdots a_n$ and any cost function $C : E \to \mathbb{R}^+$ in the following way. The set of vertices is $V = \{0, 1, \ldots, n\}$, where vertex i corresponds to a_i, i.e. the i-th character in the text T, for $1 \leq i \leq n$ and vertex 0 correspond to the position at the beginning of the text, before any characters. The empty word ϵ is associated to vertex 0 that is also called the *origin* of the graph. The set of directed edges is $E = \{(p, q) \subset (V \times V) \mid p < q \text{ and } \exists\ w = T[p+1 : q] \in D_p\}$, where $T[p+1 : q] = a_{p+1}a_{p+2} \cdots a_q$ and D_p is the dictionary relative to the processing step p-th, i.e. the step in which the algorithm either has processed the input text up to character a_p, for $0 < p$, or it has to begin, for $p = 0$. For each edge (p, q) in E, we say that (p, q) is *associated to* the dictionary phrase $w = T[p+1 : q] = a_{p+1} \cdots a_q \in D_p$. In the case of static dictionary D_i is constant along the algorithm steps, i.e. $D_i = D_j, \forall i, j = 0 \cdots n$. L is the set of edge labels $L_{p,q}$ for every edge $(p, q) \in E$, where the label $L_{p,q}$ is defined as the cost of the edge (p, q) when the dictionary D_p is in use, i.e. $L_{p,q} = C((p, q))$. When $L_{p,q}$ is always defined for each edge of the graph we say that $G_{\mathcal{A},T}$ is *well defined*.

A *dictionary-symbolwise* algorithm is a compression algorithm that uses both dictionary and symbolwise compression methods. Such compressors parse the text as a free mixture of dictionary phrases and literal characters, which are substituted by the corresponding pointers or literal codes, respectively. Therefore, the description of a dictionary-symbolwise algorithm should also include the so called *flag information*, that is the technique used to distinguish the actual compression method (dictionary or symbolwise) used for each segment or factor of the parsed text. Often, as in the case of LZSS (see [19]), an extra bit is added either to each pointer or encoded character to distinguish between them. Encoded information flag can require less (or more) space than one bit.

For instance, a dictionary-symbolwise compression algorithm with a fixed dictionary $D = \{ab, cbb, ca, bcb, abc\}$ and the static symbolwise codeword assignment $[a = 1, b = 2, c = 3]$ could compress the text *abccacbbabbcbcbb* as $F_d1F_s3F_d3F_d2F_d1F_d4F_d2$, where F_d is the information flag for dictionary pointers and F_s is the information flag for the symbolwise code.

More formally, a parsing of a text T in a dictionary-symbolwise algorithm is a pair $(parse, Fl)$ where *parse* is a sequence $(u_1, \cdots, u_s)$ of words such that

$T = u_1 \cdots u_s$ and where Fl is a boolean function that, for $i = 1, \ldots, s$ indicates whether the word u_i has to be coded as a dictionary pointer or as a symbol. See Tab. 1 for an example of dictionary-symbolwise compression.

A dictionary-symbolwise compression algorithm, analogously as done in [1] for the pure dictionary case, is specified by:

1. The dictionary description.
2. The encoding of dictionary pointers.
3. The symbolwise encoding method.
4. The encoding of the flag information.
5. The parsing method.

We can naturally extend the definition of the graph associated to an algorithm for the dictionary-symbolwise case. Given a text $T = a_1 \ldots a_n$, we denote by $T[i : j]$ the factor of T equal to $a_i \ldots a_j$. Given a dictionary-symbolwise algorithm $\mathcal{A}$, a text T and a cost function C defined on edges, the graph $G_{\mathcal{A},T} = (V, E, L)$ is defined as follows. The vertexes set is $V = \{0 \cdots n\}$, with $n = |T|$. The set of directed edges $E = E_d \bigcup E_s$, where $E_d = \{(p, q) \subset (V \times V) \mid p < q-1, \text{ and } \exists w = T[p+1 : q] \in D_p\}$ is the set of dictionary edges and $E_s = \{(q-1, q) \mid 0 < q \leq n\}$ is the set of symbolwise edges. L is the set of edge labels $L_{p,q}$ for every edge $(p, q) \in E$, where the label $L_{p,q} = C((p, q))$. Let us notice that the cost function C hereby used has to include the cost of the flag information to each edge, i.e. either $C(p, q)$ is equal to $\langle$ the cost of the encoding of $F_d\rangle$ + $\langle$ the cost of the encoded dictionary phrase $w \in D_p$ associated to the edge $(p, q)\rangle$ if $(p, q) \in E_d$ or $C(p, q)$ is equal to $\langle$ the cost of encoded $F_s\rangle$ + $\langle$ the cost of the encoded symbol $a_q\rangle$ if $(p, q) \in E_s$. Moreover, since E_d does not contain edges of length one by definition, $G_{\mathcal{A},T} = (V, E, L)$ is not a multigraph. Since this graph approach can be extended to multigraph, with a overhead of formalism, one can relax the $p < q - 1$ constrain in the definition of E_d to $p \leq q - 1$. All the results we will state in this paper, naturally extend to the multigraph case.

We call *dictionary-symbolwise scheme* a set of algorithms having in common the same first four specifics (i.e. they differ one each other for just the parsing methods). A scheme does not need to contain *all* algorithms having the same first four specifics. We notice that any of the specifics from 1 to 5 above can depend on all the others, i.e. they can be mutually interdependent. Fixed a *dictionary-symbolwise scheme*, whenever the specifics of the parsing method are given, exactly one algorithm is completely described. Notice that the word *scheme* has been used by other authors with other related meaning. For us the meaning is rigorous.

Table 1. Example of compression for the text *abccacbbabbcbcbb* by a simple Dyctionary-Symbolwise algorithm that use $D = \{ab, cbb, ca, bcb, abc\}$ as static dictionary, the identity as dictionary encoding and the mapping $[a = 1, b = 2, c = 3]$ as symbolwise encoding

Input	*ab*	*c*	*ca*	*cbb*	*ab*	*bcb*	*cbb*
Output	F_d1	F_s3	F_d3	F_d2	F_d1	F_d4	F_d2

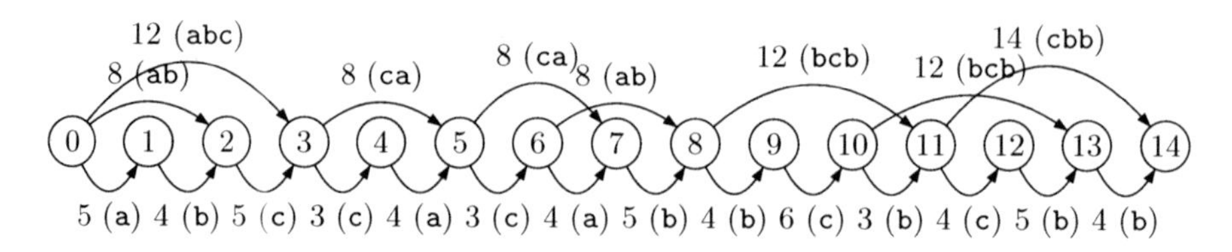

Fig. 1. Graph $G_{\mathcal{A},T}$ for the text T = abccacabbcbcbb, for the dictionary-symbolwise algorithm $\mathcal{A}$ with static dictionary $D = \{$ab, abc, bcb, ca, cbb$\}$ and cost function C as defined in the graph. The dictionary phrase or the symbol associated to an edge is reported near the edge label within parenthesis.

3 On Optimality

In this section we assume that the reader is familiar with LZ-alike dictionary encoding and with some simple statistical encodings such as the Huffman encoding.

Definition 1. *Fixed a dictionary description, a cost function C and a text T, a* dictionary (dictionary-symbolwise) algorithm *is* optimal *within a set of algorithms if the cost of the encoded text is minimal with respect to all others algorithms in the same set. The parsing of an* optimal *algorithm is called* optimal *within the same set.*

When the length in bit of the encoded dictionary pointers is used as cost function, the previous definition of optimality is equivalent to the classical well known definition of bit-optimality for dictionary algorithm. Notice that the above definition of optimality strictly depends on the text T and on a set of algorithms. A parsing can be optimal for a text and not for another one. Clearly, we are mainly interested on parsings that are optimal either for *all* texts over an alphabet or for classes of texts. Whenever it is not explicitly written, from now on when we talk about optimal parsing we mean optimal parsing for *all* texts. About the set of algorithm it make sense to find sets as large as possible.

Classically, there is a bijective correspondence between parsings and paths in $G_{\mathcal{A},T}$ from vertex 0 to vertex n, where optimal parses correspond to minimal paths and vice-versa. We say that a parse (resp. path) *induces* a path (resp. parse) to denote this correspondence. This correspondence was firstly stated in [18] only in the case of sets of algorithms sharing the same *static* dictionary and where the encoding of pointers has constant cost.

For example the path along vertexes $(0, 3, 4, 5, 6, 8, 11, 12, 13, 14)$ is the shortest path for the graph in Fig. 1. Authors of [17] were the firsts to formally extend the Shortest Path approach to dynamically changing dictionaries and variable costs.

Definition 2. *A scheme $\mathcal{S}$ has the* Schuegraf property *if, for any text T and for any pair of algorithms $\mathcal{A}, \mathcal{A}' \in \mathcal{S}$, the graph $G_{\mathcal{A},T} = G_{\mathcal{A}',T}$ with $G_{\mathcal{A},T}$ well defined.*

This property of schemes is called *property of Schuegraf* in honor to the first of the authors in [18]. In this case we define $G_{\mathcal{S},T} = G_{\mathcal{A},T}$ as the graph of (any algorithm of) the scheme. The proof of the following proposition is straightforward.

Proposition 1. *There is a bijective correspondence between optimal parsings and shortest paths in $G_{\mathcal{S},T}$ from vertex 0 to vertex n.*

Definition 3. *Let us consider an algorithm $\mathcal{A}$ and a text T and suppose that graph $G_{\mathcal{A},T}$ is well defined. We say that $\mathcal{A}$ is* dictionary *optimal (with respect to T) if its parsing induces a shortest path in $G_{\mathcal{A},T}$ from the origin (i.e. vertex 0) to vertex n, with $n = |T|$. In this case we say that its parsing is* dictionary *optimal.*

Let $\mathcal{A}$ be an algorithm such that for any text T the graph $G_{\mathcal{A},T}$ is well defined. We want to associate to it a scheme $\mathcal{SC}_A$ in the following way. Let S be the set of all algorithms $\mathcal{A}$ such that for any text T $G_{\mathcal{A},T}$ exists (i.e. it is well defined). Let $\mathcal{B}$ and $\mathcal{C}$ two algorithms in S. We say that $\mathcal{B}$ and $\mathcal{C}$ are equivalent or $\mathcal{B} \equiv \mathcal{C}$ if, for any text T, $G_{\mathcal{B},T} = G_{\mathcal{C},T}$.

We define the scheme $\mathcal{SC}_A$ to be the equivalence class that has $\mathcal{A}$ as a representative. It is easy to prove that $\mathcal{SC}_A$ has the Schuegraf property.

We can connect the definition of *dictionary* optimal parsing with the previous definition of $\mathcal{SC}_A$ to obtain the next proposition, that says, roughly speaking, that dictionary optimality implies scheme (or global) optimality within the scheme $\mathcal{SC}_A$.

Proposition 2. *Let us consider an algorithm $\mathcal{A}$ such that for any text T the graph $G_{\mathcal{A},T}$ is well defined. Suppose further that for a text T the parsing of $\mathcal{A}$ is dictionary optimal. Then the parsing of $\mathcal{A}$ of the text T is (globally) optimal within the scheme $\mathcal{SC}_A$.*

We have simple examples where a parsing of a text is dictionary optimal and the corresponding algorithm belongs to a scheme that has not the Schuegraf property and it is not (globally) optimal within the same scheme. For pure dictionary scheme having the Schuegraf Property we mean a dictionary-symbolwise scheme having the Schuegraf Property where all algorithms in the scheme are pure dictionary. We have to be a bit careful using this terminology. Indeed, LZ78, LZW, LZ77 and related algorithms often parse the text with a dictionary pointer and then add a symbol, i.e. the parse phrase is composed by a dictionary pointer and a symbol. In these cases all edges of $G_{\mathcal{A},T}$ denote parse phrases coupled to the corresponding symbol. Edges are labeled by the cost of the dictionary pointer plus the cost of the symbol. We consider these cases included in the class of "pure dictionary" algorithms and schemes.

4 Dictionary-Symbolwise Flexible Parsing Algorithm

In this section we extend the notion of flexible parsing to the dictionary-symbolwise case and we prove that it is still optimal within any scheme having the Schuegraf Property. We assume here that the dictionary must be at any moment prefix closed. The algorithm is quite different from the original Flexible Parsing but it has some analogies with it and, in the case of LZ78-alike dictionaries, it makes use of one of the main data structures used for the original flexible parsing in order to be implemented in linear time. Concerning the costs of encoding pointers, we recall that costs can vary but that they assume positive values and that they include the cost of flag information. Concerning the symbolwise compressor, the costs of symbols must be positive, including the flag information cost. They can vary depending on the position of the character in the text and on the symbol itself. We suppose further that a text T of length n is fixed and that we are considering the graph $G_{\mathcal{A},T}$, where $\mathcal{A}$ is a dictionary-symbolwise algorithm, and $G_{\mathcal{A},T}$ is well defined under our assumption. We denote by d the function that represent the distance of the vertexes of $G_{\mathcal{A},T}$ from the origin of the graph. Such a distance $d(i)$ is classically defined as the minimal cost of all possible weighted paths from the origin to the vertex i, where $d(0) = 0$. This distance obviously depends on the cost function. We say that cost function C is *prefix-non-decreasing* at any moment if for any $u, v \in D_p$ phrases associated to edges $(p, i), (p, q)$, with $p < i < q$, that implies that u is prefix of v, one has that $C((p, i)) \leq C((p, q))$.

Lemma 1. *Let $\mathcal{A}$ be a dictionary-symbolwise algorithm such that for any text T the graph $G_{\mathcal{A},T}$ is well defined. If the dictionary is always (at any moment) prefix-closed and if the cost function is always (at any moment) prefix-non-decreasing then the function d is non-decreasing monotone.*

In what follows in this paper we suppose that the graph $G_{\mathcal{A},T}$ is well defined. Let us call vertex j a *predecessor* of vertex $i \iff \exists(j, i) \in E$ such that $d(i) = d(j) + C((j, i))$. Let us define $pre(i)$ to be the smallest of the predecessors of vertex $i, 0 < i \leq n$, that is $pre(i) = \min\{j \mid d(i) = d(j) + C((j, i))\}$. In other words $pre(i)$ is the smallest vertex j that contributes to the definition of $d(i)$. Clearly $pre(i)$ has distance smaller than $d(i)$. Moreover the function pre is not necessarily injective. For instance, a vertex can be a predecessor either "via" a dictionary edge or "via" a symbol edge. It is also possible to extend previous definition to pointers having a cost smaller than or equal to a fixed c.

Definition 4. *For any cost c we define $pre_c(i) = \min\{j \mid d(i) = d(j) + C((j, i))$ and $C((j, i)) \leq c\}$. If none of the predecessor j of i is such that $C((j, i)) \leq c$ then $pre_c(i)$ is undefined.*

If all costs of the pointers are smaller than or equal to c then for any i one has obviously that $pre_c(i)$ is equal to $pre(i)$.

Analogously to the notation of [15], we want to define two boolean operations *Weighted-Extend* and *Weighted-Exist*.

Definition 5. *Given an edge (i, j) in $G_{\mathcal{A},T}$ and its associated phrase w, a cost value c and a character 'a', the operation* Weighted-Extend$((i, j), a, c)$ *finds out whether the word wa is a phrase in D_i having cost smaller than or equal to c.*

More formally, let (i, j) be such that $w = T[i + 1 : j] \in D_i$ and, then, (i, j) is in $G_{\mathcal{A},T}$. *Weighted-Extend*$((i, j), a, c)$ = *"yes"* $\iff$ $wa = T[i + 1 : j + 1] \in D_i$ *and* $C((i, j + 1)) \leq c$, where C is the cost function associated to the algorithm $\mathcal{A}$. Otherwise *Weighted-Extend*$((i, j), a, c)$ = *"no"*.

Definition 6. *Given* $0 < i, j \leq n$ *and a cost value* c*, the operation* Weighted-Exist(i, j, c) *finds out whether or not the phrase* $w = T[i+1 : j]$ *is in* D_i *and the cost of the corresponding edge* (i, j) *is smaller than or equal to* c*.*

Let us notice that doing successfully a *Weighted-Extend* operation on $((i, j), a, c)$ means that $wa \in D_i$ is the weighted extension of w and the encoding of $(i, j+1)$ has cost less or equal to c. Similarly, doing a *Weighted-Exist* operation on (i, j, c) means that an edge (i, j) exist in $G_{\mathcal{A},T}$ having cost less or equal to c.

Let E_c be the subset of all edges of the graph having cost smaller than or equal to c.

Definition 7. *Let us also define, for any cost c the set* $M_c \subseteq E_c$ *to be the set of* c-supermaximal *edges, where* $(i, j) \in M_c \iff (i, j) \in E_c$ *and* $\forall p, q \in V$*, with* $p < i$ *and* $j < q$*, the arcs* $(p, j), (i, q)$ *are not in* E_c*. For any* $(i, j) \in M_c$ *let us call* i *a c-starting point and* j *a c-ending point.*

Proposition 3. *Suppose that* (i, j) *and* (i', j') *are in* M_c*. One has that* $i < i'$ *if and only if* $j < j'$*.*

By previous proposition, if $(i, j) \in M_c$ we can think j as function of i and conversely. Therefore it is possible to represent M_c by using an array $M_c[j]$ such that if (i, j) is in M_c then $M_c[j] = i$ otherwise $M_c[j] = Nil$. Moreover the non-*Nil* values of this array are strictly increasing. The positions j having value different from *Nil* are the ending positions.

We want now to describe a simple algorithm that outputs all *c-supermaximal* edges scanning the text left-to-right. We call it *Find Supermaximal*(c). It uses the operations *Weighted-Extend* and *Weighted-Exist*. The algorithm starts with $i = 0, j = 1$ and $w = a_1$.The word w is indeed implicitely defined by the arc (i, j) and therefore it will not appear explicitely in the algorithm. At each step j is increased by one and w is set to w concatenated to $T[j]$. The algorithm executes a series of *Weighted-Extend* until this operation give a positive answer or the end of the text is reached. After a negative answer of *Weighted-Extend*, the algorithm does a series of *Weighted-Exist* increasing i by one until a positive answer. The algorithm is stated more formally in the following pseudo code.

```
FIND SUPERMAXIMAL (c)
01.  i ← 0, j ← 1
02.  WHILE j < n
03.  DO
04.        WHILE Weighted-Extend((i, j), a_{j+1}, c) = "yes" AND j < n
05.        DO
06.              j ← j + 1
07.        INSERT (i, j) in M_c, j ← j + 1
08.        DO
09.              i ← i + 1
10.        WHILE Weighted-Exist(i, j, c) = "no" AND i < j
```

We notice that when exiting from cycle of lines $4 - 6$, the cost of the edge (i, j) could still be strictly smaller than c. The function INSERT simply insert the edge (i, j) in the dynamical set M_c. If we represent M_c by an array as described after Prop. 3, function INSERT sets $M_c[j]$ equal to i. Array $M_c[j]$ was initialized by setting all its entries to *Nil*.

Proposition 4. *Above algorithm correctly computes* M_c.

Proposition 5. *For any edge* $(i, j) \in E_c$ *there exists a* c-supermaximal *edge* $(\hat{i}, \hat{j})$ *containing it, i.e. such that* $\hat{i} \leq i$ *and* $j \leq \hat{j}$.

By previous proposition for any node $v \in G_{\mathcal{A},T}$ if there exists a node $i < v$ such that $C((i, v)) = c$ and $d(v) = d(i) + c$ then there exists a *c-supermaximal* edge $(\hat{i}, \hat{j})$ containing (i, v) and such that $\hat{j}$ is the *closest* arrival point greater that v. Let us call this *c-supermaximal* edge $(\hat{i}_v, \hat{j}_v)$. We use $\hat{i}_v$ in next proposition.

Proposition 6. *Suppose that* $v \in G_{\mathcal{A},T}$ *is such that there exists a previous node* i *such that* $C((i, v)) = c$ *and* $d(v) = d(i) + c$. *Then* $\hat{i}_v$ *is a predecessor of* v, *i.e.* $d(v) = d(\hat{i}_v) + C((\hat{i}_v, v))$ *and, moreover,* $d(\hat{i}_v) = d(i)$ *and* $C((\hat{i}_v, v)) = c$.

Corollary 1. *For any vertex* v, *the edge* $(\hat{i}_v, v)$ *is the last edge of a path of minimal cost from the origin to vertex* v.

In what follows we describe a graph $G'_{\mathcal{A},T}$ that is a subgraph of $G_{\mathcal{A},T}$ and that is such that for any node $v \in G_{\mathcal{A},T}$ there exists a minimal path from the origin to v in $G'_{\mathcal{A},T}$ that is also a minimal path from the origin to v in $G_{\mathcal{A},T}$. The proof of this property, that will be stated in the subsequent proposition, is a consequence of Proposition 6 and Corollary 1.

We describe the building of $G'_{\mathcal{A},T}$ in an algorithmic way. Even if we do not give the pseudocode, algorithm BUILD $G'_{\mathcal{A},T}$ is described in a rigorous way and it makes use, as a part of it, of algorithm FIND SUPERMAXIMAL.

The set of nodes of $G'_{\mathcal{A},T}$ is the same of $G_{\mathcal{A},T}$. First of all we insert all symbolwise edges of $G_{\mathcal{A},T}$ in $G'_{\mathcal{A},T}$. Let now $\mathcal{C}$ be the set of all possible costs that any dictionary edge has. This set can be build starting from $G_{\mathcal{A},T}$ but in all known meaningful situations the set $\mathcal{C}$ is usually well known and can be ordered and stored in an array in a time that is linear in the size of the text.

For any $c \in \mathcal{C}$ we use algorithm FIND SUPERMAXIMAL to obtain the array $M_c[j]$. For any *c-supermaximal* edge (i, j), we add in $G'_{\mathcal{A},T}$ all edges of the form (i, x) where x varies from j down to (and not including) the previous arrival position j' if this position is greater than $i + 1$ otherwise down to $i + 2$. More formally, for any j such that $M_c[j] \neq Nil$ let j' be the greatest number smaller than j such that $M_c[j] \neq Nil$. For any x, such that $\max(j', i + 2) \leq x \leq j$, add (i, x) to $G'_{\mathcal{A},T}$ together with its label. This concludes the construction of $G'_{\mathcal{A},T}$.

Since (i, j) and the dictionary D_i is prefix closed then all previous arcs of the form (i, x) are also arcs of $G_{\mathcal{A},T}$ and, therefore, $G'_{\mathcal{A},T}$ is a subgraph of $G_{\mathcal{A},T}$.

Proposition 7. *For any node* $v \in G_{\mathcal{A},T}$ *there exists a minimal path from the origin to* v *in* $G'_{\mathcal{A},T}$ *that is also a minimal path from the origin to* v *in* $G_{\mathcal{A},T}$.

We can now finally describe the *Dictionary-symbolwise flexible parsing.*

The *Dictionary-symbolwise flexible parsing* firstly uses algorithm BUILD $G'_{\mathcal{A},T}$ and then uses the classical SINGLE SOURCE SHORTEST PATH algorithm to recover a minimal path from the origin to the end of graph $G_{\mathcal{A},T}$. The correctness of above algorithm is stated in the following theorem and it follows from above description and from Prop. 7.

Theorem 1. Dictionary-symbolwise flexible parsing *is dictionary optimal.*

With respect to the original Flexible Parsing algorithm we gain the fact that it can works with variable costs of pointers and that it is extended to the dictionary-symbolwise case. But we loose the fact that the original one was "on-line". A minimal path has to be recovered, starting from the end of the graph backward. But this is an intrinsic problem that cannot be eliminated. Even if the dictionary edges have just one possible cost, in the dictionary-symbolwise case it is possible that any minimal path for a text T is totally different from any minimal path for the text Ta, that is the previous text T concatenated to the symbol 'a'. Even if the cost of pointers is constant. The same can happen when we have a "pure dictionary" case with variable costs of dictionary pointers. In both cases for this reason, there cannot exists "on-line" optimal parsing algorithms, and, indeed, flexible parsing fails being optimal in the pure dictionary case when costs are variable.

On the other hand our algorithm is suitable when the text is cut in several blocks and, therefore, in practice there is not the need to process the whole text but it suffices to end the current block in order to have the optimal parsing (relative to that block). As another alternative, it is possible to keep track of just one minimal path all along the text and to use some standard tricks to arrange it if it does not reach the text end, i.e. the wished target node. In the last cases one get a suboptimal solution that is a path with a cost extremely close to the minimal path.

5 Data Structures and Time Analysis

In this subsection we analyze *Dictionary-symbolwise flexible parsing* in both LZ78 and LZ77-alike algorithms.

Concerning LZ78-alike algorithms, the dictionary is prefix closed and it is usually implemented by using a technique that is usually referred as LZW implementation. We do not enter in details of this technique. We just recall that the cost of pointers increases by one unit whenever the dictionary size is "close" to a power of 2. The moment when the cost of pointers increases is clear to both encoder and decoder. In our dictionary-symbolwise setting, we suppose that the flag information for dictionary edges is constant. We assume therefore that it takes $O(1)$ time to determine the cost of all dictionary edges outgoing node i.

The maximal cost that a pointer can assume is smaller than $\log_2(n)$ where n is the text-size. Therefore the set $\mathcal{C}$ of all possible costs of dictionary edges has logarithmic size and it is cheap to calculate.

In [15] it is used a data structure, called trie-reverse pair, that is able to perform the operation of *Extend* and *Contract* in $O(1)$ time.

Since at any position we can calculate in $O(1)$ time the cost of outgoing edges, we can use the same data structure to perform our operations of *Weighted-Extend* and of *Weighted-Exist* in constant time. In order to perform a *Weighted-Extend* we simply use the *Extend* on the same non-weight parameters and, if the answer is "yes" we perform a further check in $O(1)$ time on the cost. In order to perform a *Weighted-Exist* we simply use the *contract* on the same non-weight parameters and, if the answer is "yes" we perform a further check in $O(1)$ time on the cost.

For any cost c finding M_c and the corresponding arcs in order to build $G'_{\mathcal{A},T}$ takes then linear time. Therefore, at a first look, performing the algorithm BUILD $G'_{\mathcal{A},T}$ would take $O(n\ log(n))$. But, since there is only one cost active at any position then if $c < c'$ then $M_c \subseteq M_{c'}$ as stated in the following proposition.

Definition 8. *We say that a cost function C is* LZW-alike *if for any i the cost of all dictionary pointers in D_i is a constant c_i and that for any i, $0 \leq i < n$ one has that $c_i \leq c_{i+1}$.*

Proposition 8. *If the cost funtcion C is* LZW-alike, *one has that if $c < c'$ then $M_c \subseteq M_{c'}$.*

At this point, in order to build $G'_{\mathcal{A},T}$ it suffices to build M_c where c is the greatest possible cost. Indeed it is useless checking for the cost and one can just use the standard operation *Extend* and *Contract*. Those operation can be implemented in $O(1)$ time using the trie reverse trie data structure for LZ78 standard dictionary or for the LZW dictionary or for the FPA dictionary (see [15]). Indeed we call a dictionary *LZ78-alike* if the operations *Extend* and *Contract* can be implemented in $O(1)$ time using the trie reverse trie data structure.

We notice that previous definition of LZ78-alikeness can be relaxed by asking that the operations *Extend* and *Contract* can be implemented in $O(1)$ amortized time using any data structure, including obviously the time used for building such data structure.

The overall time for building $G'_{\mathcal{A},T}$ is therefore linear, as well as its size. The SINGLE SOURCE SHORTEST PATH over $G'_{\mathcal{A},T}$, that is a DAG topologically ordered, takes linear time.

In conclusion we state the following theorem.

Theorem 2. *Suppose that we have a dictionary-symbolwise scheme, where the dictionary is LZ78-alike and the cost function is LZW-alike. The symbolwise compressor is supposed to be, as usual, linear time. Using the trie-reverse pair data structure,* Dictionary-Symbolwise flexible parsing *is linear.*

Concerning LZ77, in [4] it has been given, with a similar shortest path approach, an optimal parsing algorithm under some assumptions on the cost function. Our prefix-non-decreasing assumption is weaker than their assumptions in the sense that it is a consequence of their assumptions (see [4, Fact 4]). The maximal cost that a pointer can have under their assumption is still $O(\log(n))$ where n is the size of the text. It seems that it is possible to use the data structure used in [4] to perform, for any cost, *Weighted-Extend* and *Weighted-Exist* in amortized

$O(1)$ time. Then the overall time for the *Dictionary-symbolwise flexible parsing* when the dictionary is LZ77-alike would be $O(n \log(n))$, extending their result to the dictionary-symbolwise case. The subgraph $G'_{\mathcal{A},T}$ of $G_{\mathcal{A},T}$ is totally different from the one used in [4]. Indeed, quite recently, we discovered a simpler data structure that allows us to perform, for any cost, *Weighted-Extend* and *Weighted-Exist* in amortized $O(1)$ time. This data structure is built by using in a clever way $O(\log(n))$ suffix trees and it will be described in the journal version of this paper.

6 Conclusions

In this paper we present some advancement on dictionary-symbolwise theory. We describe the *Dictionary-Symbolwise Flexible Parsing*, a parsing algorithm that extends in non obvious way the *Flexible Parsing* (see [16]) to *variable* and *unbounded* costs and to the dictionary-symbolwise algorithm domain. We prove its optimality for prefix-closed dynamic dictionaries under some reasonable assumption. *Dictionary-Symbolwise Flexible Parsing* is *linear* for LZ78-alike dictionaries and even if it is not able to run online it allow to easily make a block programming implementation and a near to optimal online implementation, too. In the case of LZ77-alike dictionary, we have reobtained the $O(n \log(n))$ complexity as authors of [4] recently did and we use a completely different and simpler subgraph and a simpler data structure.

Our algorithm has therefore two advantages with respect to the classical Flexible Parsing. First, it can handle *variable cost* of dictionary pointers. This fact allows to extend the range of application of Flexible Parsing to almost all LZ78-alike known algorithms of our extension. Secondly, our Dictionary-Symbolwise Flexible Parsing implemented in the case of LZ77 dictionary gives as particular case when the symbolwise is not in use, a result that is similar to the one presented in [4] that has $O(n\ log(n))$ complexity, using a completely different and simpler subgraph and a simpler data structure. Last but not least our algorithm allows to couple classical LZ-alike algorithms with several symbolwise algorithms to obtain dictionary-symbolwise algorithms that achieve better compression with prove of optimality.

It is possible to prove, and we did not do it due to space limitation, that dictionary-symbolwise compressors can be asymptotically better than optimal pure dictionary compression algorithms in compression ratio terms, with LZ78 based dictionary and the same can be proved for LZ77 based dictionary. In our case, using a simple static Huffman coding as symbolwise compressor we improved the compression ratio of the *Flexible Parsing* of $7\% - 4\%$ on texts such as prefixes of English Wikipedia data base with a negligible slow down in compressing and decompressing time. The slow down comes from the fact that we have to add to the dictionary compression and decompression time the Huffman coding and decoding time. The same experimental result holds in general when the dictionary is LZ78-alike. Indeed a dictionary-symbolwise compressor when the dictionary is LZ78-alike and the symbolwise is a simple Huffman coding with optimal parsing has a compression ratio that is is more or less 5% better than the compression ratio of a pure LZ78-alike dictionary compressor that uses an optimal parsing. In general smaller is the file greater is the gain. The 5% refers to text sizes of around 20 megabytes. Moreover, preliminary results show that using more powerful but still fast symbolwise compressor, such as an arithmetic encoder of order 1, there is a further 10% gain in compression ratio. When the dictionary is, instead, LZ77-alike the gain in compression when we use a dictionary-symbolwise compressor with optimal parsing and Huffman coding with respect to a pure dictionary compressor with optimal parsing reduces down to more or less 3% over texts of 20 megabytes, smaller the file greater the gain. The compression ratio seems to be sensibly better than in the case of LZ78-alike dictionaries when we use, in both cases, unbounded dictionaries. The distance, however, between the compression ratio of dictionary-symbolwise compressors that use LZ78-alike dictionaries and the ones that use LZ77-alike dictionaries is smaller, following our preliminary results, when we use an arithmetic encoder of order 1 instead than an Huffman encoding. We have experimental evidence that many of the most relevant commercial compressors use, following our definition, optimal parsing in the dictionary-symbolwise case where the dictionary is LZ77-alike. The method described in this paper therefore has as a consequence the possibility of optimizing the trade-off between some of the main parameters used for evaluating commercial compressors, such as compression ratio, decompression time, compression time and so on. We plan to extend our experimentation on LZ-alike dictionary algorithms and many other symbolwise algorithms, since this direction seems to be very promising.

We conclude this paper with two open problems.

1. Theoretically, LZ78 is better on memoryless sources than LZ77. Experimental results say that when optimal parsing is in use it happens the opposite. Prove this fact both in pure dictionary case and in dictionary-symbolwise case.
2. Common symbolwise compressors are based on the arithmetic coding approach. When these compressors are used, the costs in the graph are almost surely non integer and, moreover, the graph is usually not well defined. The standard workaround is to use an approximation strategy. A big goal should be to find an optimal solution for these important cases.

References

1. Bell, T.C., Witten, I.H.: The relationship between greedy parsing and symbolwise text compression. J. ACM 41(4), 708–724 (1994)
2. Cohn, M., Khazan, R.: Parsing with prefix and suffix dictionaries. In: Data Compression Conference, pp. 180–189 (1996)
3. Cormen, T.H., Leiserson, C.E., Rivest, R.L., Stein, C.: Introduction to Algorithms, 2nd edn. MIT Press, Cambridge (2001)
4. Ferragina, P., Nitto, I., Venturini, R.: On the bit-complexity of lempel-ziv compression. In: Proceedings of the Nineteenth Annual ACM -SIAM Symposium on Discrete Algorithms, SODA 2009, pp. 768–777. Society for Industrial and Applied Mathematics, Philadelphia (2009)

5. Gzip's Home Page, http://www.gzip.org
6. Hartman, A., Rodeh, M.: Optimal parsing of strings, pp. 155–167. Springer, Heidelberg (1985)
7. Horspool, R.N.: The effect of non-greedy parsing in ziv-lempel compression methods. In: Data Compression Conference (1995)
8. Katajainen, J., Raita, T.: An approximation algorithm for space-optimal encoding of a text. Comput. J. 32(3), 228–237 (1989)
9. Katajainen, J., Raita, T.: An analysis of the longest match and the greedy heuristics in text encoding. J. ACM 39(2), 281–294 (1992)
10. Katz, P.: Pkzip archiving tool (1989), http://en.wikipedia.org/wiki/pkzip
11. Kim, T.Y., Kim, T.: On-line optimal parsing in dictionary-based coding adaptive. Electronic Letters 34(11), 1071–1072 (1998)
12. Klein, S.T.: Efficient optimal recompression. Comput. J. 40(2/3), 117–126 (1997)
13. Mahoney, M.: Large text compression benchmark, http://mattmahoney.net/text/text.html
14. Martelock, C.: Rzm order-1 rolz compressor (April 2008), http://encode.ru/forums/index.php?action=vthread&forum=1&topic=647
15. Matias, Y., Rajpoot, N., Shainalp, S.C.: The effect of flexible parsing for dynamic dictionary-based data compression. ACM Journal of Experimental Algorithms 6, 10 (2001)
16. Matias, Y., Shainalp, S.C.: On the optimality of parsing in dynamic dictionary based data compression. In: SODA, pp. 943–944 (1999)
17. Della Penna, G., Langiu, A., Mignosi, F., Ulisse, A.: Optimal parsing in dictionary-symbolwise data compression schemes (2006), http://www.di.univaq.it/mignosi/ulicompressor.php
18. Schuegraf, E.J., Heaps, H.S.: A comparison of algorithms for data base compression by use of fragments as language elements. Information Storage and Retrieval 10(9-10), 309–319 (1974)
19. Storer, J.A., Szymanski, T.G.: Data compression via textural substitution. J. ACM 29(4), 928–951 (1982)
20. Wagner, R.A.: Common phrases and minimum-space text storage. ACM Commun. 16(3), 148–152 (1973)

Regular Language Constrained Sequence Alignment Revisited

Gregory Kucherov[1], Tamar Pinhas[2], and Michal Ziv-Ukelson[2]

[1] LIFL/CNRS and INRIA Lille Nord-Europe, Villeneuve d'Ascq, France
[2] Department of Computer Science, Ben-Gurion University of the Negev, Be'er Sheva, Israel

Abstract. Imposing constraints in the form of a finite automaton or a regular expression is an effective way to incorporate additional a priori knowledge into sequence alignment procedures. With this motivation, Arslan [1] introduced the Regular Language Constrained Sequence Alignment Problem and proposed an $O(n^2t^4)$ time and $O(n^2t^2)$ space algorithm for solving it, where n is the length of the input strings and t is the number of states in the non-deterministic automaton, which is given as input. Chung et al. [2] proposed a faster $O(n^2t^3)$ time algorithm for the same problem. In this paper, we further speed up the algorithms for Regular Language Constrained Sequence Alignment by reducing their worst case time complexity bound to $O(n^2t^3/\log t)$. This is done by establishing an optimal bound on the size of Straight-Line Programs solving the maxima computation subproblem of the basic dynamic programming algorithm. We also study another solution based on a Steiner Tree computation. While it does not improve the run time complexity in the worst case, our simulations show that both approaches are efficient in practice, especially when the input automata are dense.

1 Introduction

1.1 Constrained Sequence Alignment

Sequence alignment algorithms use a position independent scoring matrix, but when biologists make an alignment they favor some similarities, depending on their knowledge of the structure and/or the function of the sequences. Various extensions of the Smith-Waterman algorithm [3] modify the alignment considerations according to a priori knowledge [4–8]. One kind of a priori knowledge is about shared properties (patterns) which are expected to be preserved by the alignment. Specifically, in protein sequence alignment, it is natural to expect that functional sites be aligned together. Several studies suggested taking into account the patterns (specified by regular expressions) from the PROSITE database [9] to guide and constrain protein alignments [1, 10], since such patterns may serve as good descriptors of protein families.

In [1], Arslan introduced the Regular Expression Constrained Sequence Alignment Problem. Here, the constraint on sequence alignment is given in the form

C.S. Iliopoulos and W.F. Smyth (Eds.): IWOCA 2010, LNCS 6460, pp. 404–415, 2011.

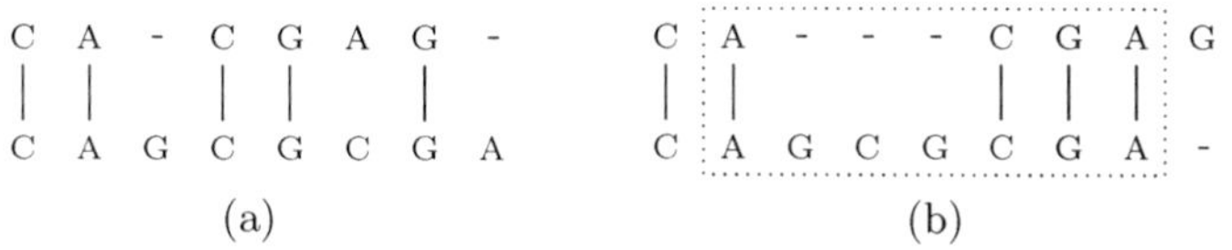

```
C A - C G A G -        C : A - - - C G A : G
| |   | | |            | : |       | | | :
C A G C G C G A        C : A G C G C G A : -
      (a)                      (b)
```

Fig. 1. Examples of a sequence alignment and a regular expression constrained sequence alignment on the two strings $CACGAG$ and $CAGCGCGA$, with a scoring matrix (-1 for mismatch/insert/delete, 1 for match). (a) The maximal score of the global alignment is 2. (b) Let R be $A(G+C)^*GA$, the constrained problem's score is 1.

of a non-deterministic finite automaton [NFA]. An alignment satisfies the constraint if a segment of it is accepted by the NFA in each aligned sequence (see Fig. 1). Arslan's dynamic programming algorithm is based on applying an NFA, with scores assigned to its states, to guide the sequence alignment. This NFA accepts all alignments of the two input strings containing a segment that belongs to the input regular language. The algorithm yields $O(n^2t^4)$ time and $O(n^2t^2)$ space complexities, where n is the sequence length and t is the number of states in the NFA expressing the constraint. The algorithm simulates copies of this automaton on alignments, updating state scores, as dictated by the underlying scoring scheme. Chung, Lu and Tang [2] proposed an improvement to the above algorithm, yielding $O(n^2t^3)$ time and $O(n^2t^2)$ space complexities, in the general case, exploiting the sparse structure of the automaton constructed in Arslan's algorithm. This improved algorithm is described in detail in section 1.3.

Our Contribution. In this paper, we further speed up the algorithms for Regular Language Constrained Sequence Alignment by reducing their worst case time complexity bound to $O(n^2t^3/\log t)$. This is done by establishing an optimal bound on the size of Straight-Line Programs [SLP] solving the maxima computation subproblem of the basic dynamic programming algorithm. We also study another solution based on a Steiner Tree computation. While it does not improve the run time complexity in the worst case, our simulations show that both approaches are efficient in practice, especially when the input automata are dense.

Roadmap. The rest of this paper proceeds as follows. In this section, we define the Regular Language Constrained Sequence Alignment problem and give an overview of previous algorithms for the problem. In Section 2, we describe and analyze two new algorithms based on Steiner Trees and SLPs; Section 2.4 includes some simulations and comparative test results.

1.2 Preliminaries and Definitions

Let Σ be a finite alphabet. Let $a, b \in \Sigma^*$ two strings over the alphabet Σ. We denote $a_{i,j}$ the substring of a from index i to index j (included) and a_i is the ith character in a. Let $\Sigma' = \Sigma \cup \{-\}$ be an extended alphabet, where $- \notin \Sigma$. Let $X, Y \in \Sigma'^*$. We denote X^-, the string result of the removal of $-$ characters from X. Let $s : \Sigma' \times \Sigma' \setminus \{-,-\} \rightarrow \Re^+$ be a scoring function over edit operations (i.e. replace, insert and delete). (X, Y) is an alignment of a and b if $|X| = |Y|$, $X^- = a$ and $Y^- = b$. The score of an alignment (X, Y) is $s((X, Y)) = \sum_{i=1}^{|X|} s(X_i, Y_i)$.

Below, we define the alignment optimization problem and its regular language constrained variant. Both problems are formalized in their score maximization version (the score minimization version is symmetric).

Definition 1 (Sequence Alignment). *Given two strings a and b, both over a fixed alphabet Σ and a scoring function s. Find an alignment of a and b with a maximal score under s.*

Let L_R be a regular language. Let $A = (Q, \Sigma, \delta, q_0, F_A)$ be an NFA with t states, such that $L(A) = L_R$. We assume that ϵ-transitions were removed from A. For convenience, we denote the automaton transition table δ as follows: $q \xrightarrow{c} p$ states that there is a transition from state q to state p by character c. In addition, we denote the number of transitions in δ as $|\delta|$.

Definition 2 (Regular Language Constrained Sequence Alignment). *Given two strings a and b, both over a fixed alphabet Σ, a scoring function s and an NFA A. Find an alignment (X, Y) of a and b such that it is the alignment with the maximal score under s which satisfies the following condition: indices i and j exist such that $X^-_{i,j}, Y^-_{i,j} \in L(A)$.*

1.3 An Overview of Previous Work

Arslan's algorithm defines an NFA M, such that the states of M are the ordered pairs of states of A, therefore, it has $O(t^2)$ reachable states. M is defined over the alphabet $\Sigma'' = \Sigma' \times \Sigma'$. For every two transitions $q_1 \xrightarrow{c_1} p_1$ and $q_2 \xrightarrow{c_2} p_2$ in A, the transitions $(q_1, q_2) \xrightarrow{(c_1,c_2)} (p_1, p_2)$, $(q_1, q_2) \xrightarrow{(c_1,-)} (p_1, q_2)$ and $(q_1, q_2) \xrightarrow{(-,c_2)} (q_1, p_2)$ exist in M. For any two final states $q_{f_1}, q_{f_2} \in F_A$ and for any characters c_1, c_2 the transitions $(q_{f_1}, q_{f_2}) \xrightarrow{(c_1,c_2)} (q_{f_1}, q_{f_2})$, $(q_{f_1}, q_{f_2}) \xrightarrow{(c_1,-)} (q_{f_1}, q_{f_2})$ and $(q_{f_1}, q_{f_2}) \xrightarrow{(-,c_2)} (q_{f_1}, q_{f_2})$ exist in M. The same addition is done for the initial state. A sequence alignment table T of size $(|a|+1) \times (|b|+1)$ is calculated. Each cell $T_{i,j}$, contains a table of scores, one for every state in M (that is, a pair of states in A). $T_{i,j}(p, q)$ is the maximal score of an alignment of $a_{1,i}$ and $b_{1,j}$, such that reading it in M ends at (p, q). The table size is clearly $O(n^2 t^2)$, since each cell holds t^2 scores.

In the following recurrence formula for $T_{i,j}$, we move from the multiplication automaton notion of Arslan to a simpler formulation. As a first step, we add to A transitions $q \xrightarrow{c} q$ where c is any symbol and q is either an initial or final state in A. The score of $T_{i,j}$ for a given state (p, q) is computed as follows.

$$T_{i,j}(q,p) = \max \begin{cases} \max\{T_{i-1,j}(q',p) | q' \in Q,\ q' \xrightarrow{a_i} q\} + s(a_i, -) \\ \max\{T_{i,j-1}(q,p') | p' \in Q,\ p' \xrightarrow{b_j} p\} + s(-, b_j) \\ \max\{T_{i-1,j-1}(q',p') | q',p' \in Q,\ q' \xrightarrow{a_i} q,\ p' \xrightarrow{b_j} p\} + s(a_i, b_j)(*) \end{cases} \quad (1)$$

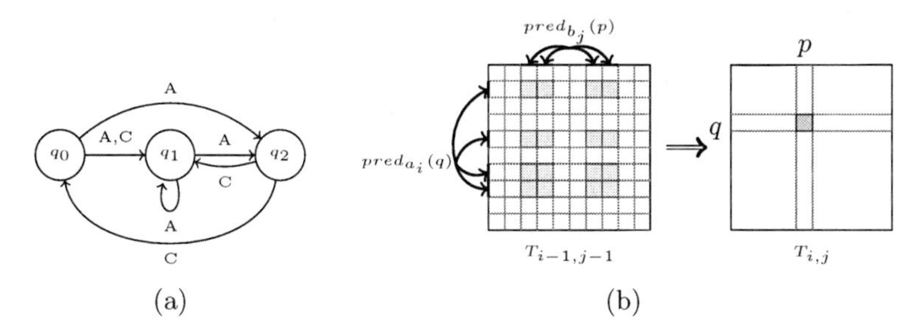

Fig. 2. (a) An example of an NFA. Its transitions yield the following *pred* sets: $pred_A(q_0) = \emptyset$, $pred_A(q_1) = pred_A(q_2) = \{q_0, q_1\}$, $pred_C(q_0) = \{q_2\}$, $pred_C(q_1) = \{q_0, q_2\}$, $pred_C(q_2) = \emptyset$.
(b) Score calculation performed by Arslan's algorithm. The green scores in $T_{i-1,j-1}$, corresponding to rows $pred_{a_i}(q)$ and columns $pred_{b_j}(p)$, are used in the calculation of $T_{i,j}$ for the state pair (q, p).

For the sake of simplicity, we define $\max \emptyset = -\infty$. The initialization consists of assigning 0 to $T_{0,0}(q_0, q_0)$ and $-\infty$ elsewhere. The optimal alignment score is $\max\{T_{|a|,|b|}(q,p)|q,p \in F_A\}$. There are a total of $O(n^2t^2)$ scores to calculate. According to Eq. 1, each score calculation (for a given i, j, q and p) involves $O(t^2)$ values in the worst case, as apparent in the third term marked $(*)$, because in an NFA there are at most t transitions $q' \xrightarrow{a_i} q$ for a single character a_i and, independently, there are at most t transitions $p' \xrightarrow{b_j} q$ for b_j (see Fig. 2(b)). The term $(*)$ is the bottleneck of the algorithm. Since A is non-deterministic, it may contain $O(t^2)$ transitions by any character c.

The algorithm of Chung et al. exploits the following redundancy: Given that M is an NFA with $|\delta| = O(t^2)$ states, and assuming no additional knowledge of M, it can be concluded that M can potentially have $O(t^4)$ transitions. Thus, Arslan's algorithm iterates over all possibilities of two states of M in each $T_{i,j}$ calculation. However, it is known, according to the way M was built, that each transition in M originates from at most two transitions in A. The iteration over the two possible transitions can be done independently of each other.

Definition 3. *Let A be an ϵ-free NFA. We denote $pred_c(q)$ the set of states with outgoing transitions labeled by character c and leading to state q.*

$$pred_c(q) = \{p | p \xrightarrow{c} q\} \tag{2}$$

Using this notation (see Fig. 2(a)), Eq. (1) can be rewritten as follows:

$$T_{i,j}(q,p) = \max \begin{cases} \max\{T_{i-1,j}(q',p)|q' \in pred_{a_i}(q)\} + s(a_i, -) \\ \max\{T_{i,j-1}(q,p')|p' \in pred_{b_j}(p)\} + s(-, b_j) \\ \max\{T_{i-1,j-1}(q',p')| \\ \qquad q' \in pred_{a_i}(q), p' \in pred_{b_j}(p)\} + s(a_i, b_j) (*) \end{cases} \tag{3}$$

Chung et al. [2] sped up Arslan's algorithm by removing redundant computations which were due to the fact that the computed value is based on two independent

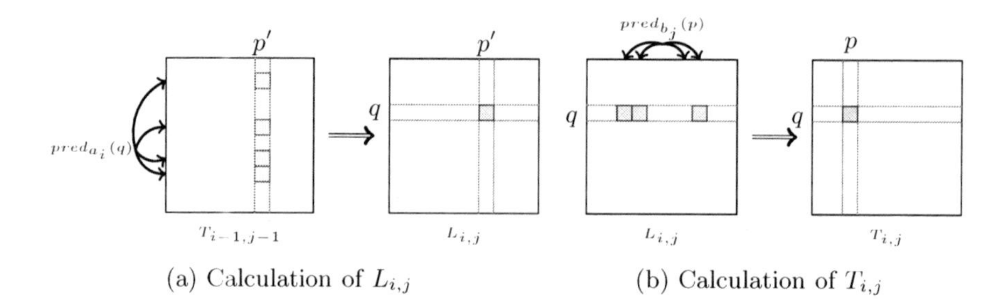

(a) Calculation of $L_{i,j}$ (b) Calculation of $T_{i,j}$

Fig. 3. Score calculation performed by Chung et al.'s algorithm

optimum calculations, one for each of the compared strings. We next describe Chung et al.'s algorithm using our own notation. The calculation of (∗) is split into two steps using an intermediate table L (see Fig. 3).

$$\begin{aligned} L_{i,j}(q,p') &= \max\{T_{i-1,j-1}(q',p')|q' \in pred_{a_i}(q)\} \\ T_{i,j}(q,p) &= \max\{L_{i,j}(q,p')|p' \in pred_{b_j}(p)\} + s(a_i,b_j) \end{aligned} \tag{4}$$

In the first step, the size of the set, over which the maximum is calculated for every pair of states of A, (q,p'), depends on the existing transitions in the automaton A with the character a_i. Since the size of the set is bounded due to $|pred_{a_i}(q)| \leq t$, the first step takes t^3 time. The same argument holds for the second step. In summary, their algorithm improved the time complexity to $O(n^2t|\delta|) = O(n^2t^3)$, while maintaining the same space complexity.

2 A Faster Algorithm

2.1 Eliminating Duplicate Computations

It is apparent from Eq. (4) that the calculation of $L_{i,j}$, for a specific value of p' and ranging over q, takes the maxima over subsets of indices of column p' of $T_{i-1,j-1}$, while the calculation of $T_{i,j}$, for a specific value of q and ranging over p, takes the optimum over subsets of indices of the qth row of $L_{i,j}$ (see Fig. 3).

The structure of the NFA transition table, namely the relations between the $pred_c$ sets, can be used to reduce the number of components required in consecutive subset maxima calculations. For instance, let us assume that a state q' is included both in $pred_c(q_1)$ and $pred_c(q_2)$ for $q_1 \neq q_2$. Then, for a given state p, the score $T_{i-1,j-1}(q',p)$ is taken into account in calculations of both $L_{i,j}(q_1,p)$ and $L_{i,j}(q_2,p)$ (see Fig. 4). By minimizing the repetition of score usage, the efficiency of the calculation of Eq. (1) and Eq. (4) can be improved.

Following the observations above, the goal of speeding up the calculation of Eq. (4) can be formulated as the following question: What is the most efficient way to calculate maximum values over given, possibly overlapping, sets of scores? Thus, the general problem underlying the speed up of these algorithms can be formulated as follows.

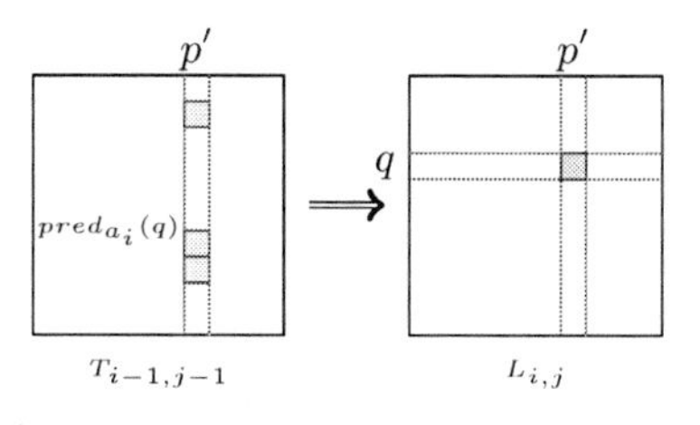

(a) The calculation of a single score in $L_{i,j}$ depends on several scores in $T_{i-1,j-1}$

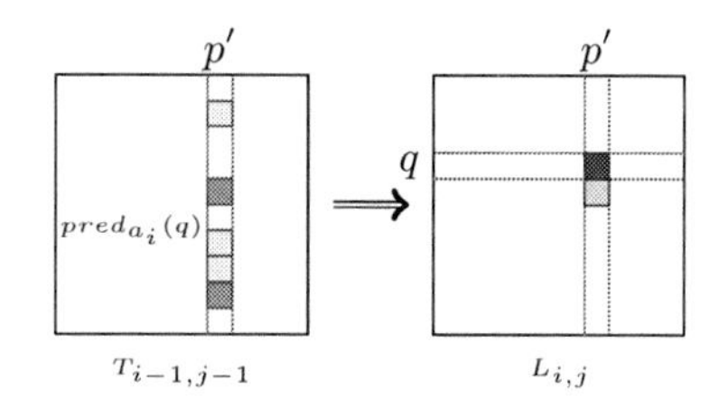

(b) The calculation of a different score in $L_{i,j}$ can be done according to the previously calculated score and some additional scores from $T_{i-1,j-1}$

Fig. 4. Similar and duplicate score calculations in Chung et al.'s algorithm can be reused

Definition 4 (Subsets Maxima Problem). *Let W be a set of scores, with $|W| = t$ and let $V = \langle v_1, ..., v_t \rangle$ be t subsets of W ($v_k \subseteq W$). Calculate* $\max v_k$ *for each $v_k \in V$.*

Thus, for the first part of Eq. 4, having fixed values of i, j and p' at a specific step, the set of scores W consists of t scores in $T_{i-1,j-1}$ and V consists of scores which correspond to all possible $pred_{a_i}$ subsets. More formally:

$$\begin{array}{l} W = \{T_{i-1,j-1}(q', p') | q' \in Q\} \\ V = \langle v_1, \ldots, v_t \rangle, v_k = \{T_{i-1,j-1}(q', p') | q' \in pred_{a_i}(q_k)\},\ q_k \in Q \end{array} \tag{5}$$

The values of W and V are similarly established for the second part of Eq. 4.

We represent each subset v_k in V by a Boolean vector, where the ith bit reflects the membership of the ith score in the subset $pred_{a_i}(q_k)$. Thus, V is represented by a tuple of Boolean vectors, denoted S. In the following sections, we discuss two alternative ways of solving the Subsets Maxima Problem: one based on Steiner trees (Section 2.2) and the other based on Straight Line programs (Section 2.3).

2.2 An Algorithm Based on a Steiner Minimal Directed Tree

In this section, we explore the possibility of employing Steiner minimal directed trees to solve the Subsets Maxima Problem. We show that the size of a Steiner minimal directed tree for a tuple of Boolean vectors S, as described above, is not greater than the number of transitions of the NFA. Thus, using a heuristic algorithm for Steiner minimal directed trees improves the run-time of our solution to Regular Language Constrained Alignment in practice, as demonstrated by our simulations (Section 2.4). But first, we give a formal definition of Steiner minimal directed trees and review related work.

There are several Steiner tree and graph problems known in the literature. The general Steiner tree problem is the problem of spanning a k-sized set S of vertices of a graph, while including the minimum number of nodes that are not in S. One variation of this problem is the Steiner minimal directed tree problem, which specifies a root node and adds the requirement that the graph be directed. These problems are known to be NP-hard [11, 12].

In this work, we are interested in the Steiner minimal directed tree problem in a specific graph, namely the Hamming hypercube. Most Steiner algorithms for graphs are not applicable to the Hamming hypercube due to its exponential size. In the Hamming hypercube, the Hamming distance between any two adjacent nodes of the tree, v and u, is 1. That is, either u has exactly one 1-valued bit more than v or vice versa.

Definition 5 (The Steiner minimal directed tree problem for Hamming hypercubes). *Given a set S of k d-dimensional points, find a rooted tree in the H^d Hamming hypercube, such that the tree spans S, has the minimal possible size N and all edges are directed away from the root (For every edge (v, u) in the tree, if v is closer to the root than u, then u has exactly one more 1-valued bit than v).*

This version of the Steiner minimal tree problem is also NP-hard [13] and have several heuristic algorithms [14–16].

Given a directed Steiner minimal tree for a tuple of Boolean vectors, S, the subsets maxima of the corresponding weight-subsets tuple, V, can be calculated by traversing the tree in a top-down fashion.

Theorem 1 (upper bound of $|\delta|$ for the size of the Steiner minimal directed tree). *Let $A = (Q, \Sigma, \delta, q_0, F_A)$ be an NFA and let S_c, for $c \in \Sigma$, be sets of Boolean vectors corresponding to δ, as described in subsection 2.1. There exist Steiner directed trees for sets S_c, such that the sum of their sizes is not greater that $|\delta| + t$.*

Proof. S_c, for a specific c, is a set of Boolean vectors representing $pred_c(A)$. Thus, the total number of 1-valued bits in all S_c sets equals $|\delta|$. For each set S_c, we build a Steiner directed tree, as follows. Let X be a Boolean vector in S_c, such that bits $x_1, x_2..., x_k$ in X are 1-valued. Starting from the zero vector, 0^t, as the root, we add a chain of nodes in the Steiner tree until X is reached. The first node connected to 0^t is the elementary vector with a 1-valued x_1 bit. Similarly, the ith node is a vector that has all bits equal to its parent node, except for the x_i bit, which is 1-valued in that vector, but is 0-valued in the parent vector. This path reaches vector X by adding at most k nodes (not including the zero vector). The total length of the tree S_c is not greater than the total number of 1-valued bits in S_c plus 1 (for the zero vector). Thus, such Steiner directed trees, for sets S_c, have the sum of their lengths not greater that $|\delta| + t$. □

It is easy to see that the size of the Steiner minimal tree is $N = \Theta(t^2)$ in the worst case. Thus, our Steiner-based algorithm, in the framework of Chung et al., runs in $O(n^2t^3)$ time.

Theorem 2 (lower bound). *For any set $S \subseteq H^t$, such that $|S| = t$, it holds that $N = \Omega(t^2)$.*

Proof. For every natural t, we show the existence of a t-sized set S, such that N is in the order of t^2. Let us assume that $t = 2^k$ for a natural number k. We select S to be any t Boolean vectors from the k-dimensional Hadamard code [17–19]. The Hadamard code contains $2t = 2^{k+1}$ vectors, each of length $t = 2^k$, such that each two vectors have a Hamming distance of at least $\frac{t}{2} = 2^{k-1}$. The Hamming distance within S is at least $\frac{t}{2}$ (the $(\frac{t}{4} - 1)$-radius ball surrounding each vector in S does not contain any other vector in S). Moreover, $\frac{t}{4} - 1$-radius balls surrounding different vectors in S, are disjoint. Thus, a tree that spans S requires at least $t(\frac{t}{4} - 1)$ Steiner nodes. □

In Section 2.4, we compare the sizes of heuristic Steiner directed trees with the size of the corresponding transition table for simulated NFAs. Our simulations show that, even though the Steiner-based algorithm does not yield the theoretical bounds obtained for SLPs, in practice it performs very well.

2.3 A Solution to Subsets Maxima via SLPs

The Subsets Maxima Problem can be reduced to the problem of finding the shortest possible SLP with Boolean operations. In order to use SLPs for the task of subsets maxima calculation, we represent V as a tuple of Boolean vectors, S, as described in subsection 2.1.

Definition 6 (SLP with Boolean operations). *We are given a tuple of t Boolean vectors $S = \langle x_1, \ldots, x_t \rangle$, $x_i \in \{0,1\}^m$. An SLP is a sequence of instructions P, of two types:*

- $\beta_i := (0, \ldots, 0, 1, 0, \ldots, 0)$ *(elementary vector),*
- $\beta_i := \beta_j \vee \beta_l$, *with $j, l < i$ (disjunction).*

An SLP computes the left-hand side vectors of its instructions $\langle \beta_1, \ldots, \beta_N \rangle$, $\beta_i \in \{0,1\}^m$. An SLP P computes S if $\langle \beta_{N-t+1}, \ldots, \beta_N \rangle = S$.

Given an SLP for S, the subsets maxima can be calculated by following the SLP in linear order: if β_h is an elementary vector, having the ith bit equal to 1, the vector is assigned the value of the ith score and if β_h is a binary disjunction of β_j and β_k, then it is assigned the value of the maximum of their assigned scores. If β_h represents a subset from V, its score is reported.

For the purpose of utilizing SLPs for the Subsets Maxima Problem, in the rest of this section we address the following goal: given a tuple S of t Boolean vectors of length t, construct an SLP for S of minimal length. This goal is achieved via the following two theorems.

Theorem 3 (upper bound). *An SLP for S can be generated such that:*
(1) $N \leq \frac{2t^2}{\log t}$, where N denotes the size of the SLP, and
(2) the time required to construct it is $O\left(\frac{t^2}{\log t}\right)$.

Proof. We will use the Four-Russians trick. A similar argument is applied in [20].

Split each vector of S into $b = t/\log t$ blocks of length $\log t$. Each block has $2^{\log t} = t$ possible values. For each $i = 1..b$, consider the set of all block vectors, say W_i, such that block i takes all the possible values and the other blocks are all zeros. All the vectors of W_i can be generated incrementally with t operations (in a bottom-up fashion): First, all the vectors in W_i which have a single 1-valued bit are generated, then all the vectors in W_i which have two 1-valued bits are generated by the disjunction of two vectors in W_i with one 1-valued bit. In general, all the vectors in W_i with $j+1$ 1-valued bits are generated by adding disjunction operations between vectors with j 1-valued bits and vectors with one 1-valued bit in W_i. Therefore, there are a total of $bt = \frac{t^2}{\log t}$ block vectors and it takes $O\left(\frac{t^2}{\log t}\right)$ time to create all the block vectors.

Each vector of S can then be generated in $b-1$ disjunction operations from pre-computed block vectors and there are t vectors in S. All the vectors of S are, therefore, computed by adding $t(b-1) \leq t^2/\log t$ operations to the SLP.

The length of the underlying SLP constructed here, equals the number of disjunction and elementary operations, summed over both stages (block vector creation plus computing S from the block vectors), which is at most $\frac{2t^2}{\log t}$. The time required for the construction of the SLP is $O\left(\frac{t^2}{\log t}\right)$. □

Remark. Note that the bound in Theorem 3 can be improved by a factor of two by taking blocks of size $\log t - \log\log t$.

The above bound is very close to the information-theoretic lower bound, as shown below.

Theorem 4 (lower bound). *An SLP for S requires* $\Omega\left(\frac{t^2}{\log t}\right)$ *operations.*

Proof. We use the standard counting argument. Again, a similar proof can be found, e.g. in [20].

There are t distinct initialization instructions and, in the minimal SLP, each of them occurs at most once. Without loss of generality, we assume that the initialization instructions form t first instructions in the SLP in any fixed order.

Let q be the number of disjunction instructions, i.e. $N = t + q$. There are at most N^2 possibilities for each disjunction instruction and, therefore, there are at most $(N^2)^q = N^{2q}$ different SLPs of length N. On the other hand, there are $(2^t)^t = 2^{t^2}$ different tuples S. We then should have $N^{2q} \geq 2^{t^2}$, i.e. $2q\log(t+q) \geq t^2$.

Resolving the inequality with respect to q gives a lower bound, matching that of Theorem 3 up to a constant factor. Specifically, it implies that, for any $\varepsilon > 0$ and for almost any tuple S, the size of the minimal SLP for S is at least $\frac{t^2}{(2+\varepsilon)\log t}$. □

Finally, we conclude that Theorems 3 and 4 improve the worst case bounds of Regular Language Constrained Alignment by a logarithmic factor.

Theorem 5. *Regular Language Constrained Alignment can be computed in* $O\left(\frac{n^2t^3}{\log t}\right)$ *time and* $O\left(n^2t^2\right)$ *space.*

Proof. The dynamic programming process described in Eq. 4 involves the calculation of $L_{i,j}$ for every $p' \in Q$, and then the calculation of $T_{i,j}$ for every $q' \in Q$, using a precomputed SLP, as described above. This takes $O(n^2 \cdot t \cdot N)$, where N denotes the maximal length of an SLP for the sets V corresponding to the given NFA. By Theorems 3 and 4, the length of such an SLP is $N = \Theta\left(\frac{t^2}{\log t}\right)$. □

Discussion. Chung et al.'s algorithm for the Regular Language Constrained Alignment Problem yields $O(n^2|Q| \cdot |\delta|) = O(n^2t^3)$ time and $O(n^2t^2)$ space. Thus, the above contribution is only interesting when the input automaton is dense, i.e. when $|\delta|$ is asymptotically larger than $O(\frac{t^2}{\log t})$.

We further note that, in the case when the input is given in the form of a regular expression rather than an automaton, the complexity analysis of the algorithm can be expressed in terms of the length r of the input regular expression. This is achieved based on recent algorithms which take as input a regular expression of length r and convert it into an ϵ-free NFA with $O(r)$ states and $O(r \log^2 r)$ transitions [21–23]. This yields $O(n^2r^2 \log^2 r)$ time and $O(n^2r^2)$ space. Note that this was not observed by Arslan [4] and Chung et al. [2].

2.4 Simulation Results

We compared the efficiency of using heuristic Steiner minimal directed trees and SLPs as a function of NFA density (see Fig. 5). To measure this, we randomly generated NFAs and constructed their corresponding data structures (Steiner minimal trees and SLPs) and measured their sizes. This simulation was repeated 500 times for each NFA size t, for $t = 10, 20, 50, 80$.

The random NFAs were constructed as follows. NFAs with a unary alphabet suffice for our purposes, since, at each step of the algorithm, transitions with a single character are used (see Eq. (4)). For a given number of states t, the transition table of an NFA was randomly generated. Only NFAs with t reachable states were considered. As shown in Fig. 5, the randomly generated NFAs range in density.

For each NFA, a corresponding heuristic Steiner minimal directed tree is constructed, having the t-dimensional Boolean zero vector, 0^t, as its root. For the construction of the heuristic Steiner minimal directed tree, we use the heuristic algorithms of [15, 16], with a minor modification that forces the constructed tree to be directed. In addition, for each NFA, a corresponding SLP is constructed. We use a greedy heuristic, which adds disjunction operations according to the bit which is set in the maximal number of Boolean vectors in the required set S. Finally, we compute the size of the SLP, according to the SLP construction described in Theorem 3, without actually computing the tree. Namely, the size of the theoretical SLP is $\min\{\frac{2t^2}{\log t}, |\delta|\}$.

We measure the efficiency of a constructed data structure as $1 - N/|\delta|$, where N is its size (i.e. the size of the constructed Steiner directed tree, the length of the constructed SLP or the upper bound of the size of the theoretical SLP construction). The efficiency of the different data structures is compared as a function of the density of the NFA. The density of the NFA equals $|\delta|/t^2$.

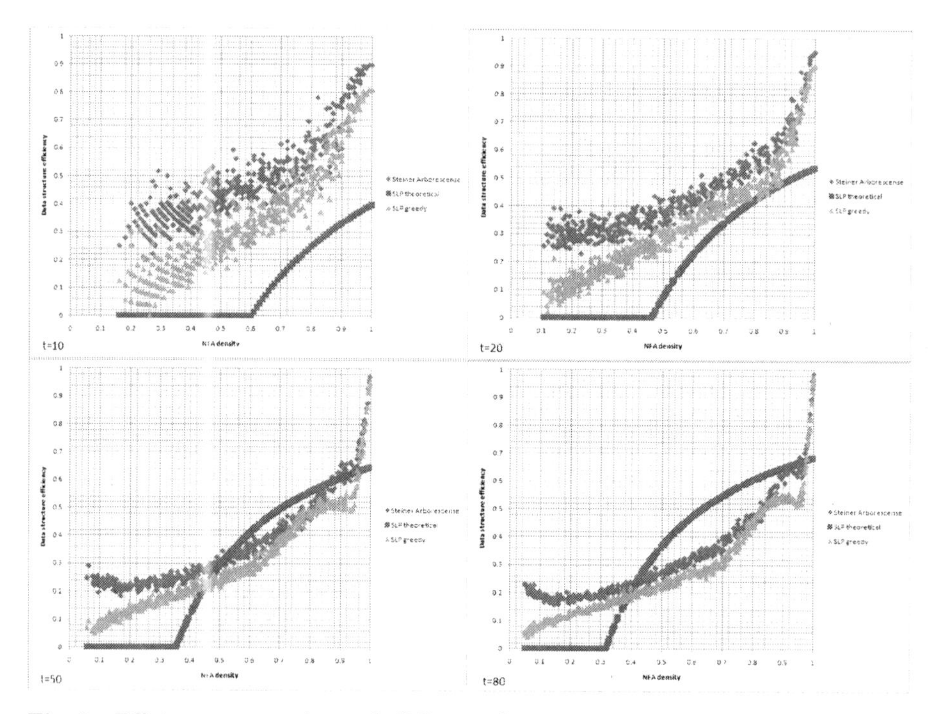

Fig. 5. Efficiency comparison of different data structures as a function of NFA density. The simulation was repeated for different automaton size (no. of states), $t = 10, 20, 50, 80$, each containing 500 randomly generated NFAs with t states and their corresponding data structures. **Legend**: blue diamond - heuristic Steiner minimal directed tree, red square - SLP theoretical construction as in Theorem 3, green triangle - SLP construction via greedy algorithm.

Our simulations show that all proposed data structures have an increased efficiency as NFA density increases. The heuristic Steiner minimal directed tree has a relatively high efficiency for low values of t, however, the SLP construction described in Theorem 3 has a relatively high efficiency for high values of t and relatively high NFA density.

Acknowledgments

We are grateful to Gregory Kabatiansky for his help on Theorem 2. The work of Tamar Pinhas and Michal Ziv-Ukelson was partially supported by the Lynn and William Frankel Center for Computer Sciences. English editing by Ethelea Katzenell.

References

1. Arslan, A.: Regular expression constrained sequence alignment. Journal of Discrete Algorithms 5(4), 647–661 (2007)
2. Chung, Y., Lu, C., Tang, C.: Efficient algorithms for regular expression constrained sequence alignment. Information Processing Letters 103(6), 240–246 (2007)

3. Smith, T., Waterman, M.: Identification of common molecular subsequences. Journal of Molecular Biology 147(1), 195–197 (1981)
4. Arslan, A., Egecioglu, O.: Algorithms for the constrained longest common subsequence problems. International Journal of Foundations of Computer Science 16(6), 1099–1110 (2005)
5. Chen, Y., Chao, K.: On the generalized constrained longest common subsequence problems. Journal of Combinatorial Optimization, 1–10 (2009)
6. Iliopoulos, C., Rahman, M.: New efficient algorithms for the LCS and constrained LCS problems. Information Processing Letters 106(1), 13–18 (2008)
7. Peng, Z., Ting, H.: Time and space efficient algorithms for constrained sequence alignment. In: Domaratzki, M., Okhotin, A., Salomaa, K., Yu, S. (eds.) CIAA 2004. LNCS, vol. 3317, pp. 237–246. Springer, Heidelberg (2005)
8. Tsai, Y.: The constrained longest common subsequence problem. Information Processing Letters 88(4), 173–176 (2003)
9. Bairoch, A.: The PROSITE dictionary of sites and patterns in proteins, its current status. Nucleic Acids Research 21(13), 3097 (1993)
10. Tang, C., Lu, C., Chang, M., Tsai, Y., Sun, Y., Chao, K., Chang, J., Chiou, Y., Wu, C., Chang, H., et al.: Constrained multiple sequence alignment tool development and its application to RNase family alignment. Journal of Bioinformatics and Computational Biology 1(2), 267–287 (2003)
11. Bern, M., Plassmann, P.: The Steiner problem with edge lengths 1 and 2. Information Processing Letters 32(4), 171–176 (1989)
12. Shi, W., Su, C.: The rectilinear Steiner arborescence problem is NP-complete. SIAM Journal on Computing 35(3), 729–740 (2006)
13. Foulds, L., Graham, R.: The Steiner problem in phylogeny is NP-complete. Advances in Applied Mathematics 3(43-49), 299 (1982)
14. Jia, W., Han, B., Au, P., He, Y., Zhou, W.: Optimal multicast tree routing for cluster computing in hypercube interconnection networks. IEICE Transactions on Information and Systems E87-D, 1625–1632 (2004)
15. Lin, X., Ni, L.: Multicast communication in multicomputer networks. IEEE Transactions on Parallel and Distributed Systems 4(10), 1105–1117 (1993)
16. Sheu, S., Yang, C.: Multicast algorithms for hypercube multiprocessors. Journal of Parallel and Distributed Computing 61(1), 137–149 (2001)
17. Dinur, I., Safra, S.: On the hardness of approximating minimum vertex cover. Annals of Mathematics 162(1), 439–486 (2005)
18. Sylvester, J.: Thoughts on inverse orthogonal matrices simultaneous sign successions, and tessellated pavements in two or more colors, with applications to Newton's rule, ornamental tile-work and the theory of numbers. Phil. Mag. 34(2), 461–475 (1867)
19. Seberry, J., Yamada, M.: Hadamard matrices, sequences, and block designs. Contemporary Design Theory: A Collection of Surveys, 431–560 (1992)
20. Savage, J.: An algorithm for the computation of linear forms. SIAM J. Comput. 3(2), 150–158 (1974)
21. Hromkoviĕc, J., Seibert, S., Wilke, T.: Translating regular expressions into small ε-free nondeterministic finite automata. Journal of Computer and System Sciences 62(4), 565–588 (2001)
22. Schnitger, G.: Regular expressions and NFAs without epsilon-transitions. In: Durand, B., Thomas, W. (eds.) STACS 2006. LNCS, vol. 3884, p. 432. Springer, Heidelberg (2006)
23. Geffert, V.: Translation of binary regular expressions into nondeterministic ε-free automata with $O(n \log n)$ transitions. Journal of Computer and System Sciences 66(3), 451–472 (2003)

Author Index